VOLUME FIVE HUNDRED AND TWENTY THREE

METHODS IN ENZYMOLOGY

Methods in Protein Design

METHODS IN ENZYMOLOGY

VOLUME FIVE HUNDRED AND TWENTY THREE

METHODS IN ENZYMOLOGY

Methods in Protein Design

Edited by

AMY E. KEATING

Department of Biology
Massachusetts Institute of Technology
Cambridge, MA, USA

ELSEVIER

AMSTERDAM • BOSTON • HEIDELBERG • LONDON
NEW YORK • OXFORD • PARIS • SAN DIEGO
SAN FRANCISCO • SINGAPORE • SYDNEY • TOKYO
Academic Press is an imprint of Elsevier

Academic Press is an imprint of Elsevier
525 B Street, Suite 1800, San Diego, CA 92101-4495, USA
225 Wyman Street, Waltham, MA 02451, USA
The Boulevard, Langford Lane, Kidlington, Oxford, OX5 1GB, UK
32, Jamestown Road, London NW1 7BY, UK
Radarweg 29, PO Box 211, 1000 AE Amsterdam, The Netherlands

First edition 2013

For information on all Academic Press publications
visit our website at store.elsevier.com

ISBN: 978-0-12-394292-0
ISSN: 0076-6879

Printed and bound in United States of America
13 14 15 16 11 10 9 8 7 6 5 4 3 2 1

CONTENTS

CONTRIBUTORS

Amy C. Anderson
Department of Pharmaceutical Sciences, University of Connecticut, Storrs, Connecticut, USA

Frances H. Arnold
Division of Chemistry and Chemical Engineering, California Institute of Technology, Pasadena, California, USA

Orr Ashenberg
Department of Biology, and Computational & Systems Biology Initiative, Massachusetts Institute of Technology, Cambridge, Massachusetts, USA

David Baker
Department of Biochemistry, University of Washington and Howard Hughes Medical Institute, Seattle, Washington, USA

Tiffany F. Chen
Department of Biological Engineering, and Koch Institute for Integrative Cancer Research, Massachusetts Institute of Technology, Cambridge, Massachusetts, USA

Cheng-Yu Chen
Department of Biochemistry, Duke University Medical Center, Durham, North Carolina, USA

Amol Date
Department of Chemical and Biomolecular Engineering, Johns Hopkins University, Baltimore, Maryland, USA

Ian W. Davis
GrassRoots Biotechnology, Durham, North Carolina, USA

Seymour de Picciotto
Department of Biological Engineering, and Koch Institute for Integrative Cancer Research, Massachusetts Institute of Technology, Cambridge, Massachusetts, USA

Bruce R. Donald
Department of Computer Science, Duke University, and Department of Biochemistry, Duke University Medical Center, Durham, North Carolina, USA

Venuka Durani
Department of Chemistry, The Ohio State University, Columbus, Ohio, USA

Ariel Erijman
Department of Biological Chemistry, The Alexander Silberman Institute of Life Sciences, The Hebrew University of Jerusalem, Jerusalem, Israel

Sarel J. Fleishman
Department of Biological Chemistry, Weizmann Institute of Science, Rehovot, Israel

James S. Fraser
Department of Cellular and Molecular Pharmacology, and California Institute for Quantitative Biosciences (QB3), University of California San Francisco, San Francisco, California, USA

Pablo Gainza
Department of Computer Science, Duke University, Durham, North Carolina, USA

Samuel H. Gellman
Department of Chemistry, University of Wisconsin, Madison, Wisconsin, USA

Ivelin Georgiev*
Department of Computer Science, Duke University, Durham, North Carolina, USA

Moshe Goldsmith
Department of Biological chemistry, Weizmann Institute of Science, Rehovot, Israel

Jeffrey J. Gray
Department of Chemical & Biomolecular Engineering, Johns Hopkins, Baltimore, Maryland, USA

Gevorg Grigoryan
Department of Computer Science, and Department of Biology, Dartmouth College, Hanover, New Hampshire, USA

Benjamin J. Hackel
Department of Chemical Engineering and Materials Science, University of Minnesota, Minneapolis, Minnesota, USA

James J. Havranek
Department of Genetics, Washington University, St. Louis, Missouri, USA

Birte Höcker
Max Planck Institute for Developmental Biology, Tübingen, Germany

Pete Heinzelman
Department of Chemical, Biological & Materials Engineering, University of Oklahoma, Norman, Oklahoma, USA

K.N. Houk
Department of Chemistry and Biochemistry, University of California, Los Angeles, California, USA

Jin Huang
Beijing Prosperous Biopharm Co., Ltd., Beijing, China

Ron Jacak
Department of Immunology and Microbial Science, The Scripps Research Institute, La Jolla, California, USA

*Current address: Vaccine Research Center, National Institute of Allergy and Infectious Diseases, National Institutes of Health (NIH), Bethesda, Maryland, USA

Lisa M. Johnson
Department of Chemistry, University of Wisconsin, Madison, Wisconsin, USA

Manu Kanwar
Department of Chemical and Biomolecular Engineering, Johns Hopkins University, Baltimore, Maryland, USA

Amy E. Keating
Computational and Systems Biology Program, and Department of Biology, Massachusetts Institute of Technology, Cambridge, Massachusetts, USA

Daniel A. Keedy
Department of Biochemistry, Duke University Medical Center, Durham, North Carolina, USA

Elizabeth H. Kellogg
Department of Biochemistry, University of Washington, Seattle, Washington, USA

Gert Kiss
Department of Chemistry, Stanford University, Stanford, California, USA

Shohei Koide
Department of Biochemistry and Molecular Biology, The University of Chicago, Chicago, Illinois, USA

Tanja Kortemme
Graduate Program in Bioinformatics; California Institute for Quantitative Biosciences (QB3), and Department of Bioengineering and Therapeutic Science, University of California San Francisco, San Francisco, California, USA

Brian Kuhlman
Department of Biochemistry, University of North Carolina, Chapel Hill, North Carolina, USA

Michael T. Laub
Department of Biology; Computational & Systems Biology Initiative, and Howard Hughes Medical Institute, Massachusetts Institute of Technology, Cambridge, Massachusetts, USA

Andrew Leaver-Fay
Department of Biochemistry, University of North Carolina, Chapel Hill, North Carolina, USA

Ryan H. Lilien
Department of Computer Science, University of Toronto, Toronto, Ontario, Canada

Sergey Lyskov
Department of Chemical & Biomolecular Engineering, Johns Hopkins, Baltimore, Maryland, USA

Thomas J. Magliery
Department of Chemistry, and Department of Biochemistry, The Ohio State University, Columbus, Ohio, USA

Megan E. McLaughlin
Department of Molecular Genetics, and Terrence Donnelly Centre for Cellular and Biomolecular Research, University of Toronto, Toronto, Ontario, Canada

Christopher Negron
Computational and Systems Biology Program, Massachusetts Institute of Technology, Cambridge, Massachusetts, USA

Matthew J. O'Meara
Department of Computer Science, University of North Carolina, Chapel Hill, North Carolina, USA

Noah Ollikainen
Graduate Program in Bioinformatics, University of California San Francisco, San Francisco, California, USA

Marc Ostermeier
Department of Chemical and Biomolecular Engineering, Johns Hopkins University, Baltimore, Maryland, USA

Roland A. Pache
Department of Bioengineering and Therapeutic Science, University of California San Francisco, San Francisco, California, USA

Vijay S. Pande
Department of Chemistry, Stanford University, Stanford, California, USA

Rama Ranganathan
Green Center for Systems Biology, Department of Pharmacology, University of Texas Southwestern Medical Center, Dallas, Texas, USA

Kimberly A. Reynolds
Green Center for Systems Biology, Department of Pharmacology, University of Texas Southwestern Medical Center, Dallas, Texas, USA

Faisal Reza*
Department of Biomedical Engineering, Duke University Medical Center, Durham, North Carolina, USA

David C. Richardson
Department of Biochemistry, Duke University Medical Center, Durham, North Carolina, USA

Jane S. Richardson
Department of Biochemistry, Duke University Medical Center, Durham, North Carolina, USA

José Arcadio Farías Rico
Max Planck Institute for Developmental Biology, Tübingen, Germany

*Current address: Department of Therapeutic Radiology, Yale University School of Medicine, New Haven, Connecticut, USA

Kyle E. Roberts
Department of Computer Science, Duke University, Durham, North Carolina, USA

Philip A. Romero
Division of Chemistry and Chemical Engineering, California Institute of Technology, Pasadena, California, USA

William P. Russ
Green Center for Systems Biology, Department of Pharmacology, University of Texas Southwestern Medical Center, Dallas, Texas, USA

Oz Sharabi
Department of Biological Chemistry, The Alexander Silberman Institute of Life Sciences, The Hebrew University of Jerusalem, Jerusalem, Israel

Julia M. Shifman
Department of Biological Chemistry, The Alexander Silberman Institute of Life Sciences, The Hebrew University of Jerusalem, Jerusalem, Israel

Sachdev S. Sidhu
Department of Molecular Genetics; Terrence Donnelly Centre for Cellular and Biomolecular Research, and Banting and Best Department of Medical Research, University of Toronto, Toronto, Ontario, Canada

Colin A. Smith*
Graduate Program in Bioinformatics, University of California San Francisco, San Francisco, California, USA

Jack Snoeyink
Department of Computer Science, University of North Carolina, Chapel Hill, North Carolina, USA

Michael Socolich
Green Center for Systems Biology, Department of Pharmacology, University of Texas Southwestern Medical Center, Dallas, Texas, USA

Yifan Song
Department of Biochemistry, University of Washington, Seattle, Washington, USA

Dan S. Tawfik
Department of Biological chemistry, Weizmann Institute of Science, Rehovot, Israel

James Thompson
Department of Biochemistry, University of Washington, Seattle, Washington, USA

Jennifer Tullman
Department of Chemical and Biomolecular Engineering, Johns Hopkins University, Baltimore, Maryland, USA

Mike Tyka
Department of Biochemistry, University of Washington, Seattle, Washington, USA

*Present address: Max Planck Institute for Biophysical Chemistry, Göttingen, Germany

Timothy A. Whitehead
Department of Chemical Engineering and Materials Science, and Department of Biosystems and Agricultural Engineering, Michigan State University, East Lansing, Michigan, USA

K. Dane Wittrup
Department of Biological Engineering; Koch Institute for Integrative Cancer Research, and Department of Chemical Engineering, Massachusetts Institute of Technology, Cambridge, Massachusetts, USA

R. Clay Wright
Department of Chemical and Biomolecular Engineering, Johns Hopkins University, Baltimore, Maryland, USA

Jian Zhang
Department of Computer Science, Dartmouth College, Hanover, New Hampshire, USA

PREFACE

It has been a privilege to edit the first *Methods in Enzymology* volume focused on protein design, a field that embraces both computational and experimental methods for creating new proteins with interesting properties. Experts who develop and use diverse approaches have generously contributed their insights and protocols to this book. There are many other talented scientists making important advances and discoveries, and I look forward to future volumes in this series that will build on what is presented here.

Protein design occupies a niche between engineering and science. As engineers, protein designers strive to make new molecules with desired characteristics. In doing so, they draw on accumulated scientific knowledge about protein structure, stability, folding, and function. The iterative process of design and redesign tests and further deepens our understanding of these relationships. Thus, even in the earliest days of computational design, when it was not clear whether this approach would ever deliver useful new proteins, the science was rich and exciting. This volume, which is focused on methods, addresses the "how" of protein design more than it does scientific questions about "why does it (or does it not) work?" Yet, in the chapters presented, the reader can find wonderful insights and ideas about ways to grow our understanding of molecular energetics, structural flexibility, and protein evolution, and how information is encoded in protein sequences.

In the past 20 years, significant progress has been made in methods for designing proteins. Advances in experimental protein library screening, for example, using phage or yeast-surface display now enable discovery efforts in many companies and have led to the development of protein therapeutics used in the clinic. Computational protein design, which was until quite recently a purely academic pursuit, is being widely used outside of the laboratories of the original developers. Chemical strategies for diversifying protein backbones, for example, making them less susceptible to proteolysis, are maturing. Applications for these methods abound and include design of novel protein interaction reagents based on non-antibody scaffolds, design of peptide-based inhibitors, design of highly stabilized proteins, design of new enzymatic functions, and prediction of the effects of mutations on stability, binding affinity and specificity, and drug resistance.

Despite progress, protein design remains very challenging. The fundamental problem of how to search vast sequence and conformational spaces efficiently, using either experimental screens/selections or computational sampling, still demands innovative solutions. And our ability to predict the relative stabilities of different proteins from their sequences is quite primitive. The chapters presented here represent the state of the art in several ways. First, they demonstrate what is possible using existing methods. In cases where methods have matured to the point that they are routinely applied in their developers' labs, formal protocols are presented so that others can use these techniques. This is the case for several powerful experimental screening and selection methods. Second, where capabilities are not as mature, important tools are presented. Several chapters describe software packages, algorithms, or scripts that are available to help users try their own versions of protein design. Third, assessment methods are presented for judging and potentially improving the quality of existing energy functions and for assessing the quality of candidate designs. Interspersed throughout, there are ideas and suggestions about how to confront challenging problems including target selection, conformational sampling, multistate design, and library design.

In the future, many of the methods presented here will be combined and applied to ever more challenging design problems. There are numerous synergies to be realized among the computational methods presented, for example, between structure-based methods and sequence-guided methods. Better integration of computational design and experimental screening is expected to dramatically improve hit rates in future design studies. Exciting innovations can be expected. For now, thanks are due to the authors who contributed to this volume, particularly those who routinely make their code and reagents widely available, thus accelerating progress in this dynamic field.

Amy E. Keating

METHODS IN ENZYMOLOGY

VOLUME XV. Steroids and Terpenoids
Edited by RAYMOND B. CLAYTON

VOLUME XVI. Fast Reactions
Edited by KENNETH KUSTIN

VOLUME XVII. Metabolism of Amino Acids and Amines (Parts A and B)
Edited by HERBERT TABOR AND CELIA WHITE TABOR

VOLUME XVIII. Vitamins and Coenzymes (Parts A, B, and C)
Edited by DONALD B. MCCORMICK AND LEMUEL D. WRIGHT

VOLUME XIX. Proteolytic Enzymes
Edited by GERTRUDE E. PERLMANN AND LASZLO LORAND

VOLUME XX. Nucleic Acids and Protein Synthesis (Part C)
Edited by KIVIE MOLDAVE AND LAWRENCE GROSSMAN

VOLUME XXI. Nucleic Acids (Part D)
Edited by LAWRENCE GROSSMAN AND KIVIE MOLDAVE

VOLUME XXII. Enzyme Purification and Related Techniques
Edited by WILLIAM B. JAKOBY

VOLUME XXIII. Photosynthesis (Part A)
Edited by ANTHONY SAN PIETRO

VOLUME XXIV. Photosynthesis and Nitrogen Fixation (Part B)
Edited by ANTHONY SAN PIETRO

VOLUME XXV. Enzyme Structure (Part B)
Edited by C. H. W. HIRS AND SERGE N. TIMASHEFF

VOLUME XXVI. Enzyme Structure (Part C)
Edited by C. H. W. HIRS AND SERGE N. TIMASHEFF

VOLUME XXVII. Enzyme Structure (Part D)
Edited by C. H. W. HIRS AND SERGE N. TIMASHEFF

VOLUME XXVIII. Complex Carbohydrates (Part B)
Edited by VICTOR GINSBURG

VOLUME XXIX. Nucleic Acids and Protein Synthesis (Part E)
Edited by LAWRENCE GROSSMAN AND KIVIE MOLDAVE

VOLUME XXX. Nucleic Acids and Protein Synthesis (Part F)
Edited by KIVIE MOLDAVE AND LAWRENCE GROSSMAN

VOLUME XXXI. Biomembranes (Part A)
Edited by SIDNEY FLEISCHER AND LESTER PACKER

VOLUME XXXII. Biomembranes (Part B)
Edited by SIDNEY FLEISCHER AND LESTER PACKER

VOLUME XXXIII. Cumulative Subject Index Volumes I-XXX
Edited by MARTHA G. DENNIS AND EDWARD A. DENNIS

VOLUME XXXIV. Affinity Techniques (Enzyme Purification: Part B)
Edited by WILLIAM B. JAKOBY AND MEIR WILCHEK

VOLUME XXXV. Lipids (Part B)
Edited by JOHN M. LOWENSTEIN

VOLUME XXXVI. Hormone Action (Part A: Steroid Hormones)
Edited by BERT W. O'MALLEY AND JOEL G. HARDMAN

VOLUME XXXVII. Hormone Action (Part B: Peptide Hormones)
Edited by BERT W. O'MALLEY AND JOEL G. HARDMAN

VOLUME XXXVIII. Hormone Action (Part C: Cyclic Nucleotides)
Edited by JOEL G. HARDMAN AND BERT W. O'MALLEY

VOLUME XXXIX. Hormone Action (Part D: Isolated Cells, Tissues, and Organ Systems)
Edited by JOEL G. HARDMAN AND BERT W. O'MALLEY

VOLUME XL. Hormone Action (Part E: Nuclear Structure and Function)
Edited by BERT W. O'MALLEY AND JOEL G. HARDMAN

VOLUME XLI. Carbohydrate Metabolism (Part B)
Edited by W. A. WOOD

VOLUME XLII. Carbohydrate Metabolism (Part C)
Edited by W. A. WOOD

VOLUME XLIII. Antibiotics
Edited by JOHN H. HASH

VOLUME XLIV. Immobilized Enzymes
Edited by KLAUS MOSBACH

VOLUME XLV. Proteolytic Enzymes (Part B)
Edited by LASZLO LORAND

VOLUME XLVI. Affinity Labeling
Edited by WILLIAM B. JAKOBY AND MEIR WILCHEK

VOLUME XLVII. Enzyme Structure (Part E)
Edited by C. H. W. HIRS AND SERGE N. TIMASHEFF

Volume XLVIII. Enzyme Structure (Part F)
Edited by C. H. W. Hirs and Serge N. Timasheff

Volume XLIX. Enzyme Structure (Part G)
Edited by C. H. W. Hirs and Serge N. Timasheff

Volume L. Complex Carbohydrates (Part C)
Edited by Victor Ginsburg

Volume LI. Purine and Pyrimidine Nucleotide Metabolism
Edited by Patricia A. Hoffee and Mary Ellen Jones

Volume LII. Biomembranes (Part C: Biological Oxidations)
Edited by Sidney Fleischer and Lester Packer

Volume LIII. Biomembranes (Part D: Biological Oxidations)
Edited by Sidney Fleischer and Lester Packer

Volume LIV. Biomembranes (Part E: Biological Oxidations)
Edited by Sidney Fleischer and Lester Packer

Volume LV. Biomembranes (Part F: Bioenergetics)
Edited by Sidney Fleischer and Lester Packer

Volume LVI. Biomembranes (Part G: Bioenergetics)
Edited by Sidney Fleischer and Lester Packer

Volume LVII. Bioluminescence and Chemiluminescence
Edited by Marlene A. DeLuca

Volume LVIII. Cell Culture
Edited by William B. Jakoby and Ira Pastan

Volume LIX. Nucleic Acids and Protein Synthesis (Part G)
Edited by Kivie Moldave and Lawrence Grossman

Volume LX. Nucleic Acids and Protein Synthesis (Part H)
Edited by Kivie Moldave and Lawrence Grossman

Volume 61. Enzyme Structure (Part H)
Edited by C. H. W. Hirs and Serge N. Timasheff

Volume 62. Vitamins and Coenzymes (Part D)
Edited by Donald B. McCormick and Lemuel D. Wright

Volume 63. Enzyme Kinetics and Mechanism (Part A: Initial Rate and Inhibitor Methods)
Edited by Daniel L. Purich

Volume 64. Enzyme Kinetics and Mechanism
(Part B: Isotopic Probes and Complex Enzyme Systems)
Edited by Daniel L. Purich

Volume 65. Nucleic Acids (Part I)
Edited by Lawrence Grossman and Kivie Moldave

Volume 66. Vitamins and Coenzymes (Part E)
Edited by Donald B. McCormick and Lemuel D. Wright

Volume 67. Vitamins and Coenzymes (Part F)
Edited by Donald B. McCormick and Lemuel D. Wright

Volume 68. Recombinant DNA
Edited by Ray Wu

Volume 69. Photosynthesis and Nitrogen Fixation (Part C)
Edited by Anthony San Pietro

Volume 70. Immunochemical Techniques (Part A)
Edited by Helen Van Vunakis and John J. Langone

Volume 71. Lipids (Part C)
Edited by John M. Lowenstein

Volume 72. Lipids (Part D)
Edited by John M. Lowenstein

Volume 73. Immunochemical Techniques (Part B)
Edited by John J. Langone and Helen Van Vunakis

Volume 74. Immunochemical Techniques (Part C)
Edited by John J. Langone and Helen Van Vunakis

Volume 75. Cumulative Subject Index Volumes XXXI, XXXII, XXXIV–LX
Edited by Edward A. Dennis and Martha G. Dennis

Volume 76. Hemoglobins
Edited by Eraldo Antonini, Luigi Rossi-Bernardi, and Emilia Chiancone

Volume 77. Detoxication and Drug Metabolism
Edited by William B. Jakoby

Volume 78. Interferons (Part A)
Edited by Sidney Pestka

Volume 79. Interferons (Part B)
Edited by Sidney Pestka

Volume 80. Proteolytic Enzymes (Part C)
Edited by Laszlo Lorand

Volume 81. Biomembranes (Part H: Visual Pigments and Purple Membranes, I)
Edited by Lester Packer

Volume 82. Structural and Contractile Proteins (Part A: Extracellular Matrix)
Edited by Leon W. Cunningham and Dixie W. Frederiksen

Volume 83. Complex Carbohydrates (Part D)
Edited by Victor Ginsburg

Volume 84. Immunochemical Techniques (Part D: Selected Immunoassays)
Edited by John J. Langone and Helen Van Vunakis

Volume 85. Structural and Contractile Proteins (Part B: The Contractile Apparatus and the Cytoskeleton)
Edited by Dixie W. Frederiksen and Leon W. Cunningham

Volume 86. Prostaglandins and Arachidonate Metabolites
Edited by William E. M. Lands and William L. Smith

Volume 87. Enzyme Kinetics and Mechanism (Part C: Intermediates, Stereo-chemistry, and Rate Studies)
Edited by Daniel L. Purich

Volume 88. Biomembranes (Part I: Visual Pigments and Purple Membranes, II)
Edited by Lester Packer

Volume 89. Carbohydrate Metabolism (Part D)
Edited by Willis A. Wood

Volume 90. Carbohydrate Metabolism (Part E)
Edited by Willis A. Wood

Volume 91. Enzyme Structure (Part I)
Edited by C. H. W. Hirs and Serge N. Timasheff

Volume 92. Immunochemical Techniques (Part E: Monoclonal Antibodies and General Immunoassay Methods)
Edited by John J. Langone and Helen Van Vunakis

VOLUME 93. Immunochemical Techniques (Part F: Conventional Antibodies, Fc Receptors, and Cytotoxicity)
Edited by JOHN J. LANGONE AND HELEN VAN VUNAKIS

VOLUME 94. Polyamines
Edited by HERBERT TABOR AND CELIA WHITE TABOR

VOLUME 95. Cumulative Subject Index Volumes 61–74, 76–80
Edited by EDWARD A. DENNIS AND MARTHA G. DENNIS

VOLUME 96. Biomembranes [Part J: Membrane Biogenesis: Assembly and Targeting (General Methods; Eukaryotes)]
Edited by SIDNEY FLEISCHER AND BECCA FLEISCHER

VOLUME 97. Biomembranes [Part K: Membrane Biogenesis: Assembly and Targeting (Prokaryotes, Mitochondria, and Chloroplasts)]
Edited by SIDNEY FLEISCHER AND BECCA FLEISCHER

VOLUME 98. Biomembranes (Part L: Membrane Biogenesis: Processing and Recycling)
Edited by SIDNEY FLEISCHER AND BECCA FLEISCHER

VOLUME 99. Hormone Action (Part F: Protein Kinases)
Edited by JACKIE D. CORBIN AND JOEL G. HARDMAN

VOLUME 100. Recombinant DNA (Part B)
Edited by RAY WU, LAWRENCE GROSSMAN, AND KIVIE MOLDAVE

VOLUME 101. Recombinant DNA (Part C)
Edited by RAY WU, LAWRENCE GROSSMAN, AND KIVIE MOLDAVE

VOLUME 102. Hormone Action (Part G: Calmodulin and Calcium-Binding Proteins)
Edited by ANTHONY R. MEANS AND BERT W. O'MALLEY

VOLUME 103. Hormone Action (Part H: Neuroendocrine Peptides)
Edited by P. MICHAEL CONN

VOLUME 104. Enzyme Purification and Related Techniques (Part C)
Edited by WILLIAM B. JAKOBY

VOLUME 105. Oxygen Radicals in Biological Systems
Edited by LESTER PACKER

VOLUME 106. Posttranslational Modifications (Part A)
Edited by FINN WOLD AND KIVIE MOLDAVE

Volume 107. Posttranslational Modifications (Part B)
Edited by Finn Wold and Kivie Moldave

Volume 108. Immunochemical Techniques (Part G: Separation and Characterization of Lymphoid Cells)
Edited by Giovanni Di Sabato, John J. Langone, and Helen Van Vunakis

Volume 109. Hormone Action (Part I: Peptide Hormones)
Edited by Lutz Birnbaumer and Bert W. O'Malley

Volume 110. Steroids and Isoprenoids (Part A)
Edited by John H. Law and Hans C. Rilling

Volume 111. Steroids and Isoprenoids (Part B)
Edited by John H. Law and Hans C. Rilling

Volume 112. Drug and Enzyme Targeting (Part A)
Edited by Kenneth J. Widder and Ralph Green

Volume 113. Glutamate, Glutamine, Glutathione, and Related Compounds
Edited by Alton Meister

Volume 114. Diffraction Methods for Biological Macromolecules (Part A)
Edited by Harold W. Wyckoff, C. H. W. Hirs, and Serge N. Timasheff

Volume 115. Diffraction Methods for Biological Macromolecules (Part B)
Edited by Harold W. Wyckoff, C. H. W. Hirs, and Serge N. Timasheff

Volume 116. Immunochemical Techniques
(Part H: Effectors and Mediators of Lymphoid Cell Functions)
Edited by Giovanni Di Sabato, John J. Langone, and Helen Van Vunakis

Volume 117. Enzyme Structure (Part J)
Edited by C. H. W. Hirs and Serge N. Timasheff

Volume 118. Plant Molecular Biology
Edited by Arthur Weissbach and Herbert Weissbach

Volume 119. Interferons (Part C)
Edited by Sidney Pestka

Volume 120. Cumulative Subject Index Volumes 81–94, 96–101

Volume 121. Immunochemical Techniques (Part I: Hybridoma Technology and Monoclonal Antibodies)
Edited by John J. Langone and Helen Van Vunakis

VOLUME 122. Vitamins and Coenzymes (Part G)
Edited by FRANK CHYTIL AND DONALD B. MCCORMICK

VOLUME 123. Vitamins and Coenzymes (Part H)
Edited by FRANK CHYTIL AND DONALD B. MCCORMICK

VOLUME 124. Hormone Action (Part J: Neuroendocrine Peptides)
Edited by P. MICHAEL CONN

VOLUME 125. Biomembranes (Part M: Transport in Bacteria, Mitochondria, and Chloroplasts: General Approaches and Transport Systems)
Edited by SIDNEY FLEISCHER AND BECCA FLEISCHER

VOLUME 126. Biomembranes (Part N: Transport in Bacteria, Mitochondria, and Chloroplasts: Protonmotive Force)
Edited by SIDNEY FLEISCHER AND BECCA FLEISCHER

VOLUME 127. Biomembranes (Part O: Protons and Water: Structure and Translocation)
Edited by LESTER PACKER

VOLUME 128. Plasma Lipoproteins (Part A: Preparation, Structure, and Molecular Biology)
Edited by JERE P. SEGREST AND JOHN J. ALBERS

VOLUME 129. Plasma Lipoproteins (Part B: Characterization, Cell Biology, and Metabolism)
Edited by JOHN J. ALBERS AND JERE P. SEGREST

VOLUME 130. Enzyme Structure (Part K)
Edited by C. H. W. HIRS AND SERGE N. TIMASHEFF

VOLUME 131. Enzyme Structure (Part L)
Edited by C. H. W. HIRS AND SERGE N. TIMASHEFF

VOLUME 132. Immunochemical Techniques (Part J: Phagocytosis and Cell-Mediated Cytotoxicity)
Edited by GIOVANNI DI SABATO AND JOHANNES EVERSE

VOLUME 133. Bioluminescence and Chemiluminescence (Part B)
Edited by MARLENE DELUCA AND WILLIAM D. MCELROY

VOLUME 134. Structural and Contractile Proteins (Part C: The Contractile Apparatus and the Cytoskeleton)
Edited by RICHARD B. VALLEE

VOLUME 135. Immobilized Enzymes and Cells (Part B)
Edited by KLAUS MOSBACH

VOLUME 136. Immobilized Enzymes and Cells (Part C)
Edited by KLAUS MOSBACH

VOLUME 137. Immobilized Enzymes and Cells (Part D)
Edited by KLAUS MOSBACH

VOLUME 138. Complex Carbohydrates (Part E)
Edited by VICTOR GINSBURG

VOLUME 139. Cellular Regulators (Part A: Calcium- and Calmodulin-Binding Proteins)
Edited by ANTHONY R. MEANS AND P. MICHAEL CONN

VOLUME 140. Cumulative Subject Index Volumes 102–119, 121–134

VOLUME 141. Cellular Regulators (Part B: Calcium and Lipids)
Edited by P. MICHAEL CONN AND ANTHONY R. MEANS

VOLUME 142. Metabolism of Aromatic Amino Acids and Amines
Edited by SEYMOUR KAUFMAN

VOLUME 143. Sulfur and Sulfur Amino Acids
Edited by WILLIAM B. JAKOBY AND OWEN GRIFFITH

VOLUME 144. Structural and Contractile Proteins (Part D: Extracellular Matrix)
Edited by LEON W. CUNNINGHAM

VOLUME 145. Structural and Contractile Proteins (Part E: Extracellular Matrix)
Edited by LEON W. CUNNINGHAM

VOLUME 146. Peptide Growth Factors (Part A)
Edited by DAVID BARNES AND DAVID A. SIRBASKU

VOLUME 147. Peptide Growth Factors (Part B)
Edited by DAVID BARNES AND DAVID A. SIRBASKU

VOLUME 148. Plant Cell Membranes
Edited by LESTER PACKER AND ROLAND DOUCE

VOLUME 149. Drug and Enzyme Targeting (Part B)
Edited by RALPH GREEN AND KENNETH J. WIDDER

VOLUME 150. Immunochemical Techniques (Part K: *In Vitro* Models of B and T Cell Functions and Lymphoid Cell Receptors)
Edited by GIOVANNI DI SABATO

Volume 166. Branched-Chain Amino Acids
Edited by Robert Harris and John R. Sokatch

Volume 167. Cyanobacteria
Edited by Lester Packer and Alexander N. Glazer

Volume 168. Hormone Action (Part K: Neuroendocrine Peptides)
Edited by P. Michael Conn

Volume 169. Platelets: Receptors, Adhesion, Secretion (Part A)
Edited by Jacek Hawiger

Volume 170. Nucleosomes
Edited by Paul M. Wassarman and Roger D. Kornberg

Volume 171. Biomembranes (Part R: Transport Theory: Cells and Model Membranes)
Edited by Sidney Fleischer and Becca Fleischer

Volume 172. Biomembranes (Part S: Transport: Membrane Isolation and Characterization)
Edited by Sidney Fleischer and Becca Fleischer

Volume 173. Biomembranes [Part T: Cellular and Subcellular Transport: Eukaryotic (Nonepithelial) Cells]
Edited by Sidney Fleischer and Becca Fleischer

Volume 174. Biomembranes [Part U: Cellular and Subcellular Transport: Eukaryotic (Nonepithelial) Cells]
Edited by Sidney Fleischer and Becca Fleischer

Volume 175. Cumulative Subject Index Volumes 135–139, 141–167

Volume 176. Nuclear Magnetic Resonance (Part A: Spectral Techniques and Dynamics)
Edited by Norman J. Oppenheimer and Thomas L. James

Volume 177. Nuclear Magnetic Resonance (Part B: Structure and Mechanism)
Edited by Norman J. Oppenheimer and Thomas L. James

Volume 178. Antibodies, Antigens, and Molecular Mimicry
Edited by John J. Langone

Volume 179. Complex Carbohydrates (Part F)
Edited by Victor Ginsburg

Volume 180. RNA Processing (Part A: General Methods)
Edited by James E. Dahlberg and John N. Abelson

Volume 181. RNA Processing (Part B: Specific Methods)
Edited by James E. Dahlberg and John N. Abelson

Volume 182. Guide to Protein Purification
Edited by Murray P. Deutscher

Volume 183. Molecular Evolution: Computer Analysis of Protein and Nucleic Acid Sequences
Edited by Russell F. Doolittle

Volume 184. Avidin-Biotin Technology
Edited by Meir Wilchek and Edward A. Bayer

Volume 185. Gene Expression Technology
Edited by David V. Goeddel

Volume 186. Oxygen Radicals in Biological Systems (Part B: Oxygen Radicals and Antioxidants)
Edited by Lester Packer and Alexander N. Glazer

Volume 187. Arachidonate Related Lipid Mediators
Edited by Robert C. Murphy and Frank A. Fitzpatrick

Volume 188. Hydrocarbons and Methylotrophy
Edited by Mary E. Lidstrom

Volume 189. Retinoids (Part A: Molecular and Metabolic Aspects)
Edited by Lester Packer

Volume 190. Retinoids (Part B: Cell Differentiation and Clinical Applications)
Edited by Lester Packer

Volume 191. Biomembranes (Part V: Cellular and Subcellular Transport: Epithelial Cells)
Edited by Sidney Fleischer and Becca Fleischer

Volume 192. Biomembranes (Part W: Cellular and Subcellular Transport: Epithelial Cells)
Edited by Sidney Fleischer and Becca Fleischer

Volume 193. Mass Spectrometry
Edited by James A. McCloskey

Volume 194. Guide to Yeast Genetics and Molecular Biology
Edited by Christine Guthrie and Gerald R. Fink

Volume 195. Adenylyl Cyclase, G Proteins, and Guanylyl Cyclase
Edited by Roger A. Johnson and Jackie D. Corbin

Volume 196. Molecular Motors and the Cytoskeleton
Edited by Richard B. Vallee

Volume 197. Phospholipases
Edited by Edward A. Dennis

Volume 198. Peptide Growth Factors (Part C)
Edited by David Barnes, J. P. Mather, and Gordon H. Sato

Volume 199. Cumulative Subject Index Volumes 168–174, 176–194

Volume 200. Protein Phosphorylation (Part A: Protein Kinases: Assays, Purification, Antibodies, Functional Analysis, Cloning, and Expression)
Edited by Tony Hunter and Bartholomew M. Sefton

Volume 201. Protein Phosphorylation (Part B: Analysis of Protein Phosphorylation, Protein Kinase Inhibitors, and Protein Phosphatases)
Edited by Tony Hunter and Bartholomew M. Sefton

Volume 202. Molecular Design and Modeling: Concepts and Applications (Part A: Proteins, Peptides, and Enzymes)
Edited by John J. Langone

Volume 203. Molecular Design and Modeling: Concepts and Applications (Part B: Antibodies and Antigens, Nucleic Acids, Polysaccharides, and Drugs)
Edited by John J. Langone

Volume 204. Bacterial Genetic Systems
Edited by Jeffrey H. Miller

Volume 205. Metallobiochemistry (Part B: Metallothionein and Related Molecules)
Edited by James F. Riordan and Bert L. Vallee

Volume 206. Cytochrome P450
Edited by Michael R. Waterman and Eric F. Johnson

Volume 207. Ion Channels
Edited by Bernardo Rudy and Linda E. Iverson

Volume 208. Protein–DNA Interactions
Edited by Robert T. Sauer

Volume 209. Phospholipid Biosynthesis
Edited by Edward A. Dennis and Dennis E. Vance

Volume 210. Numerical Computer Methods
Edited by Ludwig Brand and Michael L. Johnson

Volume 211. DNA Structures (Part A: Synthesis and Physical Analysis of DNA)
Edited by David M. J. Lilley and James E. Dahlberg

Volume 212. DNA Structures (Part B: Chemical and Electrophoretic Analysis of DNA)
Edited by David M. J. Lilley and James E. Dahlberg

Volume 213. Carotenoids (Part A: Chemistry, Separation, Quantitation, and Antioxidation)
Edited by Lester Packer

Volume 214. Carotenoids (Part B: Metabolism, Genetics, and Biosynthesis)
Edited by Lester Packer

Volume 215. Platelets: Receptors, Adhesion, Secretion (Part B)
Edited by Jacek J. Hawiger

Volume 216. Recombinant DNA (Part G)
Edited by Ray Wu

Volume 217. Recombinant DNA (Part H)
Edited by Ray Wu

Volume 218. Recombinant DNA (Part I)
Edited by Ray Wu

Volume 219. Reconstitution of Intracellular Transport
Edited by James E. Rothman

Volume 220. Membrane Fusion Techniques (Part A)
Edited by Nejat Düzgüneş

Volume 221. Membrane Fusion Techniques (Part B)
Edited by Nejat Düzgüneş

Volume 222. Proteolytic Enzymes in Coagulation, Fibrinolysis, and Complement Activation (Part A: Mammalian Blood Coagulation

Factors and Inhibitors)
Edited by Laszlo Lorand and Kenneth G. Mann

Volume 223. Proteolytic Enzymes in Coagulation, Fibrinolysis, and Complement Activation (Part B: Complement Activation, Fibrinolysis, and Nonmammalian Blood Coagulation Factors)
Edited by Laszlo Lorand and Kenneth G. Mann

Volume 224. Molecular Evolution: Producing the Biochemical Data
Edited by Elizabeth Anne Zimmer, Thomas J. White, Rebecca L. Cann, and Allan C. Wilson

Volume 225. Guide to Techniques in Mouse Development
Edited by Paul M. Wassarman and Melvin L. DePamphilis

Volume 226. Metallobiochemistry (Part C: Spectroscopic and Physical Methods for Probing Metal Ion Environments in Metalloenzymes and Metalloproteins)
Edited by James F. Riordan and Bert L. Vallee

Volume 227. Metallobiochemistry (Part D: Physical and Spectroscopic Methods for Probing Metal Ion Environments in Metalloproteins)
Edited by James F. Riordan and Bert L. Vallee

Volume 228. Aqueous Two-Phase Systems
Edited by Harry Walter and Göte Johansson

Volume 229. Cumulative Subject Index Volumes 195–198, 200–227

Volume 230. Guide to Techniques in Glycobiology
Edited by William J. Lennarz and Gerald W. Hart

Volume 231. Hemoglobins (Part B: Biochemical and Analytical Methods)
Edited by Johannes Everse, Kim D. Vandegriff, and Robert M. Winslow

Volume 232. Hemoglobins (Part C: Biophysical Methods)
Edited by Johannes Everse, Kim D. Vandegriff, and Robert M. Winslow

Volume 233. Oxygen Radicals in Biological Systems (Part C)
Edited by Lester Packer

Volume 234. Oxygen Radicals in Biological Systems (Part D)
Edited by Lester Packer

Volume 235. Bacterial Pathogenesis (Part A: Identification and Regulation of Virulence Factors)
Edited by Virginia L. Clark and Patrik M. Bavoil

Volume 236. Bacterial Pathogenesis (Part B: Integration of Pathogenic Bacteria with Host Cells)
Edited by Virginia L. Clark and Patrik M. Bavoil

Volume 237. Heterotrimeric G Proteins
Edited by Ravi Iyengar

Volume 238. Heterotrimeric G-Protein Effectors
Edited by Ravi Iyengar

Volume 239. Nuclear Magnetic Resonance (Part C)
Edited by Thomas L. James and Norman J. Oppenheimer

Volume 240. Numerical Computer Methods (Part B)
Edited by Michael L. Johnson and Ludwig Brand

Volume 241. Retroviral Proteases
Edited by Lawrence C. Kuo and Jules A. Shafer

Volume 242. Neoglycoconjugates (Part A)
Edited by Y. C. Lee and Reiko T. Lee

Volume 243. Inorganic Microbial Sulfur Metabolism
Edited by Harry D. Peck, Jr., and Jean LeGall

Volume 244. Proteolytic Enzymes: Serine and Cysteine Peptidases
Edited by Alan J. Barrett

Volume 245. Extracellular Matrix Components
Edited by E. Ruoslahti and E. Engvall

Volume 246. Biochemical Spectroscopy
Edited by Kenneth Sauer

Volume 247. Neoglycoconjugates (Part B: Biomedical Applications)
Edited by Y. C. Lee and Reiko T. Lee

Volume 248. Proteolytic Enzymes: Aspartic and Metallo Peptidases
Edited by Alan J. Barrett

Volume 249. Enzyme Kinetics and Mechanism (Part D: Developments in Enzyme Dynamics)
Edited by Daniel L. Purich

Volume 250. Lipid Modifications of Proteins
Edited by Patrick J. Casey and Janice E. Buss

VOLUME 251. Biothiols (Part A: Monothiols and Dithiols, Protein Thiols, and Thiyl Radicals)
Edited by LESTER PACKER

VOLUME 252. Biothiols (Part B: Glutathione and Thioredoxin; Thiols in Signal Transduction and Gene Regulation)
Edited by LESTER PACKER

VOLUME 253. Adhesion of Microbial Pathogens
Edited by RON J. DOYLE AND ITZHAK OFEK

VOLUME 254. Oncogene Techniques
Edited by PETER K. VOGT AND INDER M. VERMA

VOLUME 255. Small GTPases and Their Regulators (Part A: Ras Family)
Edited by W. E. BALCH, CHANNING J. DER, AND ALAN HALL

VOLUME 256. Small GTPases and Their Regulators (Part B: Rho Family)
Edited by W. E. BALCH, CHANNING J. DER, AND ALAN HALL

VOLUME 257. Small GTPases and Their Regulators (Part C: Proteins Involved in Transport)
Edited by W. E. BALCH, CHANNING J. DER, AND ALAN HALL

VOLUME 258. Redox-Active Amino Acids in Biology
Edited by JUDITH P. KLINMAN

VOLUME 259. Energetics of Biological Macromolecules
Edited by MICHAEL L. JOHNSON AND GARY K. ACKERS

VOLUME 260. Mitochondrial Biogenesis and Genetics (Part A)
Edited by GIUSEPPE M. ATTARDI AND ANNE CHOMYN

VOLUME 261. Nuclear Magnetic Resonance and Nucleic Acids
Edited by THOMAS L. JAMES

VOLUME 262. DNA Replication
Edited by JUDITH L. CAMPBELL

VOLUME 263. Plasma Lipoproteins (Part C: Quantitation)
Edited by WILLIAM A. BRADLEY, SANDRA H. GIANTURCO, AND JERE P. SEGREST

VOLUME 264. Mitochondrial Biogenesis and Genetics (Part B)
Edited by GIUSEPPE M. ATTARDI AND ANNE CHOMYN

VOLUME 265. Cumulative Subject Index Volumes 228, 230–262

Volume 266. Computer Methods for Macromolecular Sequence Analysis
Edited by Russell F. Doolittle

Volume 267. Combinatorial Chemistry
Edited by John N. Abelson

Volume 268. Nitric Oxide (Part A: Sources and Detection of NO; NO Synthase)
Edited by Lester Packer

Volume 269. Nitric Oxide (Part B: Physiological and Pathological Processes)
Edited by Lester Packer

Volume 270. High Resolution Separation and Analysis of Biological Macromolecules (Part A: Fundamentals)
Edited by Barry L. Karger and William S. Hancock

Volume 271. High Resolution Separation and Analysis of Biological Macromolecules (Part B: Applications)
Edited by Barry L. Karger and William S. Hancock

Volume 272. Cytochrome P450 (Part B)
Edited by Eric F. Johnson and Michael R. Waterman

Volume 273. RNA Polymerase and Associated Factors (Part A)
Edited by Sankar Adhya

Volume 274. RNA Polymerase and Associated Factors (Part B)
Edited by Sankar Adhya

Volume 275. Viral Polymerases and Related Proteins
Edited by Lawrence C. Kuo, David B. Olsen, and Steven S. Carroll

Volume 276. Macromolecular Crystallography (Part A)
Edited by Charles W. Carter, Jr., and Robert M. Sweet

Volume 277. Macromolecular Crystallography (Part B)
Edited by Charles W. Carter, Jr., and Robert M. Sweet

Volume 278. Fluorescence Spectroscopy
Edited by Ludwig Brand and Michael L. Johnson

Volume 279. Vitamins and Coenzymes (Part I)
Edited by Donald B. McCormick, John W. Suttie, and Conrad Wagner

Volume 280. Vitamins and Coenzymes (Part J)
Edited by Donald B. McCormick, John W. Suttie, and Conrad Wagner

VOLUME 327. Applications of Chimeric Genes and Hybrid Proteins (Part B: Cell Biology and Physiology)
Edited by JEREMY THORNER, SCOTT D. EMR, AND JOHN N. ABELSON

VOLUME 328. Applications of Chimeric Genes and Hybrid Proteins (Part C: Protein–Protein Interactions and Genomics)
Edited by JEREMY THORNER, SCOTT D. EMR, AND JOHN N. ABELSON

VOLUME 329. Regulators and Effectors of Small GTPases (Part E: GTPases Involved in Vesicular Traffic)
Edited by W. E. BALCH, CHANNING J. DER, AND ALAN HALL

VOLUME 330. Hyperthermophilic Enzymes (Part A)
Edited by MICHAEL W. W. ADAMS AND ROBERT M. KELLY

VOLUME 331. Hyperthermophilic Enzymes (Part B)
Edited by MICHAEL W. W. ADAMS AND ROBERT M. KELLY

VOLUME 332. Regulators and Effectors of Small GTPases (Part F: Ras Family I)
Edited by W. E. BALCH, CHANNING J. DER, AND ALAN HALL

VOLUME 333. Regulators and Effectors of Small GTPases (Part G: Ras Family II)
Edited by W. E. BALCH, CHANNING J. DER, AND ALAN HALL

VOLUME 334. Hyperthermophilic Enzymes (Part C)
Edited by MICHAEL W. W. ADAMS AND ROBERT M. KELLY

VOLUME 335. Flavonoids and Other Polyphenols
Edited by LESTER PACKER

VOLUME 336. Microbial Growth in Biofilms (Part A: Developmental and Molecular Biological Aspects)
Edited by RON J. DOYLE

VOLUME 337. Microbial Growth in Biofilms (Part B: Special Environments and Physicochemical Aspects)
Edited by RON J. DOYLE

VOLUME 338. Nuclear Magnetic Resonance of Biological Macromolecules (Part A)
Edited by THOMAS L. JAMES, VOLKER DÖTSCH, AND ULI SCHMITZ

VOLUME 339. Nuclear Magnetic Resonance of Biological Macromolecules (Part B)
Edited by THOMAS L. JAMES, VOLKER DÖTSCH, AND ULI SCHMITZ

VOLUME 340. Drug–Nucleic Acid Interactions
Edited by JONATHAN B. CHAIRES AND MICHAEL J. WARING

VOLUME 341. Ribonucleases (Part A)
Edited by ALLEN W. NICHOLSON

VOLUME 342. Ribonucleases (Part B)
Edited by ALLEN W. NICHOLSON

VOLUME 343. G Protein Pathways (Part A: Receptors)
Edited by RAVI IYENGAR AND JOHN D. HILDEBRANDT

VOLUME 344. G Protein Pathways (Part B: G Proteins and Their Regulators)
Edited by RAVI IYENGAR AND JOHN D. HILDEBRANDT

VOLUME 345. G Protein Pathways (Part C: Effector Mechanisms)
Edited by RAVI IYENGAR AND JOHN D. HILDEBRANDT

VOLUME 346. Gene Therapy Methods
Edited by M. IAN PHILLIPS

VOLUME 347. Protein Sensors and Reactive Oxygen Species (Part A: Selenoproteins and Thioredoxin)
Edited by HELMUT SIES AND LESTER PACKER

VOLUME 348. Protein Sensors and Reactive Oxygen Species (Part B: Thiol Enzymes and Proteins)
Edited by HELMUT SIES AND LESTER PACKER

VOLUME 349. Superoxide Dismutase
Edited by LESTER PACKER

VOLUME 350. Guide to Yeast Genetics and Molecular and Cell Biology (Part B)
Edited by CHRISTINE GUTHRIE AND GERALD R. FINK

VOLUME 351. Guide to Yeast Genetics and Molecular and Cell Biology (Part C)
Edited by CHRISTINE GUTHRIE AND GERALD R. FINK

VOLUME 352. Redox Cell Biology and Genetics (Part A)
Edited by CHANDAN K. SEN AND LESTER PACKER

VOLUME 353. Redox Cell Biology and Genetics (Part B)
Edited by CHANDAN K. SEN AND LESTER PACKER

VOLUME 354. Enzyme Kinetics and Mechanisms (Part F: Detection and Characterization of Enzyme Reaction Intermediates)
Edited by DANIEL L. PURICH

VOLUME 355. Cumulative Subject Index Volumes 321–354

VOLUME 356. Laser Capture Microscopy and Microdissection
Edited by P. MICHAEL CONN

VOLUME 357. Cytochrome P450, Part C
Edited by ERIC F. JOHNSON AND MICHAEL R. WATERMAN

VOLUME 358. Bacterial Pathogenesis (Part C: Identification, Regulation, and Function of Virulence Factors)
Edited by VIRGINIA L. CLARK AND PATRIK M. BAVOIL

VOLUME 359. Nitric Oxide (Part D)
Edited by ENRIQUE CADENAS AND LESTER PACKER

VOLUME 360. Biophotonics (Part A)
Edited by GERARD MARRIOTT AND IAN PARKER

VOLUME 361. Biophotonics (Part B)
Edited by GERARD MARRIOTT AND IAN PARKER

VOLUME 362. Recognition of Carbohydrates in Biological Systems (Part A)
Edited by YUAN C. LEE AND REIKO T. LEE

VOLUME 363. Recognition of Carbohydrates in Biological Systems (Part B)
Edited by YUAN C. LEE AND REIKO T. LEE

VOLUME 364. Nuclear Receptors
Edited by DAVID W. RUSSELL AND DAVID J. MANGELSDORF

VOLUME 365. Differentiation of Embryonic Stem Cells
Edited by PAUL M. WASSAUMAN AND GORDON M. KELLER

VOLUME 366. Protein Phosphatases
Edited by SUSANNE KLUMPP AND JOSEF KRIEGLSTEIN

VOLUME 367. Liposomes (Part A)
Edited by NEJAT DÜZGÜNEŞ

VOLUME 368. Macromolecular Crystallography (Part C)
Edited by CHARLES W. CARTER, JR., AND ROBERT M. SWEET

VOLUME 369. Combinational Chemistry (Part B)
Edited by GUILLERMO A. MORALES AND BARRY A. BUNIN

VOLUME 370. RNA Polymerases and Associated Factors (Part C)
Edited by SANKAR L. ADHYA AND SUSAN GARGES

VOLUME 371. RNA Polymerases and Associated Factors (Part D)
Edited by SANKAR L. ADHYA AND SUSAN GARGES

VOLUME 372. Liposomes (Part B)
Edited by NEJAT DÜZGÜNEŞ

VOLUME 373. Liposomes (Part C)
Edited by NEJAT DÜZGÜNEŞ

VOLUME 374. Macromolecular Crystallography (Part D)
Edited by CHARLES W. CARTER, JR., AND ROBERT W. SWEET

VOLUME 375. Chromatin and Chromatin Remodeling Enzymes (Part A)
Edited by C. DAVID ALLIS AND CARL WU

VOLUME 376. Chromatin and Chromatin Remodeling Enzymes (Part B)
Edited by C. DAVID ALLIS AND CARL WU

VOLUME 377. Chromatin and Chromatin Remodeling Enzymes (Part C)
Edited by C. DAVID ALLIS AND CARL WU

VOLUME 378. Quinones and Quinone Enzymes (Part A)
Edited by HELMUT SIES AND LESTER PACKER

VOLUME 379. Energetics of Biological Macromolecules (Part D)
Edited by JO M. HOLT, MICHAEL L. JOHNSON, AND GARY K. ACKERS

VOLUME 380. Energetics of Biological Macromolecules (Part E)
Edited by JO M. HOLT, MICHAEL L. JOHNSON, AND GARY K. ACKERS

VOLUME 381. Oxygen Sensing
Edited by CHANDAN K. SEN AND GREGG L. SEMENZA

VOLUME 382. Quinones and Quinone Enzymes (Part B)
Edited by HELMUT SIES AND LESTER PACKER

VOLUME 383. Numerical Computer Methods (Part D)
Edited by LUDWIG BRAND AND MICHAEL L. JOHNSON

VOLUME 384. Numerical Computer Methods (Part E)
Edited by LUDWIG BRAND AND MICHAEL L. JOHNSON

VOLUME 385. Imaging in Biological Research (Part A)
Edited by P. MICHAEL CONN

VOLUME 386. Imaging in Biological Research (Part B)
Edited by P. MICHAEL CONN

Volume 387. Liposomes (Part D)
Edited by Nejat Düzgüneş

Volume 388. Protein Engineering
Edited by Dan E. Robertson and Joseph P. Noel

Volume 389. Regulators of G-Protein Signaling (Part A)
Edited by David P. Siderovski

Volume 390. Regulators of G-Protein Signaling (Part B)
Edited by David P. Siderovski

Volume 391. Liposomes (Part E)
Edited by Nejat Düzgüneş

Volume 392. RNA Interference
Edited by Engelke Rossi

Volume 393. Circadian Rhythms
Edited by Michael W. Young

Volume 394. Nuclear Magnetic Resonance of Biological Macromolecules (Part C)
Edited by Thomas L. James

Volume 395. Producing the Biochemical Data (Part B)
Edited by Elizabeth A. Zimmer and Eric H. Roalson

Volume 396. Nitric Oxide (Part E)
Edited by Lester Packer and Enrique Cadenas

Volume 397. Environmental Microbiology
Edited by Jared R. Leadbetter

Volume 398. Ubiquitin and Protein Degradation (Part A)
Edited by Raymond J. Deshaies

Volume 399. Ubiquitin and Protein Degradation (Part B)
Edited by Raymond J. Deshaies

Volume 400. Phase II Conjugation Enzymes and Transport Systems
Edited by Helmut Sies and Lester Packer

Volume 401. Glutathione Transferases and Gamma Glutamyl Transpeptidases
Edited by Helmut Sies and Lester Packer

Volume 402. Biological Mass Spectrometry
Edited by A. L. Burlingame

Volume 403. GTPases Regulating Membrane Targeting and Fusion
Edited by William E. Balch, Channing J. Der, and Alan Hall

Volume 404. GTPases Regulating Membrane Dynamics
Edited by William E. Balch, Channing J. Der, and Alan Hall

Volume 405. Mass Spectrometry: Modified Proteins and Glycoconjugates
Edited by A. L. Burlingame

Volume 406. Regulators and Effectors of Small GTPases: Rho Family
Edited by William E. Balch, Channing J. Der, and Alan Hall

Volume 407. Regulators and Effectors of Small GTPases: Ras Family
Edited by William E. Balch, Channing J. Der, and Alan Hall

Volume 408. DNA Repair (Part A)
Edited by Judith L. Campbell and Paul Modrich

Volume 409. DNA Repair (Part B)
Edited by Judith L. Campbell and Paul Modrich

Volume 410. DNA Microarrays (Part A: Array Platforms and Web-Bench Protocols)
Edited by Alan Kimmel and Brian Oliver

Volume 411. DNA Microarrays (Part B: Databases and Statistics)
Edited by Alan Kimmel and Brian Oliver

Volume 412. Amyloid, Prions, and Other Protein Aggregates (Part B)
Edited by Indu Kheterpal and Ronald Wetzel

Volume 413. Amyloid, Prions, and Other Protein Aggregates (Part C)
Edited by Indu Kheterpal and Ronald Wetzel

Volume 414. Measuring Biological Responses with Automated Microscopy
Edited by James Inglese

Volume 415. Glycobiology
Edited by Minoru Fukuda

Volume 416. Glycomics
Edited by Minoru Fukuda

Volume 417. Functional Glycomics
Edited by Minoru Fukuda

Volume 418. Embryonic Stem Cells
Edited by Irina Klimanskaya and Robert Lanza

VOLUME 419. Adult Stem Cells
Edited by IRINA KLIMANSKAYA AND ROBERT LANZA

VOLUME 420. Stem Cell Tools and Other Experimental Protocols
Edited by IRINA KLIMANSKAYA AND ROBERT LANZA

VOLUME 421. Advanced Bacterial Genetics: Use of Transposons and Phage for Genomic Engineering
Edited by KELLY T. HUGHES

VOLUME 422. Two-Component Signaling Systems, Part A
Edited by MELVIN I. SIMON, BRIAN R. CRANE, AND ALEXANDRINE CRANE

VOLUME 423. Two-Component Signaling Systems, Part B
Edited by MELVIN I. SIMON, BRIAN R. CRANE, AND ALEXANDRINE CRANE

VOLUME 424. RNA Editing
Edited by JONATHA M. GOTT

VOLUME 425. RNA Modification
Edited by JONATHA M. GOTT

VOLUME 426. Integrins
Edited by DAVID CHERESH

VOLUME 427. MicroRNA Methods
Edited by JOHN J. ROSSI

VOLUME 428. Osmosensing and Osmosignaling
Edited by HELMUT SIES AND DIETER HAUSSINGER

VOLUME 429. Translation Initiation: Extract Systems and Molecular Genetics
Edited by JON LORSCH

VOLUME 430. Translation Initiation: Reconstituted Systems and Biophysical Methods
Edited by JON LORSCH

VOLUME 431. Translation Initiation: Cell Biology, High-Throughput and Chemical-Based Approaches
Edited by JON LORSCH

VOLUME 432. Lipidomics and Bioactive Lipids: Mass-Spectrometry–Based Lipid Analysis
Edited by H. ALEX BROWN

VOLUME 433. Lipidomics and Bioactive Lipids: Specialized Analytical Methods and Lipids in Disease
Edited by H. ALEX BROWN

VOLUME 434. Lipidomics and Bioactive Lipids: Lipids and Cell Signaling
Edited by H. ALEX BROWN

VOLUME 435. Oxygen Biology and Hypoxia
Edited by HELMUT SIES AND BERNHARD BRÜNE

VOLUME 436. Globins and Other Nitric Oxide-Reactive Protiens (Part A)
Edited by ROBERT K. POOLE

VOLUME 437. Globins and Other Nitric Oxide-Reactive Protiens (Part B)
Edited by ROBERT K. POOLE

VOLUME 438. Small GTPases in Disease (Part A)
Edited by WILLIAM E. BALCH, CHANNING J. DER, AND ALAN HALL

VOLUME 439. Small GTPases in Disease (Part B)
Edited by WILLIAM E. BALCH, CHANNING J. DER, AND ALAN HALL

VOLUME 440. Nitric Oxide, Part F Oxidative and Nitrosative Stress in Redox Regulation of Cell Signaling
Edited by ENRIQUE CADENAS AND LESTER PACKER

VOLUME 441. Nitric Oxide, Part G Oxidative and Nitrosative Stress in Redox Regulation of Cell Signaling
Edited by ENRIQUE CADENAS AND LESTER PACKER

VOLUME 442. Programmed Cell Death, General Principles for Studying Cell Death (Part A)
Edited by ROYA KHOSRAVI-FAR, ZAHRA ZAKERI, RICHARD A. LOCKSHIN, AND MAURO PIACENTINI

VOLUME 443. Angiogenesis: *In Vitro* Systems
Edited by DAVID A. CHERESH

VOLUME 444. Angiogenesis: *In Vivo* Systems (Part A)
Edited by DAVID A. CHERESH

VOLUME 445. Angiogenesis: *In Vivo* Systems (Part B)
Edited by DAVID A. CHERESH

VOLUME 446. Programmed Cell Death, The Biology and Therapeutic Implications of Cell Death (Part B)
Edited by ROYA KHOSRAVI-FAR, ZAHRA ZAKERI, RICHARD A. LOCKSHIN, AND MAURO PIACENTINI

Volume 447. RNA Turnover in Bacteria, Archaea and Organelles
Edited by Lynne E. Maquat and Cecilia M. Arraiano

Volume 448. RNA Turnover in Eukaryotes: Nucleases, Pathways and Analysis of mRNA Decay
Edited by Lynne E. Maquat and Megerditch Kiledjian

Volume 449. RNA Turnover in Eukaryotes: Analysis of Specialized and Quality Control RNA Decay Pathways
Edited by Lynne E. Maquat and Megerditch Kiledjian

Volume 450. Fluorescence Spectroscopy
Edited by Ludwig Brand and Michael L. Johnson

Volume 451. Autophagy: Lower Eukaryotes and Non-Mammalian Systems (Part A)
Edited by Daniel J. Klionsky

Volume 452. Autophagy in Mammalian Systems (Part B)
Edited by Daniel J. Klionsky

Volume 453. Autophagy in Disease and Clinical Applications (Part C)
Edited by Daniel J. Klionsky

Volume 454. Computer Methods (Part A)
Edited by Michael L. Johnson and Ludwig Brand

Volume 455. Biothermodynamics (Part A)
Edited by Michael L. Johnson, Jo M. Holt, and Gary K. Ackers (Retired)

Volume 456. Mitochondrial Function, Part A: Mitochondrial Electron Transport Complexes and Reactive Oxygen Species
Edited by William S. Allison and Immo E. Scheffler

Volume 457. Mitochondrial Function, Part B: Mitochondrial Protein Kinases, Protein Phosphatases and Mitochondrial Diseases
Edited by William S. Allison and Anne N. Murphy

Volume 458. Complex Enzymes in Microbial Natural Product Biosynthesis, Part A: Overview Articles and Peptides
Edited by David A. Hopwood

Volume 459. Complex Enzymes in Microbial Natural Product Biosynthesis, Part B: Polyketides, Aminocoumarins and Carbohydrates
Edited by David A. Hopwood

VOLUME 460. Chemokines, Part A
Edited by TRACY M. HANDEL AND DAMON J. HAMEL

VOLUME 461. Chemokines, Part B
Edited by TRACY M. HANDEL AND DAMON J. HAMEL

VOLUME 462. Non-Natural Amino Acids
Edited by TOM W. MUIR AND JOHN N. ABELSON

VOLUME 463. Guide to Protein Purification, 2nd Edition
Edited by RICHARD R. BURGESS AND MURRAY P. DEUTSCHER

VOLUME 464. Liposomes, Part F
Edited by NEJAT DÜZGÜNEŞ

VOLUME 465. Liposomes, Part G
Edited by NEJAT DÜZGÜNEŞ

VOLUME 466. Biothermodynamics, Part B
Edited by MICHAEL L. JOHNSON, GARY K. ACKERS, AND JO M. HOLT

VOLUME 467. Computer Methods Part B
Edited by MICHAEL L. JOHNSON AND LUDWIG BRAND

VOLUME 468. Biophysical, Chemical, and Functional Probes of RNA Structure, Interactions and Folding: Part A
Edited by DANIEL HERSCHLAG

VOLUME 469. Biophysical, Chemical, and Functional Probes of RNA Structure, Interactions and Folding: Part B
Edited by DANIEL HERSCHLAG

VOLUME 470. Guide to Yeast Genetics: Functional Genomics, Proteomics, and Other Systems Analysis, 2nd Edition
Edited by GERALD FINK, JONATHAN WEISSMAN, AND CHRISTINE GUTHRIE

VOLUME 471. Two-Component Signaling Systems, Part C
Edited by MELVIN I. SIMON, BRIAN R. CRANE, AND ALEXANDRINE CRANE

VOLUME 472. Single Molecule Tools, Part A: Fluorescence Based Approaches
Edited by NILS G. WALTER

VOLUME 473. Thiol Redox Transitions in Cell Signaling, Part A Chemistry and Biochemistry of Low Molecular Weight and Protein Thiols
Edited by ENRIQUE CADENAS AND LESTER PACKER

Volume 474. Thiol Redox Transitions in Cell Signaling, Part B Cellular Localization and Signaling
Edited by Enrique Cadenas and Lester Packer

Volume 475. Single Molecule Tools, Part B: Super-Resolution, Particle Tracking, Multiparameter, and Force Based Methods
Edited by Nils G. Walter

Volume 476. Guide to Techniques in Mouse Development, Part A Mice, Embryos, and Cells, 2nd Edition
Edited by Paul M. Wassarman and Philippe M. Soriano

Volume 477. Guide to Techniques in Mouse Development, Part B Mouse Molecular Genetics, 2nd Edition
Edited by Paul M. Wassarman and Philippe M. Soriano

Volume 478. Glycomics
Edited by Minoru Fukuda

Volume 479. Functional Glycomics
Edited by Minoru Fukuda

Volume 480. Glycobiology
Edited by Minoru Fukuda

Volume 481. Cryo-EM, Part A: Sample Preparation and Data Collection
Edited by Grant J. Jensen

Volume 482. Cryo-EM, Part B: 3-D Reconstruction
Edited by Grant J. Jensen

Volume 483. Cryo-EM, Part C: Analyses, Interpretation, and Case Studies
Edited by Grant J. Jensen

Volume 484. Constitutive Activity in Receptors and Other Proteins, Part A
Edited by P. Michael Conn

Volume 485. Constitutive Activity in Receptors and Other Proteins, Part B
Edited by P. Michael Conn

Volume 486. Research on Nitrification and Related Processes, Part A
Edited by Martin G. Klotz

Volume 487. Computer Methods, Part C
Edited by Michael L. Johnson and Ludwig Brand

Volume 488. Biothermodynamics, Part C
Edited by Michael L. Johnson, Jo M. Holt, and Gary K. Ackers

Volume 489. The Unfolded Protein Response and Cellular Stress, Part A
Edited by P. Michael Conn

Volume 490. The Unfolded Protein Response and Cellular Stress, Part B
Edited by P. Michael Conn

Volume 491. The Unfolded Protein Response and Cellular Stress, Part C
Edited by P. Michael Conn

Volume 492. Biothermodynamics, Part D
Edited by Michael L. Johnson, Jo M. Holt, and Gary K. Ackers

Volume 493. Fragment-Based Drug Design Tools, Practical Approaches, and Examples
Edited by Lawrence C. Kuo

Volume 494. Methods in Methane Metabolism, Part A
Methanogenesis
Edited by Amy C. Rosenzweig and Stephen W. Ragsdale

Volume 495. Methods in Methane Metabolism, Part B
Methanotrophy
Edited by Amy C. Rosenzweig and Stephen W. Ragsdale

Volume 496. Research on Nitrification and Related Processes, Part B
Edited by Martin G. Klotz and Lisa Y. Stein

Volume 497. Synthetic Biology, Part A
Methods for Part/Device Characterization and Chassis Engineering
Edited by Christopher Voigt

Volume 498. Synthetic Biology, Part B
Computer Aided Design and DNA Assembly
Edited by Christopher Voigt

Volume 499. Biology of Serpins
Edited by James C. Whisstock and Phillip I. Bird

Volume 500. Methods in Systems Biology
Edited by Daniel Jameson, Malkhey Verma, and Hans V. Westerhoff

Volume 501. Serpin Structure and Evolution
Edited by James C. Whisstock and Phillip I. Bird

Volume 502. Protein Engineering for Therapeutics, Part A
Edited by K. Dane Wittrup and Gregory L. Verdine

Volume 503. Protein Engineering for Therapeutics, Part B
Edited by K. Dane Wittrup and Gregory L. Verdine

Volume 504. Imaging and Spectroscopic Analysis of Living Cells
Optical and Spectroscopic Techniques
Edited by P. Michael Conn

Volume 505. Imaging and Spectroscopic Analysis of Living Cells
Live Cell Imaging of Cellular Elements and Functions
Edited by P. Michael Conn

Volume 506. Imaging and Spectroscopic Analysis of Living Cells
Imaging Live Cells in Health and Disease
Edited by P. Michael Conn

Volume 507. Gene Transfer Vectors for Clinical Application
Edited by Theodore Friedmann

Volume 508. Nanomedicine
Cancer, Diabetes, and Cardiovascular, Central Nervous System, Pulmonary and Inflammatory Diseases
Edited by Nejat Düzgüneş

Volume 509. Nanomedicine
Infectious Diseases, Immunotherapy, Diagnostics, Antifibrotics, Toxicology and Gene Medicine
Edited by Nejat Düzgüneş

Volume 510. Cellulases
Edited by Harry J. Gilbert

Volume 511. RNA Helicases
Edited by Eckhard Jankowsky

Volume 512. Nucleosomes, Histones & Chromatin, Part A
Edited by Carl Wu and C. David Allis

Volume 513. Nucleosomes, Histones & Chromatin, Part B
Edited by Carl Wu and C. David Allis

Volume 514. Ghrelin
Edited by Masayasu Kojima and Kenji Kangawa

Volume 515. Natural Product Biosynthesis by Microorganisms and Plants, Part A
Edited by David A. Hopwood

VOLUME 516. Natural Product Biosynthesis by Microorganisms and Plants, Part B
Edited by DAVID A. HOPWOOD

VOLUME 517. Natural Product Biosynthesis by Microorganisms and Plants, Part C
Edited by DAVID A. HOPWOOD

VOLUME 518. Fluorescence Fluctuation Spectroscopy (FFS), Part A
Edited by SERGEY TETIN

VOLUME 519. Fluorescence Fluctuation Spectroscopy (FFS), Part B
Edited by SERGEY TETIN

VOLUME 520. G Protein Coupled Receptors
Structure
Edited by P. MICHAEL CONN

VOLUME 521. G Protein Coupled Receptors
Trafficking and Oligomerization
Edited by P. MICHAEL CONN

VOLUME 522. G Protein Coupled Receptors
Modeling, Activation, Interactions and Virtual Screening
Edited by P. MICHAEL CONN

VOLUME 523. Methods in Protein Design
Edited by AMY E. KEATING

CHAPTER ONE

Computational Design of Novel Protein Binders and Experimental Affinity Maturation

Timothy A. Whitehead[*,†,1], **David Baker**[‡], **Sarel J. Fleishman**[§,1]
[*]Department of Chemical Engineering and Materials Science, Michigan State University, East Lansing, Michigan, USA
[†]Department of Biosystems and Agricultural Engineering, Michigan State University, East Lansing, Michigan, USA
[‡]Department of Biochemistry, University of Washington and Howard Hughes Medical Institute, Seattle, Washington, USA
[§]Department of Biological Chemistry, Weizmann Institute of Science, Rehovot, Israel
[1]Corresponding authors: e-mail address: taw@msu.edu; sarel@weizmann.ac.il

Contents

Abstract

Computational design of novel protein binders has recently emerged as a useful technique to study biomolecular recognition and generate molecules for use in biotechnology, research, and biomedicine. Current limitations in computational design methodology have led to the adoption of high-throughput screening and affinity maturation techniques to diagnose modeling inaccuracies and generate high activity binders. Here, we scrutinize this combination of computational and experimental aspects and propose areas for future methodological improvements.

Methods in Enzymology, Volume 523
ISSN 0076-6879
http://dx.doi.org/10.1016/B978-0-12-394292-0.00001-1

1. INTRODUCTION

Molecular recognition underlies all of biological function including signaling, immune recognition, and catalysis. Molecular structures of thousands of naturally occurring protein interactions illuminate the physical basis for biomolecular recognition. These structures reveal very high shape complementarity between the interacting surfaces and energetically optimized interactions, including van der Waals, electrostatic, and hydrogen-bonding contacts. Computational modeling has been able to recapitulate some of these structural features to design novel protein–protein interactions (e.g., Huang, Love, & Mayo, 2007; Jha et al., 2010; Karanicolas et al., 2011; Liu et al., 2007), but until recently, the ability to design high-affinity and specific protein binders of naturally occurring biomolecules without recourse to existing cocrystal structures was not demonstrated, signifying gaps in our understanding of biomolecular recognition and frustrating attempts to program new molecular interactions that impact biological processes.

Recently, we described a new computational method for the design of protein binders, which focused on designing the surfaces of natural proteins of diverse folds to incorporate a region of high affinity for interaction with the target protein, and used this method to generate binders of the highly conserved stem region on influenza hemagglutinin (HA; Fleishman, Whitehead, Ekiert, et al., 2011; Fig. 1.1). The designs were found to interact specifically with the desired site, but initial binding affinities were low. We therefore combined computational design with *in vitro* affinity maturation to generate high-affinity binders of influenza HA that inhibited its cell-invasion function. The affinity maturation process also diagnosed inaccuracies in the energy function which underlies the computational design process, thereby suggesting a future route to improvements in our understanding of molecular recognition through the iterative application of *de novo* design and affinity maturation (Whitehead et al., 2012). Here, we provide an in-depth description of the computational and experimental techniques, focusing on what appear to us as the most fruitful ways to combine the computational and experimental aspects. We suggest areas where additional methodological developments are necessary for robust and reproducible design of protein binding to become routine.

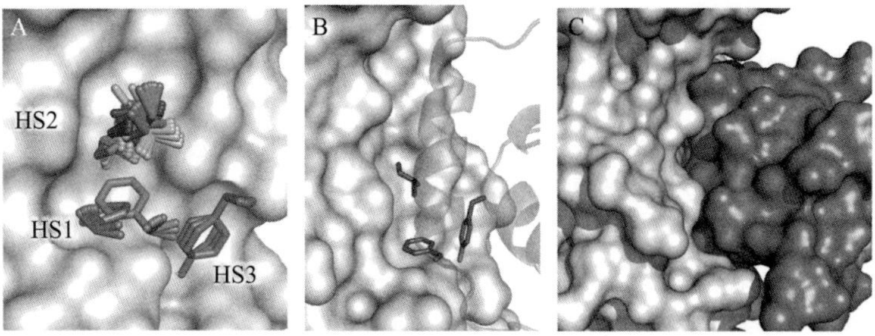

Figure 1.1 The computational design procedure realizes three features of natural protein–protein interactions: cores of high-affinity interactions with the target surface (A), favorable interactions among core residues (B), and high shape complementarity (C). (A) Three hotspot residue libraries (HS1, HS2, and HS3) were computed to form the idealized core of the interaction with the influenza hemagglutinin (HA) surface (gray). HS1 comprises two major configurations for a Phe aromatic ring and is supported by HS2, which contains the hydrophobic residues Phe, Leu, Ile, Met, and Val (green, purple, navy blue, cyan, and light brown, respectively), and HS3 comprises Tyr conformations. In the specific case of design of HA binders, the geometric constraint on HS2 is laxer than on the other hotspot positions and many different residues can be accommodated there. Residues from each hotspot-residue library interact favorably both with the target HA surface and the other hotspot-residue libraries, recapitulating two features common to many natural complexes: a core of highly optimized interactions with the target and internally stabilized contacts between key sidechains. (B) Cocrystal structure of HB80 and the HA surface. The hotspot residues are shown in dark green, realizing one of the energetically favorable combinations seen in panel a (comprising a Phe, Ile, and Tyr, for HS1, HS2, and HS3, respectively). (C) Common to many natural protein interactions, the surfaces of the designed and target proteins fit together snugly in a high shape complementary configuration. All molecular graphics were generated using PyMol (DeLano, 2002). (See Color Insert.)

2. COMPUTATIONAL DESIGN OF BINDERS USING NOVEL SCAFFOLDS

Surveys of the molecular structures of protein–protein interactions have underscored the importance of high shape complementarity at the interface with many molecular structures showing interface-packing densities as high as those seen in protein cores (Lo Conte, Chothia, & Janin, 1999). Another feature of many protein–protein interfaces is energetically highly optimized interactions at the core of the interface, typically comprising long sidechains such as Tyr, Gln, and Leu. Such interaction hotspot regions contribute a large share of the binding energy (Bogan & Thorn,

1998; Clackson & Wells, 1995). In our preliminary design attempts (where the above features of optimized cores and high shape complementarity were the main selection criteria), we noted that the resulting designs still failed to embody another key feature of natural binders: when compared to natural binding surfaces, predicted hotspot residues on designed surfaces did not form appreciable stabilizing interactions with other structural elements in their host monomer (Fleishman, Khare, Koga, & Baker, 2011). The relative lack of stabilizing structural features in designed surfaces suggested that the designed surfaces were conformationally less rigid. We suggested that restricted sidechain plasticity was an important feature of binding surfaces, reducing entropy loss upon binding and precluding the reorganization of the binding surface into configurations that are incompatible with binding the target (Fleishman, Khare, Koga, et al., 2011). The organization of many natural hotspot regions into spatial clusters is potentially a negative design feature, disfavoring alternative conformations of the binding surface; due to spatial clustering, these alternative conformations are likely to introduce rotameric strain, voids, or clashes in the unbound protein. These three structural features—high shape complementarity, energetically optimized interactions between core residues on the binder with the target, and the clustering of these core residues—serve as the basis for the computational design method (Fleishman, Corn, Strauch, et al., 2011), and in the following, we discuss how each feature is realized in a computational design framework (Fig. 1.1).

The computational method was implemented as an extension of the Rosetta software suite for macromolecular modeling (Das & Baker, 2008). Rosetta provides implementations of many key functionalities in biomolecule structure prediction and design, providing a straightforward means to access sophisticated computational methods. Rosetta and the methods described here can be freely obtained by academic users through the RosettaCommons agreement.

3. TARGET SELECTION

In selecting a target for designing inhibitors, a number of structural and experimental considerations need to be taken into account. The combined computational–experimental approach that we describe requires that the protein that is targeted for binding is characterized with high-resolution molecular structures, is stable, and can be produced in good yield and purity for binding measurements. It is important to have a protein/small molecule that binds at the target surface. Such preexisting binders provide valuable information on whether the target surface forms correctly in experiment, allowing confirmation that the designed binders target the intended site by serving as positive binding controls and as competitive inhibitors of the designed binders. Such preexisting binders might not be available for all desired applications. In such cases, amino acid substitutions at the target surface that disable binding to the designed proteins yet preserve the overall structure of the target can provide an alternative control for binding at the target site. Our computational strategy, which generates exposed hotspot sites of interaction, is best equipped to target concave protein surfaces. The conserved epitope of the soluble ectodomain of HA common to Group I influenza viruses exhibited all of these features: influenza HA is well characterized biochemically and structurally, it can be produced recombinantly in insect cells, and there exist multiple antibodies that bind at or near the epitope to serve as positive controls in experiments (Corti et al., 2011; Ekiert et al., 2009, 2011; Sui et al., 2009). It is also an important target for drug design, as binding in this region has been linked to preventing HA-mediated fusion of the viral and host endosomal membranes, thereby blocking viral entry into the cytoplasm of the host cell (Ekiert & Wilson, 2012).

4. GENERATING AN IDEALIZED CONCEPT OF THE HOTSPOT

A central element of the computational method is the construction of a spatial region in which high affinity, sidechain-mediated interactions are formed between the designed binder and the target protein; the designed sidechains should also be stabilized through intrachain interactions on the designed protein. As different surfaces on scaffold proteins for design present different ways in which to incorporate such key residues, we start the design process by precomputing a spatially clustered set of residue combinations (Fig. 1.1A). To date, we have generated binders with two to four hotspot positions, but the methods described below could be used to specify any number of hotspot positions. We previously provided examples based on natural protein–protein interactions for how to generate a hotspot conception when molecular structures of bound components are available (Fleishman, Corn, Strauch, et al., 2011). In the following, we describe in detail how to generate a hotspot region for a site that is known to serve in protein–protein interactions, but with minimal or no recourse to the natural binding mode; this approach was used to generate the hotspot region for

HA binding (Fleishman, Whitehead, Ekiert et al., 2011) and provides a way to generate protein binders with desired structural and biophysical properties without the limitations of naturally occurring binders (Fig. 1.1). To accomplish this, we docked aromatic residues against the hemagglutinin Trp21 on chain 2 of the HA (HA2) (HA residue numbering corresponds to the H3 subtype sequence-numbering convention) and isolated two major conformations of a Phe residue (other aromatic residue identities failed to produce energetically optimized configurations with respect to the target surface). For each Phe residue, we computed positions for backbone and Cβ atoms such that the aromatic ring moieties of the computed Phe residues intersected with the energetically optimized configurations computed in the previous step (inverse rotamers) (Fig. 1.1A); this step produces approximately a dozen different conformations for each of the two major configurations of the aromatic ring. All of these spatially clustered residues are saved in a hotspot-residue library for use in subsequent scaffold design steps. The HA target site is quite hydrophobic, and we extended this hotspot with hydrophobic residues (Phe, Leu, Ile, Met, Val) all of which formed favorable interactions both with the target HA surface and the previously computed hotspot Phe. A third Tyr hotspot position was extracted from antibody-bound structures of HA (Ekiert et al., 2009). This last hotspot residue was used in the design calculations that resulted in the binder codenamed HB80 (Fleishman, Whitehead, Ekiert, et al., 2011), but this design strategy, which included three hotspot residues, was found to be very restrictive (eliminating many potentially favorable designs), and the binder HB36 was computed without the Tyr hotspot residue. Once residue identities for the hotspot region were defined, the rigid-body conformations of these residues with respect to the target HA surface and to one another were computationally optimized by subjecting them to rigid-body docking and minimization simulations using Rosetta, and the lowest-energy conformations were isolated. Some of the hotspot-residue combinations, which appeared feasible and favorable at the initial hotspot construction phase, such as Val and Phe at HS2, failed to produce designed proteins with favorable energetic and structural characteristics, underscoring the importance of formulating a diverse hotspot concept. In summary, a hotspot region can be computed based on existing bound structures (Fleishman, Corn, Strauch, et al., 2011) or based solely on the molecular structure of the target protein (Fleishman, Whitehead, Ekiert et al., 2011). It is important to note that at this point of method development producing a hotspot concept requires human intervention in choice of target site and residue identities. Part of the reason why this crucial step has not been automated stems from the fact that the energetics of residue binding to the target surface within the context of an entire protein are poorly modeled when these residues are dismembered from a protein as in the case of building a hotspot. We overcame this difficulty by testing diverse hotspot concepts as explained above. Recently, a method has succeeded in recapitulating natural hotspot residues and may serve in future design studies to automatically generate hotspot regions from scratch (Ben-Shimon & Eisenstein, 2010).

5. SELECTING SHAPE COMPLEMENTARY SCAFFOLD SURFACES FOR DESIGN

Shape complementarity is a key feature of biological protein–protein interactions (Lo Conte et al., 1999). Although certain protein families recur in biology as protein interaction modules (e.g., SH3, ankyrin, immunoglobulins; Pawson & Nash, 2003), we reasoned that using more scaffolds for design would increase the chances of designing proteins that exhibit high shape complementarity. We therefore used a set of more than 800 unique protein structures deposited in the Protein DataBank as scaffolds for design (Fleishman, Whitehead, Ekiert, et al., 2011). This library of protein structures was selected to improve the chances that the proteins would be easy to experimentally express and test; they contain no disulfide bridges, are relatively small (<250 amino acid residues), contain no small molecule ligands, and are predicted to be monomeric. This scaffold library could be periodically updated with newly deposited structures. Another potential extension may be to include structures of proteins containing disulfide bridges, as those may present stable scaffolds for design. A number of computational methods have been described that capitalize on the high shape complementarity of interacting molecular surfaces to predict the proteins' mode of interaction (e.g., Gabb, Jackson, & Sternberg, 1997; Katchalski-Katzir et al., 1992). In a step independent of the hotspot construction step above, we use an efficient docking method called PatchDock (Schneidman-Duhovny, Inbar, Nussinov, & Wolfson, 2005) to isolate hundreds of unique configurations for each scaffold protein that show high shape complementarity to the target. The PatchDock software is run externally to Rosetta as a precomputation step and the output files from PatchDock are read by the RosettaScripts (Fleishman, Leaver-Fay, Corn, et al., 2011) module of Rosetta to set up the configurations with which design simulations start.

6. INTERFACE DESIGN

At this point in the protocol, we have obtained hundreds of thousands of coarse-grained binding configurations of the scaffolds in the library docked against the target epitope. The next step is to computationally determine which scaffold surfaces can incorporate hotspot residues. We developed two approaches for placing the hotspot residues on the scaffold protein (Fleishman, Corn, Strauch, et al., 2011). The approaches either translate the scaffold protein such that residues in the vicinity of the precomputed hotspot residues align their backbone N—C and Cα–Cβ vectors perfectly with those of the hotspot residue (scaffold placement) or by starting from the PatchDock configuration and replacing a scaffold position of appropriate geometry with one of the hotspot residues in the library (hotspot placement). The scaffold placement approach reproduces with high fidelity the geometric relationships of the precomputed hotspot residue with the target surface, whereas the hotspot placement strategy allows for more slack in incorporating the hotspot residue. Choice of which algorithm to use in placement depends on the physicochemical nature of the hotspot positions: residues that form geometrically constrained interactions, such as hydrogen bonds should be positioned with scaffold placement, whereas sidechains that form hydrophobic interactions can be incorporated using hotspot placement. When applied sequentially, these two algorithms can be used to test the energetic compatibility of every combination of hotspot residues from all hotspot-residue libraries with the scaffold protein. This is an exhaustive approach, but one that is poorly scalable and we have found that for three or more hotspot positions a computationally less demanding strategy is needed. We developed an alternative simultaneous-placement procedure. This procedure focuses design calculations on a set of scaffold residues that provide the most optimal geometric compatibility with the hotspot-residue libraries and designs this set of residues simultaneously. Since this procedure does not iterate over each combination of hotspot residues defined in the hotspot-residue libraries, it is much more scalable and allows the design of, in principle, as many hotspot positions as needed. In practice, combinations of these three methods can be used for any design task. For instance, the accurate scaffold placement procedure would be applied to hotspot positions that are geometrically very constrained (e.g., HS3 in Fig. 1.1A), whereas all other hotspot positions would be designed with the simultaneous-placement strategy.

Every target surface has unique structural features that demand a tailored computational design protocol. We implemented all of the above algorithms within the RosettaScripts framework allowing the protein designer to define specific constraints and filters in a versatile XML-style scripting language (Fleishman, Leaver-Fay, Corn, et al., 2011). For example, the HA target surface contains a recessed hydrophobic residue (Trp21) and an exposed backbone carbonyl. Following hotspot-residue placement, we found that many of the designs do not pack well against Trp21 or hydrogen bond with the backbone carbonyl. In the RosettaScripts framework, this problem was easily put right, without additional programming of the underlying C++ source code, by adding filters that prune modeling trajectories if these two constraints were not satisfied immediately following the hotspot placement steps. The RosettaScripts for running the HA design protocol and additional scripts for recapitulating a diverse set of natural complexes have been published providing computational design programs for a broad spectrum of molecular surfaces that could be used or modified to target other desired surfaces (Fleishman, Corn, Strauch, et al., 2011; Fleishman, Whitehead, Ekiert, et al., 2011).

Following the design of the core hotspot region, we use iterations of RosettaDesign and minimization (Kuhlman & Baker, 2000) to optimize the sequence of the scaffold protein for binding the target, while keeping the hotspot region fixed. The computational design strategy described here has been found to recapitulate natural binding interfaces with high sequence and conformational recovery rates (Fleishman, Corn, Strauch, et al., 2011). In the design of HA binders, we used small backbone motions to further optimize the binding interface, but in retrospect found that designed proteins that bound the target in experiment had very rigid backbones, where these motions had little effect (Fleishman, Whitehead, Ekiert, et al., 2011). In unpublished design work, we similarly found that all designs that bind their targets as intended contain a very high fraction of rigid secondary structural elements at the binding surface. The question of whether and what type of backbone motions are useful for interface design is the subject of ongoing research (Humphris & Kortemme, 2008; Zhang & Lai, 2012) and will impact our design capabilities as well as our understanding of the intimate connections between protein stability, conformational flexibility, and molecular function.

7. YEAST CELL-SURFACE DISPLAY AS A SCREENING METHOD FOR DESIGNED BINDERS

The advent of custom and affordable DNA synthesis allows the testing of scores of potential designs and greatly expands the diversity of designs that can be considered. This added capability necessitates a matching experimental method to screen for interactions. While every screening method to identify putative binders has its advantages and pitfalls, the ideal features of a screening method would be the following: that the throughput of the experimental screening method be matched to that of the computational design process, that diverse proteins can be robustly expressed, that weak binders (dissociation constants in the micromolar range; designs with fast kinetic dissociation rates) be positively identified, and that false negatives and false positives be minimized. Screening methods previously used by design groups in published and unpublished work included pull-down approaches, ELISA with phage display (Gu et al., 1995), cell extracts, or purified proteins (Karanicolas et al., 2011), and purification followed by binding verification using surface plasmon resonance or fluorescence polarization. After critically evaluating these alternatives, we found yeast cell-surface display coupled to flow cytometry as best matched to our screening needs. Yeast cell-surface display is well documented methodologically (Chao et al., 2006). There is a demonstrated correspondence between dissociation constants (K_D) measured using yeast display and *in vitro* measurements on purified proteins up to $K_D = 100$ nM (Colby et al., 2004). It is possible to improve detection limits by increasing the effective affinity using avidity between the yeast surface and naturally oligomeric targets (like the trimeric influenza HA) or by creating oligomeric complexes of the target protein using a biotin-conjugated target protein bound to streptavidin (Chao et al., 2006). There are published protocols for screening as well as directed evolution approaches for affinity maturation. Finally, using the yeast display format described below, we have found that more than 80% of the designed proteins express robustly. In a side-by-side comparison, only 50% of these proteins could be solubly expressed in BL21 bacterial cells under standard expression protocols (Studier, 2005). Yeast display requires access to a flow cytometer, and affordable cytometers have been marketed in recent years, with many academic institutions having dedicated flow cytometry facilities. We recommend using flow cytometry for monitoring binding events. Although there is a literature on identifying weak interactions using yeast display coupled to magnetic bead screening (Ackerman et al., 2009), in our hands these systems were less robust for screening than flow cytometry. For reproducibility by other laboratories, we have made yeast display expression plasmids encoding 71 different designs available through AddGene (www.AddGene.com).

As *S. cerevisiae* surface-expressed proteins could be glycosylated by the cellular machinery, care must be taken to remove potential N-glycosylation consensus sequences near the designed surface. Unpaired surface exposed cysteines are also removed from the designs. We note that yeast are able to express complicated multidomain and disulfide-linked proteins on their cell surface; thus the scaffold set used in the HA design effort could be expanded in future design efforts to include more diverse scaffolds.

One consideration of any screening method is the choice of a positive control to monitor binding events. Positive binding controls are important to ensure that the screen is run correctly; an ideal positive control would bind the targeted surface at the weak limits of detection. In our implementation, we chose the CR6261 antibody fragment (Fab) that was previously found to bind the targeted HA surface (Throsby et al., 2008). We created a CR6261 scFv using splice overlap extension PCR, displayed this construct on the cell surface, and verified binding against biotinylated HA. HA was biotinylated either through a genetically encoded Avi-tag or chemically using NHS-Ester chemistry (Pierce). We used alanine-scanning mutagenesis on multiple antibody residues directly contacting HA and combined several mutations to significantly reduce the affinity of the CR6261–HA interaction.

With this experimental setup in place, screening could be carried out at a high pace. Four days after DNA-encoding designs arrived, designs could be tested for binding. As the method is efficient, dozens of designs could be tested in a single afternoon by following the screening procedure in Chao et al. (2006). Once we isolated designs that bound the target, the yeast display system allowed rapid implementation of controls, including testing the nondesigned wild-type scaffold for binding and competitive binding experiments with the soluble CR6261 Fab. Combined, these controls were used to rapidly screen out those scaffolds that bound natively to the target and to cull designs that bound the target but not at the desired (and designed) location (Fig. 1.2).

8. AFFINITY MATURATION

Limitations in the energy function and design method resulted in designs that initially bound specifically but rather weakly to the target site. We wanted to identify amino acid substitutions that conferred tighter binding to the target. Identification of such substitutions is important for three reasons: by improving the affinity of our designs toward the target, we can improve biologically relevant function; consistency between the substitutions uncovered through affinity maturation and the designed model structures can affirm the designed binding mode, even in the absence of experimentally determined molecular structures of complexes, which are sometimes difficult to obtain; identifying substitutions that are clearly better

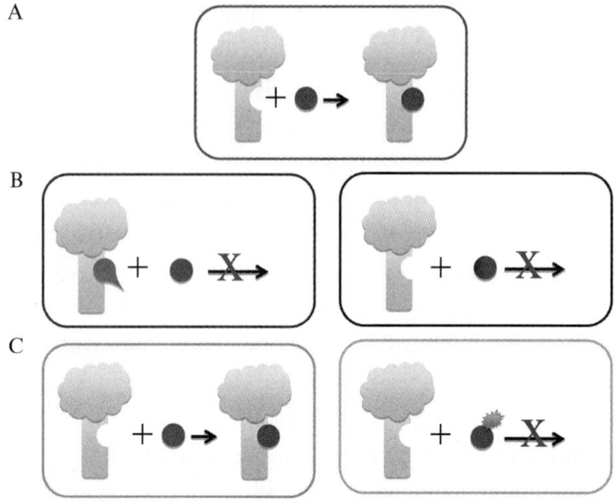

Figure 1.2 Yeast cell-surface display can be used to experimentally screen and test the precision of designed protein binders. (A) Screening for interactions is accomplished by incubation of a yeast-displayed designed protein (circle) with a purified, biotinylated target protein (gray stalk and head). After secondary labeling with a streptavidin-linked fluorophore, binding is measured by increased fluorescence of yeast cells as monitored by flow cytometry. (B) Precision of the designed binders can be tested using (left panel) competitive inhibition of the target surface with a small molecule/protein binder (teardrop) of the target surface and (right panel) coincubation of the target protein with a yeast-displayed scaffold from which the design was derived. The designed protein binds the targeted surface only if both experiments result in no increase in fluorescence. (C) Affinity maturation using yeast cell-surface display can further validate the accuracy of the design. Identification of mutations conferring affinity increases at the designed surface (left panel) suggests precision of the design if the improvements can be rationalized posteriori. Conversely, mutations conferring affinity increases distal to the designed surface (right panel) provide important clues that the protein is not binding the target as designed. (For color version of this figure, the reader is referred to the online version of this chapter.)

than the starting design helps, by retroactive rationalization, to improve the design process and identify inaccuracies in the energy function (Fig. 1.2).

We found yeast display coupled to flow cytometry to be well equipped for the affinity maturation process, as our binding screen could be readily reconfigured for cell sorting. In the affinity maturation protocol we implemented, single substitutions conferring large increases in affinity were isolated. Libraries encoding design variants could be generated either through site saturation mutagenesis (SSM) by the method of Kunkel (1985) or by error prone PCR (epPCR) (Genemorph II random mutagenesis kit, Stratagene) at low mutational loads (2–4 mutations/kb DNA), followed by high efficiency transformation into *S. cerevisiae* EBY100 cells (Benatuil, Perez, Belk, & Hsieh, 2010). Constructs were combined with the yeast display backbone by homologous recombination, obviating a subcloning step. Retrospectively, the choice of mutagenesis procedure was not essential, as both SSM and epPCR approaches yielded the same consensus mutations. This may reflect the small size and high backbone rigidity of the designed scaffold proteins; with other scaffolds, the choice of library construction method may be significant. We recommend epPCR as it is considerably less laborious than SSM, and mutations are located throughout the protein sequence, not just at the designed epitope. This feature is particularly important for removing potential designs from consideration, as affinity-enhancing substitutions distal from the designed binding site provide clues that the design might not bind in the intended mode (Fig. 1.2). Care must be taken to sort the library under conditions where a single clone cannot dominate the library because in this case the mutation(s) responsible for increased affinity cannot be unequivocally determined. In our implementation, we used two (at most three) sorts of epPCR mutagenesis libraries at a sorting threshold of 5%. Sorting gates were set to collect cells with the tightest binding to the target protein. Following sorting, we ensured that the library was improved relative to the starting sequence by measuring dissociation constants using yeast cell-surface display against the soluble target protein. We then plated yeast cells and sequenced individual clones using yeast colony PCR to extract the DNA. Once affinity-enhancing substitutions were verified, they were directly incorporated into the original design framework, eliminating substitutions that might have arisen through genetic drift, for example, on protein surfaces away from the binding surface. The entire procedure of generating a library, selecting and identifying beneficial mutations, and testing them in a clean background took a little over 2 weeks.

9. WHAT WORKS, WHAT FAILS, AND WHAT IT MEANS

De novo design of protein binders is in its infancy. We have implemented a computational design strategy, which produced two different designed proteins that bound with atomic precision at the desired protein surface (Fleishman, Whitehead, Ekiert, et al., 2011). In unpublished work using the same methodology, we generated several binders of other protein targets, demonstrating the method's robustness and scope. By comparing the relatively small number of designed proteins which bind their targets with the much larger number of designs which fail to do so, we have learned important lessons on biomolecular recognition (Fleishman, Whitehead, Strauch, et al., 2011). Similarly, by exhaustive characterization of working designs, we have been able to uncover weaknesses in the energy function used in design. We anticipate that the scrutiny of our design efforts by such computational and experimental methods will advance the ability to design new interactions in the future, and we have published the coordinates of all designed complexes that were experimentally tested (Fleishman, Whitehead, Ekiert, et al., 2011; Fleishman, Whitehead, Strauch, et al., 2011). In the following paragraphs, we describe two complementary approaches that we undertook in order to improve our understanding of molecular recognition and diagnose areas for future improvements (Fig. 1.3).

To get an unbiased view of structural features that distinguish computational designs from naturally occurring protein–protein interfaces, we compiled 88 designed proteins that expressed well and were tested for binding against three target proteins (HA, the human IgG1 Fc region, and an acyl carrier protein from *Mycobacterium tuberculosis*) (Fleishman, Whitehead, Strauch, et al., 2011). These designs did not bind their targets detectably. This set of protein structures was provided to 28 research groups that participated in the Critical Assessment of PRedicted Interactions experiment, and each group was asked to develop and disclose a computational metric for discriminating the nonbinding designs from naturally occurring interfaces. One of the surprising results came from an analysis by Haliloglu and coworkers revealing that many of the failed designs had binding surfaces that were predicted to be highly mobile and suggesting that these surfaces would not form as designed in experiment. This result underscores the importance of rigidity in functional surfaces: where such rigidity is not provided by the protein's secondary structure, the design effort was largely in vain. This insight has been translated to a simple computational filter, which tests whether the scaffold surface for design is embedded within the scaffold protein to stabilize it, demonstrating how *de novo* design, experimental characterization, and posteriori analysis can be used to diagnose and improve our understanding and ability to design binders. A recent cocrystal structure from another *de novo* protein binding study showed that plasticity either at the level of the sidechains or the backbone can lead to proteins which recognize their targets through quite different binding modes than those that were planned (Karanicolas et al., 2011) and might also have a role in the reduced effectiveness of *de novo* designed enzymes (Fleishman & Baker, 2012). Two directions for future research are consequently how to stabilize backbones that are poorly embedded in the scaffold protein, such as

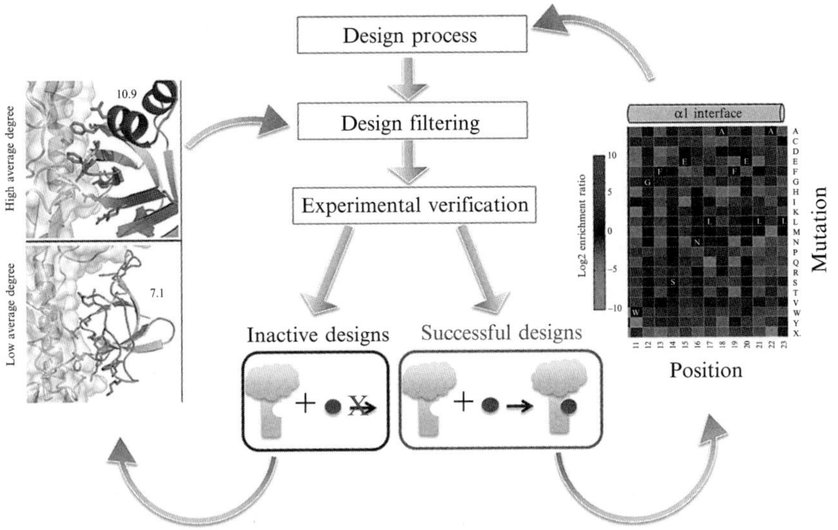

Figure 1.3 A schematic for how computation and experiment have been integrated to probe the physical basis for molecular recognition and to generate novel proteins of biomedical potential. Experimental screening of designs using yeast cell-surface display leads to their classification as active or inactive. Inactive designs can be compared to native protein interfaces (arrow pointing left): metrics that discriminate inactive designs from natural proteins can be used to formulate automated computational filters to prune unpromising designs in future design efforts and highlight areas for future methodological improvements in design. By contrast, successful designs (arrow pointing right) can be experimentally probed for detailed structure–function relationships using newly developed next generation sequencing technologies. Here, a plot showing enrichment ratio (a proxy for affinity) as a function of point mutation is shown for the interface stretch of one of the designs. This wealth of information can be used to identify limiting approximations in the energetic potential underlying the design calculations. By identifying mutations that clearly improve binding, this dataset can also be used to program high affinity and specificity binders from these initial designs that could subsequently be used as therapeutics. *The left-hand side of panel was reproduced with permission from Fleishman, Whitehead, Strauch, et al. (2011), and right-hand side with permission from Whitehead et al. (2012).* (See Color Insert.)

unstructured regions, and how to predict the stability of designed surfaces. This is particularly relevant for efforts to design antibodies or antibody-like scaffolds for therapeutically relevant targets.

In a second approach, we sought to fully characterize the effects on binding affinity and specificity of substitutions on the designed binders. Using the affinity maturation approach mentioned above, we had identified a handful of substitutions that increased the affinity of our designs to HA (Fleishman,

Whitehead, Ekiert, et al., 2011). The substitutions that increased affinity delineated potential shortcomings in the energy model that undergirds the design calculations, yet the data were too sparse to use our findings to improve the computational design process. To understand more completely the shortcomings of our energy model, we extended a recently described approach for experimentally mapping the affinity contributions of residues at binding interfaces using high-throughput sequencing to encompass much larger sets of positions (Araya & Fowler, 2011; Fowler et al., 2010). Very briefly, we transformed SSM libraries encoding all possible single point mutants of our designs into yeast and used fluorescence-activated cell sorting to select variants that bound the target protein HA (Whitehead et al., 2012) We then used Illumina DNA sequencing to sequence the entire population of design variants before and after selection. In so doing, a comprehensive sequence–function map for nearly every possible single point substitution in HB36 and HB80 was generated. As these maps were generated using selections for binding of H1 and H5 HA subtypes and at differing selection stringencies, for both designs we were able to determine the sequence determinants for affinity and subtype specificity.

For both designs, the sequence–function map identified many affinity-enhancing substitutions, and computational modeling indicated that a large fraction of these improved the long-range electrostatic complementarity of the designed binders with the HA surface. Long-range electrostatics effects are notoriously difficult to model (Fleishman & Baker, 2012), and the unprecedented amount of experimental data generated by this experimental approach provided an opportunity to test a variety of different electrostatic models, ultimately leading us to develop a rapidly computable static Poisson–Boltzmann electrostatics model that can be used as a final step in the design process to ensure that the designed proteins exhibit high charge complementarity with the target molecular surface. These maps enabled us to improve dissociation constants of both designs against HA to picomolar levels through the combination of many of these beneficial substitutions. The best design variant neutralized two different H1 influenza strains at doses comparable to the CR6261 antibody and is currently being tested for influenza abatement in live animal models. Thus, the combination of *de novo* protein design with comprehensive sequence–function mapping by deep sequencing can be used to generate proteins of potential therapeutic relevance.

Computational design of interactions holds great promise for extending our understanding of biomolecular recognition and ability to design novel proteins with useful molecular functions (Fleishman & Baker, 2012).

Although the field is young, computational design has already generated proteins with antiviral potential. Pressing areas for future development include a more general method for modeling of a hotspot region, the ability to design conformationally stable loop regions in functional sites, and improvements in the accuracy of the energy function. These abilities will open the way to routine and robust generation of novel biomolecules for biomedicine, biotechnology, and research.

ACKNOWLEDGMENTS

The authors thank past and present members of the Baker lab for many contributions to the methods described here, and Eva-Maria Strauch, Jim Stapleton, and Ravit Netzer for comments. Research in the Whitehead lab is supported by the National Science Foundation. Research in the Baker lab is supported by the National Institutes of Health, the Defense Advanced Research Projects Agency, the Defense Threat Reduction Agency, and the Howard Hughes Medical Institute. Research in the Fleishman lab is supported by the Geffen Fund, the Yeda-Sela Center, a donation from Sam Switzer and Family, a Marie Curie Integration Grant, an Alon Fellowship, the Israel Science Foundation, and the Human Frontier Science Program. SJF is the incumbent of the Martha S Sagon Career Development Chair.

REFERENCES

Ackerman, M., Levary, D., Tobon, G., Hackel, B., Orcutt, K. D., & Wittrup, K. D. (2009). Highly avid magnetic bead capture: An efficient selection method for de novo protein engineering utilizing yeast surface display. *Biotechnology Progress*, *25*, 774–783.

Araya, C. L., & Fowler, D. M. (2011). Deep mutational scanning: Assessing protein function on a massive scale. *Trends in Biotechnology*, *29*, 435–442.

Benatuil, L., Perez, J. M., Belk, J., & Hsieh, C. M. (2010). An improved yeast transformation method for the generation of very large human antibody libraries. *Protein Engineering, Design and Selection*, *23*, 155–159.

Ben-Shimon, A., & Eisenstein, M. (2010). Computational mapping of anchoring spots on protein surfaces. *Journal of Molecular Biology*, *402*, 259–277.

Bogan, A. A., & Thorn, K. S. (1998). Anatomy of hot spots in protein interfaces. *Journal of Molecular Biology*, *280*, 1–9.

Chao, G., Lau, W. L., Hackel, B. J., Sazinsky, S. L., Lippow, S. M., & Wittrup, K. D. (2006). Isolating and engineering human antibodies using yeast surface display. *Nature Protocols*, *1*, 755–768.

Clackson, T., & Wells, J. A. (1995). A hot spot of binding energy in a hormone-receptor interface. *Science*, *267*, 383–386.

Colby, D. W., Kellogg, B. A., Graff, C. P., Yeung, Y. A., Swers, J. S., & Wittrup, K. D. (2004). Engineering antibody affinity by yeast surface display. *Methods in Enzymology*, *388*, 348–358.

Corti, D., Voss, J., Gamblin, S. J., Codoni, G., Macagno, A., Jarrossay, D., et al. (2011). A neutralizing antibody selected from plasma cells that binds to group 1 and group 2 influenza A hemagglutinins. *Science*, *333*, 850–856.

Das, R., & Baker, D. (2008). Macromolecular modeling with rosetta. *Annual Review of Biochemistry*, *77*, 363–382.

DeLano, W. L. (2002). *The PyMol molecular graphics systems*. Palo Alto, CA: DeLano Scientific.

Ekiert, D. C., Bhabha, G., Elsliger, M. A., Friesen, R. H., Jongeneelen, M., Throsby, M., et al. (2009). Antibody recognition of a highly conserved influenza virus epitope. *Science, 324*, 246–251.

Ekiert, D. C., Friesen, R. H., Bhabha, G., Kwaks, T., Jongeneelen, M., Yu, W., et al. (2011). A highly conserved neutralizing epitope on group 2 influenza A viruses. *Science, 333*, 843–850.

Ekiert, D. C., & Wilson, I. A. (2012). Broadly neutralizing antibodies against influenza virus and prospects for universal therapies. *Current Opinion in Virology, 2*, 134–141.

Fleishman, S. J., & Baker, D. (2012). Role of the biomolecular energy gap in protein design, structure, and evolution. *Cell, 149*, 262–273.

Fleishman, S. J., Corn, J. E., Strauch, E. M., Whitehead, T. A., Karanicolas, J., & Baker, D. (2011). Hotspot-centric de novo design of protein binders. *Journal of Molecular Biology, 413*, 1047–1062.

Fleishman, S. J., Khare, S. D., Koga, N., & Baker, D. (2011). Restricted sidechain plasticity in the structures of native proteins and complexes. *Protein Science, 20*, 753–757.

Fleishman, S. J., Leaver-Fay, A., Corn, J. E., Strauch, E. M., Khare, S. D., Koga, N., et al. (2011). RosettaScripts: A scripting language interface to the Rosetta macromolecular modeling suite. *PLoS One, 6*, e20161.

Fleishman, S. J., Whitehead, T. A., Ekiert, D. C., Dreyfus, C., Corn, J. E., Strauch, E.-M., et al. (2011). Computational design of proteins targeting the conserved stem region of influenza hemagglutinin. *Science, 332*, 816–821.

Fleishman, S. J., Whitehead, T. A., Strauch, E. M., Corn, J. E., Qin, S., Zhou, H. X., et al. (2011). Community-wide assessment of protein-interface modeling suggests improvements to design methodology. *Journal of Molecular Biology, 414*, 289–302.

Fowler, D. M., Araya, C. L., Fleishman, S. J., Kellogg, E. H., Stephany, J. J., Baker, D., et al. (2010). High-resolution mapping of protein sequence-function relationships. *Nature Methods, 7*, 741–746.

Gabb, H. A., Jackson, R. M., & Sternberg, M. J. (1997). Modelling protein docking using shape complementarity, electrostatics and biochemical information. *Journal of Molecular Biology, 272*, 106–120.

Gu, H., Yi, Q., Bray, S. T., Riddle, D. S., Shiau, A. K., & Baker, D. (1995). A phage display system for studying the sequence determinants of protein folding. *Protein Science, 4*, 1108–1117.

Huang, P. S., Love, J. J., & Mayo, S. L. (2007). A de novo designed protein protein interface. *Protein Science, 16*, 2770–2774.

Humphris, E. L., & Kortemme, T. (2008). Prediction of protein-protein interface sequence diversity using flexible backbone computational protein design. *Structure, 16*, 1777–1788.

Jha, R. K., Leaver-Fay, A., Yin, S., Wu, Y., Butterfoss, G. L., Szyperski, T., et al. (2010). Computational design of a PAK1 binding protein. *Journal of Molecular Biology, 400*, 257–270.

Karanicolas, J., Corn, J. E., Chen, I., Joachimiak, L. A., Dym, O., Peck, S. H., et al. (2011). A de novo protein binding pair by computational design and directed evolution. *Molecular Cell, 42*, 250–260.

Katchalski-Katzir, E., Shariv, I., Eisenstein, M., Friesem, A. A., Aflalo, C., & Vakser, I. A. (1992). Molecular surface recognition: Determination of geometric fit between proteins and their ligands by correlation techniques. *Proceedings of the National Academy of Sciences of the United States of America, 89*, 2195–2199.

Kuhlman, B., & Baker, D. (2000). Native protein sequences are close to optimal for their structures. *Proceedings of the National Academy of Sciences of the United States of America, 97*, 10383–10388.

Kunkel, T. A. (1985). Rapid and efficient site-specific mutagenesis without phenotypic selection. *Proceedings of the National Academy of Sciences of the United States of America, 82*, 488–492.

Liu, S., Zhu, X., Liang, H., Cao, A., Chang, Z., & Lai, L. (2007). Nonnatural protein-protein interaction-pair design by key residues grafting. *Proceedings of the National Academy of Sciences of the United States of America, 104*, 5330–5335.

Lo Conte, L., Chothia, C., & Janin, J. (1999). The atomic structure of protein-protein recognition sites. *Journal of Molecular Biology, 285*, 2177–2198.

Pawson, T., & Nash, P. (2003). Assembly of cell regulatory systems through protein interaction domains. *Science, 300*, 445–452.

Schneidman-Duhovny, D., Inbar, Y., Nussinov, R., & Wolfson, H. J. (2005). PatchDock and SymmDock: Servers for rigid and symmetric docking. *Nucleic Acids Research, 33*, W363–W367.

Studier, F. W. (2005). Protein production by auto-induction in high density shaking cultures. *Protein Expression and Purification, 41*, 207–234.

Sui, J., Hwang, W. C., Perez, S., Wei, G., Aird, D., Chen, L. M., et al. (2009). Structural and functional bases for broad-spectrum neutralization of avian and human influenza A viruses. *Nature Structural and Molecular Biology, 16*, 265–273.

Throsby, M., van den Brink, E., Jongeneelen, M., Poon, L. L., Alard, P., Cornelissen, L., et al. (2008). Heterosubtypic neutralizing monoclonal antibodies cross-protective against H5N1 and H1N1 recovered from human IgM+ memory B cells. *PLoS One, 3*, e3942.

Whitehead, T. A., Chevalier, A., Song, Y., Dreyfus, C., Fleishman, S. J., De Mattos, C., et al. (2012). Optimization of affinity, specificity and function of designed influenza inhibitors using deep sequencing. *Nature Biotechnology, 30*, 543–548.

Zhang, C., & Lai, L. (2012). Automatch: Target-binding protein design and enzyme design by automatic pinpointing potential active sites in available protein scaffolds. *Proteins, 80*, 1078–1094.

CHAPTER TWO

Mining Tertiary Structural Motifs for Assessment of Designability

Jian Zhang*, Gevorg Grigoryan*,†,1
*Department of Computer Science, Dartmouth College, Hanover, New Hampshire, USA
†Department of Biology, Dartmouth College, Hanover, New Hampshire, USA
1Corresponding author: e-mail address: gevorg.grigoryan@dartmouth.edu

Contents

Abstract

The observation of a limited secondary-structural alphabet in native proteins, with significant sequence preferences, has profoundly influenced the fields of protein design and structure prediction (Simons, Kooperberg, Huang, & Baker, 1997; Verschueren et al., 2011). In the era of structural genomics, as the size of the structural dataset continues to grow rapidly, it is becoming possible to extend this analysis to tertiary structural motifs and their sequences. For a hypothetical tertiary motif, the rate of its utilization in natural proteins may be used to assess its designability—the ease with which the motif can be realized with natural amino acids. This requires a structural similarity search methodology, which rather than looking for global topological agreement (more appropriate for categorization of full proteins or domains), identifies detailed geometric matches. In this chapter, we introduce such a method, called MaDCaT, and demonstrate its use by assessing the designability landscapes of two tertiary structural motifs. We also show that such analysis can establish structure/sequence links by providing the sequence constraints necessary to encode designable motifs. As logical extension of their secondary-structure counterparts, tertiary structural preferences will likely prove extremely useful in *de novo* protein design and structure prediction.

Methods in Enzymology, Volume 523
ISSN 0076-6879
http://dx.doi.org/10.1016/B978-0-12-394292-0.00002-3

1. INTRODUCTION

The universe of natural protein structures appears to be degenerate, with many frequently repeating structural motifs (Vanhee et al., 2010; Verschueren et al., 2011). This is certainly apparent on the level of secondary structure as the majority of structured residues in folded proteins are found in either α-helices or β-strands (Joosten et al., 2011). However, the structural degeneracy goes beyond that. For example, clear preferences have been found for the ways in which secondary-structural elements come together in folded proteins. Helix–helix interactions in trans-membrane (TM) proteins (Walters & DeGrado, 2006) as well as overall topologies of TM proteins (Fuchs & Frishman, 2010) have been effectively classified and shown to have strong geometric biases. In soluble proteins, helix–helix crossings represent a classical example of a structural motif with strong geometric preferences (Grigoryan & Degrado, 2011; Kallblad & Dean, 2004; Moutevelis & Woolfson, 2009; Testa, Moutevelis, & Woolfson, 2009). Other well-established biases in super-secondary arrangements include packing of α-helices against β-sheets (Hu & Koehl, 2010), shearing and twisting of β-sheets (Ho & Curmi, 2002), β-turn, and α–α linking geometries (Engel & DeGrado, 2005; Hutchinson & Thornton, 1994), and others (Platt, Guerra, Zanotti, & Rigoutsos, 2003). In fact, when Fernandez-Fuentes, Dybas, and Fiser (2010)structurally classified all motifs consisting of two secondary-structural elements (α-helices or β-strands) connected by a loop, they found the different classes to occur at very different frequencies in the Protein Data Bank (PDB). The structural degeneracy of proteins is further evident at the level of domains (i.e., separable globular segments of structure), which are highly reused in nature (Marchler-Bauer et al., 2011), domain–domain and domain–peptide interaction interfaces (London, Movshovitz-Attias, & Schueler-Furman, 2010; Stein, Ceol, & Aloy, 2011; Vanhee et al., 2009, 2010), and even at the level of full-length protein structures, which are amenable to systematic hierarchical classification (Andreeva et al., 2008; Greene et al., 2007).

There can be several explanations for why some seemingly reasonable geometries appear to be very rare in proteins while others are very frequent. This may, in part, be due to incomplete coverage of the protein structural universe by the PDB, though at its present size the database is believed to have nearly saturated many structural features (Baeten et al., 2008; Fernandez-Fuentes et al., 2010; Zhang & Skolnick, 2005). Stochasticity in early evolution may have also contributed to higher prevalence of some types of structures over others. However, an important reason is likely that some structures are simply more difficult to realize using the 20 naturally occurring amino acids. This concept has been referred to as the *designability* of a protein structure, loosely defined as the number of amino-acid sequences capable of folding into it (England, Shakhnovich, & Shakhnovich, 2003; Govindarajan & Goldstein, 1996; Grigoryan & Degrado, 2011; Helling et al., 2001; Li, Helling, Tang, & Wingreen, 1996; Wingreen, Li, & Tang, 2004). Designability is a complex property that combines many physical factors. Certainly, designable structures must be able to accommodate a variety of amino acid combinations in an energetically favored fashion. In fact, Koehl and Levitt (2002) have shown that the magnitude of the sequence space compatible with a natural protein backbone correlates well with its natural sequence diversity. But designability is also related to less easily measurable properties such as fold specificity—that is, whether for a given sequence the structure is optimal from within the continuum of possible folds. Designable structures should represent such an optimum for many sequences and several investigators have demonstrated this to be highly structure dependent (Govindarajan & Goldstein, 1996; Wingreen et al., 2004). Additional factors contributing to the natural utilization of structural motifs may include folding/unfolding rates and robustness to small changes in environmental conditions.

As a result of combining many desirable but difficult-to-compute properties, designability is of significant utility in fields such as computational protein design or structure prediction. In design, one would like to *a priori* limit oneself to considering only highly designable structural templates. In structure prediction, designability would be a useful filter for discarding likely nonnative structures. Natural structures are certainly expected to be at least somewhat designable. As a consequence, many methods in computational protein design have relied on using native backbone structures (Kortemme et al., 2004; Reina et al., 2002), building novel structures from combinations of native segments (Azoitei et al., 2011; Kuhlman et al., 2003), or incorporating measures of native-like structural arrangements into scoring functions (Simons et al., 1997). Related approaches have also shown significant promise in structure prediction (Rohl, Strauss, Misura, & Baker, 2004; Zhang, Liang, & Zhang, 2011).

The natural abundance of a structural motif and its designability are related. Thus, a potential approach for evaluating the designability of a motif is to measure the degree of its recurrence in natural structures. Larger structural motifs, which contain pairs of segments not in contact and free to evolve independently, may not be sampled well either in the PDB or indeed in nature. However, this concern is greatly diminished for compact structural motifs, whose possible geometries are more likely to be well represented in the known structural universe (Fernandez-Fuentes et al., 2010; Grigoryan & Degrado, 2011; Vanhee et al., 2009; Verschueren et al., 2011). Further, even without any assumptions on the saturation of the PDB, if we do observe a motif to be highly recurrent, it is very likely designable. That is, we do not expect many false positives. On the other hand, false negatives— designable motifs that are labeled undesignable owing to poor representation in the PDB, are possible due to limited database size (especially for large motifs). This type of an error is reasonably tolerable in the context of protein design, as long as designable structures can still be identified that meet all other design criteria. Finally, one may often need to compare the designabilities of different motifs of the same size (e.g., different specific geometries of a given topology). For this purpose, only relative recurrence rates are important, which are expected to be more robust to database size and bias effects.

To enable an abundance-based metric of designability, an efficient method of searching for protein structural similarity is required. Many computational approaches have been proposed under the general category of protein structural comparison and search (Budowski-Tal, Nov, & Kolodny, 2010; Choi, Kwon, & Kim, 2004; Hasegawa & Holm, 2009; Holm & Rosenstrom, 2010). As designability is likely highly sensitive to the precise local geometry, one needs a method for finding matches to the detailed arrangement of atoms in the query structure. Here, we present such a method, which we term MaDCaT (*Ma*pping of *D*istances for the *Ca*tegorization of *T*opology), and provide examples of its usage for describing the designability landscape of structural motifs. A C++ implementation of MaDCaT, along with a web-based tool for structural similarity searching, can be found at: http://www.grigoryanlab.org/madcat/.

Quantification of designability provides a systematic filter for engineering novel protein structures. This is particularly useful in *de novo* design applications, where there is no guarantee that the hypothesized structure is easily achievable with natural amino acids. Further, it is known that many apparently feasible structural templates are in fact nondesignable (Grigoryan & Degrado, 2011). Recently, MaDCaT was used to impose designability in engineering peptide assemblies around single-walled carbon nanotubes (Grigoryan et al., 2011). Because the designed structure was entirely unprecedented in nature, there was not a clear basis for the choice of assembly geometry. Imposing high designability via MaDCaT provided such a basis, reducing the very large space of apparently reasonable geometries down to the single most appropriate structure. MaDCaT can be similarly used to provide a designability filter in other *de novo* design applications, provided that the desired structure is partitioned into motifs small enough to be well sampled in nature and the PDB, but large enough to capture important tertiary structural information. Designability may also provide a useful filter in structure prediction, where a localized density of nondesignable motifs could serve as an indicator of a poorly predicted region.

2. MaDCaT

MaDCaT relies on a distance-map representation of protein structure. A distance map is a two-dimensional matrix that stores distances between atoms of a protein (Choi et al., 2004). This representation is essentially lossless in that there is a one-to-one correspondence between three-dimensional structures and distance maps, to within chirality (i.e., mirror-image structures produce the same distance matrices; Dattorro, 2012). In its current implementation, MaDCaT considers distances between only C_α atoms (see Fig. 2.1). Though some structural information is lost in this way, the overall backbone geometry is preserved (Gront, Kmiecik, & Kolinski, 2007). This also comes with the convenience of representing a structure in an amino acid independent manner, which is useful for relating structure and sequence for designable motifs (see Section 3). On the other hand, the search methodology does not assume that only C_α atoms are used, so it is easy to extend the approach to deal with additional backbone atoms, side-chain atoms, or pseudo-atoms (e.g., side-chain centroids). Distance maps are a particularly convenient representation for structural similarity searches because (1) they contain enough information to identify detailed matches and (2) two distance maps can be compared in a computationally efficient manner (see below), without having to invoke optimal structural super-positions, as with other similarity metrics such as root mean squared deviation (RMSD).

2.1. Similarity score

Consider two protein structures (or structural segments), S_1 and S_2, each with n residues. Let $r_1(i, j)$ and $r_2(i, j)$ be the distances between the C_α atoms of i-th and j-th residues in S_1 and S_2 respectively. A reasonable metric of similarity between the two structures, assuming a linear correspondence between residues, is the Euclidian norm:

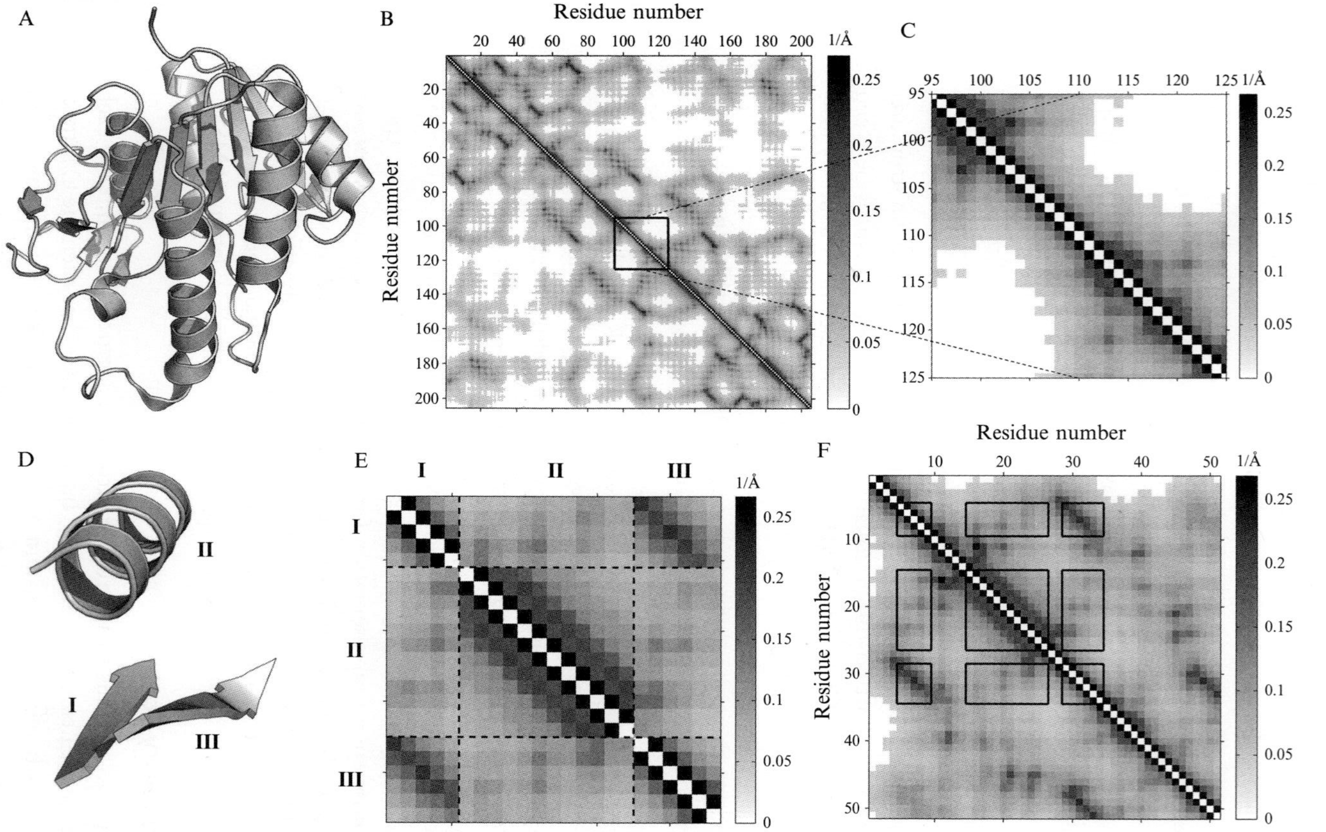

Figure 2.1 Distance-map representation of protein structure. (A) PDB entry 1HE4 used to demonstrate distance-map-based representation. (B) The matrix of inverse distances corresponding to 1HE4 (values below $1/25\,\text{Å}^{-1}$ are set to zero and shown in white). When searching this structure for a match against a single-segment query, only diagonal alignments of the query map need to be considered, with an example alignment outlined in black. (C) Magnified version of the diagonal alignment region. (D) and (E) are a multisegment query structure and its corresponding distance map, respectively. Dotted lines in (E) denote breaks between adjacent segments and roman numerals illustrate the correspondence between query segments in (D) and submaps in (E). (F) A potentially matching alignment of the query map onto a database map (outlined in black) may have gaps between adjacent segments of the query. Further, the sequence order of segments is not guaranteed to be the same in the query and the match.

$$d_{1,2} = \sqrt{\sum_{i<j} [r_1(i,j) - r_2(i,j)]^2} \qquad [2.1]$$

A possible issue with this score is that most of its magnitude is likely to arise from far-away C_α atoms, because larger distances imply larger potential deviations in two related structures. On the other hand, for the purpose of assessing designability and linking sequence to structure, it is the closely contacting residue pairs that are likely most important. For this reason, in its current implementation MaDCaT uses inverse distances. Given S_1 and S_2, distance maps are stored as, respectively:

$$M_1(i,j) = \frac{1}{r_1(i,j)}; \quad M_2(i,j) = \frac{1}{r_2(i,j)} \qquad [2.2]$$

with the corresponding score:

$$s_{1,2} = ||M_1 - M_2|| = \sqrt{\sum_{i<j} [M_1(i,j) - M_2(i,j)]^2} \qquad [2.3]$$

a better indicator of local structural similarity. Though this score has worked well for our applications, the search method in MaDCaT is very general so any other functions of distance can be used. Hereafter, distance maps will refer to matrices of inverse distances, as in Eq. (2.2).

An added benefit of using maps of inverse distances is that they are well suited for sparse representation. This is because the less "important" distances above a suitably chosen cutoff r_{cut}, corresponding to map entries below $1/r_{\text{cut}}$, can be replaced with zeros in the map. Such sparse matrices not only reduce storage and memory requirements but also result in significant speedups of the search procedure (see below). Finally, because they resemble interatomic interaction potentials, inverse distances (and their powers) are perhaps more natural basis functions for expressing structural similarity than distances themselves.

2.2. The algorithm

As the input query, MaDCaT takes any structure composed of an arbitrary number of disjoint segments. The query is converted to its distance-map representation and used to search a database of proteins with precomputed distance maps. Each segment within the query is considered as a whole and is only matched against segments of consecutive residues in database structures. The goal of the algorithm is to find alignments of segments in the query onto regions of database structures in a way that optimizes the score in Eq. (2.3). Because it is usually not know *a priori* what scores are good for a given query structure, MaDCaT finds the L best scoring alignments, where L is a user-specified value. In cases where an appropriate score cutoff does exist, it can be specified and will speed up the search. To introduce the algorithm, we shall first consider the case when the query is composed of a single segment and then generalize to multisegment structures.

2.2.1 Single-segment queries

In this case, one only needs to consider alignments of the query distance map on the main diagonal of database maps (Fig. 2.1B). Although this is a straightforward computation, its time cost of $O(n^2 \cdot N \cdot m)$ (where n is the length of the query structure, N is that for an average database structure, and m is the number of database structures) can get high in practice, especially for large query structures. MaDCaT mitigates this by taking advantage of the sparsity of distance maps. Consider a particular alignment of the query map Q onto a database map M, the score for which is $s(Q,M,k)=\sum_{i<j}[Q(i,j)-M(i+k,j+k)]^2$, where i and j iterate over the elements of Q, and k is the offset defining the alignment (for simplicity, the square root in Eq. (2.3) will be omitted hereafter; this does not change the relative ordering of matches, and the root function can always be applied as a last stage in the calculation). Many elements $M(i+k, j+k)$ may be zero, especially for large query maps (white cells in Fig. 2.1C). The contribution to the total score of these elements is dependent only on Q, such that for a given Q a default score can be computed, which assumes all corresponding elements in the database map to be zero, i.e. $s_d=\sum_{i<j}Q(i,j)^2$. Then, in order to find the score for a specific alignment, $s(Q, M, K)$, s_d needs to be updated to reflect only the nonzero elements of M corresponding to Q in the alignment:

$$s(Q,M,k) = s_d + \sum_{\substack{i<j \\ M(i+k,j+k)\neq 0}} \left([Q(i,j)-M(i+k,j+k)]^2 - Q(i,j)^2\right) \quad [2.4]$$

Zero values in our distance maps, which store inverse distances, correspond to atom pairs farther apart than a given cutoff r_{cut} (by default, 25 Å is used in MaDCaT). Because the number of atom pairs within a certain distance cutoff grows at most linearly with the number of residues in the structure, this modification gives an asymptotic time of $O(n \cdot N \cdot m)$ and results in significant speedups in practice.

It is easy to imagine how simple heuristics can be used to significantly cut down on the number of alignments that need to be considered for a given query map/database map pair. These could be based on secondary-structure matching or other local structural properties. Whereas heuristics are reasonably safe for query structures with good matches in the database, they can present significant artifacts in cases of rare on unusual queries. Since one of the envisioned uses of MaDCaT was the ability to start with an implausible hypothetical structure and progressively move toward a nearby more designable one, no heuristic prefilters are currently available in MaDCaT, though the implementation does support them.

2.2.2 Multisegment queries

In cases where the query structure consists of multiple disjoint segments, the query map can be thought of as composed of submaps (Fig. 2.1D and E). Each of these submaps represents either a contiguous segment of structure or an interface between two segments (diagonal and off-diagonal submaps in Fig. 2.1E, respectively). An alignment of the query structure onto a database structure involves the placement of each segment of the query onto an equally sized contiguous region in the database structure. In distance-map terms, this means that diagonal maps of the query line up along the diagonal of the database map (without overlaps), and off-diagonal maps align at the resulting intersection points (see Fig. 2.1F). Because the alignment of individual segments is independent, the number of potential alignments grows exponentially with the number of segments. In fact, finding the optimal alignment is known to be NP-hard (Lathrop, 1994).

MaDCaT applies a branch-and-bound approach to solving this combinatorial problem, reminiscent of the approach first introduced by Holm and Sander (1996). The algorithm represents the space of possible alignments as a search tree. At each level i of the tree, a choice has to be made as to the alignment of the i-th segment. This tree is traversed, top to bottom, making a specific choice for the alignment of the i-th segment, and moving onto the $(i+1)$-st. The key part of the algorithm, which enables it to give up on unproductive combinations of segment alignments early on, is the computation of a lower bound on the score of an incomplete alignment. Submap alignments are visited in the order of best to worst lower bound, such that as soon as the bound becomes larger than the L-th worst match found so far (where L is the desired number of top matches), the current branch can be safely terminated. When the top branch is terminated, the search tree is completely pruned. The lower bound is based on prescoring each individual submap of the query against the database map, in all relevant alignments (e.g., diagonal submaps are scored only in diagonal alignments). Based on these scores, lower bounds for incomplete alignments are estimated by relaxing some constraints on the remainder of the alignment (e.g., allowing submaps in the same column of the query map to align in different columns of the database map).

To aid in searching for larger structures or those with more than a few segments, MaDCaT has an optional greedy setting that enables it to give up on segment alignments purely based on the score of the diagonal submap. Using this filter eliminates the optimality guarantee that MaDCaT otherwise provides, and may lead to significantly different results for queries without well-matching structures in the database. The greedy filter requires that the similarity score in Eq. (2.3) accumulate from throughout the query matrix, rather than originating heavily from a particular submap. For a given user-specified greediness level g, the filter requires that the score originating from each submap s be not worse than $g \cdot m_s \cdot (w_L/m)$, where w_L is the worst score among the top L solutions currently found, m is the total mass of the query matrix (the sum of all elements), and m_s is the mass originating from submap s. Although any value can be specified for g, values above 1.0 make most sense, with larger values corresponding to less greediness.

As described above, the search algorithm will consider alignments of query segments that map arbitrarily far apart in sequence of database structures. In fact, all sequence-order permutations of segments in the query structure are automatically considered by MaDCaT (e.g., the motif in Fig. 2.1D will match similar motifs in the database, even if the order of the three secondary-structure elements in the database structure is different). This is useful when one only cares to find matches to the segments themselves (e.g., segments represent discontinuous chains) or when all the ways of bridging the gaps between the segments are of interest. MaDCaT additionally enables one to limit the number of residues that map between two consecutive segments, by establishing lower and/or upper bounds. This can be useful when one aims to investigate motifs of a certain length for bridging two or more structural segments.

2.2.3 Interfacial searches

In some applications, the intersegment interfacial geometry may be of more interest than the segments themselves. For example, this may be the case where one looks for starting structures to mimic one side of an existing protein–protein interface. For such cases, MaDCaT allows one to search for intersegment portions of distance maps (e.g., the submap at the intersection of segments I and II in Fig. 2.1E). From the standpoint of computational efficiency, interfacial searches have the advantage of requiring fewer independent submaps, but also have the disadvantage that they can be aligned almost anywhere in database maps. Overall running times are thus comparable in practice.

2.2.4 Dali versus MaDCaT

The algorithm underlying MaDCaT bares resemblance to the structure search technique by Holm and Sander, now part of the Dali search suite (Holm & Rosenstrom, 2010; Holm & Sander, 1996). However, there are important differences between MaDCaT and Dali. With MaDCaT, the query represents an exact specification of the structure of interest (e.g., precisely defined contiguous segments and locations of allowed gaps), and the results are provably optimal matches from the given database. On the other hand, the aim of Dali is to discover close matches to substructures of the query. These substructures are not fixed *a priori*, and although they may cover the entire query in some cases, they are determined by a series of heuristic techniques that try to identify larger conserved regions but avoid visiting the entire search space (Holm & Sander, 1993, 1994, 1996). These differences arise primarily from different intended uses of the methods. Dali is very well suited for identifying overall structural similarities between proteins or larger protein fragments (in fact, it requires a minimum chain length of 30 residues to perform a search). For example, Dali has been used to identify topological "attractors" in protein structure space (Holm & Sander, 1996). On the other hand, with MaDCaT we aim to find close matches to precisely defined tertiary structural motifs, aiming to quantify their designabilities. The provable optimality of MaDCaT matches is particularly critical for the latter goal.

2.2.5 Obtaining MaDCaT

MaDCaT is implemented as a C++ suite, freely available under the terms of the GNU General Public License (see http://www.grigoryanlab.org/madcat/). Inquiries about commercial licensing should be directed to the corresponding author. Support programs for MaDCaT (e.g., for building database and query maps and analysis of results) utilize the Molecular Software Library (MSL; Kulp et al., 2012), freely available at http://msl-libraries.org. The Web interface for MaDCaT is currently limited to searching over a subsample of the PDB, which does not find the best available matches to a motif, but in our experience, it can still be used to grossly estimate its designability.

3. QUANTIFYING DESIGNABILITY

Several investigators have shown that the universe of sequences compatible with a native protein backbone (or close structural variations thereof) in an *in silico* protein design experiment correlates with evolutionary sequence profiles of the protein (Koehl & Levitt, 2002; Kuhlman & Baker, 2000; Smith & Kortemme, 2011). So, when a structure is designable, computational protein design can often identify some of the sequence space compatible with it, albeit much room for improvement remains (Boas & Harbury, 2007; Pantazes, Grisewood, & Maranas, 2011). However, presently it is not easy to recognize that a structure is not designable using computation protein design. This is a particular limitation for *de novo* protein design, where novel protein structures are proposed and are not guaranteed to be designable. Thus, a method for quantifying designability is sorely needed.

3.1. Motif usage in nature varies significantly

We expect different local geometries of protein structure to have different designabilities and thus to have been sampled at different rates in nature. To illustrate the sensitivity of this effect, we shall consider abundance as a function of small perturbations in local geometry for two structural motifs—the parallel dimeric α-helical coiled coil and an α-helix packed against a parallel two-strand β-sheet, αββ (see Fig. 2.2).

The backbone of a coiled-coil structure is well described with simple parametric equations modeling the α-helix wrapping around a larger superhelix (Crick, 1953; Grigoryan & Degrado, 2011). For an ideal parallel structure, critical parameters are R_o—the radius of the superhelix, α—the pitch angle of the superhelical curve with respect to the interface axis, and φ_1—the helical phase defining helical sides facing each other (see Fig. 2.2A). Whereas it is easy to imagine how superhelical radius may affect the designability of a structure (e.g., packing preferences of interface-lining amino acids at least partially explain R_o variations, Grigoryan & Degrado, 2011), it is less clear that certain pitch angles and phases should necessarily be selectively preferred. To look at how α and φ_1 affect designability, we systematically varied these parameters in the context of a 12-residue parallel dimeric coiled-coil fragment, using MaDCaT to find all nonredundant structural matches in the PDB for each sampled structure. Both phase and pitch angle were varied around their previously found canonical values (Grigoryan & Degrado, 2011) in 30 increments (between $-20°$ and $+10°$ for α, and $-24°$ and

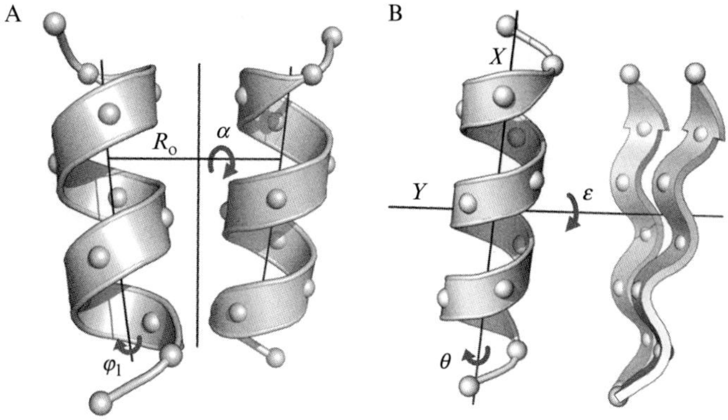

Figure 2.2 Some of the structural parameters defining a parallel α-helical coiled coil (in A) and an αββ motif (in B). (A) For formal definitions of superhelical radius R_o, pitch angle α, and helical phase φ_1 see references (Crick, 1953; Grigoryan & Degrado, 2011). In this work, R_o was fixed at 4.88 Å. (B) To model the ideal αββ motif, an initial structure was generated by fitting a naturally occurring αββ motif (taken from PDB entry 3EGD, residue ranges 505–511, 615–628, and 632–638) with ideal secondary-structure elements (i.e., with exactly repeating backbone φ/ψ angles). Helical phase θ was defined as a rotation around the helical axis X. The crossing angle ε was encoded as a rotation around axis Y defined to be orthogonal to X and in the plane formed by X and the third principal axis of the β-sheet (the "out-of-plane" component). The two parameters were taken to be zero for the initially fit ideal αββ motif.

+6° for φ_1), resulting in 900 structures. A nonredundant subset of the PDB, generated by taking the first member of each sequence cluster produced by blastclust (Camacho et al., 2009) at 30% sequence identity, was used as the search database. Figure 2.3A shows a contoured heatmap of the number of close structural hits as a function of structural parameters. The significant bias for specific geometries is obvious in this motif. Though both pitch angle and phase contribute to designability, changes in the latter are much more tolerable and many phases can be accommodated (see also Fig. 2.3B and C). The most designable region corresponds to the canonical range of values identified in an earlier analysis (Grigoryan & Degrado, 2011). A less designable but still populated region of positive pitch angles, corresponding to right-handed crossings, is also apparent (Fig. 2.3A and B).

Figure 2.4 illustrates the results of a similar analysis for the αββ motif. Here, the varied parameters were the helix-sheet crossing angle ε and the helical phase θ (see Fig. 2.2B). Both parameters were varied between −30° and +30°, in 31 steps, for a total of 961 structures. The heatmap in Fig. 2.4A describes the designability landscape of this motif. Once again,

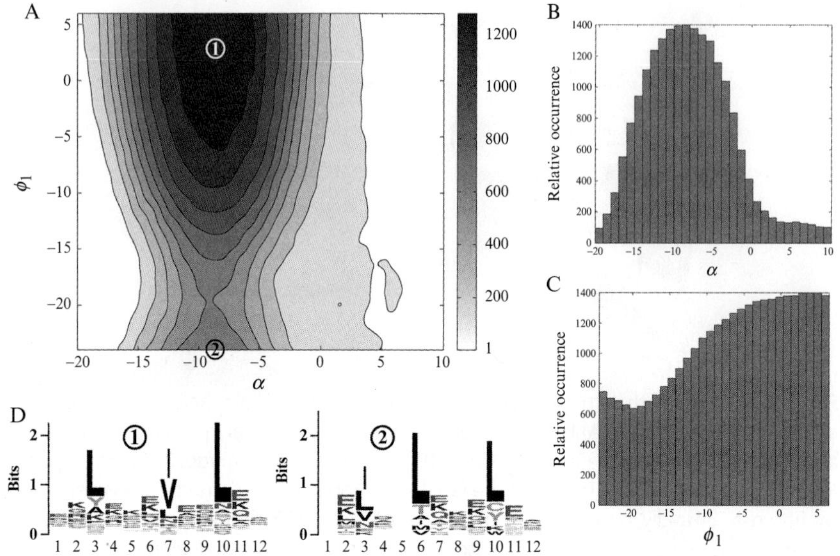

Figure 2.3 The designability landscape of a parallel dimeric coiled-coil motif. (A) The number of matches identified by MaDCaT within 1.0 Å RMSD of the query structure, as a function of pitch angle α and helical phase ϕ_1. For each value of α (B) plots the number of matches maximized over all values of ϕ_1 sampled. The equivalent for ϕ_1 is plotted in (C). For two highly designable structures (marked with circled numbers in (A)), (D) shows sequence logos originating from close matches. RMSD cutoffs of 0.7 Å and 0.4 Å were used for generating the left and the right sequence logos, respectively.

we see that the phase is a weaker determinant of designability than crossing angle (see also Fig. 2.4B and C).

With both motifs, very drastic changes in designability can result upon seemingly small perturbations to structure. Figure 2.5 shows structures with very different designabilities for the αββ motif. It is difficult to tell *a priori* which structure is more designable. On the other hand, MaDCaT enables rapid quantification of designability in a systematic manner for an arbitrary motif.

3.2. Connection between structure and sequence

An important additional piece of information revealed by finding close matches to a structural motif are sequence features needed to realize the given motif. As the different matches come from different structure/sequence contexts, significantly conserved amino acids are likely important for stabilizing the motif itself or for encoding its structural specificity. On the other hand, positions with weak conservation can tolerate many different amino acids and are likely important for adjusting to the specific context. Such insight is highly useful

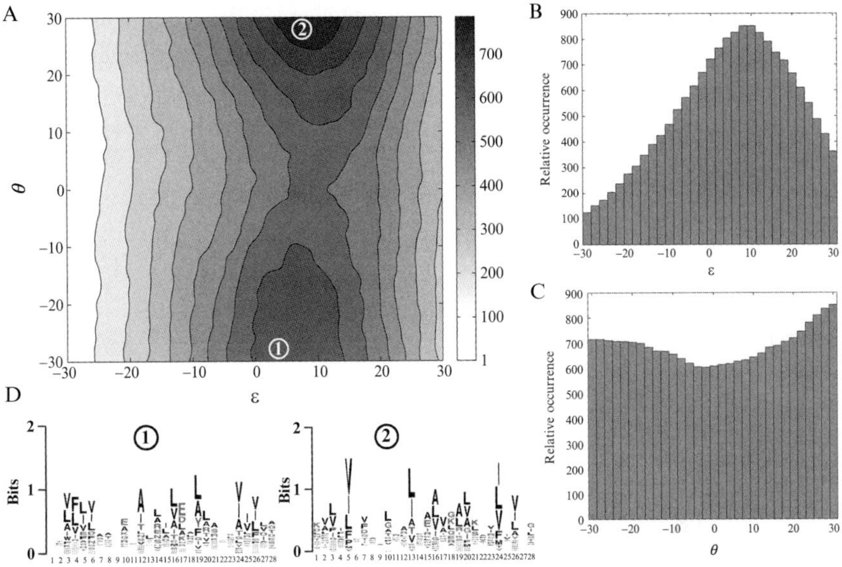

Figure 2.4 The designability landscape of the αββ motif. (A) The number of matches identified by MaDCaT within 1.5 Å RMSD of the query structure, as a function of its crossing angle ε and helical phase θ. For each value of ε (B) plots the number of matches maximized over all values of θ sampled. The equivalent for θ is plotted in (C). For two highly designable structures (marked with circled numbers in (A)), (D) shows sequence logos originating from close matches with RMSD to the query below 1.0 Å.

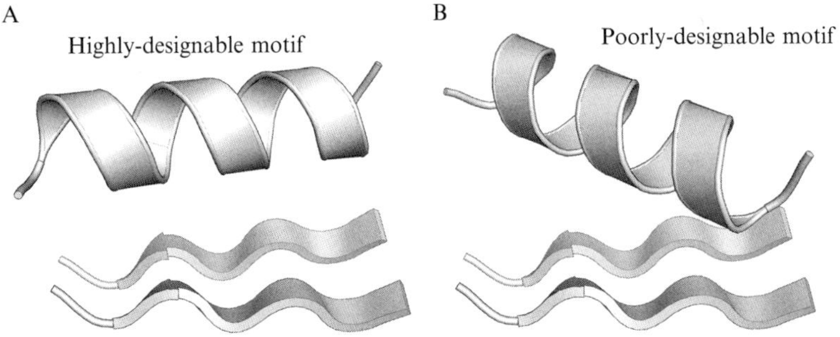

Figure 2.5 Examples of designable and nondesignable instances of the αββ motif. The structure in (A) has 44 unique examples, within 1.0 Å RMSD, in the nonredundant subset of the PDB used for searching, compared to 0 such examples for the structure in (B).

in protein design as it significantly constraints the productive sequence space (Grigoryan et al., 2011). This information can also, in principle, be used for structure prediction to assure that all local structural motifs in the predicted model are consistent with their corresponding sequences.

Figure 2.3D shows the amino acid distributions in the sequences of close matches corresponding to two different designable coiled-coil geometries (Crooks, Hon, Chandonia, & Brenner, 2004). Canonical coiled coils exhibit a seven-residue sequence repeat, designated with letters **abcdefg**, with **a** and **d** positions generally occupied with hydrophobic amino acids. The main difference between the two motifs is the helical phase and this is clearly reflected in the sequence logos as a register shift. Whereas the first motif starts with position **b**, such that residues 3 (**d**), 7 (**a**), and 10 (**d**) are hydrophobic, the second motif starts with an **f**, leading to residues 3 (**a**), 6 (**d**), and 10 (**a**) being hydrophobic. Though decades of study have led to a very good understanding of coiled-coil position-specific amino acid preferences, the above analysis can be performed for any structural motif, rapidly revealing sequence preferences in a geometry-specific manner.

Figure 2.4D shows a similar analysis for two designable geometries of the αββ motif. Though the amino acid preferences here are less pronounced than for the coiled coil, clear trends are still evident and the differences between the two geometries are apparent. Such sequence trends can be used to encode a specific geometry in design.

The sequence logos above capture only position-specific distributions, ignoring inter-position correlations. In principle, the latter can also be extracted from sequence alignments of matches, provided enough sequences are available, and these data can be equally useful in design or structure prediction. In fact, significant inter-positional correlations can flatten individual (marginalized) position distributions, leading to apparently lower information content by the standard sequence-logo analysis. Many methods for extracting meaningful mutual correlations between alignment positions have been proposed and can be employed here (for a recent example, see Morcos et al., 2011).

4. FURTHER DEVELOPMENTS

Though at present MaDCaT is efficient enough for many practical applications, further improvements in computational speed need to be pursued. Because structural searching is an example of an "embarrassingly parallel" problem, leveraging the massive parallelism offered by GPU technologies is one potential direction. Heuristics-based prefiltering or preclassification of database structures, already explored in the literature (Budowski-Tal et al., 2010; Hasegawa & Holm, 2009; Kolodny, Koehl, & Levitt, 2005), may offer another fruitful avenue for efficiency gains, though it must be performed carefully not to bias search accuracy based on motif type. In the limit of very rapid (lookup-like) structural searching, designability analysis may be incorporated as a routine step in such applications as structural sampling for structure prediction, in alternating between sequence and structure selection for *de novo* protein design, or in automatic refinement of experimentally determined structures.

5. SUMMARY

Protein structural comparison, classification, and searching for structural similarity are problems that have received considerable attention in the past several decades (Hasegawa & Holm, 2009). It has been shown that such methodologies can be used for establishing evolutionary and functional relationships between proteins (Ouzounis, Coulson, Enright, Kunin, & Pereira-Leal, 2003). Here and in prior work (Grigoryan & Degrado, 2011; Grigoryan et al., 2011), we have demonstrated that structural similarity, when considered at a detailed local level, can also shed light on the designability of different structural motifs comprising proteins. It can provide a connection between structure and sequence, invaluable in *de novo* computational protein design, and potentially in structure prediction. MaDCaT is a tool particularly well suited for establishing such links, as its definition of similarity is focused on the precise local geometry, with particular emphasis on close contacts. By making MaDCaT freely available, we hope to stimulate its application in protein design and structure prediction, as well as its further development.

ACKNOWLEDGMENTS

This work was supported by NIH grant 5F32GM084631-02 and startup funds from Dartmouth College (G. G.).

REFERENCES

Andreeva, A., Howorth, D., Chandonia, J. M., Brenner, S. E., Hubbard, T. J., Chothia, C., et al. (2008). Data growth and its impact on the SCOP database: New developments. *Nucleic Acids Research*, *36*, D419–D425.

Azoitei, M. L., Correia, B. E., Ban, Y. E., Carrico, C., Kalyuzhniy, O., Chen, L., et al. (2011). Computation-guided backbone grafting of a discontinuous motif onto a protein scaffold. *Science*, *334*, 373–376.

Baeten, L., Reumers, J., Tur, V., Stricher, F., Lenaerts, T., Serrano, L., et al. (2008). Reconstruction of protein backbones from the BriX collection of canonical protein fragments. *PLoS Computational Biology*, *4*, e1000083.

Boas, F. E., & Harbury, P. B. (2007). Potential energy functions for protein design. *Current Opinion in Structural Biology*, *17*, 199–204.

Budowski-Tal, I., Nov, Y., & Kolodny, R. (2010). FragBag, an accurate representation of protein structure, retrieves structural neighbors from the entire PDB quickly and accurately. *Proceedings of the National Academy of Sciences of the United States of America*, *107*, 3481–3486.
Camacho, C., Coulouris, G., Avagyan, V., Ma, N., Papadopoulos, J., Bealer, K., et al. (2009). BLAST+: Architecture and applications. *BMC Bioinformatics*, *10*, 421.
Choi, I. G., Kwon, J., & Kim, S. H. (2004). Local feature frequency profile: A method to measure structural similarity in proteins. *Proceedings of the National Academy of Sciences of the United States of America*, *101*, 3797–3802.
Crick, F. H. C. (1953). The Fourier transform of a coiled-coil. *Acta Crystallographica*, *6*, 685.
Crooks, G. E., Hon, G., Chandonia, J. M., & Brenner, S. E. (2004). WebLogo: A sequence logo generator. *Genome Research*, *14*, 1188–1190.
Dattorro, J. (2012). *Convex optimization & euclidean distance geometry*. USA: Meboo.
Engel, D. E., & DeGrado, W. F. (2005). Alpha-alpha linking motifs and interhelical orientations. *Proteins*, *61*, 325–337.
England, J. L., Shakhnovich, B. E., & Shakhnovich, E. I. (2003). Natural selection of more designable folds: A mechanism for thermophilic adaptation. *Proceedings of the National Academy of Sciences of the United States of America*, *100*, 8727–8731.
Fernandez-Fuentes, N., Dybas, J. M., & Fiser, A. (2010). Structural characteristics of novel protein folds. *PLoS Computational Biology*, *6*, e1000750.
Fuchs, A., & Frishman, D. (2010). Structural comparison and classification of alpha-helical transmembrane domains based on helix interaction patterns. *Proteins*, *78*, 2587–2599.
Govindarajan, S., & Goldstein, R. A. (1996). Why are some proteins structures so common? *Proceedings of the National Academy of Sciences of the United States of America*, *93*, 3341–3345.
Greene, L. H., Lewis, T. E., Addou, S., Cuff, A., Dallman, T., Dibley, M., et al. (2007). The CATH domain structure database: New protocols and classification levels give a more comprehensive resource for exploring evolution. *Nucleic Acids Research*, *35*, D291–D297.
Grigoryan, G., & Degrado, W. F. (2011). Probing designability via a generalized model of helical bundle geometry. *Journal of Molecular Biology*, *405*, 1079–1100.
Grigoryan, G., Kim, Y. H., Acharya, R., Axelrod, K., Jain, R. M., Willis, L., et al. (2011). Computational design of virus-like protein assemblies on carbon nanotube surfaces. *Science*, *332*, 1071–1076.
Gront, D., Kmiecik, S., & Kolinski, A. (2007). Backbone building from quadrilaterals: A fast and accurate algorithm for protein backbone reconstruction from alpha carbon coordinates. *Journal of Computational Chemistry*, *28*, 1593–1597.
Hasegawa, H., & Holm, L. (2009). Advances and pitfalls of protein structural alignment. *Current Opinion in Structural Biology*, *19*, 341–348.
Helling, R., Li, H., Melin, R., Miller, J., Wingreen, N., Zeng, C., et al. (2001). The designability of protein structures. *Journal of Molecular Graphics & Modelling*, *19*, 157–167.
Ho, B. K., & Curmi, P. M. (2002). Twist and shear in beta-sheets and beta-ribbons. *Journal of Molecular Biology*, *317*, 291–308.
Holm, L., & Rosenstrom, P. (2010). Dali server: Conservation mapping in 3D. *Nucleic Acids Research*, *38*, W545–W549.
Holm, L., & Sander, C. (1993). Protein structure comparison by alignment of distance matrices. *Journal of Molecular Biology*, *233*, 123–138.
Holm, L., & Sander, C. (1994). Parser for protein folding units. *Proteins*, *19*, 256–268.
Holm, L., & Sander, C. (1996). Mapping the protein universe. *Science*, *273*, 595–603.
Hu, C., & Koehl, P. (2010). Helix-sheet packing in proteins. *Proteins*, *78*, 1736–1747.
Hutchinson, E. G., & Thornton, J. M. (1994). A revised set of potentials for beta-turn formation in proteins. *Protein Science*, *3*, 2207–2216.
Joosten, R. P., te Beek, T. A., Krieger, E., Hekkelman, M. L., Hooft, R. W., Schneider, R., et al. (2011). A series of PDB related databases for everyday needs. *Nucleic Acids Research*, *39*, D411–D419.

Kallblad, P., & Dean, P. M. (2004). Backbone-backbone geometry of tertiary contacts between alpha-helices. *Proteins*, *56*, 693–703.

Koehl, P., & Levitt, M. (2002). Protein topology and stability define the space of allowed sequences. *Proceedings of the National Academy of Sciences of the United States of America*, *99*, 1280–1285.

Kolodny, R., Koehl, P., & Levitt, M. (2005). Comprehensive evaluation of protein structure alignment methods: Scoring by geometric measures. *Journal of Molecular Biology*, *346*, 1173–1188.

Kortemme, T., Joachimiak, L. A., Bullock, A. N., Schuler, A. D., Stoddard, B. L., & Baker, D. (2004). Computational redesign of protein-protein interaction specificity. *Nature Structural & Molecular Biology*, *11*, 371–379.

Kuhlman, B., & Baker, D. (2000). Native protein sequences are close to optimal for their structures. *Proceedings of the National Academy of Sciences*, *97*, 10383–10388.

Kuhlman, B., Dantas, G., Ireton, G. C., Varani, G., Stoddard, B. L., & Baker, D. (2003). Design of a novel globular protein fold with atomic-level accuracy. *Science*, *302*, 1364–1368.

Kulp, D. W., Subramaniam, S., Donald, J. E., Hannigan, B. T., Mueller, B. K., Grigoryan, G., et al. (2012). Structural informatics, modeling, and design with an open-source molecular software library (MSL). *Journal of Computational Chemistry*, *33*, 1645–1661.

Lathrop, R. H. (1994). The protein threading problem with sequence amino acid interaction preferences is NP-Complete. *Protein Engineering*, *7*, 1059–1068.

Li, H., Helling, R., Tang, C., & Wingreen, N. (1996). Emergence of preferred structures in a simple model of protein folding. *Science*, *273*, 666–669.

London, N., Movshovitz-Attias, D., & Schueler-Furman, O. (2010). The structural basis of peptide-protein binding strategies. *Structure*, *18*, 188–199.

Marchler-Bauer, A., Lu, S., Anderson, J. B., Chitsaz, F., Derbyshire, M. K., DeWeese-Scott, C., et al. (2011). CDD: A conserved domain database for the functional annotation of proteins. *Nucleic Acids Research*, *39*, D225–D229.

Morcos, F., Pagnani, A., Lunt, B., Bertolino, A., Marks, D. S., Sander, C., et al. (2011). Direct-coupling analysis of residue coevolution captures native contacts across many protein families. *Proceedings of the National Academy of Sciences of the United States of America*, *108*, E1293–E1301.

Moutevelis, E., & Woolfson, D. N. (2009). A periodic table of coiled-coil protein structures. *Journal of Molecular Biology*, *385*, 726–732.

Ouzounis, C. A., Coulson, R. M., Enright, A. J., Kunin, V., & Pereira-Leal, J. B. (2003). Classification schemes for protein structure and function. *Nature Reviews. Genetics*, *4*, 508–519.

Pantazes, R. J., Grisewood, M. J., & Maranas, C. D. (2011). Recent advances in computational protein design. *Current Opinion in Structural Biology*, *21*, 467–472.

Platt, D. E., Guerra, C., Zanotti, G., & Rigoutsos, I. (2003). Global secondary structure packing angle bias in proteins. *Proteins*, *53*, 252–261.

Reina, J., Lacroix, E., Hobson, S. D., Fernandez-Ballester, G., Rybin, V., Schwab, M. S., et al. (2002). Computer-aided design of a PDZ domain to recognize new target sequences. *Nature Structural Biology*, *9*, 621–627.

Rohl, C. A., Strauss, C. E., Misura, K. M., & Baker, D. (2004). Protein structure prediction using Rosetta. *Methods in Enzymology*, *383*, 66–93.

Simons, K. T., Kooperberg, C., Huang, E., & Baker, D. (1997). Assembly of protein tertiary structures from fragments with similar local sequences using simulated annealing and Bayesian scoring functions. *Journal of Molecular Biology*, *268*, 209–225.

Smith, C. A., & Kortemme, T. (2011). Predicting the tolerated sequences for proteins and protein interfaces using Rosettabackrub flexible backbone design. *PloS One*, *6*, e20451.

Stein, A., Ceol, A., & Aloy, P. (2011). 3did: Identification and classification of domain-based interactions of known three-dimensional structure. *Nucleic Acids Research*, *39*, D718–D723.

Testa, O. D., Moutevelis, E., & Woolfson, D. N. (2009). CC+: A relational database of coiled-coil structures. *Nucleic Acids Research*, *37*, D315–D322.

Vanhee, P., Reumers, J., Stricher, F., Baeten, L., Serrano, L., Schymkowitz, J., et al. (2010). PepX: A structural database of non-redundant protein-peptide complexes. *Nucleic Acids Research*, *38*, D545–D551.

Vanhee, P., Stricher, F., Baeten, L., Verschueren, E., Lenaerts, T., Serrano, L., et al. (2009). Protein-peptide interactions adopt the same structural motifs as monomeric protein folds. *Structure*, *17*, 1128–1136.

Verschueren, E., Vanhee, P., van der Sloot, A. M., Serrano, L., Rousseau, F., & Schymkowitz, J. (2011). Protein design with fragment databases. *Current Opinion in Structural Biology*, *21*, 452–459.

Walters, R. F., & DeGrado, W. F. (2006). Helix-packing motifs in membrane proteins. *Proceedings of the National Academy of Sciences of the United States of America*, *103*, 13658–13663.

Wingreen, N. S., Li, H., & Tang, C. (2004). Designability and thermal stability of protein structures. *Polymer*, *45*, 699–705.

Zhang, J., Liang, Y., & Zhang, Y. (2011). Atomic-level protein structure refinement using fragment-guided molecular dynamics conformation sampling. *Structure*, *19*, 1784–1795.

Zhang, Y., & Skolnick, J. (2005). The protein structure prediction problem could be solved using the current PDB library. *Proceedings of the National Academy of Sciences of the United States of America*, *102*, 1029–1034.

CHAPTER THREE

Computational Methods for Controlling Binding Specificity

Oz Sharabi, Ariel Erijman, Julia M. Shifman[1]

Department of Biological Chemistry, The Alexander Silberman Institute of Life Sciences, The Hebrew University of Jerusalem, Jerusalem, Israel

[1]Corresponding author: e-mail address: jshifman@cc.huji.ac.il

Contents

Abstract

Learning to control, protein-binding specificity is useful for both fundamental and applied biology. In fundamental research, better understanding of complicated signaling networks could be achieved through engineering of regulator proteins to bind to only a subset of their effector proteins. In applied research such as drug design, non-specific binding remains a major reason for failure of many drug candidates. However, developing antibodies that simultaneously inhibit several disease-associated pathways are a rising trend in pharmaceutical industry. Binding specificity could be manipulated experimentally through various display technologies that allow us to select desired binders from a large pool of candidate protein sequences. We developed an alternative approach for controlling binding specificity based on computational protein design. We can enhance binding specificity of a protein by computationally optimizing its

Methods in Enzymology, Volume 523
ISSN 0076-6879
http://dx.doi.org/10.1016/B978-0-12-394292-0.00003-5

sequence for better interactions with one target and worse interaction with alternative target(s). Moreover, we can design multispecific proteins that simultaneously interact with a predefined set of proteins. Unlike combinatorial techniques, our computational methods for manipulating binding specificity are fast, low cost and in principle are able to consider an unlimited number of desired and undesired binding partners.

1. INTRODUCTION

Computational protein design is a growing strategy for predicting protein sequences that fold into a particular target structure and/or perform a specific function. A vast number of potential amino acid sequences are evaluated with an atomic-based energy function and the best (lowest energy) sequence is selected for a particular protein structure. Computationally predicted sequences are then generated experimentally and assayed for the desired property.

While the initial methods for protein design were developed for monomeric proteins, more recently great progress has been made in design of protein–protein interactions. Computational design presents a future alternative to combinatorial techniques that involve selecting binders from large combinatorial libraries of protein mutants using phage, ribosome, or other display technologies. Interestingly, computational strategies for controlling binding specificity are conceptually similar to positive and negative selection used in experimental approaches.

1.1. Positive and negative design in manipulating binding specificity

Proteins are dynamical entities that can assume different structures during their functional cycle. Hence, protein design frequently requires considering several protein structures (i.e., design states). This introduces the concepts of positive and negative design that mimic positive and negative selection in protein evolution (Fig. 3.1). In positive design, we search for the amino acid sequence that is optimal for one or several design states. Hence, the sum of the sequence energy in all the designed states is minimized in positive design. In negative design, we would like to eliminate sequences compatible with undesirable states. Hence, the energy of the sequence is maximized in the negative states. The concepts of positive and negative design become especially important when enhancing binding specificity since binding specificity is essentially an affinity ratio of a protein for one target (positive state) versus alternative target(s) (negative states). In the design calculations that

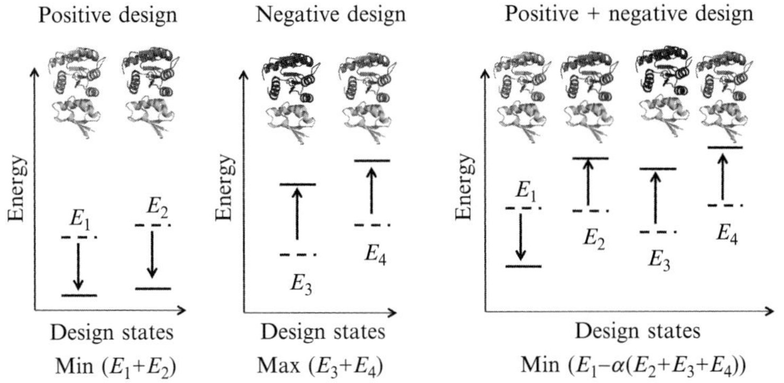

Figure 3.1 Positive versus negative design. Initially, one protein can bind to several targets with similar binding energy (dashed lines). During the positive design procedure, a common binding interface sequence is selected to minimize the sum of the energies in different design states. The energies of the designed sequences are lowered with respect to the initial energy for all the design states (solid lines). During the negative design procedure, a common binding interface sequence is selected to maximize the energy in the negative design states. The energies of the designed sequences are raised with respect to the initial energy for all the design states (solid lines). When both positive and negative design procedures are used, a common sequence is selected that minimizes the difference between the energy of the positive design state(s) and the negative design states. The optimization lowers the energy of the positive design state and raises the energy of the negative design states. (For color version of this figure, the reader is referred to the online version of this chapter.)

incorporate both positive and negative design, the difference between the sequence energy in the positive and the negative states should be minimized. Unfortunately, this is not a straightforward procedure since minimizing the difference in energies could produce solutions with high energy in both positive and negative states. Such protein sequences in reality will not fold and hence are not physiologically relevant. This problem could be solved by scaling down the contribution to the scoring function of the negative-state energy in comparison to the positive-state energy (Bolon, Grant, Baker, & Sauer, 2005; Grigoryan, Reinke, & Keating, 2009; Havranek & Harbury, 2003; Leaver-Fay, Jacak, Stranges, & Kuhlman, 2011). This procedure, however, involves introduction of an arbitrary weight that is likely to be different for each experimental system. Due to the described difficulties, our group has avoided simultaneous use of positive and negative design when manipulating binding specificity. Instead, we suggest two alternative approaches for selecting against undesired binders. One approach involves optimizing a large number of binding interface positions using only positive design. The second strategy involves introduction of single mutations while computing their energetic contribution to the binding energy in both positive and negative design states. Below, we include discussion of both protocols and give examples of their application to various experimental systems.

2. NARROWING DOWN BINDING SPECIFICITY

2.1. Optimizing a large number of residues with positive design

Several groups working on coiled coils argued for the necessity of both positive and negative design when designing highly specific proteins (Grigoryan et al., 2009; Havranek & Harbury, 2003). We, on the contrary, demonstrated that at least in some biological systems (e.g., calmodulin (CaM)–target interactions), it is possible to achieve great increase in binding specificity without the explicit use of negative design. In order for this to work, two requirements have to be satisfied: (1) binding residues on the desired and undesired targets should be sufficiently different and (2) a large number of positions on the binder protein should be computationally optimized (Fig. 3.2, upper panel). The rational for neglecting negative design is the following. During the design procedure, we introduce mutations that are likely to improve interactions with respect to the desired target but are random with respect to alternative targets. Since each random mutation is likely to decrease affinity to alternative targets, several random mutations will almost certainly produce a destabilizing effect for alternative complexes (Fig. 3.2, lower panel). When studying CaM–target interactions, we demonstrated that significant enhancement in binding specificity (up to 900-fold) could be achieved toward different CaM targets without implicit consideration of negative design (Shifman & Mayo, 2002, 2003; Yosef, Politi, Choi, & Shifman, 2009). Large-scale computational analysis of CaM––target interactions further supported our hypothesis that negative design is not necessary for this experimental system (Fromer & Shifman, 2009). Below, we list a protocol for narrowing down binding specificity of a multispecific binder without the explicit use of negative design.

2.1.1 Structure preparation

1. Download from PDB the structure of your protein in complex with the desired target (your positive design state). X-ray structures are preferred to NMR structures since they were shown to work better on several protein design problems (Schneider, Fu, & Keating, 2009). Structures with resolution of less than 2.5 Å are preferred.

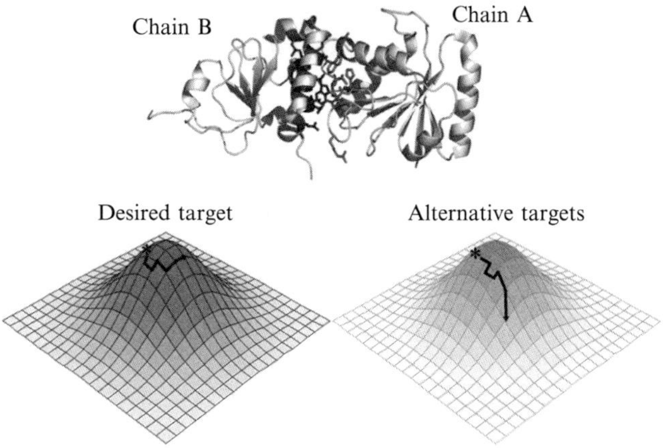

Figure 3.2 Optimizing multiple residues using only positive design. Upper panel: Using the structure of the complex between chains A and B (the positive design state), all binding interface residues on chain A are simultaneously optimized while binding interface residues on chain B are allowed to repack. Binding interface residues are shown as sticks. Lower panel: Fitness landscape for interactions of chain A with the desired target, chain B, and with alternative targets that are not considered in the optimization. The WT sequence lies near the respective maxima for both the desired target and the alternative targets as indicated by *. Arrows show the change in fitness due to mutations in chain A predicted by the optimization. *The lower panel is reproduced from Shifman & Mayo (2003).* (For color version of this figure, the reader is referred to the online version of this chapter.)

2. Remove water molecules and heteroatoms.
3. Add hydrogen atoms to the structure using MolProbity (http://molprobity.biochem.duke.edu/; Chen et al., 2010).
4. Minimize the structure (including both backbone and side chains) to relax clashes and to allow optimal hydrogen bond conformations. We perform 50 steps of atom displacement using the conjugated gradient minimization. From this step on, protein backbone remains fixed throughout the design calculations.
5. Classify the residues as surface, boundary, or core according to their surface accessibility (Dahiyat & Mayo, 1996; Dahiyat, Sarisky, & Mayo, 1997). This is done to calculate the solvation energy term, which depends on the surface accessibility of the residue.
6. Define the binding interface on both chains. We classify a residue as interfacial if any of its atoms is found within 4 Å of the other chain in the complex. Alternatively, interfacial residues could be defined according to their Cα–Cα distance or the percentage of surface area buried upon complex formation.

2.1.2 Set up the design calculation

1. Choose the rotamer library for the design calculations. We suggest using the medium-size rotamer library (Shapovalov & Dunbrack, 2011) that adds subrotamers at ± 1 standard deviation around χ_1 angles. The library could be further expanded to reflect the nature of the binding interface by using additional rotamers (with expansion around χ_1 and χ_2 angles) for amino acids that frequently appear in the particular binding interface (i.e., aromatics or positively charged residues).
2. Define the energy function for calculating interaction energies. The energy function should be specifically optimized for designing protein–protein interfaces. In our previous work, we reported such an energy function that was optimized by trying to reproduce side-chain conformations in a large dataset of protein–protein complexes with available structures (Sharabi, Dekel, & Shifman, 2011). This energy function does not penalize burial of polar atoms in the interface and enhances electrostatic interactions in comparison to the energy function used for design of monomeric proteins.
3. (Optional) Bias the scoring energy function toward prediction of favorable inter- rather than intramolecular interactions. The incentive for doing this is to favor mutations that improve intermolecular interactions rather than stabilize the single chain. This is done by simply multiplying the contribution of the intermolecular part of the total energy by a factor $\alpha \geq 1$ and the contribution of the intramolecular part by a factor $\beta \leq 1$. Since values of α and β are determined empirically, we recommend testing a few combinations of different factors α and β by constructing the designed sequences and verifying their binding specificities as described in our previous work (Yosef et al., 2009).

2.1.3 Perform the design calculation

1. Simultaneously optimize all the binding interface positions on chain A (the chain whose specificity should be enhanced). All 20 amino acids could be used as possible candidates for the design calculation. We recommend, however, excluding Pro and Gly because they might cause protein misfolding and excluding Cys because it might result in formation of undesired disulphide bonds. At the same time, allow for repacking of the binding interface residues on chain B (Fig. 3.2, upper panel). Find the best binding interface sequence by minimizing the total energy of the complex using the dead-end elimination (DEE) algorithm (Desmet, De Maeyer, Hazes, & Lasters, 1992) or an alternative energy

minimization algorithm (Allen & Mayo, 2006). Here, the total energy of the complex rather than the intermolecular energy is minimized since minimization of the latter quantity is likely to produce significant destabilization of the protein under design (see Section 1 for discussion).

2. (Optional) Calculation is repeated with different bias factors α and β to obtain sequences with various weights for inter- and intramolecular contributions.

2.1.4 Analyze the results

1. Computationally verify the reduction in binding affinity for those alternative targets whose structures in complex with the multispecific binder protein (chain A) are available. Starting from the structure of the binder in complex with an alternative target, introduce mutations designed in Section 2.1.3 into the binder. Allow for rotamer repacking. Compute the total energy of the binder–target complex using the unbiased energy function. Perform the same calculation for the WT binder–target complex. Verify that the energy of the designed binder sequence is higher (less favorable) than that of the WT binder sequence when interacting with an alternative target (Fig. 3.1).
2. Compute an evolutionary profile of the binding interface sequences for natural homologous proteins using the homology-derived secondary structure of proteins (HSSP) database (Dodge, Schneider, & Sander, 1998). Compare your designed binding interface sequences with the HSSP profile. Mutations that are common to your designs and to the evolutionary profile are likely to enhance affinity to the desired target. Mutations that are different from the HSSP profile are likely to be specificity determining, although they might decrease affinity to the desired target (Fromer & Shifman, 2009).

2.1.5 Example: Enhancing binding specificity of CaM to CaM-dependent protein kinase II (CaMKII) relative to calcineurin (CaN)

CaM is a Ca^{2+}-sensor protein that in nature exhibits very low-binding specificity by recognizing and activating hundreds of different target proteins. We redesigned CaM to increase its specificity toward CaM-dependent protein kinase II (CaMKII) relative to calcineurin (CaN) (Yosef et al., 2009). Starting from the structure of CaM in complex with the CaMKII peptide, 18 positions on the CaM-binding interface were optimized, while the residues on CaMKII were allowed to change their conformation. No negative design or design against CaN was incorporated into the calculations. We experimentally tested six CaM sequences designed with different bias factors. These mutants contained 5–9 mutations with respect to WT CaM. All but one CaM mutant demonstrated a small increase in affinity for the CaMKII peptide ($\Delta\Delta G_{\text{bind}}$ of up to -0.9 kcal/mol). Moreover, all CaM mutants exhibited substantial decreases in binding to the CaN peptide ($\Delta\Delta G_{\text{bind}}$ ranging from 1.3 to 4.6 kcal/mol). All CaM variants showed an increase in binding specificity toward CaMKII, ranging from a factor of 23 to a factor of 880. Interestingly, the largest enhancement in binding specificity was obtained by sacrificing some affinity to the desired target CaMKII.

The described strategy of using only the positive design protocol and simultaneously optimizing all binding interface residues has an advantage of being algorithmically simple, and it does not require structural information for the negative design states, which in many cases is not available. However, the strategy is likely to perform poorly when positive and negative design states are highly similar. In addition, due to accumulation of error in our predictions, simultaneous introduction of multiple mutations into the binder sequence has high probability of reducing binding affinity to the desired target. It is subsequently difficult to dissect which of the predicted mutations decreased affinity. Hence, more recently, we developed a different algorithm for prediction of single affinity- and specificity-enhancing mutations. This algorithm involves scanning each position of the binding interface with all possible amino acids and computing the intermolecular energy for this residue in the context of both the positive and the negative target (Fig. 3.3). The specificity-enhancing mutations are defined as those that form specific intermolecular interactions such as side chain–side chain hydrogen bonds, pi–pi, and cation–pi interactions only in the positive design state. The effect of the selected mutations is then tested experimentally for affinity to both the positive and the negative targets. Verified specificity-enhancing mutations can be subsequently combined into one sequence to produce an additive effect.

2.2. Designing single specificity-enhancing mutations with positive and negative design

2.2.1 Structure preparation

Select two or more structures of complexes that contain the same protein as chain A and different proteins as chain B. Follow steps 1–6 in Section 2.1.1. In step 4, residue surface classification should be performed for both the bound complex and for the two unbound chains.

For each interfacial position X on chain A, follow Sections 2.2.2 and 2.2.3:

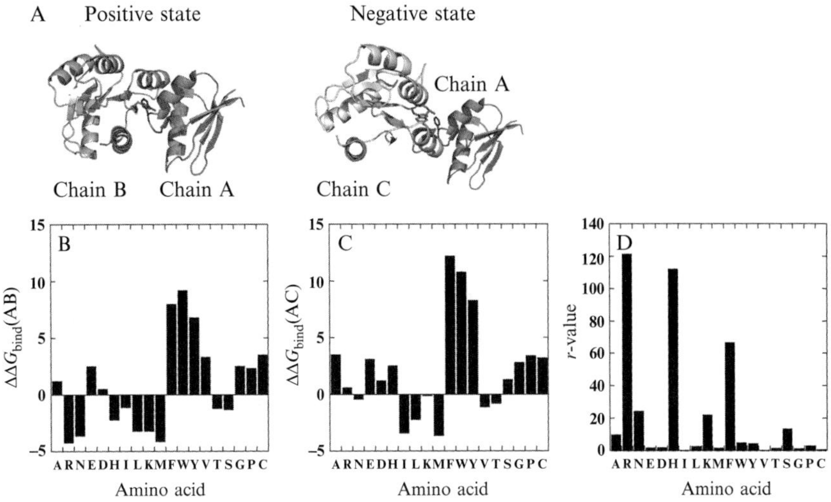

Figure 3.3 Saturated mutagenesis protocol for identifying specificity-enhancing mutations. (A) Positive and negative states are selected. All amino acids are substituted at a single interface position, the surrounding residues are repacked and $\Delta\Delta G_{\text{bind}}$ is calculated for both positive and negative design states. (B) $\Delta\Delta G_{\text{bind}}$ values are plotted for each amino acid substitution at a certain position in the context of the positive design state (chain A interacting with chain B). (C) $\Delta\Delta G_{\text{bind}}$ values are plotted for each amino acid substitution at the same position in the context of the negative design state (chain A interacting with chain C). (D) Specificity ratio *r* for each amino acid. Mutations to R and to H show the highest specificity shift. Mutation to F shows a high specificity shift but reduced affinity to both targets B and C. (For color version of this figure, the reader is referred to the online version of this chapter.)

2.2.2 Define the design problem

1. Define a shell around the interfacial residue. The residues in the shell are allowed to repack their side-chain conformations upon introduction of a mutation in chain A. We suggest using cutoffs of 3 Å for the shell residues belonging to chain A and 3.5 Å for the shell residues belonging to chain B.
2. Choose the rotamer library for the design calculations. Due to the small combinatorial complexity of the problem, we suggest using the largest possible rotamer library such as the one that adds subrotamers at ±1 standard deviation around both χ_1 and χ_2 angles. We also recommend using backbone-independent library rather than backbone-dependent library since the latter might lack rotamers for certain amino acids at some of the design positions.
3. Define the energy function. Use the same energy function as described in step 2 in Section 2.1.2.

2.2.3 Perform the design calculation

Probe position X with all twenty amino acids, including the WT amino acid. For each amino acid, Y placed at position X follow steps 1–3.

1. Find the total energy of the complex E_{tot} when placing amino acid Y at position X (denoted further as $X \rightarrow Y$) in the context of the protein–protein complex structure.

   a. Set the amino acid identity at the considered position X to Y.

   b. Allow for repacking of all the shell residues around position X.

   c. Search for the Global Minimum Energy Conformation (GMEC) of rotamers using DEE (Desmet et al., 1992) or an alternative energy minimization algorithm (Allen & Mayo, 2006).

   d. Define the energy of the complex, E_{tot}, as the energy of the GMEC.

2. Calculate the energies of the single chains E_A and E_B for the mutation $X \rightarrow Y$.

   a. Split the GMEC structure generated in step 1 in Section 2.2.3 into two separate PDB files containing coordinates for chain A and chain B.

   b. Calculate the energies for the single chains A and B. Use all the same parameters as in the calculation performed for the complex (step 1 in Section 2.2.3) except the surface accessibility becomes different for single chains in comparison to the complex. Two options are possible to perform this calculation. One does not allow for additional rotamer repacking, that is, the rotameric conformations are frozen in the GMEC obtained for the complex structure. The second option is to allow for rotamer repacking when the single chains are separated. Our unpublished results as well as results by the Kuhlman group (Sammond et al., 2007) suggest that rotamer repacking at this step results in reduction in accuracy of predicting free energy of binding, $\Delta\Delta G_{bind}$. Hence, no rotamer repacking is a preferred and faster option for this calculation.

3. Calculate $\Delta\Delta G_{bind}$ for the mutation $X \rightarrow Y$.

   a. To obtain the intermolecular energy, subtract the energies of the single chains from the complex energy:

$$E_{inter} = E_{tot} - E_B - E_A$$

   b. To calculate $\Delta\Delta G_{bind}$, subtract the intermolecular energy of the mutated complex from that of the WT complex:

$$\Delta\Delta G_{bind} = E_{inter}^{WT} - E_{inter}^{X \rightarrow Y} = \Delta E_{tot} - \Delta E_A - \Delta E_B,$$

   where Δ denotes the difference between the WT and the mutated state.

In most cases, the energy of the nonmutated chain E_B is the same for the WT and the mutated complexes. However, in a few cases, we noticed that this energy is different, giving rise to nonzero ΔE_B. This is obviously an artifact arising from not repacking rotamers during the single-chain calculations. Since theoretically the energy of the nonmutated chain should not change upon introduction of mutation in the opposite chain, we recommend setting ΔE_B to zero, hence simplifying the formula:

$$\Delta\Delta G_{\text{bind}} = \Delta E_{\text{tot}} - \Delta E_A$$

Such simplification gave an improvement in the accuracy of $\Delta\Delta G_{\text{bind}}$ prediction using a large set of mutational data (our unpublished results).

2.2.4 Identify specificity-enhancing mutations

Assume that protein A binds to proteins B and C. You would like to find single mutations in protein A that shift its binding specificity toward protein B relative to protein C.

1. Repeat the saturated mutagenesis protocol (Sections 2.2.1–2.2.3) for the complex between proteins A and C.
2. Calculate the predicted specificity ratio for each position and each amino acid:

$$r = \exp[-\Delta\Delta G_{\text{bind}}(\text{AB}) + \Delta\Delta G_{\text{bind}}(\text{AC})]$$

   where $\Delta\Delta G_{\text{bind}}$ (AB) and $\Delta\Delta G_{\text{bind}}$ (AC) denote the predicted free energy of binding of protein A to proteins B and C, respectively.
3. Select specificity-enhancing mutations with the highest r value (Fig. 3.3). Pay attention that some mutations might increase binding specificity but decrease affinity to both proteins B and C. If such mutations are not desirable, select only mutations that have negative $\Delta\Delta G_{\text{bind}}$ (AB) and the highest r value. Avoid mutations that are predicted to destabilize chain A by a large amount since such mutations are likely to result in unfolding of protein A and hence are not likely to produce the expected effects on binding affinity and specificity.

2.2.5 Examples: Specificity-enhancing mutations identified by the saturated mutagenesis protocol

1. Specificity and affinity enhancement of Raf kinase (Raf) toward Ras^{GDP}

   Ras is a small GTPase that is an essential molecular switch for a wide variety of signaling pathways. In the GTP-bound state, Ras can interact with its effector proteins and activate downstream signaling pathways.

In the GDP-bound state, Ras affinity for its effectors is reduced by several orders of magnitude. Using the saturated mutagenesis protocol, we searched for mutations that increase the affinity of one of the Ras effectors Raf for the usually inactive state of Ras, Ras-GDP, while at the same time preserving or destabilizing the complex between Raf and the active Ras state, Ras-GTP (Filchtinski et al., 2010). Experimental testing showed that most of the designed mutations narrowed the gap between Raf's affinity for Ras-GTP and Ras-GDP, producing the desired shift in the binding specificity. A combination of our best designed mutation, N71R, with another mutation, A85K, yielded a Raf mutant with a 100-fold improvement in affinity toward Ras-GDP and a 12-fold shift in binding specificity.

2. Enhancing binding specificity of a snake toxin Fasciculin (Fas) toward *Torpedo californica* acetylcholinesterase.

Fas is a snake toxin that binds very tightly and inhibits synaptic enzyme acetylcholinesterase from Humans (*h*AChE) and some other species such as *T. californica* (*Tc*AChE). Structures of Fas in complex with *h*AChE and *Tc*AChE enzymes reveal a very similar binding mode where Fas seals the entrance to the active site, thus preventing the access of the substrate acetylcholine. Although the Fas-binding surfaces on the two enzymes are quite similar, Fas binds to *h*AChE with about four times stronger affinity compared to *Tc*AChE. To identify mutations that reverse Fas-binding specificity toward *Tc*AChE, we performed the *in silico* saturated mutagenesis protocol (Aizner et al., 2013). Several specificity-enhancing mutations were identified. For example, at position 29, mutation H29R resulted in predicted $\Delta\Delta G_{\text{bind}}$ values of −8.2 and −2.7 for Fas–*Tc*AChE and Fas–*h*AChE interactions, respectively, and an *r* value of 244. Experimentally, the given mutant showed ~fivefold enhancement in binding specificity.

3. BROADENING BINDING SPECIFICITY

Multispecific proteins play a central role in the organization of protein interaction networks. These proteins, frequently referred to as protein hubs, are critical for cell survival and exhibit slower rates of evolution. The requirement of binding to multiple partners imposes constraints on the amino acid sequences of multispecific proteins, especially in their binding interface region. The binding interface sequences of such proteins is a compromise obtained to satisfy interaction requirements with all partner proteins.

Sequence analysis of multispecific proteins is an interesting problem in evolutionary biology that could be addressed by computational protein design.

Recently, binding multispecificity has also been explored for therapeutic applications. Multispecific antibodies are promising drug candidates since they can simultaneously target a number of disease-associated proteins and hence be more effective in battling the disease. An example of such a bispecific antibody that binds both human epidermal growth factor receptor 2 (HER2) and the vascular endothelial growth factor (VEGF) was reported by Bostrom et al. (2009). Multispecific binders are usually engineered through combinatorial approaches by alternating target proteins in subsequent rounds of selection. Alternatively, multispecific binders could be selected *in silico*, using the multistate design procedure described below. The advantages of using this procedure over the combinatorial approaches is that in principle there is no limitation on the number of targets incorporated in the calculation, while experimental selection for binding to a large number of targets is problematic.

Several algorithms for design of multispecific binders have been proposed (Allen & Mayo, 2010; Fromer & Shifman, 2009; Fromer, Yanover, & Linial, 2010; Humphris & Kortemme, 2007; Yanover, Fromer, & Shifman, 2007). In such studies, the binding interface sequence(s) of the multispecific protein is designed to provide optimal interactions with a predefined set of targets (Fig. 3.4). This process is referred to as multistate design. Below, we provide a detailed protocol for such design.

3.1. Structure preparation for multistate design

1. Select your design states: two or more structures of protein–protein complexes where chain A is the multispecific binder and chain B is a variable protein target.
2. Perform steps 2–6 in the Section 2.1.1 to prepare structures for design.
3. Define the common binding interface for all the structures. We suggest including all positions that appear in the binding interface of at least 75% of the selected design states. Binding interface positions that are common to only one or a few design states should not be included since they are not constrained for multispecificity.

3.2. Set up the design calculation for each of the selected design states

1. Select the rotamer library for design. Due to the larger combinatorial complexity of the calculation, we suggest using a medium-size rotamer

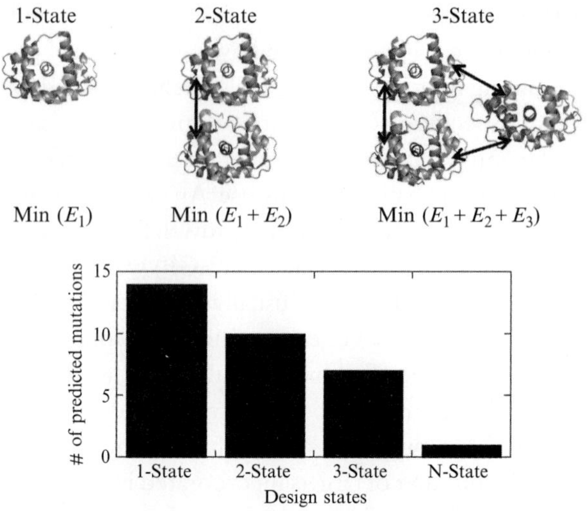

Figure 3.4 Multistate design protocol. Upper panel: A multispecific binder is optimized to bind to each single target (1-state design), pairs of targets (2-state design), and triplets of targets (3-state design). Multiple target design is implemented by minimizing the sum of the binder sequence energies in each structure, with the constraint (denoted by arrows) that the same amino acid sequence be predicted for all structures. Lower panel: Average number of mutations with respect to WT protein predicted for different numbers of design states. Note that the number of mutations decreases as more design states are incorporated and approaches zero for large Ns. *This figure is reproduced with changes from Fromer and Shifman (2009).* (For color version of this figure, the reader is referred to the online version of this chapter.)

library that adds subrotamers at ±1 standard deviation around χ_1 angle. The library could be expanded to reflect the nature of the binding interface by using additional rotamers (with expansion around χ_1 and χ_2 angles) for amino acids that frequently appear in the particular binding interface (i.e., aromatics or positively charged residues).

2. Define the energy function for calculating interaction energies. See step 2 in Section 2.1.2.
3. Allow all the binding interface residues on the multispecific protein (chain A) to mutate to all natural amino acids.
4. Allow all the binding interface residues on the variable chain B to repack their side-chain conformations.

3.3. Design of multispecific binding interface sequences

1. Design a single-binding interface sequence that minimizes the sum of the energies in n design states: $\min \sum_{i=0}^{n} E_i$ (Fig. 3.4).

Here, different search algorithms could be applied. While we utilized a combination of an algorithm based on the (DEE) theorem (Gordon, Hom, Mayo, & Pierce, 2003) and an algorithm based on belief propagation for probabilistic graphical models (type-specific Best Max-Marginal First (tBMMF)) (Fromer & Yanover, 2008, 2009) other groups used a simple Monte Carlo Simulated Annealing (MCSA) search (Humphris & Kortemme, 2007; Kirckpatrick, Gelatt, & Vecchi, 1983).

2. One multispecific sequence could be further expanded by designing a small number of near-optimal sequences. Here, the tBBMF algorithm becomes particularly useful since it is guaranteed to find a desired number of the best-energy sequences (Fromer & Yanover, 2009).
3. To better visualize the results generate a WebLogo of the best hundred-binding interface sequences using a WebLogo server (Crooks, Hon, Chandonia, & Brenner, 2004).

An interesting insight could be obtained by comparing sequences designed for simultaneous binding to multiple targets to those designed to best satisfy interactions with only one target at a time. Below, we provide a protocol for such a comparison.

3.4. Design sequences with single specificity

1. For each target included in the multistate design, perform the calculation considering only interactions with this particular target (1-state design, Fig. 3.4).
2. Expand the results by finding the hundred lowest energy interfacial sequences for each of the targets.
3. To better visualize the results generate a WebLogo of the best hundred-binding interface sequences using a WebLogo server for all of the design scenarios.

Below are some types of analysis that can provide insight into differences between single-state and multistate designs.

3.5. Analyze the results

1. Calculate the number of designed mutations with respect to WT protein for each single-state and multistate design calculation. Compare the average number of mutations predicted in the single-state and multistate design calculations. The number of predicted mutations should decrease as more targets are incorporated in the design procedure (Fig. 3.4).
2. Compute the evolutionary (HSSP) profile of the binding interface sequences (Dodge et al., 1998). This is done by calculating probabilities

of each amino acid to occur at each of the design positions among natural homologous sequences.

3. Compare the designed sequences resulting from either multispecific or single-state design scenarios with the evolutionary profile. This is done by computing the Jensen–Shannon Divergence (JSD) score for two distributions P and Q containing frequencies of each amino acid observed at a certain position:

$$\mathrm{JSD}(P,Q) = \frac{1}{2} D_{\mathrm{KL}}(P,R) + \frac{1}{2} D_{\mathrm{KL}}(Q,R)$$

where $R = 1/2(P+Q)$, $D_{\mathrm{KL}}(A,B) = \sum_x a(x)\log_2(a(x)/b(x))$, and $a(x)$ and $b(x)$ are probabilities of an amino acid x observed at a certain position.

The JSD score is equal to 0 for the same distributions and is equal to 1 for very distant distributions. The mean JSD score for the whole sequence profile is calculated by averaging JSD scores over all the designed positions. The JSD score between designed and natural profiles should decrease with increasing the number of design states incorporated into the calculation.

4. Classify the binding interface positions into the affinity-defining and the specificity-defining. For this purpose, calculate the JSD score for all the pairs of single-state design profiles for each design position. Affinity-determining positions are defined as those that exhibit similar amino acid identities in all single-state calculations (JSD scores close to 0) Specificity-determining positions are defined as those that exhibit higher diversity (JSD scores close to 1). We used cutoffs of 0.25 and 0.75 to define affinity- and specificity-determining positions, respectively.

5. Analyze the sequence compromise needed to achieve multispecificity. For a particular 2-state design scenario, compare the profile based on its 100 lowest energy sequences to those profiles designed for the same two single targets (1-state designs) for each of the design positions. Calculate the JSD score between the 2-state design and each of the 1-state designs and the JSD score between the 1-state designs. Compromise at each position can then be categorized into five intuitive scenarios according to four calculated dissimilarity scores. "Kept same"—the 1-state designs predicted similar results (pairwise $\mathrm{JSD} < 0.3$) and the 2-state design was similar to both of them (both pairwise $\mathrm{JSD} < 0.5$); "Combined"—the 1-state designs were dissimilar (pairwise $\mathrm{JSD} \geq 0.3$), but the 2-state design was similar to both of them ($\mathrm{JSD} < 0.5$); "Preferred

one"—2-state design was similar to only one of the 1-state designs (JSD <0.5); "New aa"—the 1-state designs predicted dissimilar results (pairwise JSD ≥ 0.3), and the 2-state design was different from both of them (both pairwise JSD ≥ 0.5); "despite same"—despite the 1-state designs predicting similar results (pairwise JSD <0.3), the 2-state design was different from both of them (JSD ≥ 0.5). The latter scenario can occur if there are contributions from correlated mutations. Note that similar analysis could be performed for analyzing sequences designed for interaction with more than two targets.

6. Investigate whether imposing a requirement for multispecificity brings the designed sequence profiles closer to the evolutionary profile. To do this, calculate the per-position JSD scores comparing the 2-state design profiles to the HSSP profile. Then, construct the profile resulting from averaging the two 1-state design profiles and calculate its per-position JSD scores from HSSP. For a particular position, calculate the difference between these JSD values (*d*) to define the effect of multistate compromise. Here, the results could be categorized into three groups: "No Change"—$|d| \leq 0.1$; "Benefit"—$d < -0.1$; "Loss"—$d > 0.1$. According to our study (Fromer & Shifman, 2009), most of the results fall into the benefit category, showing that multispecificity requirements bring the designed sequence profiles closer to those of natural multispecific binders.

3.6. Example: Multistate protein design in CaM–target interactions

To investigate how multispecificity is acquired in nature, we studied one prototypical mutispecific protein CaM, which has evolved to interact with hundreds of different targets (Fromer & Shifman, 2009). Starting from 16 structures of CaM–target complexes, we computationally predicted the hundred best sequences of the CaM-binding interface for each individual CaM target. We then designed CaM sequences optimal for interactions with all possible combinations of two, three, and finally all 16 targets simultaneously. By comparing these sequences and their energies, we gained insight into how nature has managed to find the compromise between the need for favorable interaction energies and the need for multispecificity. We observed that designing for more partners simultaneously yields CaM sequences that better match natural sequence profiles, thus emphasizing the importance of such strategies in nature. Furthermore, we showed that the CaM-binding interface could be nicely partitioned into positions that are critical for the affinity of all CaM–target complexes and those that are molded to provide interaction specificity. We revealed several basic categories of sequence-level tradeoffs that enable the compromises necessary for the promiscuity of this protein. We also thoroughly quantified the tradeoff between interaction energetics and multispecificity and found that facilitating seemingly competing interactions requires only a small deviation from optimal energies.

4. SUMMARY

Here, we present three computational protocols for controlling binding specificity in proteins. The first two protocols aim at enhancing binding specificity, while the third protocol intends to broaden binding specificity. The first protocol for enhancing specificity optimizes a large number of positions and uses only information about interactions with the desired target. The second protocol introduces single mutations that increase binding specificity toward one of the considered targets relative to another. The third protocol presents an approach for designing proteins that are optimal for interactions with multiple targets. Such an approach could be applied to study evolution of hub proteins in nature and to design novel mutispecific binders with therapeutical potential.

ACKNOWLEDGMENTS

This work was supported by the ISF Grant 1372/10, Deutsche Forschungsgemeinschaft Grant EI 423/2-1, and the Abisch Frenkel foundation.

REFERENCES

Aizner, Y., Sharabi, O., Dakwar, G., Shirian, J., Risman, M., Lewinson, O., et al. (2013). Walking the energy landscape of a picomolar protein-protein complex. Submitted.

Allen, B. D., & Mayo, S. L. (2006). Dramatic performance enhancements for the FASTER optimization algorithm. *Journal of Computational Chemistry*, *27*(10), 1071–1075.

Allen, B. D., & Mayo, S. L. (2010). An efficient algorithm for multistate protein design based on FASTER. *Journal of Computational Chemistry*, *31*(5), 904–916.

Bolon, D. N., Grant, R. A., Baker, T. A., & Sauer, R. T. (2005). Specificity versus stability in computational protein design. *Proceedings of the National Academy of Sciences of the United States of America*, *102*(36), 12724–12729.

Bostrom, J., Yu, S. F., Kan, D., Appleton, B. A., Lee, C. V., Billeci, K., et al. (2009). Variants of the antibody herceptin that interact with HER2 and VEGF at the antigen binding site. *Science*, *323*(5921), 1610–1614.

Chen, V. B., Arendall, W. B., 3rd., Headd, J. J., Keedy, D. A., Immormino, R. M., Kapral, G. J., et al. (2010). MolProbity: All-atom structure validation for macromolecular crystallography. *Acta Crystallographica. Section D, Biological Crystallography*, *66*(Pt 1), 12–21.

Crooks, G. E., Hon, G., Chandonia, J. M., & Brenner, S. E. (2004). WebLogo: A sequence logo generator. *Genome Research*, *14*(6), 1188–1190.

Dahiyat, B. I., & Mayo, S. L. (1996). Protein design automation. *Protein Science*, *5*(5), 895–903.

Dahiyat, B. I., Sarisky, C. A., & Mayo, S. L. (1997). De novo protein design: Towards fully automated sequence selection. *Journal of Molecular Biology*, *273*(4), 789–796.

Desmet, J., De Maeyer, M., Hazes, B., & Lasters, I. (1992). The dead-end elimination theorem and its use in side chain packing problem. *Nature*, *356*, 539–542.

Dodge, C., Schneider, R., & Sander, C. (1998). The HSSP database of protein structure sequence alignments and family profiles. *Nucleic Acids Research*, *26*(1), 313–315.

Filchtinski, D., Sharabi, O., Rüppel, A., Vetter, I. R., Herrmann, C., & Shifman, J. M. (2010). What makes Ras an efficient molecular switch: A computational, biophysical, and structural study of Ras-GDP interactions with mutants of Raf. *Journal of Molecular Biology*, *399*(3), 422–435.

Fromer, M., & Shifman, J. M. (2009). Tradeoff between stability and multispecificity in the design of promiscuous proteins. *PLoS Computational Biology*, *5*(12), e1000627.

Fromer, M., & Yanover, C. (2008). A computational framework to empower probabilistic protein design. *Bioinformatics*, *24*(13), i214–i222.

Fromer, M., & Yanover, C. (2009). Accurate prediction for atomic-level protein design and its application in diversifying the near-optimal sequence space. *Proteins*, *75*(3), 682–705.

Fromer, M., Yanover, C., & Linial, M. (2010). Design of multispecific protein sequences using probabilistic graphical modeling. *Proteins*, *78*(3), 530–547.

Gordon, D. B., Hom, G. K., Mayo, S. L., & Pierce, N. A. (2003). Exact rotamer optimization for protein design. *Journal of Computational Chemistry*, *24*(2), 232–243.

Grigoryan, G., Reinke, A. W., & Keating, A. E. (2009). Design of protein-interaction specificity gives selective bZIP-binding peptides. *Nature*, *458*(7240), 859–864.

Havranek, J. J., & Harbury, P. B. (2003). Automated design of specificity in molecular recognition. *Nature Structural Biology*, *10*(1), 45–52.

Humphris, E. L., & Kortemme, T. (2007). Design of multi-specificity in protein interfaces. *PLoS Computational Biology*, *3*(8), e164.

Kirckpatrick, S., Gelatt, C. D., & Vecchi, M. P. (1983). Optimization by simulating annealing. *Science*, *220*(4598), 671–680.

Leaver-Fay, A., Jacak, R., Stranges, P. B., & Kuhlman, B. (2011). A generic program for multistate protein design. *PLoS One*, *6*(7), e20937.

Sammond, D. W., Eletr, Z. M., Purbeck, C., Kimple, R. J., Siderovski, D. P., & Kuhlman, B. (2007). Structure-based protocol for identifying mutations that enhance protein-protein binding affinities. *Journal of Molecular Biology*, *371*(5), 1392–1404.

Schneider, M., Fu, X., & Keating, A. E. (2009). X-ray vs. NMR structures as templates for computational protein design. *Proteins*, 77(1), 97–110.

Shapovalov, M.S., & Dunbrack, R.L., Jr. (2011). A smoothed backbone-dependent rotamer library for proteins derived from adaptive kernel density estimates and regressions. *Structure, 19,* 844–858.

Sharabi, O., Dekel, A., & Shifman, J. M. (2011). Triathlon for energy functions: Who is the winner for design of protein-protein interactions? *Proteins*, *79*(5), 1487–1498.

Shifman, J. M., & Mayo, S. L. (2002). Modulating calmodulin specificity through computational protein design. *Journal of Molecular Biology*, *323*, 417–423.

Shifman, J. M., & Mayo, S. L. (2003). Exploring the origins of binding specificity through the computational redesign of calmodulin. *Proceedings of the National Academy of Sciences of the United States of America*, *100*(23), 13274–13279.

Yanover, C., Fromer, M., & Shifman, J. M. (2007). Dead-end elimination for multistate protein design. *Journal of Computational Chemistry*, *28*(13), 2122–2129.

Yosef, E., Politi, R., Choi, M. H., & Shifman, J. M. (2009). Computational design of calmodulin mutants with up to 900-fold increase in binding specificity. *Journal of Molecular Biology*, *385*(5), 1470–1480.

CHAPTER FOUR

Flexible Backbone Sampling Methods to Model and Design Protein Alternative Conformations

Noah Ollikainen[*], **Colin A. Smith**[*,1], **James S. Fraser**[†,‡,2], **Tanja Kortemme**[*,‡,§,2]

[*]Graduate Program in Bioinformatics, University of California San Francisco, San Francisco, California, USA
[†]Department of Cellular and Molecular Pharmacology, University of California San Francisco, San Francisco, California, USA
[‡]California Institute for Quantitative Biosciences (QB3), University of California San Francisco, San Francisco, California, USA
[§]Department of Bioengineering and Therapeutic Science, University of California San Francisco, San Francisco, California, USA
[1]Present address: Max Planck Institute for Biophysical Chemistry, Göttingen, Germany
[2]Corresponding authors: e-mail address: james.fraser@ucsf.edu; kortemme@cgl.ucsf.edu

Contents

Abstract

Sampling alternative conformations is key to understanding how proteins work and engineering them for new functions. However, accurately characterizing and modeling protein conformational ensembles remain experimentally and computationally challenging. These challenges must be met before protein conformational heterogeneity can be exploited in protein engineering and design. Here, as a stepping stone, we describe methods to detect alternative conformations in proteins and strategies to model these near-native conformational changes based on backrub-type Monte Carlo moves in

Methods in Enzymology, Volume 523
ISSN 0076-6879
http://dx.doi.org/10.1016/B978-0-12-394292-0.00004-7

Rosetta. We illustrate how Rosetta simulations that apply backrub moves improve modeling of point mutant side-chain conformations, native side-chain conformational heterogeneity, functional conformational changes, tolerated sequence space, protein interaction specificity, and amino acid covariation across protein–protein interfaces. We include relevant Rosetta command lines and RosettaScripts to encourage the application of these types of simulations to other systems. Our work highlights that critical scoring and sampling improvements will be necessary to approximate conformational landscapes. Challenges for the future development of these methods include modeling conformational changes that propagate away from designed mutation sites and modulating backbone flexibility to predictively design functionally important conformational heterogeneity.

1. INTRODUCTION

Proteins are constantly fluctuating between alternative conformations (Frauenfelder, Sligar, & Wolynes, 1991). Processes including folding (Korzhnev, Religa, Banachewicz, Fersht, & Kay, 2010), ligand binding (Boehr, Nussinov, & Wright, 2009), and enzymatic catalytic cycles (Nagel & Klinman, 2009) depend on the movement across the energy landscape. While protein folding is generally driven by a large energy gap between the "native" state and the unfolded ensemble, functionally essential conformations within the "native" state are often separated by smaller energy differences (Fleishman & Baker, 2012).

Computational modeling of the "native" state can result in either a representative single structure or a limited ensemble of conformations. Several straightforward global and local metrics have been developed to compare computational predictions of a representative single structure to an experimentally derived X-ray structure (MacCallum et al., 2011). In contrast, modeling conformational heterogeneity within the "native" state presents significant complications. For example, conformational heterogeneity present in NMR structural ensembles can result from a lack of restraints, limitations in sampling methods, or genuine heterogeneity (Rieping, Habeck, & Nilges, 2005; Schneider, Brunger, & Nilges, 1999). Additionally, comparisons to simulations are often necessary to distinguish between multiple motional models suggested by NMR dynamics observables including residual dipolar couplings (Meiler, Prompers, Peti, Griesinger, & Bruschweiler, 2001), CPMG relaxation dispersion (Bouvignies et al., 2011), and side-chain order parameters (S^2) (Li, Raychaudhuri, & Wand, 1996). X-ray crystallography, which is traditionally interpreted in terms of a single static structure, can also contain information about protein conformational heterogeneity (Best, Lindorff-Larsen, DePristo, & Vendruscolo, 2006; Furnham, Blundell, DePristo, &

Terwilliger, 2006; Lang et al., 2010; Levin, Kondrashov, Wesenberg, & Phillips, 2007). All of these experimental data types can be integrated to improve the computational modeling of protein conformational ensembles.

Ultimately, to connect protein conformational dynamics to function, simulations must be leveraged to provide structural mechanisms consistent with the experimental data. Molecular dynamics simulations present the most obvious solution to identify the structural mechanisms of conformational heterogeneity (Maragakis et al., 2008). However, other than in exceptional cases (Kelley, Vishal, Krafft, & Pande, 2008; Shaw et al., 2010), the timescales accessible to molecular dynamics often preclude sampling functional motions. The computational requirements of molecular dynamics simulations also make it prohibitive to simultaneously sample sequence space for protein design.

Monte Carlo simulations, for example, as used in Rosetta (Leaver-Fay et al., 2011), can also be used to sample protein conformations but rely on having moves that result in energetically accessible conformations. Fixed backbone Monte Carlo simulations, where side-chain conformations are sampled based on a rotamer library, can provide some indications of local flexibility (DuBay & Geissler, 2009). However, it is clear that both side-chain and backbone flexibility are necessary to describe and design protein conformational heterogeneity (Friedland, Linares, Smith, & Kortemme, 2008; Mandell & Kortemme, 2009). Backbone conformations are less easily discretized compared to side-chain rotamers, leading to problems in both creating and validating Monte Carlo backbone moves. Many strategies to efficiently search through backbone space have been implemented in Rosetta including fragment insertion (Simons, Kooperberg, Huang, & Baker, 1997), loop closure with cyclic coordinate decent (Canutescu & Dunbrack, 2003), local torsion sampling with kinematic loop closure (Mandell, Coutsias, & Kortemme, 2009), and backrub (Davis, Arendall, Richardson, & Richardson, 2006; Smith & Kortemme, 2008). Additionally, these moves can be combined and iterated with sequence design to enrich for proteins with desired conformational and functional properties that could not be explored without backbone flexibility.

Here, we describe the application and validation of the backrub sampling move, which was initially inspired by observations using high-resolution X-ray crystallography (Davis et al., 2006), in Rosetta (Smith & Kortemme, 2008; Fig. 4.1A–C). To test whether these moves accurately represent protein conformational heterogeneity, we provide example command lines and scripts that can be run using Rosetta version 3.5. These commands and scripts examine how backrub sampling affects predictions of mutant structures, alternative conformations observed by X-ray crystallography, peptide-ligand binding specificities, and evolutionary properties (Fig. 4.1D). The continued development of

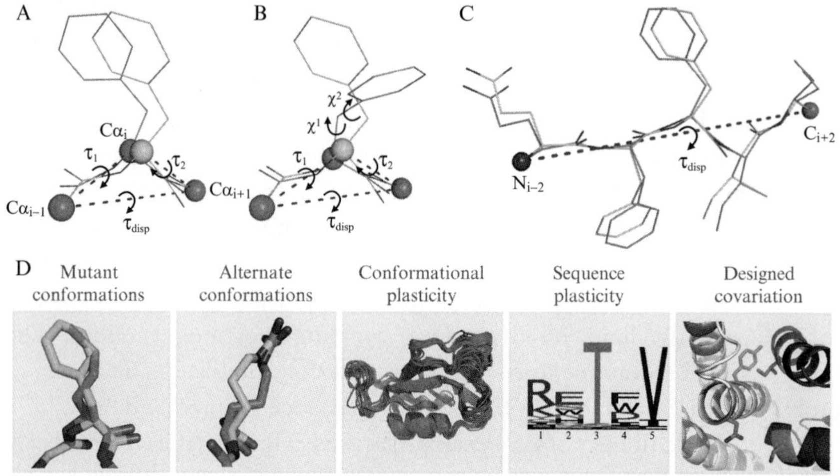

Figure 4.1 The backrub move and its applications in Rosetta. (A) The Richardson group originally described the "backrub" move as a rotation around the $C\alpha_{i-1}$ and $C\alpha_{i+1}$ axis by τ_{disp}, along with simultaneous peptide plane rotations (τ_1 and τ_2), without disturbing other surrounding atom coordinates. (B) By changing the position of the $C\alpha_i$–$C\beta_i$ bond vector, this move can couple side-chain rotameric changes with small local backbone adjustments. (C) In Rosetta, the generalized backrub move is a single rotation that can also include longer intervals and other backbone atom types as pivots for the rotations. (D) Implementing backrubs as a Monte Carlo move in Rosetta enables a variety of flexible backbone prediction and design applications that are described in this chapter: predicting mutant conformations (Fig. 4.2), modeling alternative conformations (Figs. 4.3 and 4.4), coupling conformational and sequence plasticity (Fig. 4.5), and designing amino acid covariation at protein interfaces (Fig. 4.6). (See Color Insert.)

flexible backbone sampling methods that agree with diverse experimental and evolutionary data will improve our ability to design and engineer new protein functions that depend on and exploit conformational heterogeneity.

2. ROSETTA MOVES TO MODEL ALTERNATIVE CONFORMATIONS IN X-RAY DENSITY

2.1. Modeling the Richardson backrub in Rosetta

Unlike other methods for flexible backbone sampling implemented in Rosetta, which are based on fragment insertion or geometric constraints, the backrub move derives its motional model from conformational variation observed in high-resolution X-ray data (Davis et al., 2006). The Richardson group observed electron density consistent with a concerted backbone reorientation that moves a central side-chain perpendicular to the main-chain direction for 3% of total residues in a dataset of ultra-high-resolution crystal structures. They noted that this move changed the accessible side-chain conformations while leaving flanking structure undisturbed.

In Rosetta, the backrub consists of a rotation about an axis defined by the flanking backbone atoms that changes six internal backbone degrees of freedom in the protein, namely the ϕ, ψ, and the N–Cα–C bond (α) angles at both pivots (Smith & Kortemme, 2008). In the Richardson formulation, the pivots were Cα atoms surrounding a single residue (Fig. 4.1A), but the move can be performed over any backbone atom type over varying length scales (Fig. 4.1C). Bond angle, rotational angle, and Cβ/Hα placement constraints are included to eliminate the need for costly minimization steps after every move. In addition, the backrub move in Rosetta can be adapted so that it obeys detailed balance (Smith & Kortemme, 2008). One notable aspect of the backrub move is that it makes certain side-chain conformations, which would not be accessible in the starting backbone conformation, accessible in the newly accepted backbone conformation (Fig. 4.1B). Such moves alter the potential to accommodate new side-chain conformations and mutations at the "backrubbed" position and its local neighbors.

2.2. Modeling the response to mutations

Subtle backbone adjustments are often necessary to accommodate differences between the wild-type and the mutant side-chain. An initial test of the backrub move in Rosetta was to compare its performance with fixed backbone sampling in predicting the conformation of mutated side-chains (Fig. 4.2). Based on a template of the wild-type structure, a successful prediction of a mutant structure would generate both the conformation of the mutant side-chain and any changes that propagate away from the mutated residue. In general, backrub moves decrease the RMSD between the prediction and conformation observed in the mutant crystal structure (Fig. 4.2). Particularly, dramatic successes are achieved when there would be a clash to neighboring atoms that is relieved by a small backbone adjustment or when a local backbone move changes the probability of accessing a new conformation from a backbone dependent rotamer library.

Due to the broad utility of this application for predicting the results of single or multiple point mutations, we have created a Web server that automates this task: https://kortemmelab.ucsf.edu/backrub/ (Lauck, Smith, Friedland, Humphris, & Kortemme, 2010). On the server, the user must enter the desired PDB, the site of mutation, and the new amino acid identity. By default, 10 independent simulations are performed but this can be

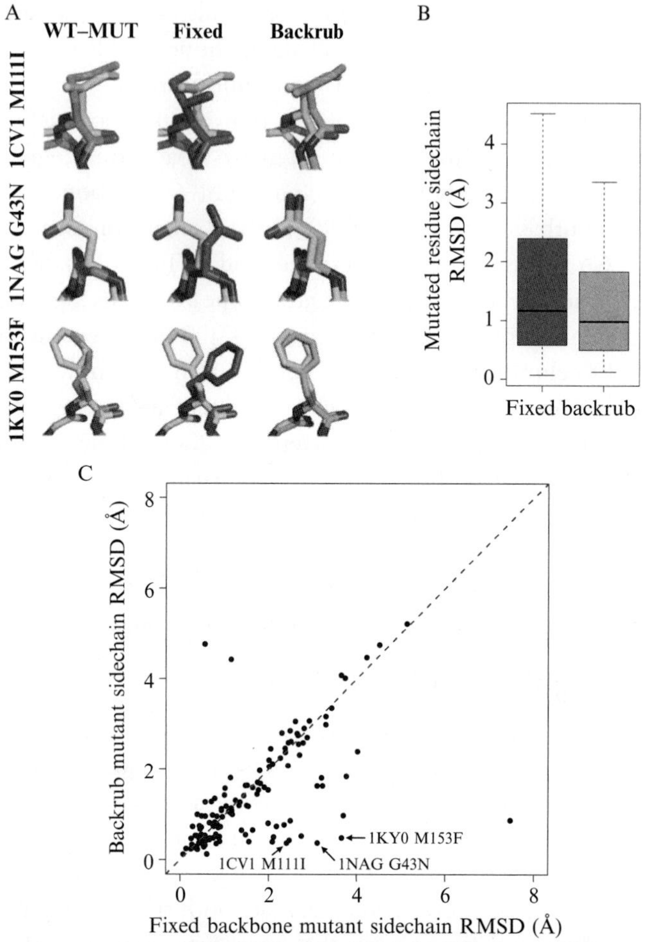

Figure 4.2 Backrub sampling improves the prediction of mutant side-chain conformation compared to fixed backbone simulations. (A) Example predictions of mutant conformations given the wild-type structure, with the indicated PDB codes and point mutations. The mutant crystal structure is shown in yellow compared to the wild-type crystal structure (green, left), prediction based on fixed backbone simulations (magenta, center), or prediction based on backrub flexible backbone sampling (cyan, right). (B) The overall quantification of the results of fixed backbone and backrub predictions over a set of 136 buried (SASA < 5%) side-chains with conformations differing by more than 0.2 Å between mutant and wild type. The median RMSD decreases from 1.17 to 0.98 Å. Shown are box plots with the median as a black line and the 25–75th percentiles in the shaded box with outlier-corrected extreme values as dashed lines. (C) A scatter plot representation of the data in (B) shows that for many mutant structure predictions backrub leads to large improvements compared to fixed backbone simulations. (For interpretation of the references to color in this figure legend, the reader is referred to the online version of this chapter.)

adjusted to 2–50 simulations. For each simulation, the server will attempt 10,000 moves that include backrub rotations of various lengths and angles centered on the mutated residue, and side-chain moves in a 6 Å shell around the mutated residue. The resulting conformations are scored with the Rosetta scoring function and accepted or rejected according to the Metropolis criterion using a kT of 0.6. The lowest scoring conformation out of all the simulations is returned to the user as the best prediction. Advanced users can exert greater control over these parameters by using the "backrub" Rosetta command line program or RosettaScripts (Fleishman et al., 2011). An example command line for point mutation prediction is as follows:

```
~/rosetta/rosetta_source/bin/backrub.linuxgccrelease -database
~/rosetta/rosetta_database/ -s 1CV1.pdb -ex1 -ex2 -extrachi_cutoff
0 -use_input_sc -backrub:ntrials 10000 -nstruct 10 -resfile
1CV1_M111I.resfile -pivot_residues 84 99 102 103 106 107 108 109 110
111 112 113 114 115 118
```

where the "resfile" sets positions to be mutated and repacked (allowing rotamer changes), and the pivot residues denote pivots allowed for backrub moves (necessary files to run the command line with Rosetta version 3.5 are included as example S1 at http://kortemmelab.ucsf.edu/resources/MIE_Supplement.tar.gz; results shown in Fig. 4.2 were obtained with Rosetta revision 18013). For details about command line flags and the resfile syntax, see the Rosetta manual at http://www.rosettacommons.org/.

2.3. Discovering and modeling alternative conformations from X-ray data

The backrub move was inspired by manual examination of ultra-high (sub 1 Å) resolution electron density maps (Davis et al., 2006) suggesting that conformational heterogeneity in X-ray data can be used to develop and validate new sampling methods. Subsequently, Alber and colleagues developed a method, Ringer (Lang et al., 2010), to automate the discovery of alternative side-chain conformations in high (sub 2 Å) resolution electron density maps by sampling around side-chain dihedral angles (Fig. 4.3A). Despite the limitation that Ringer uses a fixed backbone to define the sampling radius, they showed that 18% of side-chains have evidence for unmodeled alternative conformations at electron density levels of 0.3–1σ. Concurrently, van den Bedem, Dhanik, Latombe, and Deacon (2009) developed a complementary method, qFit, which includes local backbone and side-chain flexibility to compute an optimal fit to the electron density for each residue. The

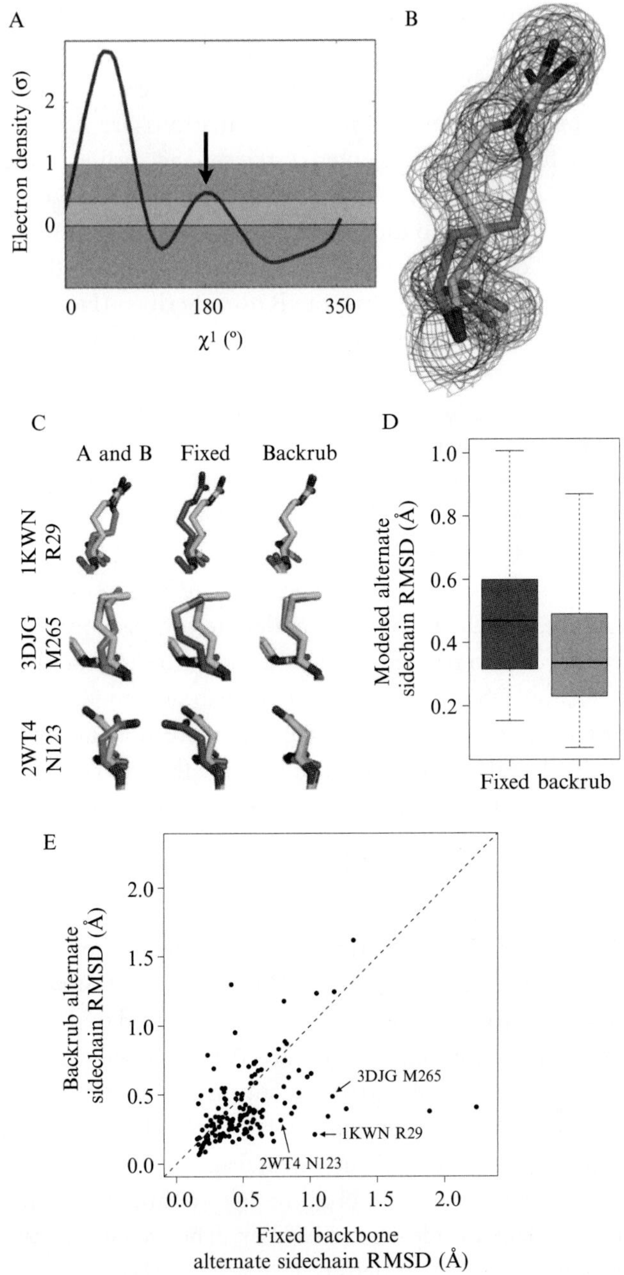
A
B
Electron density (σ)
2
1
0
0
180
350
χ1 (°)
C
D
A and B
Fixed
Backrub
1KWN R29
3DJG M265
2WT4 N123
1.0
0.8
0.6
0.4
0.2
Modeled alternate sidechain RMSD (Å)
Fixed backrub
E
2.0
1.5
1.0
0.5
0.0
Backrub alternate sidechain RMSD (Å)
3DJG M265
1KWN R29
2WT4 N123
0.0
0.5
1.0
1.5
2.0
Fixed backbone alternate sidechain RMSD (Å)

resulting 1–3 backbone and side-chain conformations per residue are merged together in a multiconformer qFit model that improves R/Rfree and maintains excellent geometry statistics. Remarkably, despite an entirely different search procedure from the original observation by the Richardson group, many alternative conformations identified by qFit can be related by backrub-like moves (Fig. 4.3B). Collectively, these studies suggest that electron density maps can provide a more informative representation of the "native" state than traditionally offered by static X-ray structures.

By examining 30 pairs of matched room temperature and cryogenic X-ray datasets, Fraser et al. used Ringer and qFit to show that room temperature X-ray data increase the evidence for alternative conformations compared to data collected at conventional cryogenic temperatures (Fraser et al., 2011). We reasoned that sampling between these experimentally visualized alternative conformations would assess the ability of fixed backbone or backrub simulations to access a representative set of conformations that are significantly populated in the "native" state. To test this idea, we considered the A and B alternative conformations (Fig. 4.3C) from 30 room temperature X-ray multiconformer models refined by qFit.

We focused our analysis on alternative conformations with Cβ deviations of 0.2 Å or greater, relative SASAs less than or equal to 30%, and different $\chi 1$ rotameric bins (152 side-chains). First, we split the multiconformer model into two separate PDB files, containing all residues without alternative

Figure 4.3 Backrub sampling improves the prediction of alternative side-chain conformations observed in protein crystal structures. (A) Electron density sampling by Ringer around the $\chi 1$ of R29 from PDB 1KWN reveals high electron density for the primary conformation 60° and a secondary peak (indicated by the black arrow), above the 0.3σ threshold that enriches for alternative conformations (shaded green area), near the 180° rotameric bin. (B) 2mFo-DFc electron density surrounding R29 from PDB 1KWN contoured at 1σ (blue mesh) and 0.3σ (cyan mesh). The original PDB model is shown in yellow, with an alternative conformation identified by Ringer and modeled with qFit at 25% occupancy shown in green. (C) Example predictions with Rosetta, with the indicated PDB codes and residues. Sampling of side-chain conformations (yellow) starting from alternative conformations (green, right) is improved by flexible backbone backrub moves (cyan, right) compared to fixed backbone side-chain only sampling (magenta, center). (D) The overall quantification of the results, showing that backrub sampling increases identification of discrete side-chain local minima modeled as alternative conformations by qFit compared to fixed backbone models over a set of 152 side-chains with solvent accessibility less than 30%. The median RMSD decreases from 0.47 to 0.33. Box plots are shown as in Fig. 4.2C. (E) A scatter plot representation of the data in (D) shows that backrub leads to large improvements compared to fixed backbone for many alternative conformation predictions. (See Color Insert.)

conformations and either the "A" conformations or "B" conformations (e.g., 1kwn_A.pdb and 1kwn_B.pdb). Next, we ran a RosettaScripts protocol that moves between the starting conformation specified by the flag –s (in this example, 1kwn_A.pdb) and the target alternative conformation specified by the flag –in:file:native (in this example, 1kwn_B.pdb). At the beginning of the protocol, the χ angles of a central side-chain are switched from the starting conformation to the target conformation. This script tests whether changes in the surrounding side-chains (fixed backbone) or both the backbone and surrounding side-chains (backrub) better accommodate the new χ angles and find a side-chain conformation close in RMSD to the target conformation. We tested this protocol in both directions between the A and B conformations. In the simple example below, the variables $\chi1$ through $\chi4$ provide the target χ angles (here, the χ angles of conformation B), and piv1 through piv3 define the positions around the central residue allowed to be pivots for the backrub move.

The command to run the protocol is

```
~/rosetta/rosetta_source/bin/rosetta_scripts.linuxgccrelease-database ~/rosetta/rosetta_database -s 1kwn_A.pdb-in:file:native 1kwn_B.pdb -parser:protocol model_alternate_conformation.xml -parser:script_vars pos=29 chi1=-145.522 chi2=-160.509 chi3=81.9108 chi4=175.816 piv1=28 piv2=29 piv3=30
```

The contents of model_alternate_conformation.xml are

```
<ROSETTASCRIPTS>
  <SCOREFXNS>
    Include the bond angle potential scoring term
    <score12_backrub weights=score12_full>
      <Reweight scoretype=mm_bend weight=1/>
    </score12_backrub>
  </SCOREFXNS>
  <TASKOPERATIONS>
    Define the restrictions on the sidechain moves that will occur in
    the simulation
    <ExtraRotamersGeneric name=extra_rot ex1=1 ex2=2 extrachi_
    cutoff=0/>
    <IncludeCurrent name=input_sc/>
    <DesignAround name=neighbors_only allow_design=0 design_shell
    =0 repack_shell=6.0 resnums=%%pos%%/>
    <RestrictToRepacking name=repack_only/>
    <PreventRepacking name=fix_central_residue resnum=%%pos%%/>
  </TASKOPERATIONS>
```

```
<FILTERS>
  Calculates the side-chain RMSD before and after simulation
  <SidechainRmsd name=rmsd threshold=10 include_backbone=0
  res1_res_num=%%pos%% res2_res_num=%%pos%%/>
</FILTERS>
<MOVERS>
  Set the chi angles of the residue of interest
  <SetChiMover name=setchi1 chinum=1 resnum=%%pos%% angle=%%
  chi1%%/>
  <SetChiMover name=setchi2 chinum=2 resnum=%%pos%% angle=%%
  chi2%%/>
  <SetChiMover name=setchi3 chinum=3 resnum=%%pos%% angle=%%
  chi3%%/>
  <SetChiMover name=setchi4 chinum=4 resnum=%%pos%% angle=%%
  chi4%%/>
  Set backrub moves to only occur near residue of interest
  <Backrub name=backrub pivot_residues=%%piv1%%,%%piv2%%,%%
  piv3%% min_atoms=3 min_atoms=7/>
  Set side-chain moves to only include residues within 6 angstrom shell
  <Sidechain name=sidechain task_operations=extra_rot,input_sc,
  fix_central_residue,
  neighbors_only,repack_only/>
  During Monte Carlo, alternate between backrub moves (75%) and side-
  chain moves (25%)
  <ParsedProtocol name=backrub_protocol mode=single_random>
  <Add mover_name=backrub apply_probability=0.75/>
  <Add mover_name=sidechain apply_probability=0.25/>
  </ParsedProtocol>
  Set up Monte Carlo simulation with 10,000 steps and kT=0.6
  <GenericMonteCarlo name=backrub_mc mover_name=backrub_protocol
  scorefxn_name=score12_backrub trials=10000 temperature=0.6
  preapply=0/>
</MOVERS>
<PROTOCOLS>
  Set the residue of interest to the desired chi angles
  <Add mover_name=setchi1/>
  <Add mover_name=setchi2/>
  <Add mover_name=setchi3/>
  <Add mover_name=setchi4/>
  Calculate RMSD before simulation
```

```
      <Add filter_name=rmsd/>
      Run backrub simulation
      <Add mover_name=backrub_mc/>
      Calculate RMSD after simulation
      <Add filter_name=rmsd/>
  </PROTOCOLS>
</ROSETTASCRIPTS>
```

Necessary files to run this script with Rosetta version 3.5 are included as example S2 at http://kortemmelab.ucsf.edu/resources/MIE_Supplement.tar.gz; results shown in Fig. 4.3 were obtained with Rosetta revision 48648.

Similarly to the mutation data set (Fig. 4.2), backrub moves significantly improve the predictions (Fig. 4.3D and E). Additionally, we tested the effect of including larger backrub moves or using C and N atoms as pivots in place of the normal Cα pivot. Applying these larger moves or using additional pivot atoms did not significantly affect the modeled alternate side-chain RMSD over the dataset. However, there may be certain residue types, secondary structures, or local environments that benefit from distinct move sets.

These results suggest that backrub moves help to model "native" state heterogeneity. While above, we have described the validation of this procedure on high-resolution room temperature X-ray data, similar strategies can be applied to cryogenic data, where models will likely contain fewer alternative conformations, or to low-resolution data, where the electron density maps do not reveal discrete alternative conformations. Therefore, flexible backbone sampling strategies in Rosetta may help to improve the description of the "native" state offered by conventional or low-resolution X-ray crystallography experiments (Tyka et al., 2011). Such sampling methodologies will have many applications including flexible receptor docking in drug discovery (Sherman, Day, Jacobson, Friesner, & Farid, 2006).

2.4. Sampling functional alternative conformations in Cyclophilin A

While the preceding examples utilized backbone flexibility centered on a single residue, many protein motions require movement of a neighborhood of residues that may potentially spread across multiple elements of secondary structure. These movements can create loop, rigid body domain, or side-chain rearrangements that are crucial for the biological mechanism.

The difficulty of discovering and simulating correlated motions is exemplified in the intrinsic conformational exchange of the proline isomerase Cyclophilin A (CypA; Fraser et al., 2009; Fig. 4.4A). Previous NMR studies by the Kern group identified a collective exchange process extending from

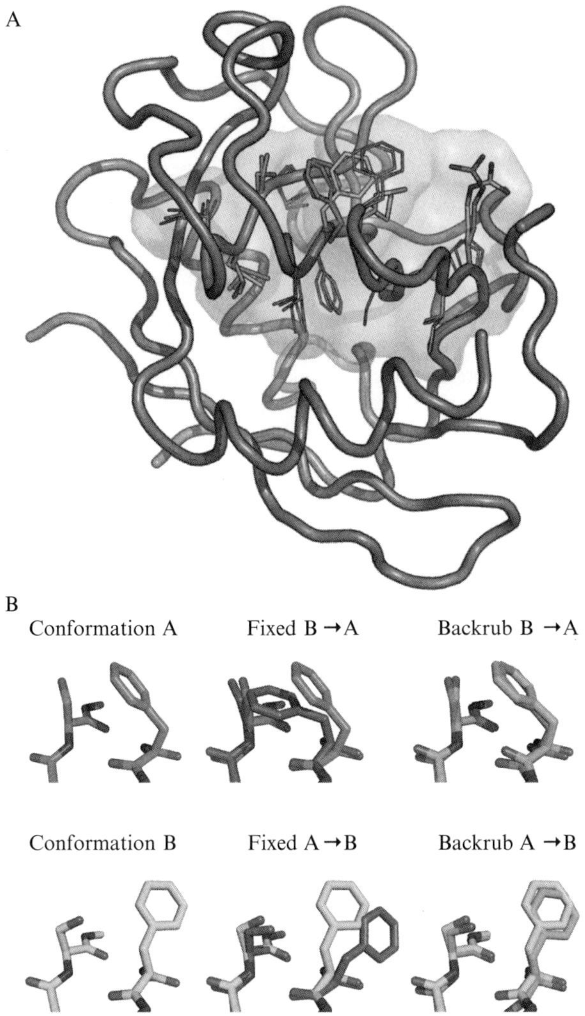

Figure 4.4 Backrub sampling can be used to model functionally relevant alternative conformations. (A) NMR relaxation experiments detect that residues in a dynamic network (cyan transparent surface) undergo a collective exchange between a major and minor conformation with and without substrate present (Eisenmesser, Bosco, Akke, & Kern, 2002; Eisenmesser et al., 2005; Fraser et al., 2009). Room temperature X-ray data collection and qFit multiconformer refinement identify a major (green) and minor (yellow) conformation providing a structural basis for the NMR observations. Additional alternative conformations are shown in orange. (B) Rosetta simulations can access the alternative conformation starting from either state (yellow/green) using backrub (right, cyan) but not fixed backbone (middle, magenta) sampling methods. (For interpretation of the references to color in this figure legend, the reader is referred to the online version of this chapter.)

the active site into the core of the protein and established a link between the rate of conformational exchange and the catalytic cycle of the enzyme (Eisenmesser et al., 2002, 2005). Room temperature X-ray crystallography and electron density interpretation using a combination of Ringer and manual inspection were used to reveal that the exchange was due to a coupled network of alternative side-chain conformations (Fraser et al., 2009). The functional importance of the alternative conformation was tested by demonstrating a parallel reduction in dynamics and catalysis upon mutation of a residue outside the active site (Fraser et al., 2009). Intriguingly, the backbone movement of Phe113 renders its alternative conformation undetectable by Ringer, which is limited to fixed backbone sampling (Fig. 4.4B).

Due to the millisecond timescale of this correlated motion, the side-chain conformational changes cannot be sampled in conventional molecular dynamics simulations. However, recent accelerated molecular dynamics simulations that reduce torsional barriers have recapitulated several key elements of the conformational dynamics during catalysis (Doshi, McGowan, Ladani, & Hamelberg, 2012). As an initial test of the ability of Rosetta to model correlated motions between neighboring side-chains in CypA, we modified the RosettaScripts protocol used to sample alternative conformations. We defined multiple positions that are allowed to be pivots for backrub and included a call to a "resfile" that specifies the positions whose side-chains can be repacked. This protocol improves sampling of the alternative conformation over fixed backbone approaches (Fig. 4.4B).

The command to run the protocol is

```
~/rosetta/rosetta_source/bin/rosetta_scripts.linuxgccrelease-database ~/rosetta/rosetta_database/ -s 3KON_A.pdb -in:file:native 3KON_B.pdb -parser:protocol model_alternate_conformation_F113.xml -resfile F113.resfile -parser:script_vars chi1=-53.763 chi2=-41.4727 chi3=0 chi4=0
```

The contents of model_alternate_conformation_F113.xml are:

```
<ROSETTASCRIPTS>
<SCOREFXNS>
Include the bond angle potential scoring term
<score12_backrub weights=score12_full>
<Reweight scoretype=mm_bend weight=1/>
</score12_backrub>
</SCOREFXNS>
<TASKOPERATIONS>
<ExtraRotamersGeneric name=extra_rot ex1=1 ex2=2 extrachi_cutoff=0/>
```

```
<ReadResfile name=read_resfile filename="F113.resfile"/>
    </TASKOPERATIONS>
    <FILTERS>
Calculates the side-chain RMSD before and after simulation
<SidechainRmsd name=rmsd threshold=10 include_backbone=0
res1_pdb_num=113A res2_pdb_num=113A/>
</FILTERS>
<MOVERS>
Set the chi angles of the residue of interest
<SetChiMover name=setchi1 chinum=1 resnum=113A angle=%%chi1%%/>
<SetChiMover name=setchi2 chinum=2 resnum=113A angle=%%chi2%%/>
<SetChiMover name=setchi3 chinum=3 resnum=113A angle=%%chi3%%/>
<SetChiMover name=setchi4 chinum=4 resnum=113A angle=%%chi4%%/>
Set backrub moves to only occur near residue of interest
<Backrub name=backrub
pivot_residues=54A,55A,56A,59A,60A,61A,
62A,63A,64A,65A,90A,91A,92A,97A,98A,99A,100A,101A,102A,103A,110A,
111A,112A,113A,114A,115A,116A,118A,119A,120A,121A,122A,123A,125A,
126A,127A,128A,129A,130A min_atoms=3 min_atoms=7/>
Set side-chain moves to only include residues within 6 angstrom shell
<Sidechain name=sidechain task_operations=read_resfile,extra_
rot/>
During Monte Carlo, alternate between backrub moves (75%) and side-
chain moves (25%)
<ParsedProtocol name=backrub_protocol mode=single_random>
<Add mover_name=backrub apply_probability=0.75/>
<Add mover_name=sidechain apply_probability=0.25/>
</ParsedProtocol>
Set up Monte Carlo simulation with 10,000 steps and kT=0.6
<GenericMonteCarlo name=backrub_mc mover_name=backrub_protocol
scorefxn_name=score12_backrub trials=10000 temperature=0.6 preapply=0/>
</MOVERS>
<PROTOCOLS>
Set the residue of interest to the desired chi angles
<Add mover_name=setchi1/>
<Add mover_name=setchi2/>
<Add mover_name=setchi3/>
<Add mover_name=setchi4/>
Calculate RMSD before simulation
<Add filter_name=rmsd/>
```

```
Run backrub simulation
<Add mover_name=backrub_mc/>
Calculate RMSD after simulation
<Add filter_name=rmsd/>
</PROTOCOLS>
</ROSETTASCRIPTS>
```

Necessary files to run this script with Rosetta version 3.5 are included in the example S3 at http://kortemmelab.ucsf.edu/resources/MIE_Supplement.tar.gz; results shown in Fig. 4.4 were obtained with Rosetta revision 48648.

Here, we have specified neighboring residues that can undergo backrub moves. Similarly, the ability of backrub to sample functionally important loop conformations has been demonstrated for triosephosphate isomerase (TIM; Smith & Kortemme, 2008). To efficiently sample these enzymatic motions, we used prior knowledge of residues that need conformational adjustments. Therefore, these strategies present an immediate challenge: sampling large correlated motions without prior knowledge of what residues are involved in the motion. One approach is to use unbiased simulations to identify flexible regions. In the case of TIM, unbiased simulations identify that the catalytically important loop region is highly flexible, but only simulations that focus on the loop have been shown to sample the entire range of motion. Another intermediate on the road to this goal is to include constraints from NMR relaxation dispersion experiments, which specify residues that are experiencing an exchange in chemical environment but do not provide direct structural information about the exchange. Recent work using T4 Lysozyme (Bouvignies et al., 2011) suggests that Rosetta fragment insertion methods biased by experimental chemical shifts can generate structural descriptions of alternative conformations discovered by NMR. The success of backrub moves in sampling the enzyme motions of CypA and TIM indicate that "native" state sampling using backrub moves can likely be exploited in a similar fashion to link conformational dynamics discovered by NMR relaxation dispersion experiments with structural mechanisms.

3. SEQUENCE PLASTICITY AND CONFORMATIONAL PLASTICITY ARE INTERTWINED

The improvements offered by backrub moves in predicting point mutant structures (Fig. 4.2) suggest that subtle backbone rearrangements can significantly alter the prediction of tolerated mutations. It follows that conformational ensembles created through backrub moves would also change the potential for sequences predicted to be consistent with a given protein fold. Indeed, incorporating backbone flexibility increased the overlap between sequences predicted to be consistent with the ubiquitin fold and the evolutionary record (Friedland, Lakomek, Griesinger, Meiler, & Kortemme, 2009). These results provided further evidence that the relationship between sequence and structural variability can be leveraged to develop and validate new conformational sampling methods. Both sequence alignments of orthologous proteins (natural selection) and sequences enriched in high-throughput binding experiments, such as phage display or peptide arrays (artificial selection), can be used to define the sequence variability that design methods can target.

3.1. Modeling peptide binding specificity

In addition to predicting sequences tolerated by a single protein fold, flexible backbone methods can improve the prediction of binding specificity in peptide binding domains such as PDZ, SH3, and WW domains. Phage display coupled with next-generation sequencing techniques can generate experimental position weight matrices (PWMs) based on large numbers of potential sequences (Huang & Sidhu, 2011). A challenge for interpreting these datasets is to define the structural basis for specificity in binding pockets that are quite similar. To test how well Rosetta can recapitulate the binding specificities discovered by these experiments, we sampled conformations of both the peptide and receptor protein using backrub moves. As observed previously for sampling the ubiquitin fold family sequences, the temperature parameter is key for controlling the conformational diversity sampled by the ensemble (Fig. 4.5A). For applications where there is a larger degree of backbone flexibility and corresponding sequence variability, higher temperatures can be explored. For PDZ domain–peptide interactions, a temperature of 0.6 kT was used (Smith & Kortemme, 2010, 2011). After generating an ensemble using backrub moves, design can be used to sample sequence changes of either the receptor or the peptide.

We have automated the sequence tolerance protocol for using flexible backbone ensembles and sequence design for predicting peptide binding specificity on a Web server: https://kortemmelab.ucsf.edu/backrub/ (Lauck et al., 2010). Users can also download a "protocol capture" of the sequence tolerance method, complete with example input/output and scripts, in the Supplementary Materials accompanying, http://www.elsevierdirect.com/companions/9780123942920 (Smith & Kortemme, 2011). Given a peptide-bound structure, backrub sampling methods are used

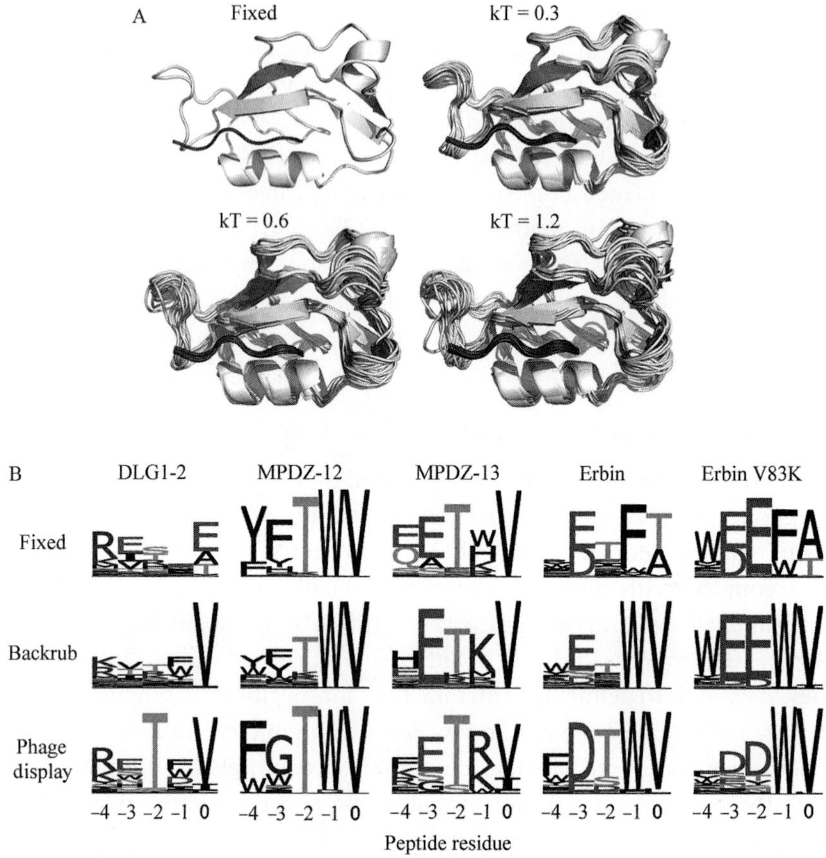

Figure 4.5 Rosetta generates near-native ensembles using backrub sampling. (A) Cα cartoons of Rosetta generated conformational ensembles using backrub sampling at different temperatures, compared to the fixed backbone (top left). Higher temperatures increase the conformational diversity and can increase agreement with experimental data. The PDZ domain structure is shown in white and peptide in gray. (B) Example results from the sequence tolerance protocol to predict peptide specificity for four PDZ domains (DLG1-2, MPDZ-12, MPDZ-13, and Erbin) and one PDZ domain point mutant (Erbin V83K); peptide positions are indicated using the standard nomenclature for PDZ domain motifs, with 0 denoting the C-terminal residue, followed by −1, −2, etc. Without backbone flexibility, Rosetta fails to predict important residue preferences observed in experimental phage display selections, such as valine at the 0 position or tryptophan at the −1 position for DLG1-2 and Erbin. (For color version of this figure, the reader is referred to the online version of this chapter.)

to generate an ensemble of conformations. For each conformation, a genetic algorithm is used to design sequences for high-affinity binding. In this protocol, the interface energy is given greater weight (Smith & Kortemme, 2010, 2011). To compare these predictions to experimental data, we generate a sequence logo based on the positional frequencies in the resulting designed sequences (Fig. 4.5B). Compared to fixed backbone methods, backrub sampling increases the agreement at several positions. Given the adaptable nature of many protein–protein interfaces, it is clear that flexible backbone methods will provide great insight into the structural and energetic basis for binding specificity. Additionally, as more datasets on mutant binding domains are collected, there is potential to look for covariation between the sequences tolerated between receptor and peptide positions (Ernst et al., 2010).

3.2. Covariation and interface design in two-component signaling

Testing computational protein design methods based on comparison with experimental PWMs is informative. However, it involves evaluating amino acid positions independently from each other and therefore may overlook some of the intricate details of pair-wise interactions between designed residues. In order to assess how well flexible backbone design protocols capture dependencies between designed residues, we directly compared designed residue covariation to native residue covariation. We chose to examine covariation within the bacterial two-component signaling system, since it has previously been shown that sensor histidine kinases (HKs) and their cognate response regulators (RRs) exhibit significant intermolecular covariation at their protein–protein interface (White, Szurmant, Hoch, & Hwa, 2007).

Designed sequences were obtained by generating a backrub conformational ensemble of 500 structures starting from the cocrystal structure of HK853 and RR468 from *Thermotoga maritima* (Casino, Rubio, & Marina, 2009; PDB 3DGE). Since bacterial HK and RR sequences are highly divergent, we used a temperature of 1.2 kT to produce a conformational ensemble that would yield sufficiently diverse designed sequences. We then performed sequence design using Monte Carlo simulated annealing on each structure, which resulted in 500 designed HK and RR sequences.

The command lines for this protocol are as follows:

Backrub ensemble generation

```
~/rosetta/rosetta_source/bin/backrub.linuxgccrelease -database ~/
rosetta/rosetta_database/ -s 3DGE.pdb -resfile NATAA.res -ex1 -ex2 -
extrachi_cutoff 0 -backrub:mc_kt 1.2 -backrub:ntrials 10000 -nstruct
500 -backrub:initial_pack
```

Sequence design

```
~/rosetta/rosetta_source/bin/fixbb.linuxgccrelease -database ~/
rosetta/rosetta_database/ -s 3DGE_0001_last.pdb -resfile ALLAA.res
-ex1 -ex2 -extrachi_cutoff 0 -nstruct 1 -overwrite -linmem_ig 10 -
no_his_his_pairE -minimize_sidechains
```

Necessary files to run these command lines with Rosetta version 3.5 are included as example S4 at http://kortemmelab.ucsf.edu/resources/MIE_Supplement.tar.gz; results shown in Fig. 4.6 were obtained with Rosetta revision 39284.

To compare the sequence features from interface design with those observed in naturally interacting proteins, we collected alignments of natural HK and RR sequences from Pfam (PF000512 for HK and PF00072 for RR) and concatenated all pairs of HKs and RRs that were adjacent in a particular genome (i.e., pairs with GI numbers differing by 1). To avoid bias from closely related sequences, we filtered the joint HK/RR alignment for redundancy using an 80% sequence identity cutoff. We quantified residue covariation of all intermolecular pairs of amino acid positions in designed and natural sequences using a mutual-information-based statistic (Dickson, Wahl, Fernandes, & Gloor, 2010).

We observed significant overlap between the designed and natural highly covarying intermolecular pairs within the HK/RR complex (Fig. 4.6A). Mapping the residue pairs that were highly covarying in both designed and natural sequences onto the structure of the complex revealed that all of these pairs are localized to the HK/RR interface (Fig. 4.6B). A closer examination of these pairs shows that each pair forms a physical interaction across the HK/RR interface (Fig. 4.6C), suggesting that these pairs may be important for determining specificity in bacterial two-component signaling systems. Indeed, several of these positions have previously been mutated to alter the specificity of HK–RR interactions: HK-Thr in Pair 1, HK-Tyr in Pair 9, and HK-Val in Pair 11 (Skerker et al., 2008). The remaining pairs, including those that highly covary in designed sequences but not natural sequences, represent potential opportunities for rewiring two-component signaling specificity using computational protein design.

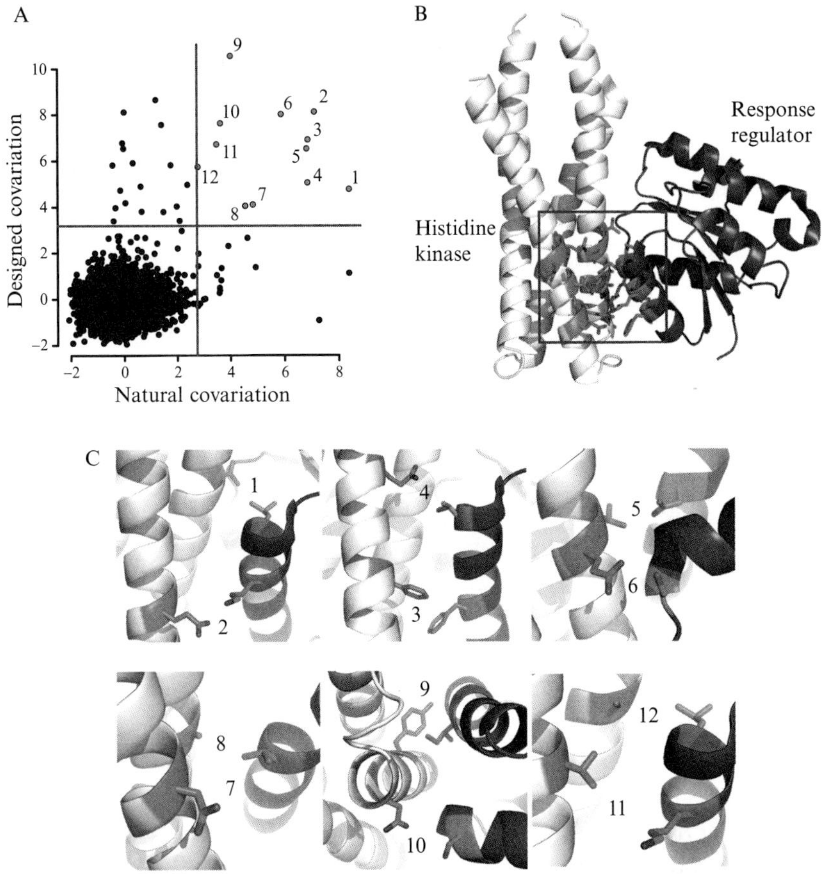

Figure 4.6 Rosetta backrub design methods capture features of evolutionary amino acid covariation. (A) Comparison between designed and natural intermolecular amino acid covariation for histidine kinases (HKs) and their cognate response regulators (RRs). Each point represents a pair of amino acid positions. Natural covariation was quantified using a mutual-information-based metric for all pairs of positions in a multiple sequence alignment of HKs concatenated to their cognate RRs. A backrub ensemble of 500 structures was generated for a HK/RR complex (PDB ID 3DGE) and RosettaDesign was used to predict one low-energy sequence for each structure in the ensemble. Designed covariation was quantified for all pairs of positions in the resulting multiple sequence alignment of 500 sequences. The red lines indicate the threshold cutoff for the top 30 designed covarying intermolecular pairs (horizontal) and the top 30 natural covarying pairs (vertical). The 12 intermolecular pairs of positions that are highly covarying in both designed and natural sequences are highlighted in green. (B) The structure of a HK/RR complex with amino acids that are involved in highly covarying intermolecular pairs in both natural and designed sequences are shown in green and stick representation. (C) Close-up of the 12 intermolecular covarying pairs. Each of these 12 pairs of amino acids forms a physical interaction across the interface of the complex. (For interpretation of the references to color in this figure legend, the reader is referred to the online version of this chapter.)

4. FUTURE CHALLENGES

The success of flexible backbone sampling methods in predicting mutant side-chain (Fig. 4.2) and alternative (Fig. 4.3) conformations indicates the broad utility of these methods in designing sequences compatible with a target "native" structure. Previous studies have used Rosetta to provide structural mechanisms for NMR measures of protein dynamics (Friedland et al., 2008, 2009) and to design mutations that stabilize specific conformations from a dynamic ensemble (Babor & Kortemme, 2009; Bouvignies et al., 2011). Additionally, the comparisons to naturally and artificially selected sequence data suggest that flexible backbone methods can be leveraged to design libraries for generating proteins with new or improved functions (Friedland & Kortemme, 2010).

Despite these successes, exploiting backbone flexibility to design conformational heterogeneity, in contrast to design of a single target structure, remains largely unaddressed. A major challenge in the coming years will be to adapt these methods to design functionally important protein conformational dynamics. Examples of these design challenges include: designing loops to sample multiple conformations that exclude water and permit substrate flux during an enzymatic catalytic cycle, creating peptide binding domains where specificity is encoded by distinct binding modes, or generating coupled networks of side-chain conformations that respond to an allosteric binding event.

To meet these lofty challenges, scoring functions must be sensitive to the small gaps that separate these conformations on the energy landscape (Fleishman et al., 2011). In addition, to avoid having populations biased by the sampling algorithm and provide better estimates of conformational entropy, the Monte Carlo move sets must obey detailed balance (Hastings, 1970). In addition to improvements in scoring and thermodynamics, more sophisticated sampling protocols will likely be needed. Here, we have primarily focused on backrub moves around Cα. However, sampling the "native" state of some protein environments may benefit from different strategies or iterations through a combination of sampling moves. Indeed, we have recently had success at modeling conformational changes that propagate away from a designed mutation by iteratively switching between different sampling and scoring strategies during the course of a single simulation (Kapp et al., 2012). Learning from the successes and failures of these new strategies will be essential to improve both protein design and our understanding of the relationship between protein conformational dynamics and function.

ACKNOWLEDGMENTS

The Fraser lab is supported by the National Institutes of Health Early Independence Award (DP5 OD009180) and QB3. Rosetta development in the Kortemme lab is supported by awards from the National Science Foundation (NSF) to T. K. (NSF CAREER MCB-0744541; NSF EF-0849400) and the Synthetic Biology Engineering Research Center (NSF EEC-0540879; PI Keasling). N. O. was additionally supported by an NSF graduate fellowship. We thank Russell Goodman and Henry van den Bedem for helpful discussions.

REFERENCES

Babor, M., & Kortemme, T. (2009). Multi-constraint computational design suggests that native sequences of germline antibody H3 loops are nearly optimal for conformational flexibility. *Proteins*, *75*, 846–858.

Best, R. B., Lindorff-Larsen, K., DePristo, M. A., & Vendruscolo, M. (2006). Relation between native ensembles and experimental structures of proteins. *Proceedings of the National Academy of Sciences of the United States of America*, *103*, 10901–10906.

Boehr, D. D., Nussinov, R., & Wright, P. E. (2009). The role of dynamic conformational ensembles in biomolecular recognition. *Nature Chemical Biology*, *5*, 789–796.

Bouvignies, G., Vallurupalli, P., Hansen, D. F., Correia, B. E., Lange, O., Bah, A., et al. (2011). Solution structure of a minor and transiently formed state of a T4 lysozyme mutant. *Nature*, *477*, 111–114.

Canutescu, A. A., & Dunbrack, R. L., Jr. (2003). Cyclic coordinate descent: A robotics algorithm for protein loop closure. *Protein Science*, *12*, 963–972.

Casino, P., Rubio, V., & Marina, A. (2009). Structural insight into partner specificity and phosphoryl transfer in two-component signal transduction. *Cell*, *139*, 325–336.

Davis, I. W., Arendall, W. B., 3rd., Richardson, D. C., & Richardson, J. S. (2006). The backrub motion: How protein backbone shrugs when a sidechain dances. *Structure*, *14*, 265–274.

Dickson, R. J., Wahl, L. M., Fernandes, A. D., & Gloor, G. B. (2010). Identifying and seeing beyond multiple sequence alignment errors using intra-molecular protein covariation. *PloS One*, *5*, e11082.

Doshi, U., McGowan, L. C., Ladani, S. T., & Hamelberg, D. (2012). Resolving the complex role of enzyme conformational dynamics in catalytic function. *Proceedings of the National Academy of Sciences of the United States of America*, *109*, 5699–5704.

DuBay, K. H., & Geissler, P. L. (2009). Calculation of proteins' total side-chain torsional entropy and its influence on protein-ligand interactions. *Journal of Molecular Biology*, *391*, 484–497.

Eisenmesser, E. Z., Bosco, D. A., Akke, M., & Kern, D. (2002). Enzyme dynamics during catalysis. *Science*, *295*, 1520–1523.

Eisenmesser, E. Z., Millet, O., Labeikovsky, W., Korzhnev, D. M., Wolf-Watz, M., Bosco, D. A., et al. (2005). Intrinsic dynamics of an enzyme underlies catalysis. *Nature*, *438*, 117–121.

Ernst, A., Gfeller, D., Kan, Z., Seshagiri, S., Kim, P. M., Bader, G. D., et al. (2010). Coevolution of PDZ domain-ligand interactions analyzed by high-throughput phage display and deep sequencing. *Molecular BioSystems*, *6*, 1782–1790.

Fleishman, S. J., & Baker, D. (2012). Role of the biomolecular energy gap in protein design, structure, and evolution. *Cell*, *149*, 262–273.

Fleishman, S. J., Leaver-Fay, A., Corn, J. E., Strauch, E. M., Khare, S. D., Koga, N., et al. (2011). RosettaScripts: A scripting language interface to the Rosetta macromolecular modeling suite. *PloS One*, *6*, e20161.

Fraser, J. S., Clarkson, M. W., Degnan, S. C., Erion, R., Kern, D., & Alber, T. (2009). Hidden alternative structures of proline isomerase essential for catalysis. *Nature*, *462*, 669–673.

Fraser, J. S., van den Bedem, H., Samelson, A. J., Lang, P. T., Holton, J. M., Echols, N., et al. (2011). Accessing protein conformational ensembles using room-temperature X-ray crystallography. *Proceedings of the National Academy of Sciences of the United States of America*, *108*, 16247–16252.

Frauenfelder, H., Sligar, S. G., & Wolynes, P. G. (1991). The energy landscapes and motions of proteins. *Science*, *254*, 1598–1603.

Friedland, G. D., & Kortemme, T. (2010). Designing ensembles in conformational and sequence space to characterize and engineer proteins. *Current Opinion in Structural Biology*, *20*, 377–384.

Friedland, G. D., Lakomek, N. A., Griesinger, C., Meiler, J., & Kortemme, T. (2009). A correspondence between solution-state dynamics of an individual protein and the sequence and conformational diversity of its family. *PLoS Computational Biology*, *5*, e1000393.

Friedland, G. D., Linares, A. J., Smith, C. A., & Kortemme, T. (2008). A simple model of backbone flexibility improves modeling of side-chain conformational variability. *Journal of Molecular Biology*, *380*, 757–774.

Furnham, N., Blundell, T. L., DePristo, M. A., & Terwilliger, T. C. (2006). Is one solution good enough? *Nature Structural & Molecular Biology*, *13*, 184–185. Discussion 185.

Hastings, W. K. (1970). Monte-Carlo sampling methods using Markov chains and their applications. *Biometrika*, *57*, 97–109.

Huang, H., & Sidhu, S. S. (2011). Studying binding specificities of peptide recognition modules by high-throughput phage display selections. *Methods in Molecular Biology*, *781*, 87–97.

Kapp, G. T., Liu, S., Stein, A., Wong, D. T., Remenyi, A., Yeh, B. J., et al. (2012). Control of protein signaling using a computationally designed GTPase/GEF orthogonal pair. *Proceedings of the National Academy of Sciences of the United States of America*, *109*, 5277–5282.

Kelley, N. W., Vishal, V., Krafft, G. A., & Pande, V. S. (2008). Simulating oligomerization at experimental concentrations and long timescales: A Markov state model approach. *The Journal of Chemical Physics*, *129*, 214707.

Korzhnev, D. M., Religa, T. L., Banachewicz, W., Fersht, A. R., & Kay, L. E. (2010). A transient and low-populated protein-folding intermediate at atomic resolution. *Science*, *329*, 1312–1316.

Lang, P. T., Ng, H. L., Fraser, J. S., Corn, J. E., Echols, N., Sales, M., et al. (2010). Automated electron-density sampling reveals widespread conformational polymorphism in proteins. *Protein Science*, *19*, 1420–1431.

Lauck, F., Smith, C. A., Friedland, G. F., Humphris, E. L., & Kortemme, T. (2010). RosettaBackrub—A web server for flexible backbone protein structure modeling and design. *Nucleic Acids Research*, *38*, W569–W575.

Leaver-Fay, A., Tyka, M., Lewis, S. M., Lange, O. F., Thompson, J., Jacak, R., et al. (2011). ROSETTA3: An object-oriented software suite for the simulation and design of macromolecules. *Methods in Enzymology*, *487*, 545–574.

Levin, E. J., Kondrashov, D. A., Wesenberg, G. E., & Phillips, G. N., Jr. (2007). Ensemble refinement of protein crystal structures: Validation and application. *Structure*, *15*, 1040–1052.

Li, Z., Raychaudhuri, S., & Wand, A. J. (1996). Insights into the local residual entropy of proteins provided by NMR relaxation. *Protein Science*, *5*, 2647–2650.

MacCallum, J. L., Perez, A., Schnieders, M. J., Hua, L., Jacobson, M. P., & Dill, K. A. (2011). Assessment of protein structure refinement in CASP9. *Proteins*, *79*(Suppl. 10), 74–90.

Mandell, D. J., Coutsias, E. A., & Kortemme, T. (2009). Sub-angstrom accuracy in protein loop reconstruction by robotics-inspired conformational sampling. *Nature Methods, 6*, 551–552.

Mandell, D. J., & Kortemme, T. (2009). Backbone flexibility in computational protein design. *Current Opinion in Biotechnology, 20*, 420–428.

Maragakis, P., Lindorff-Larsen, K., Eastwood, M. P., Dror, R. O., Klepeis, J. L., Arkin, I. T., et al. (2008). Microsecond molecular dynamics simulation shows effect of slow loop dynamics on backbone amide order parameters of proteins. *The Journal of Physical Chemistry. B, 112*, 6155–6158.

Meiler, J., Prompers, J. J., Peti, W., Griesinger, C., & Bruschweiler, R. (2001). Model-free approach to the dynamic interpretation of residual dipolar couplings in globular proteins. *Journal of the American Chemical Society, 123*, 6098–6107.

Nagel, Z. D., & Klinman, J. P. (2009). A 21st century revisionist's view at a turning point in enzymology. *Nature Chemical Biology, 5*, 543–550.

Rieping, W., Habeck, M., & Nilges, M. (2005). Inferential structure determination. *Science, 309*, 303–306.

Schneider, T. R., Brunger, A. T., & Nilges, M. (1999). Influence of internal dynamics on accuracy of protein NMR structures: Derivation of realistic model distance data from a long molecular dynamics trajectory. *Journal of Molecular Biology, 285*, 727–740.

Shaw, D. E., Maragakis, P., Lindorff-Larsen, K., Piana, S., Dror, R. O., Eastwood, M. P., et al. (2010). Atomic-level characterization of the structural dynamics of proteins. *Science, 330*, 341–346.

Sherman, W., Day, T., Jacobson, M. P., Friesner, R. A., & Farid, R. (2006). Novel procedure for modeling ligand/receptor induced fit effects. *Journal of Medicinal Chemistry, 49*, 534–553.

Simons, K. T., Kooperberg, C., Huang, E., & Baker, D. (1997). Assembly of protein tertiary structures from fragments with similar local sequences using simulated annealing and Bayesian scoring functions. *Journal of Molecular Biology, 268*, 209–225.

Skerker, J. M., Perchuk, B. S., Siryaporn, A., Lubin, E. A., Ashenberg, O., Goulian, M., et al. (2008). Rewiring the specificity of two-component signal transduction systems. *Cell, 133*, 1043–1054.

Smith, C. A., & Kortemme, T. (2008). Backrub-like backbone simulation recapitulates natural protein conformational variability and improves mutant side-chain prediction. *Journal of Molecular Biology, 380*, 742–756.

Smith, C. A., & Kortemme, T. (2010). Structure-based prediction of the peptide sequence space recognized by natural and synthetic PDZ domains. *Journal of Molecular Biology, 402*, 460–474.

Smith, C. A., & Kortemme, T. (2011). Predicting the tolerated sequences for proteins and protein interfaces using RosettaBackrub flexible backbone design. *PloS One, 6*, e20451.

Tyka, M. D., Keedy, D. A., Andre, I., Dimaio, F., Song, Y., Richardson, D. C., et al. (2011). Alternate states of proteins revealed by detailed energy landscape mapping. *Journal of Molecular Biology, 405*, 607–618.

van den Bedem, H., Dhanik, A., Latombe, J. C., & Deacon, A. M. (2009). Modeling discrete heterogeneity in X-ray diffraction data by fitting multi-conformers. *Acta Crystallographica Section D: Biological Crystallography, 65*, 1107–1117.

White, R. A., Szurmant, H., Hoch, J. A., & Hwa, T. (2007). Features of protein-protein interactions in two-component signaling deduced from genomic libraries. *Methods in Enzymology, 422*, 75–101.

CHAPTER FIVE

OSPREY: Protein Design with Ensembles, Flexibility, and Provable Algorithms

Pablo Gainza[*,3], Kyle E. Roberts[*,3], Ivelin Georgiev[*,1], Ryan H. Lilien[†], Daniel A. Keedy[‡], Cheng-Yu Chen[‡], Faisal Reza[§,2], Amy C. Anderson[¶], David C. Richardson[‡], Jane S. Richardson[‡], Bruce R. Donald[*,‡,4]

[*]Department of Computer Science, Duke University, Durham, North Carolina, USA
[†]Department of Computer Science, University of Toronto, Toronto, Ontario, Canada
[‡]Department of Biochemistry, Duke University Medical Center, Durham, North Carolina, USA
[§]Department of Biomedical Engineering, Duke University Medical Center, Durham, North Carolina, USA
[¶]Department of Pharmaceutical Sciences, University of Connecticut, Storrs, Connecticut, USA
[1]Current address: Vaccine Research Center, National Institute of Allergy and Infectious Diseases, National Institutes of Health (NIH), Bethesda, Maryland, USA
[2]Current address: Department of Therapeutic Radiology, Yale University School of Medicine, New Haven, Connecticut, USA
[3]These authors contributed equally to this work.
[4]Corresponding author: e-mail address: brd@cs.duke.edu

Contents

Methods in Enzymology, Volume 523
ISSN 0076-6879
http://dx.doi.org/10.1016/B978-0-12-394292-0.00005-9

Abstract

Summary: We have developed a suite of protein redesign algorithms that improves realistic *in silico* modeling of proteins. These algorithms are based on three characteristics that make them unique: (1) *improved flexibility* of the protein backbone, protein side-chains, and ligand to accurately capture the conformational changes that are induced by mutations to the protein sequence; (2) modeling of proteins and ligands as *ensembles* of low-energy structures to better approximate binding affinity; and (3) a globally optimal protein design search, guaranteeing that the computational predictions are optimal with respect to the input model. Here, we illustrate the importance of these three characteristics. We then describe OSPREY, a protein redesign suite that implements our protein design algorithms. OSPREY has been used prospectively, with experimental validation, in several biomedically relevant settings. We show in detail how OSPREY has been used to predict resistance mutations and explain why improved flexibility, ensembles, and provability are essential for this application.

Availability: OSPREY is free and open source under a Lesser GPL license. The latest version is OSPREY 2.0. The program, user manual, and source code are available at www.cs.duke.edu/donaldlab/software.php. *Contact*: osprey@cs.duke.edu

1. INTRODUCTION

Technological advances in protein redesign could revolutionize therapeutic treatment. With these advances, proteins and other molecules can be designed to act on today's undruggable proteins or tomorrow's drug-resistant diseases. One of the most promising approaches in protein redesign is structure-based computational protein redesign (SCPR). SCPR programs model a protein's three-dimensional structure and predict mutations to the native protein sequence that will have a desired effect on its biochemical properties and function, such as improving the affinity of a drug-like protein for a disease target. In this chapter, we describe OSPREY (*O*pen *S*ource *P*rotein *Re*design for *Y*ou), a free, open-source SCPR program. We have prospectively used OSPREY, with experimental validation, to redesign enzymes (Chen, Georgiev, Anderson, & Donald, 2009), design new drugs (Gorczynski et al., 2007), predict drug resistance (Frey, Georgiev, Donald, & Anderson, 2010), design peptide inhibitors of protein–protein interactions (Roberts, Cushing, Boisguerin, Madden, & Donald, 2012), and design epitope-specific antibody probes (Georgiev, Acharya, et al., 2012).

Predicting mutations that result in a desired protein structure and enable novel function or new biochemical properties presents four main protein design challenges. First, as the number of mutated residues to the native sequence increases, the number of unique protein sequences, or the size of *sequence space*, increases exponentially. Second, mutating a protein sequence induces conformational changes to the protein structure. Thus, the most stable, lowest energy conformations of one sequence can differ significantly from those of another sequence. A protein's potential flexibility occurs over many degrees of freedom. This results in an astronomically large, continuous space over which SCPR algorithms must search. A third challenge is that, for each protein sequence, an ensemble of low-energy states exists, which contributes to protein–ligand binding (Gilson, Given, Bush, & McCammon, 1997). Thus, each binding partner's conformational ensemble must be considered to compute the binding energy of the protein and ligand (Donald, 2011; Lilien, Stevens, Anderson, & Donald, 2005). Finally, the fourth challenge in protein design is calculating the energy that drives protein structure and function at the molecular level. The most accurate models would require computationally expensive quantum mechanical simulations of the protein and solvent, which is intractable for SCPR problems.

These challenges require SCPR programs to make approximations in their *input model*. The input model defines (i) the initial protein structure, (ii) the sequence space to which the protein can mutate, (iii) the allowed protein flexibility, and (iv) the energy function to rank the generated conformations. The input model must be carefully chosen to minimize the error that stems from its approximations, while at the same time ensuring that the SCPR algorithm can efficiently search the protein conformational space.

The accuracy of an SCPR program largely depends on how it addresses the protein design challenges. OSPREY's approach is based on three main protein design principles: (1) realistic, yet efficient, models of flexibility; (2) ensembles of low-energy conformations; and (3) provable optimality with respect to the input model. In Section 2, we describe these three principles and their importance. Section 3 details specific design problems where OSPREY has been applied. Section 4 describes the OSPREY program and its input, algorithms, and expected output. In Section 5, we show how to use OSPREY to predict drug resistance-conferring mutations.

2. OSPREY DESIGN PRINCIPLES

2.1. Protein flexibility

Proteins are dynamic and can exist in many different low-energy, relatively near-native conformations at physiological conditions. The ability of a ligand to select or induce protein conformations demonstrates the requirement of SCPR algorithms to accurately model flexibility (Teague, 2003). However, SCPR algorithms often must limit protein flexibility during the design search in the interest of computational feasibility.

One common SCPR approximation is to limit the allowed side-chain conformations to search. Protein amino acid side-chains appear in clusters at low-energy regions of χ-angle space, known as rotamers (Lovell, Word, Richardson, & Richardson, 2000). Many SCPR programs use discrete rotamers to represent each cluster as only a single point in χ-angle space. However, protein energetics are sensitive to small changes in atom coordinates; so, the reduction of a cluster to a single discrete conformation cannot fully describe a continuous region of side-chain conformation space.

To improve upon the limitations of discrete rotamers, OSPREY implements *continuous rotamers* (Gainza, Roberts, & Donald, 2012; Georgiev, Lilien, & Donald, 2008). In contrast to discrete rotamers, each continuous rotamer is a region in χ-angle space that more accurately reflects the empirically discovered side-chain clusters. A large-scale study of protein core designs using continuous rotamers versus discrete rotamers demonstrated the benefits of continuous rotamers in protein design (Gainza et al., 2012). Importantly, continuous rotamers were able to find conformations that were both lower in energy and had different sequences than the conformations found using discrete rotamers, even when more expansive discrete rotamer libraries were used. This means that discrete rotamers do not accurately quantize conformation space and will likely result in less than optimal design predictions. Also, using continuous rotamers improves the biological accuracy of the designs. Specifically, sequences found using continuous rotamers were significantly more similar to native sequences than sequences found with rigid rotamers. The accuracy improvements are comparable to gains achieved when incorporating sophisticated energy terms such as solvation (Hu & Kuhlman, 2006). Therefore, continuous rotamers are likely required to accurately search conformation space to find the true low-energy protein structures.

While the large-scale study in Gainza et al. (2012) was conducted for side-chain flexibility, OSPREY can also be used to search over local backbone flexibility (Georgiev, Lilien, & Donald, 2008), or continuous global backbone flexibility (Georgiev & Donald, 2007). Extrapolating from the benefits obtained by using continuous rotamers, similar benefits were shown (Hallen, Keedy, & Donald, 2013) when using OSPREY's flexible backbone models instead of traditional backbone models (that use only a fixed backbone or discrete backbone conformers). The benefits of continuous rotamers and continuous backbone flexibility have been experimentally demonstrated by Chen et al. (2009), Frey et al. (2010), and Roberts et al. (2012).

2.2. Ensemble-based design

Traditional protein design methods often focus on finding the single global minimum energy conformation (GMEC) for a design. However, this simplification ignores the reality that proteins in solution exist as a thermodynamic ensemble of conformations, and not just a single low-energy structure (Fig. 5.1). In fact, current nuclear magnetic resonance (NMR) techniques can now estimate relative populations of side-chain rotamers in folded proteins (Chou, Case, & Bax, 2003). It is the nature of this thermodynamic ensemble that governs protein–ligand binding (Gilson et al., 1997). Therefore, if several low-energy conformations contribute to protein–ligand binding, a model that only considers a single GMEC is likely to incorrectly predict binding.

OSPREY uses the K^* algorithm (Donald, 2011; Georgiev, Lilien, et al., 2008) to efficiently approximate the association constant, K_A, of a protein–ligand complex using structural ensembles. K^* considers ensembles of only the most probable low-energy conformations and discards the majority of conformations that are rarely populated by the protein or ligand. K^*'s ability to accurately rank protein sequences by weighting ensembles of low-energy conformations relies heavily on OSPREY's provable guarantees

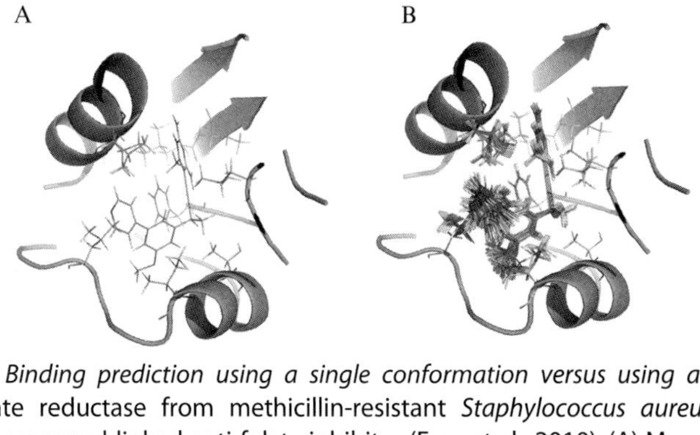

Figure 5.1 *Binding prediction using a single conformation versus using an ensemble.* Dihydrofolate reductase from methicillin-resistant *Staphylococcus aureus* is shown bound to a propargyl-linked anti-folate inhibitor (Frey et al., 2010). (A) Many SCPR algorithms use a single low-energy conformation to model a protein–ligand complex. The GMEC for the protein–ligand complex is shown. (B) OSPREY's MinDEE/A^*/K^* pipeline models the most populated conformations in which binding occurs. Members of an ensemble of bound low-energy conformations are superimposed. (For color version of this figure, the reader is referred to the online version of this chapter.)

(discussed below). Since OSPREY can guarantee that it finds all low-energy conformations for a protein sequence, the generated ensembles do not lack any critical conformations and can be accurately ranked for each sequence. We found K^* to be more accurate and reliable than GMEC-based designs when applying OSPREY to biologically relevant protein design systems (Chen et al., 2009; Roberts et al., 2012).

2.3. Provable guarantees

SCPR requires searching over a very large protein conformation space. Even when searching over a relatively small rotamer library (152 rotamers; Lovell et al., 2000), redesigning 10 residues results in approximately 10^{21} possible rotamer combinations. To handle this large space, heuristic search methods, such as Monte Carlo, are often used. However, when using heuristic methods, it is impossible to know when the design search is complete and how close the computed protein conformation is to the GMEC. Therefore, OSPREY uses provable techniques that guarantee that it finds all low-energy conformations with respect to the input model.

As discussed above, the protein design input model contains many assumptions that can potentially cause errors in the protein design predictions. Ultimately, experimental validation is required to determine whether these assumptions are sufficiently accurate. If the experimentally tested SCPR predictions are successful, the input model is considered sufficiently accurate. However, if the designs fail, it is crucial to ascertain why they did not function as designed. One key advantage of provable SCPR is that there is no error or inaccuracy arising from the search; so, all error can be attributed to the input model. Specifically, if a design prediction fails, one can be confident that improvements should be made to the input model. In contrast, if a heuristic approach were used, it is impossible to disambiguate inaccuracies in the input model from inaccuracies resulting from an insufficient search of the input model.

Misattributing heuristic SCPR search inaccuracies as flaws in the input model could have dire consequences when trying to improve protein energy models. If energy term weights are recalculated or additional terms are added to the energy function based on this misinformation, overfitting is likely to occur. The overfitting is worsened because the energies are not fit to the actual GMECs but rather to the local minima found by the heuristic search. Therefore, training an energy function and improving an input model is more straightforward when using provable SCPR techniques (Roberts et al., 2012).

2.4. Significance of design principles in positive/negative design

Most applications of SCPR focus on stabilizing a target protein fold or binding capability (positive design). When trying to design specificity for a single target, it is also important to *prevent* unwanted folds or binding events from occurring (negative design). A successful positive design only requires finding at least one protein sequence with the desired properties. However, in negative design, the SCPR algorithm must be confident that no off-target binding occurs. Therefore, negative design is much more sensitive to false negatives and requires a more thorough search of the conformation space. Missing low-energy conformations is more detrimental to a negative design than to a positive design. All of the main OSPREY design principles focus on accurately and completely searching the low-energy protein conformation space, which will likely be a great advantage for negative design efforts (Donald, 2011; Frey et al., 2010; Georgiev, Acharya, et al., 2012; Roberts et al., 2012). We further explore positive and negative design with OSPREY in Section 5.2.

3. APPLICATIONS OF OSPREY

We have used OSPREY in several successful prospective designs. In this section, we summarize these designs and mention which protein design algorithms were used for each design. All of these algorithms are explained in detail in Section 4.2.

OSPREY was used to switch the specificity of the phenylalanine adenylation domain of the nonribosomal peptide synthetase enzyme gramicidin S synthetase toward a set of substrates for which the wild-type enzyme had little or no specificity (Chen et al., 2009). The K^* algorithm, with both Minimized DEE (MinDEE) and BD, predicted mutations to the catalytic active site that would switch the substrate specificity. The OSPREY self-consistent mean field (SCMF) module was then used to find residue positions distal from the active site that could bolster the stability of the redesigned enzymes. The chosen distal positions were analyzed with MinDEE to determine the most stabilizing mutations. The mutant enzyme with the highest activity toward its noncognate substrate (L-Leu) showed 1/6 of the wild-type protein/substrate activity. This mutant showed a 2168-fold switch in specificity from the cognate (L-Phe) to the noncognate (L-Leu) substrate.

In Georgiev, Acharya, et al. (2012), OSPREY used a positive/negative design approach to design epitope-specific antibody probes. OSPREY predicted HIV-1 gp120 mutations that would eliminate the binding of specific, undesired antibodies. ELISA assays confirmed that the designed probes maintained binding to their target antibodies, and had weak or only moderate binding toward undesired antibodies. A set of these designed probes is currently being used to isolate antibodies that target epitopes of interest. In Section 5, we describe how OSPREY's positive/negative design approach was used to predict bacterial resistance mutations (Frey et al., 2010).

In Roberts et al. (2012), MinDEE and K^* were used to design peptides that inhibited the cystic fibrosis-related interaction between the proteins CAL and cystic fibrosis transmembrane conductance regulator (CFTR). Specifically, OSPREY redesigned the protein–protein interface between the CFTR C-terminus and CAL to produce a competitive peptide inhibitor of the interaction. The top-ranked peptide bound the CAL protein with a sevenfold better affinity than the previous best-known hexamer peptide and 170-fold more tightly than the CFTR C-terminus. The top-ranked peptide was also shown to rescue chloride flux in human airway epithelial cells containing the ΔF508-CFTR mutation.

In Gorczynski et al. (2007), OSPREY screened inhibitors of a leukemia-associated protein–protein interaction. An earlier version of K^* (Georgiev, Lilien, et al., 2008) used ensembles of protein structures from NMR to rank small molecules that could bind to CBF-β and disrupt its interaction with the protein Runx1. The small molecules allosterically inhibited the interaction and prevented proliferation of cancerous cells.

Finally, we have also used OSPREY in combination with sparse NMR data to determine protein structures. In Zeng, Zhou, and Donald (2011), the DEE/A^* algorithms enabled the computation of side-chain resonance assignments and backbone structure with less NMR data than traditional structure determination methods. In Zeng, Roberts, Zhou, and Donald (2011), side-chain conformations were inferred using a modified version of OSPREY that incorporates unassigned distance restraints data into side-chain placement optimization.

4. PROTEIN DESIGN IN OSPREY

OSPREY redesigns a protein's function and biochemical properties. To perform a redesign, OSPREY requires as input: (a) a 3D structure of the protein to be redesigned; (b) the sequence space, as defined by the allowed mutations to the redesigned protein; (c) the allowed protein flexibility, defined by both an empirical database of favored side-chain conformations (a *rotamer library*) and the type of allowed flexibility (e.g., see below and Fig. 5.2); and (d) an all-atom pairwise energy function to score protein conformations. The 3D structure, sequence space, and allowed flexibility define the conformation search space. A suite of algorithms with mathematical guarantees then computes the GMEC and, optionally, a gap-free list of the other lowest energy conformations. Finally, sequences are ranked using either the GMEC for each sequence or a binding constant prediction based on the computed ensemble of low-energy conformations. Here, we give a brief overview of the input and algorithms. Detailed explanations can be found in Donald (2011), Gainza et al. (2012), Georgiev, Lilien, and Donald (2006), Georgiev and Donald (2007), Georgiev, Keedy, Richardson, Richardson, and Donald (2008), Georgiev, Lilien, et al. (2008), and Lilien et al. (2005) and in the OSPREY user manual (Georgiev, Roberts, Gainza, & Donald, 2012).

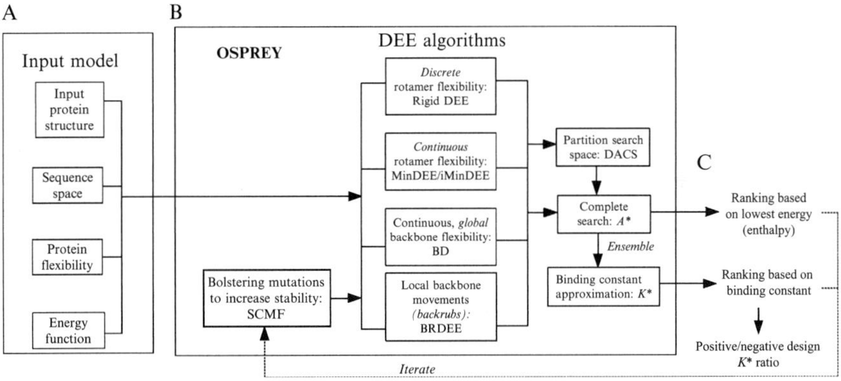

Figure 5.2 *SCPR with OSPREY.* (A) The input model (see Section 4.1). (B) According to the type of flexibility allowed, a specific pruning algorithm is run. The output from the pruning algorithm is directed to either a divide-and-conquer algorithm or directly to the A^* algorithm for a full conformational search. According to the user's selection, the A^* output can then be used to generate a ranking for each sequence based on either the lowest energy structure or on an ensemble of structures generated by the K^* algorithm. (C) If the goal is to find sequences that have a high affinity for one ligand (positive design) while having a low affinity for another (negative design), a ranking can be produced based on the ratio of K^* scores (i.e., positive design score/negative design score; Frey et al., 2010). In addition, if desired, predicted mutants can be improved by finding bolstering mutations that can increase the stability of a mutant. The bolstering mutations can be designed using any of the DEE variants and A^*.

4.1. Input model

OSPREY requires a protonated protein structure in PDB format that has no residues with missing atoms. A single molecule can be modeled as a flexible ligand, while all other water and nonamino acid molecules present in the PDB file must be either specified as rigid bodies or removed. When a ligand is present, OSPREY can use distinct or identical structures for both the protein and ligand's unbound (*apo*) states and for the bound (*holo*) state.

By default, OSPREY includes the Richardsons' Penultimate Rotamer Library (Lovell et al., 2000) and is extensible to other rotamer libraries. Rotamer libraries are available for all natural amino acids, but they are rare for nonamino acid molecules. Thus, in cases where a nonamino acid small molecule is used as the ligand, the user must define its low-energy conformations (called the small molecule's *generalized rotamers*). Within OSPREY, the small molecule's generalized rotamers consist of different conformations of the molecule's flexible dihedral-torsion angles. These conformations are defined by specifying the angle value for each flexible dihedral torsion in the molecule. For a specific example of a generalized rotamer, see Frey et al. (2010) and Georgiev, Roberts, et al. (2012).

OSPREY relies on empirical pairwise-decomposable energy functions to rank protein conformations. OSPREY includes both the Amber96 (Pearlman et al., 1995) and CHARMM (Brooks et al., 2009) energy functions for electrostatics and repulsive–attractive van der Waals (vdW) forces. EEF1 (Lazaridis & Karplus, 1999) is used to score solvation penalties, which are the cost to bury hydrophilic amino acids and/or solvate hydrophobic residues. OSPREY includes charges for some nonamino acid molecules, such as DNA and RNA nucleotides (Reza, 2010), as well as waters. Charges for other organic molecules can be precomputed with a program such as Antechamber (Chen et al., 2009; Frey et al., 2010; Wang, Wang, Kollman, & Case, 2001). All energy parameters, vdW, solvation, and electrostatics, can be scaled and weighted by the user from the defaults provided. Other pairwise-decomposable energy functions can be incorporated into OSPREY; in fact, users are encouraged to improve designs by modifying the energy function.

The user can also specify other design parameters that can significantly improve the accuracy of OSPREY. Amino acid reference energies (Lippow, Wittrup, & Tidor, 2007) account for the energy of a residue in the unfolded state of the protein. These reference energies are important in GMEC-based designs but are not necessary in K^* designs (see following section). Dihedral energy penalties can also be used to prevent continuous flexibility algorithms from minimizing away from the most frequently observed protein conformations. These and several other design parameters are thoroughly explained in Georgiev, Roberts, et al. (2012).

4.2. Protein design algorithms

The search space in protein design is large, and grows exponentially with the number of protein residues and side-chain rotamers. To search in an efficient manner, OSPREY first reduces the size of the search space through a suite of algorithms based on extensions and generalizations of the dead-end elimination (DEE) algorithm (Desmet, de Maeyer, Hazes,.& Lasters, 1992). These algorithms prune the rotamers that, even in the presence of backbone or continuous side-chain flexibility, would not lead to the GMEC or one of the lowest energy conformations (Gainza et al., 2012; Georgiev & Donald, 2007; Georgiev, Keedy, et al., 2008; Georgiev et al., 2006; Georgiev, Lilien, et al., 2008; Lilien et al., 2005). A branch-and-bound algorithm based on A^* (Georgiev, Lilien, et al., 2008; Leach & Lemon, 1998) then traverses the remaining search space and outputs the GMEC and, if desired, a gap-free list of low-energy conformations. The K^* algorithm uses the gap-free list of low-energy conformations to approximate the protein–ligand binding constant. After a design iteration, the results can be reintroduced into OSPREY to search for mutations distal from the active site that will increase the stability of the design (Chen et al., 2009; Fig. 5.2).

Many SCPR algorithms restrict the backbone to a single rigid conformation and the side-chains to discrete, rigid rotameric conformations. The original DEE algorithm (Desmet et al., 1992) falls into this category, and we will refer to it as *rigid DEE* because the rotamers are discrete, rigid geometries that do not include the continuous χ-angle space that immediately surrounds them. OSPREY includes rigid DEE as well as improved variations of DEE that search the continuous χ-angle space that surrounds side-chain rotamers and the continuous ϕ- and ψ-angle space that surrounds the protein backbone (Fig. 5.3, B–E). These continuous-flexibility algorithms compute upper and lower *bounds* on the energy that a backbone or a rotamer could reach after minimization and use these bounds for pruning instead of the rigid energies.

The MinDEE (Georgiev, Lilien, et al., 2008) algorithm extends rigid DEE by including in the search the continuous χ-angle space that immediately surrounds rotamers, and guarantees that no rotamer that can minimize and be part of a minimized GMEC (minGMEC) will be pruned. The *iMinDEE* algorithm (Gainza et al., 2012) improves over MinDEE by

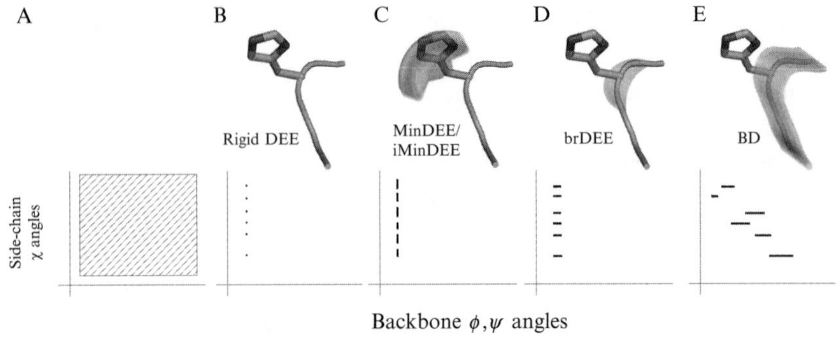

Figure 5.3 *Conceptual illustration of the protein flexibility modeled by* OSPREY*'s algorithms.* Each panel portrays a different algorithm, with the flexibility it models shown above ("blurs" denote movement) and a plot illustrating which dihedral angles it changes shown below. (A) Theoretical "complete" flexibility that could be induced by the introduction of one or more mutations. OSPREY does not yet model "complete" protein flexibility. (B) Rigid DEE: discrete side-chain flexibility with a rigid backbone. (C) MinDEE: continuous side-chain flexibility with a rigid backbone. (D) brDEE: discrete side-chain flexibility and local backbone (backrub) moves. (E) BD: discrete side-chain flexibility and continuous global backbone moves. (For color version of this figure, the reader is referred to the online version of this chapter.)

pruning orders of magnitude more rotamers with close to the same efficiency as rigid DEE, and also guarantees to find the minGMEC. The backrub DEE (brDEE) algorithm (Georgiev, Keedy, et al., 2008) allows mutants to undergo *backrub motions*, which are entirely local backbone movements that each change the orientation of one C_α–C_β bond vector by performing a small rotation of the surrounding dipeptide (Davis, Arendall, Richardson, & Richardson, 2006). The Backbone DEE (BD) algorithm (Georgiev & Donald, 2007) prunes only rotamers that cannot be part of the GMEC after allowing continuous global backbone movements.

When a ligand is present, it can rotate and translate with respect to the protein (i.e., rigid-body motions), and continuous rotamers can be defined for the ligand. If the ligand is a polypeptide, both the ligand and protein can mutate and rigid DEE, MinDEE, iMinDEE, brDEE, or BD can be used on both molecules.

Each OSPREY algorithm (rigid DEE, MinDEE, iMinDEE, brDEE, and BD) reduces the search space through a stage of DEE-based pruning. In addition, several extensions to the DEE algorithm implemented in OSPREY further improve its pruning capabilities, including generalized DEE (Goldstein, 1994), split flags (Pierce, Spriet, Desmet, & Mayo, 2000), bounds pruning (Gordon, Hom, Mayo, & Pierce, 2003), and a divide-and-conquer strategy called DACS (Georgiev et al., 2006).

Once the DEE algorithms prune the majority of the conformational space, the remaining space must be searched to find the lowest energy conformation(s). We have implemented a branch-and-bound algorithm based on the A^* algorithm (Georgiev, Lilien, et al., 2008; Leach & Lemon, 1998) that searches conformations in a tree and traverses only the branches that might lead to the lowest energy structure, even in the presence of flexibility. A^* searches the space completely and guarantees to find the optimal answer. When rigid DEE is used, our extension of A^* can also enumerate conformations in order of the lowest energy, which makes it possible to enumerate a gap-free list of low-energy conformations and their sequences. When MinDEE, brDEE, or BD is used, our version of A^* enumerates conformations in order of the lower *bounds* on the energy of each conformation, and can also enumerate a gap-free list of low-energy conformations.

The K^* algorithm (Georgiev, Lilien, et al., 2008; Lilien et al., 2005) uses this list of low-energy conformations to compute a provable ε-approximation to the binding constant (a K^* *score*) with respect to the input model (Fig. 5.2A). A provable ε-approximation algorithm guarantees that the computed binding constant is mathematically accurate up to a user-specified percentage error of ε, with respect to the input model. The ε parameter is specified by the user as the desired accuracy and all computed solutions are guaranteed to be ε-accurate. K^* is efficient because it uses A^* to compute only the reduced set of low-energy conformations that are most likely to be taken on by the protein, the ligand, and the protein–ligand complex. These low-energy conformations are then Boltzmann-weighted and used to approximate the partition function for the unbound and bound states. Because A^* enumerates conformations in order of their low-energy bound, K^* can calculate exactly when each partition function is within an ε-factor of the exact solution and stop the computation. Since the energy of each low-energy conformation is Boltzmann-weighted, K^* must only compute a small percentage of the total number of conformations (Georgiev, Lilien, et al., 2008). Once each partition function is computed, the K^* score is computed by dividing the partition function of the bound state (i.e., the protein–ligand complex) by the partition functions of both unbound states (Georgiev, Lilien, et al., 2008).

After redesigning a protein core, boundary, surface, or active site, it can be beneficial to increase the mutant's stability by further mutating residues that are distal to the redesigned region. However, since proteins can be large, searching at these distal positions for potential bolstering mutations using algorithms with mathematical guarantees can be an expensive process. To address this issue, OSPREY uses a heuristic SCMF algorithm to find residue positions, that when mutated, might increase the stability of the engineered protein. After SCMF identifies distal residue positions, OSPREY's MinDEE variants and A^* can be used to find mutant residues at those distal positions that stabilize the fold. We have used this approach to increase the stability and catalytic efficiency of a redesigned enzyme using MinDEE/A^* (Chen et al., 2009).

5. EXAMPLE: PREDICTING DRUG RESISTANCE MUTATIONS USING OSPREY

Pharmaceutical companies periodically release new, effective drugs to treat the world's most dangerous infectious diseases. After these drugs are first introduced, pathogens that cause these diseases, such as methicillin-resistant *Staphylococcus aureus* (MRSA), recede temporarily, only to reemerge months or years later as drug-resistant strains. When novel drugs are first discovered, little is known about how pathogens will develop drug resistance. Without that information, drug designers cannot improve existing drugs or develop new ones to target resistant strains until after they spread in the community.

Fortunately, SCPR programs can be used to predict drug resistance that could arise through mutations in enzyme active sites as soon as a drug is developed. This type of application exemplifies the next frontier in SCPR: the design of proteins not only for activity but for specificity. We have used OSPREY to predict resistance mutations in enzyme active sites that confer resistance to competitive inhibitors. Competitive inhibitors are drugs that compete with the natural substrate for binding, and thus inhibit a critical enzyme in a pathogen. Organisms can often evolve active site mutations that maintain catalysis of the substrate but reduce the affinity for the competitive inhibitor. In effect, this resistance mechanism allows the substrate to outcompete the inhibitor for binding to the enzyme. To perform this kind of design, OSPREY must design specificity for the natural substrate over the drug, which is a relatively new and attractive goal for SCPR algorithms.

We have developed a methodology that uses OSPREY to accurately predict resistance-conferring mutations. In Frey et al. (2010), we showed that this methodology can successfully predict resistance mutations to a new antibiotic, UCP111D26M (termed D26M, and shown in Fig. 5.4) that inhibits the MRSA DHFR enzyme. In this section, we describe in depth the methods, empirical rationale, and experimental validation used in Frey et al. (2010).

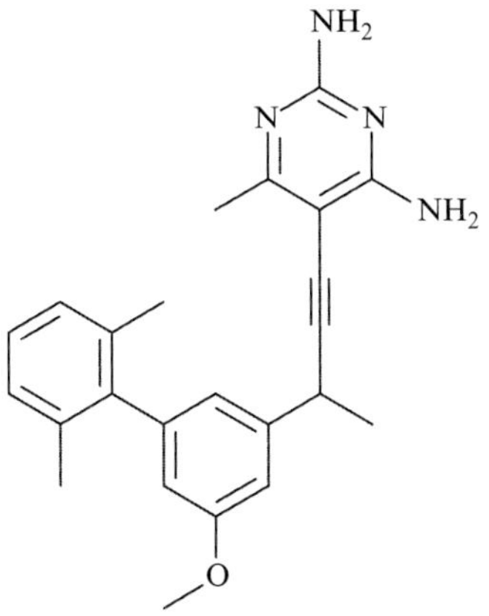

Figure 5.4 *D26M compound.* This compound belongs to a new class of propargyl-linked antifolates. D26M is an effective antibiotic against MRSA.

5.1. Input model in a resistance prediction problem

5.1.1 Initial protein structure

Our approach assumes that mutant sequences that bind the natural substrate well (*positive design*) while binding the competitive inhibitor poorly (*negative design*) will confer drug resistance. This positive/negative design approach needs both a structure of the wild-type enzyme bound to its natural substrate and a structure of the wild-type enzyme bound to the competitive inhibitor drug. In general, higher quality input structures should lead to more accurate results. However, if the relevant high-quality structure is not available, modeling based on a related structure can be used. In the case of MRSA DHFR resistance prediction, a structure of the wild-type enzyme bound to D26M (Fig. 5.4) was determined (PDB ID: 3F0Q) for negative design, and the structure of the F98Y DHFR mutant bound to folate was used to generate a model of the wild type for positive design.

5.1.2 Protein flexibility

The structural changes caused by resistance mutations must preserve the catalytic activity of the enzyme. Thus, we expect that successful resistance mutations to the active site of MRSA DHFR will cause small conformational changes in the protein structure. The BD, brDEE, and MinDEE algorithms can model these conformational changes. However, in the case of DHFR, we chose to model continuous side-chain flexibility (MinDEE) in addition to ligand flexibility (described below) because we expected that the protein backbone surrounding the active site would remain relatively rigid after the introduction of resistance mutations to maintain catalytic activity. Thus, we assumed that the protein's side-chain interactions with the ligand would determine resistance.

The ligand (both D26M and dihydrofolate) can potentially bind DHFR in a large number of conformations (see Section 4.1). However, choosing the D26M negative design rotamers is especially important because missing low-energy binding conformations between D26M and DHFR could result in a failure of the negative design (see Section 2.4). We chose 512 generalized rotamers for D26M over four rotatable dihedrals. Based on structural information, in addition to the reasons described in Section 2.4, we believe that the catalytic activity of DHFR is highly optimized for a few specific binding conformations of dihydrofolate. We chose 12 generalized rotamers over 10 rotatable dihedrals for dihydrofolate. Each of these generalized rotamers was treated as a continuous rotamer, meaning that each flexible dihedral was allowed to minimize by rotating its torsional dihedrals $\pm 9°$. Additionally, continuous rigid-body motions (rotation and translation) were allowed for each ligand within the active site. The allowed conformation space, C_I, for the inhibitor D26M is completely described by these three modeling choices. Namely, C_I is defined by the chosen generalized rotamers, the allowed continuous minimization around these rotamers, and the allowed rigid-body rotation and translation. The conformation space for DHFR and dihydrofolate, C_P, and C_S, respectively, were similarly defined.

5.1.3 Sequence space

For this study, we only allowed residues in the active site with direct contact to the drug to mutate. Specifically, only residues L5, V6, L20, D27, L28, V31, T46, I50, L54, and F92 were allowed to mutate and/or change conformation. The allowed amino acid mutations were selected based on the wild-type amino acid and correlation with other DHFR species. Residues 5, 6, 20, 28, 31, 50, and 92 were allowed to mutate to Ala, Val, Leu, Ile, Met, Phe, Trp, and Tyr. Residue D27 was allowed to maintain its identity or mutate to Glu, while residues T46 and L54 were allowed to change their conformation but not mutate. Only mutant sequences representing single- or double-point amino acid mutations were allowed to mimic resistance mutations that could evolve naturally. This resulted in a total sequence space of 1173 mutants to the wild type.

5.2. Results

OSPREY computed the K^* score for the 1173 DHFR single- and double-point mutants bound to dihydrofolate (positive design) or D26M (negative design). Positive and negative design computations were performed separately and then combined.

Each sequence was ranked by its K^* ratio: the K^* score of the positive design divided by the K^* score of the negative design. Sequences with a negative design score of zero were ranked solely by the positive design score, namely by their binding affinity for dihydrofolate. Since the K^* score approximates a K_A for the protein–ligand complex, a higher K^* score represents better binding. Therefore, a mutant DHFR sequence with a high positive design score and negative design score of zero is predicted to destabilize binding to D26M versus dihydrofolate. All of the top 10 mutations were predicted to bind dihydrofolate, while disrupting the binding of D26M in the conformations specified by C_I for negative design. Four of the top 10 predicted resistance mutants were tested experimentally: (1) V31Y/F92I, (3) V31I/F92S, (7) V31F/F92L, and (9) I50W/F92S. Mutants (1), (3), and (7) yielded biologically successful results. Specifically, these mutants maintained catalytic activity and had a lower affinity for the D26M drug. The top-ranked mutant sequence, V31Y/F92I, conferred the greatest decrease in binding to D26M, an 18-fold loss.

In addition to confirming that the top mutants were biologically successful, it is important to evaluate the success of the underlying computational predictions. The success of the computational prediction relies entirely on the accuracy of the input model's definition of conformation space and energy function because OSPREY guarantees to find the optimal conformation(s) given the input model. Moreover, the goal of the computational negative design was to find protein sequences that cannot bind any conformation in the D26M ligand's conformational space, C_I. Note that if, for example, C_I does not accurately represent the conformational energy landscape, it is possible that the computational prediction would successfully exclude binding in C_I but result in biologically failed designs because the protein could bind to a D26M conformation outside of C_I. Of course, the issue of defining the input conformational space C_I arises in any protein design algorithm. But since K^*'s search guarantees completeness, we can rule out failures of optimization and attribute any discrepancies between predictions and experimental measurements exclusively to the input model, which includes C_I. This guarantee is crucial because if any low-energy conformation was missed, there would likely be a low-energy D26M binding mode resulting in designs that fail both computationally and biologically.

To analyze the binding mechanism of the top resistant DHFR mutant, V31Y/F92I, the crystal structure of this mutant bound to D26M was determined (PDB ID: 3LG4). The structure showed that D26M occupies the active site of mutant V31Y/F92I with 50% occupancy. In contrast, wild-type DHFR binds D26M with full occupancy, which suggests that poor occupancy in V31Y/F92I is caused by reduced ligand binding. D26M binds the V31Y/F92I mutant weakly in a conformation that was not in C_I, which demonstrates the difficulty for the user to determine *a priori* all conformations of a drug to input into the design protocol. However, the predicted energy of this new conformation bound to DHFR V31Y/F92I was very similar to the lowest energy conformations in the predicted K^* ensemble. This demonstrates that C_I accurately covered the energy landscape and OSPREY successfully found a mutant protein sequence that could destabilize binding to D26M. Since OSPREY uses provable algorithms, it can guarantee that no conformations in C_I would bind with a better energy than what was found in the K^* ensemble. This is confirmed by the conformation of D26M in the V31Y/F92I crystal structure.

5.2.1 Effect of limiting flexibility on mutation predictions

We have argued that ensembles, flexibility, and provability are essential for both positive design and negative design (see Section 2). We have also shown that using ensembles can have important consequences on binding affinity rankings (Roberts et al., 2012), and that provably modeling continuous flexibility in protein core redesign is critical for accuracy (Gainza et al., 2012). Similarly, we now show the importance of improving flexibility in an additional example. For this example, we performed predictions almost identical to those in Frey et al. (2010) with one crucial change: we limited protein flexibility to discrete rotamers for both the rotamer and ligands, and disabled continuous rigid-body motions. Using such a discrete, rigid model is very common in the protein design field.

Rigid DEE/A^*/K^* was used to compute a positive and a negative design K^* score for all 1173 DHFR mutant sequences. We found that all of the top 10 ranked mutants predicted by MinDEE/K^* in Frey et al. (2010) now received radically different scores. They all had a positive design score of 0 in the rigid, discrete designs because rigid DEE/A^*/K^* could not find any low-energy binding conformation of dihydrofolate for these sequences. Thus, the rigid model incorrectly predicted that the experimentally tested mutants of Frey et al. (2010) would not bind dihydrofolate. These results suggest that rigid rotamers not only fail to cover the entire rotamer space but also are sensitive to small changes in torsional dihedrals and rigid-body motions. Consequently, few mutants are predicted by rigid DEE/A^*/K^* to bind dihydrofolate or D26M, which is manifestly wrong in light of the MinDEE/A^*/K^* results and our experimental validation.

6. FUTURE DIRECTIONS AND AVAILABILITY

We have presented an overview of OSPREY, a comprehensive open-source SCPR suite. OSPREY has been in continuous development over the last decade and both the algorithms and functionality will continue to improve. The variety of the prospective designs where OSPREY has been applied, and the suite of sophisticated algorithms with which they were created, suggest that OSPREY can be adapted to facilitate protein engineering in a number of settings. Several enhancements of OSPREY are planned, including support for explicit water-mediated hydrogen bonds, concerted backbone and side-chain continuous flexibility (Hallen, Keedy, & Donald, 2013), protein loop modeling (Tripathy, Zeng, Zhou, & Donald, 2012), and RNA rotamers.

OSPREY is available under a GNU Lesser General Public License. As such, the source code is provided as part of the distribution. We encourage users to customize and/or improve it. Specifically, OSPREY provides a platform for the development of new algorithms and new protein design methodology, beyond the features we have presented here. All software is implemented in Java, with parallel computing capabilities provided by mpiJava (Baker, Carpenter, Hoon Ko, & Li, 1998). OSPREY can run on any operating system that supports Java, but we recommend a computing cluster to run OSPREY to distribute and reduce the computation time.

ACKNOWLEDGMENTS

This work was supported by the following NIH grants to B. R. D.: R01 GM-78031, R01 GM-65982, and T32 GM-71340. D. A. K., J. S. R., and D. C. R. were supported by grant NIH R01-GM073930 to D. C. R.

REFERENCES

Baker, M., Carpenter, B., Hoon Ko, S., & Li, X. (1998). mpiJava: A Java interface to MPI. First UK Workshop on Java for High Performance Network Computing.

Brooks, B. R., Brooks, C. L., III, Mackerell, A. D., Jr., Nilsson, L., Petrella, R. J., Roux, B., et al. (2009). CHARMM: The biomolecular simulation program. *Journal of Computational Chemistry*, *30*, 1545–1614.

Chen, C. Y., Georgiev, I., Anderson, A. C., & Donald, B. R. (2009). Computational structure-based redesign of enzyme activity. *Proceedings of the National Academy of Sciences of the United States of America*, *106*, 3764–3769.

Chou, J. J., Case, D. A., & Bax, A. (2003). Insights into the mobility of methyl-bearing side chains in proteins from (3)J(CC) and (3)J(CN) couplings. *Journal of the American Chemical Society, 125*, 8959–8966.

Davis, I. W., Arendall, W. B., Richardson, D. C., & Richardson, J. S. (2006). The backrub motion: How protein backbone shrugs when a sidechain dances. *Structure, 14*, 265–274.

Desmet, J., de Maeyer, M., Hazes, B., & Lasters, I. (1992). The dead-end elimination theorem and its use in protein side chain positioning. *Nature, 356*, 539–542.

Donald, B. R. (2011). *Algorithms in structural molecular biology*. Cambridge, MA: MIT Press.

Frey, K. M., Georgiev, I., Donald, B. R., & Anderson, A. C. (2010). Predicting resistance mutations using protein design algorithms. *Proceedings of the National Academy of Sciences of the United States of America, 107*, 13707–13712.

Gainza, P., Roberts, K. E., & Donald, B. R. (2012). Protein design using continuous rotamers. *PLoS Computational Biology, 8*(1), e1002335.

Georgiev, I., Acharya, P., Schmidt, S. D., Li, Y., Wycuff, D., Ofek, G., et al. (2012). Design of epitope-specific probes for sera analysis and antibody isolation. *Retrovirology, 9* (Suppl. 2):P50. PMC id: PMC3442034.

Georgiev, I., & Donald, B. R. (2007). Dead-end elimination with backbone flexibility. *Bioinformatics, 23*, i185–i194.

Georgiev, I., Keedy, D., Richardson, J., Richardson, D., & Donald, B. R. (2008). Algorithm for backrub motions in protein design. *Bioinformatics, 24*, i196–i1204.

Georgiev, I., Lilien, R. H., & Donald, B. R. (2006). Improved pruning algorithms and divide-and-conquer strategies for dead-end elimination, with application to protein design. *Bioinformatics, 22*, e174–e183.

Georgiev, I., Lilien, R. H., & Donald, B. R. (2008). The minimized dead-end elimination criterion and its application to protein redesign in a hybrid scoring and search algorithm for computing partition functions over molecular ensembles. *Journal of Computational Chemistry, 29*, 1527–1542.

Georgiev, I., Roberts, K.E., Gainza, P., & Donald, B.R. (2012). OSPREY v2.0 manual. Dept. of Computer Science, Duke University http://www.cs.duke.edu/donaldlab/software/osprey/osprey2.0.pdf.

Gilson, M. K., Given, J. A., Bush, B. L., & McCammon, J. A. (1997). The statistical-thermodynamic basis for computation of binding affinities: A critical review. *Biophysical Journal, 72*, 1047–1069.

Goldstein, R. (1994). Efficient rotamer elimination applied to protein side-chains and related spin glasses. *Biophysical Journal, 66*, 1335–1340.

Gorczynski, M. J., Grembecka, J., Zhou, Y., Kong, Y., Roudaia, L., Douvas, M. G., et al. (2007). Allosteric inhibition of the protein–protein interaction between the leukemia-associated proteins RUNX1 and CBF β. *Chemistry & Biology, 14*, 1186–1197.

Gordon, D. B., Hom, G. K., Mayo, S. L., & Pierce, N. A. (2003). Exact rotamer optimization for protein design. *Journal of Computational Chemistry, 24*, 232–243.

Hallen, M. A., Keedy, D. A., & Donald, B. R. (2013). Dead-End Elimination with Perturbations ('DEEPer'): A provable protein design algorithm with continuous sidechain and backbone flexibility. *Proteins, 81*(1), 18–39.

Hu, X., & Kuhlman, B. (2006). Protein design simulations suggest that side-chain conformational entropy is not a strong determinant of amino acid environmental preferences. *Proteins, 62*, 739–748.

Lazaridis, T., & Karplus, M. (1999). Discrimination of the native from misfolded protein models with an energy function including implicit solvation. *Journal of Molecular Biology, 288*, 477–487.

Leach, A. R., & Lemon, A. P. (1998). Exploring the conformational space of protein side chains using dead-end elimination and the A^* algorithm. *Proteins, 33*, 227–239.

Lilien, R. H., Stevens, B. W., Anderson, A. C., & Donald, B. R. (2005). A novel ensemble-based scoring and search algorithm for protein redesign and its application to modify the substrate specificity of the gramicidin synthetase a phenylalanine adenylation enzyme. *Journal of Computational Biology, 12*, 740–761.

Lippow, S. M., Wittrup, K. D., & Tidor, B. (2007). Computational design of antibody-affinity improvement beyond in vivo maturation. *Nature Biotechnology, 25*, 1171–1176.

Lovell, S. C., Word, J. M., Richardson, J. S., & Richardson, D. C. (2000). The penultimate rotamer library. *Proteins, 40*, 389–408.

Pearlman, D. A., Case, D. A., Caldwell, J. W., Ross, W. S., Cheatham, T. E., DeBolt, S., et al. (1995). AMBER, a package of computer programs for applying molecular mechanics, normal mode analysis, molecular dynamics and free energy calculations to simulate the structural and energetic properties of molecules. *Computer Physics Communications, 91*, 1–41.

Pierce, N. A., Spriet, J. A., Desmet, J., & Mayo, S. L. (2000). Conformational splitting: A more powerful criterion for dead-end elimination. *Journal of Computational Chemistry, 21*, 999–1009.

Reza, F. (2010). Computational molecular engineering of nucleic acid binding proteins and enzymes. Doctoral Dissertation, Duke University.

Roberts, K. E., Cushing, P. R., Boisguerin, P., Madden, D. R., & Donald, B. R. (2012). Design of a PDZ domain peptide inhibitor that rescues CFTR activity. *PLoS Computational Biology, 8*(4), e1002477. http://dx.doi.org/10.1371/journal.pcbi.1002477.

Teague, S. J. (2003). Implications of protein flexibility for drug discovery. *Nature Reviews Drug Discovery, 2*, 527–541.

Tripathy, C., Zeng, J., Zhou, P., & Donald, B. R. (2012). Protein loop closure using orientational restraints from NMR data. *Proteins, 80*, 433–453.

Wang, J., Wang, W., Kollman, P., & Case, D. (2001). Antechamber, an accessory software package for molecular mechanical calculations. *Abstracts of Papers of the American Chemical Society, 222*, U403.

Zeng, J., Roberts, K. E., Zhou, P., & Donald, B. R. (2011). A Bayesian approach for determining protein side-chain rotamer conformations using unassigned NOE data. *Journal of Computational Biology, 18*, 1661–1679.

Zeng, J., Zhou, P., & Donald, B. R. (2011). Protein side-chain resonance assignment and NOE assignment using RDC-defined backbones without TOCSY data. *Journal of Biomolecular NMR, 50*, 371–395.

CHAPTER SIX

Scientific Benchmarks for Guiding Macromolecular Energy Function Improvement

Andrew Leaver-Fay[*,1,2], Matthew J. O'Meara[†,1,2], Mike Tyka[‡], Ron Jacak[§], Yifan Song[‡], Elizabeth H. Kellogg[‡], James Thompson[‡], Ian W. Davis[¶], Roland A. Pache[||], Sergey Lyskov[#], Jeffrey J. Gray[#], Tanja Kortemme[||], Jane S. Richardson[**], James J. Havranek[††], Jack Snoeyink[†], David Baker[‡], Brian Kuhlman[*]

[*]Department of Biochemistry, University of North Carolina, Chapel Hill, North Carolina, USA
[†]Department of Computer Science, University of North Carolina, Chapel Hill, North Carolina, USA
[‡]Department of Biochemistry, University of Washington, Seattle, Washington, USA
[§]Department of Immunology and Microbial Science, The Scripps Research Institute, La Jolla, California, USA
[¶]GrassRoots Biotechnology, Durham, North Carolina, USA
[||]Department of Bioengineering and Therapeutic Science, University of California San Francisco, San Francisco, California, USA
[#]Department of Chemical & Biomolecular Engineering, Johns Hopkins, Baltimore, Maryland, USA
[**]Department of Biochemistry, Duke University, Durham, North Carolina, USA
[††]Department of Genetics, Washington University, St. Louis, Missouri, USA
[1]These authors contributed equally to this work.
[2]Corresponding authors: e-mail address: leaverfa@email.unc.edu; momeara@cs.unc.edu

Contents

Methods in Enzymology, Volume 523
ISSN 0076-6879
http://dx.doi.org/10.1016/B978-0-12-394292-0.00006-0

Abstract

Accurate energy functions are critical to macromolecular modeling and design. We describe new tools for identifying inaccuracies in energy functions and guiding their improvement, and illustrate the application of these tools to the improvement of the Rosetta energy function. The feature analysis tool identifies discrepancies between structures deposited in the PDB and low-energy structures generated by Rosetta; these likely arise from inaccuracies in the energy function. The optE tool optimizes the weights on the different components of the energy function by maximizing the recapitulation of a wide range of experimental observations. We use the tools to examine three proposed modifications to the Rosetta energy function: improving the unfolded state energy model (reference energies), using bicubic spline interpolation to generate knowledge-based torisonal potentials, and incorporating the recently developed Dunbrack 2010 rotamer library (Shapovalov & Dunbrack, 2011).

1. INTRODUCTION

Scientific benchmarks are essential for the development and parameterization of molecular modeling energy functions. Widely used molecular mechanics energy functions such as Amber and OPLS were originally parameterized with experimental and quantum chemistry data from small molecules and benchmarked against experimental observables such as intermolecular energies in the gas phase, solution phase densities, and heats of vaporization (Jorgensen, Maxwell, & Tirado-Rives, 1996; Weiner et al., 1984). More recently, thermodynamic measurements and high-resolution structures of macromolecules have provided a valuable testing ground for energy function development. Commonly used scientific tests include discriminating the ground state conformation of a macromolecule from higher energy conformations (Novotný, Bruccoleri, & Karplus, 1984; Park & Levitt, 1996; Simons et al., 1999), and predicting amino acid sidechain conformations (Bower, Cohen, & Dunbrack, 1997; Jacobson, Kaminski, Friesner, & Rapp, 2002) and free energy changes associated with protein mutations (Gilis & Rooman, 1997; Guerois, Nielsen, & Serrano, 2002; Potapov, Cohen, & Schreiber, 2009).

Many studies have focused on optimizing an energy function for a particular problem in macromolecular modeling, for instance, the FoldX energy function was empirically parameterized for predicting changes to the free energy of a protein when it is mutated (Guerois et al., 2002). Often, these types of energy functions are well suited only to the task they have been trained for. Kellogg, Leaver-Fay, and Baker (2011) showed that an energy function explicitly trained to predict energies of mutation did not produce native-like sequences when redesigning proteins. For many projects, it is advantageous to have a single energy function that can be used for diverse modeling tasks. For example, protocols in the molecular modeling program Rosetta for ligand docking (Meiler & Baker, 2003), protein design (Kuhlman et al., 2003), and loop modeling (Wang, Bradley, & Baker, 2007) share a common energy function, which allowed Murphy, Bolduc, Gallaher, Stoddard, and Baker (2009) to combine them to shift an enzyme's substrate specificity.

Sharing a single energy function between modeling applications presents both opportunities and challenges. Researchers applying the energy function to new tasks sometimes uncover deficiencies in the energy function. The opportunities are that correcting the deficiencies in the new tasks will result in improvements in the older tasks—after all, nature uses only one energy function. Sometimes, however, modifications to the energy function that improve its performance at one task degrade its performance at others. The challenges are then to discriminate beneficial from deleterious modifications and reconcile task-specific objectives.

To address these challenges, we have developed three tools based on benchmarking Rosetta against macromolecular data. The first tool (Section 3), a suite we call "feature analysis," can be used to contrast ensembles of structural details from structures in the PDB and from structures generated by Rosetta. The second tool (Section 4), a program we call "optE," relies on fast, small-scale benchmarks to train the weights in the energy function. These two tools can help identify and fix flaws in the energy function, facilitating the process of integrating a proposed modification. We follow (Section 5) with a curated set of large-scale benchmarks meant to provide sufficient coverage of Rosetta's applications. The use of these benchmarks will provide evidence that a proposed energy function modification should be widely adopted. To conclude (Section 6), we demonstrate our tools and benchmarks by evaluating three incremental modifications to the Rosetta energy function.

Alongside this chapter, we have created an online appendix, which documents usage of the tools, input files, instructions for running the benchmarks, and current testing results: http://rosettatests.graylab.jhu.edu/guided_energy_function_improvement.

2. ENERGY FUNCTION MODEL

The Rosetta energy function is a linear combination of terms that model interactions between atoms, solvation effects, and torsion energies. More specifically, *Score12* (Rohl, Strauss, Misura, & Baker, 2004), the default fullatom energy function in Rosetta, consists of a Lennard–Jones term, an implicit solvation term (Lazaridis & Karplus, 1999), an orientation-dependent H-bond term (Kortemme, Morozov, & Baker, 2003), sidechain and backbone torsion potentials derived from the PDB, a short-ranged knowledge-based electrostatic term, and reference energies for each of the 20 amino acids that model the unfolded state. Formally, given a molecular conformation, C, the total energy of the system is given by

$$E(C|w,\Theta) = \sum_{j}^{|T|} w_j T_j\left(C|\Theta_j\right) \qquad [6.1]$$

where each energy term T_j has parameters Θ_j and weight w_j. The feature analysis tool is meant to aid the refinement of the parameters Θ. The optE tool is meant to fit the weights, w.

3. FEATURE ANALYSIS

We aim to facilitate the analysis of distributions of measurable properties of molecular conformations, which we call "feature analysis." By formalizing the analysis process, we are able to create a suite of tools and benchmarks that unify the collection, visualization, and comparison of feature distributions. After motivating our work, we describe the components (Section 3.1) and illustrate how they can be integrated into a workflow (Section 3.2) by investigating the distribution of the lengths of H-bonds with hydroxyl donors.

Feature distributions, broadly construed, have long held a prominent role in structural biochemistry. The Boltzmann equation—relating probability with energy—has been used to justify creating knowledge-based potentials from distribution of features from crystal structures (Miyazawa & Jernigan, 1985; Sippl, 1990); for example, rotamer libraries are often based on feature distributions (Dunbrack & Karplus, 1993; Ponder & Richards, 1987). The Boltzmann equation also motivates comparing feature distributions from predicted structures against those observed in nature: for an energy function to generate geometries that are rarely observed means the energy function is wrongly assigning low energies to high-energy geometries. Many structure validation tools, such as MolProbity (Chen et al., 2010), identify outlier features as indications of errors in a structure.

The Rosetta community has also relied on feature analysis: for example, the derivation of low-resolution (Simons et al., 1999) and high-resolution (Kortemme et al., 2003) knowledge-based potentials, and the analysis of core-packing quality (Sheffler & Baker, 2009) and surface hydrophobic patches (Jacak, Leaver-Fay, & Kuhlman, 2012). However, each feature analysis foray has been *ad hoc*, limiting reuse and reproducibility. Our primary goal was to create a unified framework for feature analysis.

Feature analysis also provides a means to tune the parameters of an energy function. Recently, Song, Tyka, Leaver-Fay, Thompson, and Baker (2011) observed peaks in the backbone φ, ψ distributions of Rosetta-generated structures, absent in those of crystal structures, which they attributed to double counting in knowledge-based potentials. In one case, a commonly observed loop motif forms an H-bond between the asparagine sidechain of residue i and the backbone of residue $i+2$, constraining ψ_i to 120°. The proliferation of this motif caused an artifact in Rosetta's knowledge-based Ramachandran energy term, leading it to favor asparagines with a ψ of 120° irrespective of H-bond formation. To correct this, they considered the Ramachandran term as a parametric model and tuned the parameters until the predicted asparagine ψ distribution matched the distribution from crystal structures. Hamelryck et al. (2010) have also observed that this tuning process is a useful way to improve an energy function. Our second major goal for the feature analysis tool is to facilitate this process of parameter tuning.

3.1. Feature analysis components

The feature analysis framework consists of two components: Feature reporters take a *batch* of structures and populate a relational database with feature data. Next, feature analysis scripts select, estimate, visualize, and compare feature distributions from the database.

3.1.1 Feature databases

To facilitate the analysis of a diversity of feature distributions, we have created a relational database architecture for feature data. Typically, when analyzing feature distributions, we decompose a basic feature distribution (e.g., H-bond length) into many conditional feature distributions (e.g., H-bond length for carboxylate-guanidine residues in beta-sheets). By putting basic features into a relational database along with other supporting data, we can perform the expensive task of extracting features from some input batch of structures once, while retaining the ability to examine arbitrary conditional feature distributions in the future.

The database schema is a hierarchy with high-level tables holding the batch of structures and low-level tables holding the basic features. For example, the HBond feature reporter manages H-bond properties with foreign-key references to the higher-level `residues` and `structures` tables (Fig. 6.1).

Each feature database holds features for a single batch of conformations. Once a batch and relevant feature reporters (Table 6.1) have been selected, the features are extracted to the database using the `ReportToDB` mover in the `RosettaScripts` XML-based protocol language (Fleishman et al., 2011). Feature extraction is robust in that it supports multiple database backends (`SQLite`, `PostgreSQL`, and `MySQL`), and incremental extraction and merging. See the online appendix for feature analysis tutorials and details about implementing new `FeatureReporters`.

3.1.2 Distribution analysis

The second component of the feature analysis suite provides tools to query feature databases, to transform features in order to correctly estimate feature distributions, and to plot those distributions. Community-created feature analysis scripts are released with Rosetta and may be found in the `rosetta/rosetta_tests/features` directory.

To run a features analysis, the user provides a configuration file specifying a set of feature databases (each extracted from a batch of structures), the analysis scripts to run, and the plot output formats. Each feature analysis script typically consists of three parts: an SQL query to retrieve features from the input databases, kernel density estimation (KDE) on the extracted features (or transformed features), and the creation of a plot using the `ggplot2` grammar-of-graphics package in *R*.

Feature analysis scripts begin by querying the input sample sources using one or more SQL statements, ending with a `SELECT` statement. These SQL queries can join multiple tables to compile arbitrarily complicated conditional feature distributions. The resulting table has rows that represent

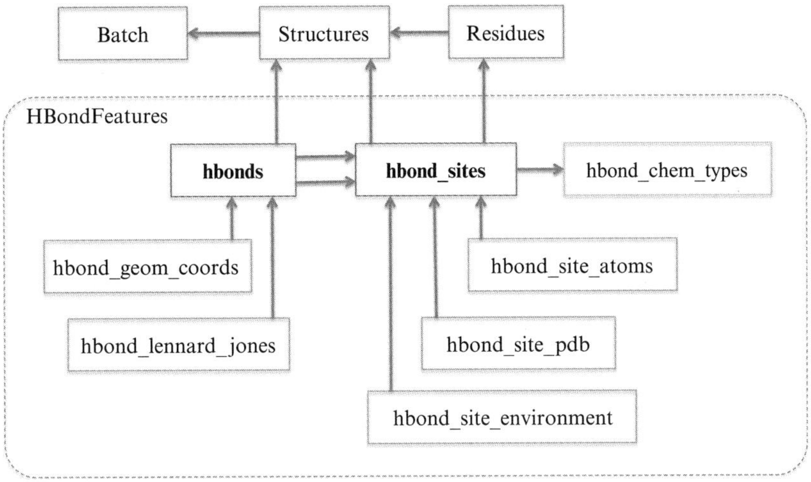

Figure 6.1 HBondFeatures database schema. The `HBondFeatures` class populates these tables with H-bond data. For each H-bond site (acceptor atom or polar hydrogen), atomic coordinates, experimental data, and solvent environment are reported. For each H-bond that forms between two H-bond sites, the geometric coordinates (i.e., distances and angles) and the sum of the Lennard–Jones energies for H-bonding atoms are reported. (For the color version of this figure, the reader is referred to the online version of this chapter.)

feature instances, where some of the columns *identify* the feature (including which batch it came from along with other covariates), and the remaining columns *measure* the feature in the feature space.

Once the features have been retrieved, density distributions can be computed. To do this, a feature analysis script can use the split-apply-combine strategy (Wickham, 2011): feature instances are grouped by their identifying columns, and for each group, a KDE is computed over the measure columns. When computing density estimations over feature spaces, care must be taken to apply appropriate transformations to normalize the space and to handle boundary conditions. Our framework provides support for common transformations and kernel bandwith selection strategies, which control smoothness of the estimated distribution. Once a collection of feature distributions have been estimated, they can be visualized through the declaritive grammar-of-graphics method (Wickham, 2010; Wilkinson, 1999).

The scripting framework also provides support for other types of feature analysis tasks. For example, the results from prediction benchmarks (e.g., from RotamerRecovery) can be regressed against various feature types. Feature instances can also be aligned and exported to a PyMOL session for interactive inspection.

Table 6.1 Each `FeatureReporter` is responsible for extracting a particular structural feature from a structure and reporting it to a relational database

Meta	One body	Two body	Multibody
Protocol	Residue	Pair	Structure
Batch	ResidueConformation	AtomAtomPair	PoseConformation
JobData	ProteinResidueConformation	AtomInResidue–AtomInResiduePair	RadiusOfGyration
PoseComments	ProteinBackboneTorsionAngle		SecondaryStructure
	ResidueBurial	ProteinBackbone–AtomAtomPair	HydrophbicPatch
Experimental Data	ResidueSecondaryStructure		Cavity
PdbData	GeometricSolvation	HBond	GraphMotif
PdbHeaderData	BetaTurn	Orbital	SequenceMotif
DDG	RotamerBoltzmannWeight	SaltBridge	Rigidity
NMR	ResidueStrideSecondaryStructure	LoopAnchor	VoronoiPacking
DensityMap	HelixCapping	DFIREPair	InterfaceAnalysis
MultiSequenceAlignment	BondGeometry	ChargeCharge	
HomologyAlignment	ResidueLazaridisKarplusSolvation		**Energy Function**
	ResidueGeneralizedBornSolvation	**Multistructure**	ScoreFunction
Chemical	ResiduePoissonBoltzmannSolvation	ProteinRMSD	ScoreType
AtomType	Pka	ResidueRecovery	StructureScores
ResidueType	ResidueCentroids	ResiduePairRecovery	ResidueScores
		ResidueClusterRecovery	HBondParameters
		Cluster	⟨EnergyTerm⟩ Parameters

The `FeatureReporters` that are currently implemented are in black while some `FeatureReporters` that would be interesting to implement in the future are in gray.

3.2. Feature analysis workflow

Feature analysis has three common stages: sample generation, feature extraction, and distribution comparison. Feature analysis can be used to optimize the energy function parameters by iteratively modifying the energy function and comparing feature distributions from structures generated with the new energy function against those from crystal structures (Fig. 6.2).

For a demonstration, we consider the hydrogen-acceptor distance of H-bonds with nonaromatic ring hydroxyl donors (i.e., serine or theonine). In X-ray crystal structures of proteins, the most common distance for this type of H-bond is 1.65 Å, which is ~0.2 Å shorter than that for H-bonds

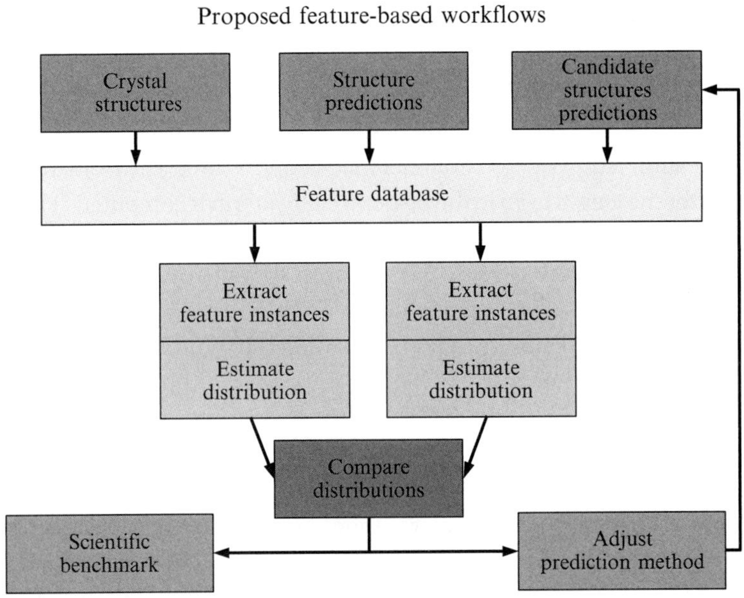

Figure 6.2 Example usage workflow for the feature analysis tool. Layer 1: Each *batch* consists of a set of molecular conformations, for example, experimentally determined or predicted conformations. Layer 2: Features from each batch are extracted into a relational database. Layer 3: Conditional feature distributions are estimated from feature instances queried from the database. Layer 4: The distributions are compared graphically. Layer 5: The comparison results are used as scientific benchmarks or to inform modifications to the structure prediction protocol and energy function, where the cycle can begin again. (For the color version of this figure, the reader is referred to the online version of this chapter.)

involving amide donors. Rosetta does not correctly recapitulate the distance distribution for hydroxyl donors, because previously to avoid the challenge of inferring the hydroxyl hydrogen locations, the hydroxyl donor parameters were taken from the sidechain amide and carboxylamide donor parameters (Kortemme et al., 2003).

To start, we compared structures generated from Rosetta's existing energy function against native structures to verify that Rosetta does not generate the correct distribution of hydroxyl H-bond distances. We used as our reference source a subset of the `top8000` chains dataset (Keedy et al., 2012, Richardson, Keedy, & Richardson, 2013) with a maximum sequence identity of 70%, which gave 6563 chains. Hydrogen atom coordinates were optimized using Reduce (Word, Lovell, Richardson, & Richardson, 1999). We further restricted our investigation to residues with B-factors less than 30. Then, for each candidate energy function, we relaxed each protein chain with the `FastRelax` protocol in Rosetta (Khatib et al., 2011).

The analysis script has three parts: First, for each batch, the lengths of all H-bonds with serine or theonine donors to protein-backbone acceptors are extracted with an SQL query. Second, for the instances associated with each batch, a one-dimensional KDE is constructed normalizing for equal volume per unit length. Also, for each nonreference sample source, the Boltzmann distribution for the length term in the H-bond energy term is computed. Third, the density distributions are plotted. This script is available in the online appendix.

With this distribution analysis script in place, we evaluated two incremental modifications to the standard *Score12* H-bond term. The first, `NewHB`, adjusts parameters for the length term for these H-bond interactions so that the optimal length is consistent with the peak in the native distribution (panel A in Fig. 6.3). This shifts the distribution of predicted H-bonds toward shorter interactions. However, the predicted distribution does not move far enough to recapitulate the observed distribution (panels B to C in Fig. 6.3). In *Score12*, the Lennard–Jones energy term between the donor and acceptor oxygen atoms is optimal at 3 Å. However, the peak in the O—O distance distribution is at 2.6 Å (see the online appendix). Thus H-bonds with the most favorable distance according to the `NewHB` parametrization experience strong repulsion from the Lennard–Jones term. To reduce correlation between the H-bond and the Lennard–Jones energy terms, we decreased the optimal distance for the Lennard–Jones term for these specific interactions to 2.6 Å. With this second modification, Rosetta recapitulates the native distribution (panel D in Fig. 6.3).

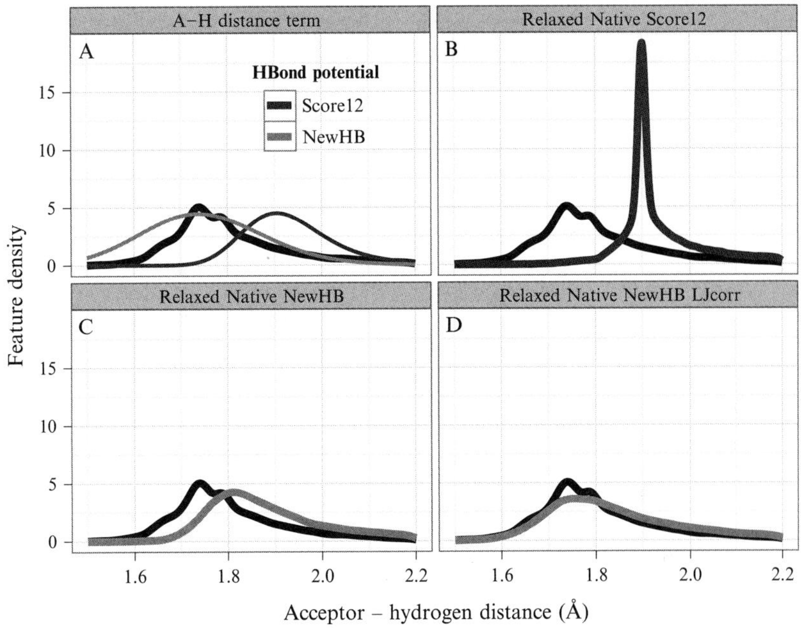

Figure 6.3 H-bond length distributions for hydroxyl donors (SER/THR) to backbone oyxgens. The thick curves are kernel density estimations from observed data normalized for equal volume per unit distance. The black curve in the background of each panel represents the Native sample source. (A) Boltzmann distribution for the length term in the Rosetta H-bond model with the *Score12* and NewHB parameterizations. (B) Relaxed Natives with the *Score12* energy function. The excessive peakiness is due to a discontinuity in the *Score12* parametrization of the H-bond model. (C) Relaxed Natives with the NewHB energy function. (D) Relaxed Natives with the NewHB energy function and the Lennard–Jones minima between the acceptor and hydroxyl heavy atoms adjusted from 3.0 to 2.6 Å, and between the acceptor and the hydrogen atoms adjusted from 1.95 to 1.75 Å.

4. MAXIMUM LIKELIHOOD PARAMETER ESTIMATION WITH optE

Recall that the Rosetta energy function is a weighted linear combination of energy terms that capture different aspects of molecular structure, as defined in Eq. (6.1). The weights, w, balance the contribution of each term to give the overall energy. Because the weights often need adjusting after modifying an energy term, we have developed a tool called "optE" to facilitate fitting them against scientific benchmarks. The benchmarks are small, tractable tests of Rosetta's ability to recapitulate experimental observations given a particular assignment of weights. Although the weight sets that optE generates have not proven to be good replacements for the existing weights in Rosetta, we have found optE useful at two tasks: identifying problems in the Rosetta energy function and fitting the 20 amino-acid-reference-energy weights.

In the next section, we give a formula for generic likelihood-based loss functions and then describe the scientific benchmarks that are available in optE.

4.1. Loss function models

To jointly optimize an energy function's performance at the scientific benchmarks, we require success at each benchmark to be reported as a single number, which is called the loss. For scientific benchmarks based on recapitulating experimental observations, a common method of defining the loss is, given the weights, the negative log probability of predicting the observed data. If the observed data are assumed to be independently sampled, the loss is the sum of the negative log-probability over all observations. Thinking of the loss as a function of the weights for a fixed set of observations, it is called the negative log-likelihood of the weights. An ideal prediction protocol will generate predictions according to the Boltzmann distribution for the energy function.[1] Therefore, the probability of an observation o is

$$p(o|w) = \mathrm{e}^{-E(o|w)/kT} / Z(w) \qquad [6.2]$$

$$Z(w) = \sum_{a \in \{A \cup o\}} \mathrm{e}^{-E(a|w)/kT} \qquad [6.3]$$

where the partition function, $Z(w)$, includes o and all possible alternatives, A. Because of the vast size of conformation space, computing Z is often intractable; this is a common problem for energy-based loss functions (LeCun & Jie, 2005). To address this problem, we rely on loss functions that do not consider all possible alternatives.

4.1.1 Recovering native sequences

Within the Rosetta community, we have observed that improvements to the energy function, independently conceived to fix a particular aspect of Rosetta's behavior, have produced improvements in sequence recovery when redesigning naturally occurring proteins (Kuhlman & Baker, 2000;

Morozov & Kortemme, 2005). Therefore, we attempt to increase sequence recovery to improve the energy function.

The standard sequence-recovery benchmark (described in Section 5.2) looks at the fraction of the amino acids recovered after performing complete protein redesign using Rosetta's Monte Carlo optimization technique. To turn this benchmark into a loss function like Eq. (6.2) would require us to compute the exact solution to the NP-Complete sidechain optimization problem (Pierce & Winfree, 2002) for the numerator, and, for an N residue protein, to repeat that computation 20^N times for each possible sequence for the denominator. This is not feasible.

Instead, we approached the benchmark as a one-at-a-time optimization, maximizing the probability of the native amino acid at a single position given some fixed context. This introduces a split between the energy function whose weights are being modified to fit the native amino acid into a particular environment and the energy function which is holding that environment together. Upweighting one term may help distinguish the native amino acid from the others, but it might also cause the rest of the environment to relax into some alternate conformation in which the native amino acid is no longer optimal. To build consistency between the two energy functions, we developed an iterative protocol (additional details given in Section 4.2.1) that oscillates between loss-function optimization and full-protein redesign. Briefly, the protocol consists of a pair of nested loops. In the outer loop, the loss function is optimized to produce a set of candidate weights. In the inner loop, the candidate weights are mixed in various proportions with the weights from the previous iteration through the outer loop, and for each set of mixed weights, complete protein redesign is performed. The redesigned structures from the last iteration through the inner loop are then used to define new loss functions for the next iteration through the outer loop.

We define the loss function for a single residue by the log-likelihood of the native amino acid defined by a Boltzmann distribution of possible amino acids at that position. We call this the p_{NatAA} loss function.

$$p_{\text{NatAA}}(w) = \frac{e^{-E(\text{nat}|w)/kT}}{\sum_{\text{aa}} e^{-E(\text{aa}|w)/kT}} \quad [6.4]$$

$$L_{p_{\text{NatRot}}}(w) = -\ln p_{\text{NatAA}}(w) \quad [6.5]$$

where $E(\text{nat} \mid w)$ is the energy of the best rotamer for the native amino acid and $E(\text{aa} \mid w)$ is the energy of the best rotamer for amino acid aa. Rotamers are sampled from Roland Dunbrack's backbone-dependent rotamer library from 2002 (Dunbrack, 2002), with extra samples taken at $\pm\sigma$ for both χ_1 and χ_2. The energies for the rotamers at a particular position are computed in the context of a fixed surrounding. The contexts for the outer-loop iteration i are the designed structures from iteration $i-1$; the first round's context comes from the initial crystal structures.

4.1.2 Recovering native rotamers

As is the case for the full-fledged sequence-recovery benchmark, the rotamer-recovery benchmark (described in Section 5.1) would be intractably expressed as a generic log-likelihood loss function as given in Eq. (6.2). Instead, we again approach the recovery test as a one-at-a-time benchmark to maximize the probability of the native rotamer at a particular position when considering all other rotamer assignments.

For residue j, the probability of the native rotamer is given by

$$p_{\mathrm{NatRot}}(w) = \frac{\mathrm{e}^{-E(\mathrm{nat}|w)/kT}}{\sum_{i \in \mathrm{rots}} \mathrm{e}^{-E(i|w)/kT}} \qquad [6.6]$$

where $E(\mathrm{nat} \mid w)$ is the energy of the native rotamer, and the set rots contains all other rotamers built at residue j. We define a loss function, Lp_{NatRot}, for residue j as the negative log of this probability.

4.1.3 Decoy discrimination

Benchmarking the ability of Rosetta to correctly predict protein structures from their sequences is an incredibly expensive task. In the high-resolution refinement benchmark (described in Section 5.4), nonnative structures, *decoys*, are generated using the energy function being tested so that each decoy and each near-native structure will lie at a local minimum, but this takes ~20 K CPU hours. Within optE, we instead test the ability of Rosetta to discriminate near-native structures from decoys looking only at static structures; as optE changes the weights, the property that each structure lies at a local minimum is lost.

Given a set N of relaxed, near-native structures for a protein, and a set D of relaxed decoy structures, we approximate the probability of the native structure for that protein as

$$p_{\mathrm{NatStruct}}(w) = \frac{1}{\sum_j n_j} \sum_{j \in N} \mathrm{e}^{-\sigma n_j E_j(w)/kT} / Z(w) \qquad [6.7]$$

$$Z(w) = \sum_{j \in D} e^{-\left(\sigma E_j(w)/kT\right) - d_j\left(R \ln \alpha - \sum_k d_k\right)} - \sum_{j \in N} e^{-\sigma E_j(w)/kT} \quad [6.8]$$

and similarly define a decoy-discrimination loss function, $L_{\text{pNatStruct}}$, as the negative log of this probability. Here, n_j is the "nativeness" of conformation j, which is 1 if j is below 1.5 Å Cα Root Mean Squared Deviation (RMSD) from the crystal structure and 0 if it is above 2 Å RMSD. The nativeness decreases linearly from 1 to 0 in the range between 1.5 and 2. Similarly, d_j is the "decoyness" of conformation j, which is 0 if j is below 4 Å RMSD and 1 otherwise. σ is a dynamic-range-normalization factor that prevents the widening of an energy gap between the natives and the decoys by scaling all of the weights; it is defined as $\sigma = \sigma(D,w)/\sigma_0$ where $\sigma(D,w)$ is the computed standard deviation for the decoy energies for a particular assignment of weights and σ_0 is the standard deviation of the decoy energies measured at the start of the simulation.

In the partition function, the $R \ln \alpha - \sum_k d_k$ term approximates the entropy of the decoys, an aspect that is otherwise neglected in a partition function that does not include all possible decoy conformations. This term attempts to add shadow decoys to the partition function, and the number of extra decoys added scales exponentially with the length of the chain, R. We chose 1.5 as the scale factor, α, which is relatively small given the number of degrees of freedom (DOFs) each residue adds. Counting torsions alone, there are between three (glycine) and seven (arginine) extra DOFs per residue. Our choice of a small α is meant to reflect the rarity of low-energy conformations. To normalize between runs which contain differing numbers of far-from-native decoys, we added the $-\sum_k d_k$ term; doubling the number of decoys between two runs should not cause the partition function to double in value.

4.1.4 ΔΔG of mutation

The full benchmark for predicting ΔΔGs of mutation (described in Section 5.3) is computationally expensive, and so, similar to the decoy-discrimination loss function, we define a ΔΔG loss function which relies on static structures. This loss function is given by:

$$L_{\Delta\Delta G}(w) = \left(\Delta\Delta G_{\text{exp}} - \left(\min_{\text{mut}\in\text{muts}} E(\text{mut}|w) - \min_{\text{wt}\in\text{wts}} E(\text{wt}|w)\right)\right)^2 \quad [6.9]$$

where muts is a set of structures for the mutant sequence, wts is a set of structures for the wild-type sequence, and the experimentally observed ΔΔG is defined such that it is positive if the mutation is destabilizing and negative if it is stabilizing. Note that this loss function is convex if there is only one structure each in the muts and wts sets. This is very similar to the linear least-squares fitting, except that it is limited to a slope of one and a y-intercept of zero. The slope can be fit by introducing a scaling parameter to weight optimization, as described in Section 4.2.2 below.

4.2. Loss function optimization

OptE uses a combination of a particle swarm minimizer (Chen, Liu, Huang, Hwang, & Ho, 2007) and gradient-based minimization to optimize the loss function. Nonconvexity of the loss function prevents perfect optimization, and independent runs sometimes result in divergent weight sets that have similar loss function values. In spite of this problem, optE tends to converge on very similar weight sets (for an example, see the online appendix).

4.2.1 Iterative protocol

As described above, training with the p_{NatAA} loss function effectively splits the energy function into two which we attempt to merge with an iterative procedure where we oscillate between loss-function optimization in an outer loop and complete protein redesign in an inner loop. Between rounds of the outer loop, the weights fluctuate significantly, and so, in each iteration of the inner loop, we create a weight set that is a linear combination of the weights resulting from round i's loss-function optimization and the weight set selected at the end of round $i-1$. The weight set used for design during outer-loop iteration i, inner-loop iteration j is given by

$$w(i,j) = \alpha w_l(i) + (1-\alpha)w(i-1) \qquad [6.10]$$

$$\alpha = \frac{1}{i+j} \qquad [6.11]$$

where $w_l(i)$ is the weight set generated by minimizing the loss function in round i, and $w(i-1)$ is the final weight set from round $i-1$. In the first round, $w(6.1)$ is simply assigned $w_l(1)$. This inner loop is exited if the sequence-recovery rate improves over the previous round or if six iterations through this loop are completed. The weight set $w(i,j)$ for the last iteration through the inner loop taken as the weight set $w(i)$ for round i and is written to disk. The set of designed structures from this iteration are taken to serve as the context for the p_{NatAA} loss function in the next round.

4.2.2 Extra capabilities

OptE provides extra capabilities useful for exploring weight space. For instance, it is possible to weigh the contributions of the various loss functions differently. Typically, we upweight the decoy-discrimination loss function by 100 when optimizing it along with the p_{NatAA} and p_{NatRot} loss functions. We typically train with ~100 native-set/decoy-set pairs, compared to several thousand residues from which we can define p_{NatAA} and p_{NatRot} loss functions. Upweighting the $p_{NatStruct}$ loss function prevents it from being drowned out.

OptE also offers the ability to fit two or more terms with the same weight, or to obey an arithmetic relationship as specified in an input text file (see the online appendix). This feature was used to scale the *Score12* terms by a linear factor but otherwise keep them fixed to optimize the $\Delta\Delta G$ of the mutation loss function (Kellogg et al., 2011). Finally, OptE allows the definition of restraints for the weights themselves to help hold them to values the user finds reasonable. This is often useful because the loss functions often prefer negative weights.

4.3. Energy function deficiencies uncovered by OptE

OptE is particularly good at two tasks: refitting reference energies (described in Section 4.5) and uncovering areas where the existing Rosetta energy function falls short. Rosetta's efficiency in searching through conformation space means it often finds decoys with lower energies than the native. Such decoys surely point to flaws in Rosetta's energy function, however, it is not always easy to see why the natives are not at lower energy than these decoys. OptE allows efficient hypothesis-driven testing: hypothesize what kind of term is absent from the Rosetta energy function, implement that term, and test that term in optE using the $p_{NatStruct}$ loss function. If the value of the loss function improves after the new term is added, that is strong evidence the term would improve the energy function. There are caveats: because the $p_{NatStruct}$ loss function relies on Rosetta-relaxed native structures, some of the features present in the crystal structures might have already been erased before optE gets started, and optE might fail to identify terms that would improve the energy function.

Using optE, we found two terms that improved the decoy-discrimination loss function over *Score12*. The first was a carbon-H-bond potential added to Rosetta (but not included as part of *Score12*) to model RNA (Das, Karanicolas, & Baker, 2010). This potential was derived from a set of protein crystal structures and a set of decoy protein structures as the log of the difference in the probability of an oxygen being observed at a particular distance in crystal structures versus Rosetta-generated structures. OptE identified this term as strongly improving decoy discrimination. We followed this lead by splitting the potential into backbone/backbone, backbone/sidechain, and sidechain/sidechain contributions. Here, optE preferred to set the weight on the sidechain/sidechain and backbone/sidechain components to zero while keeping the weight on the backbone/backbone interactions high. This left exactly one interaction: the Hα hydrogen interacting with the carbonyl oxygen. This contact is observed principally in β-sheets and has been reported previously as giving evidence for a carbon H-bond (Fabiola, Krishnaswarmy, Nagarajan, & Pattabhi, 1997; Taylor & Kennard, 1982).

The original CH-bond potential proved to be a poor addition to the energy function: when used to generate new structures or to relax natives, previously observed deep minima at low RMSD (<1.5 Å) from the crystal structure were lost and were instead replaced by broad, flat, near-native minima that reached out as far as 4 Å. To improve the potential, Song et al. (2011) iteratively adjusted the parameters to minimize the difference between the observed and predicted structures. This iterative process resulted in better Hα—O distance distributions; unexpectedly, it also resulted in better distance distributions for other atom-pairs in β-sheets.

A second term identified by optE was a simple Coulombic electrostatic potential with a distance-dependent dielectric (Yanover & Bradley, 2011). After an initial signal from optE suggested the importance of this term, we again separated the term into backbone/backbone, backbone/sidechain, and sidechain/sidechain components. Again, the backbone/backbone portion showed the strongest signal. From there, we separated each term into attractive and repulsive components, and optE suggested that the repulsive backbone/backbone interaction contributed the most toward improved decoy discrimination. OptE also identified which low-energy decoys in the training set were problematic in the absence of the electrostatic term. By hand, we determined that the loops in these decoys contained extremely close contacts between backbone carbonyl oxygens: they were only 3.4 Å apart, which is closer than is commonly observed in crystal structures.

4.4. Limitations

Though our original goal was to fit all the weights simultaneously, optE has not proven exceptionally useful at that task. There are a number of factors that contribute to optE's failures here. For one, the loss function that we use is not convex, and therefore its optimization cannot be guaranteed. This sometimes leads to a divergence of the weight sets in independent trajectories, which makes interpreting the results somewhat tricky.

The biggest problem facing optE, however, is its inability to see all of the conformation space. This is true for all of the loss functions we try to optimize, but the $p_{NatStruct}$ loss function, in particular, can see only the decoys we give it, and typically we cannot computationally afford to give optE more than a few hundred decoys per protein. Thus, a weight set optE generates may reflect artifacts of our decoy selection. For example, optE typically tries to assign a negative weight to the Ramachandran term (described briefly in Section 6.2), suggesting that this term is overoptimized in our decoys compared to native structures; optE finds that the easiest way to discriminate natives from decoys is turning the weight negative. In general, the weights that optE produces are not good for protein modeling (see the online appendix for benchmark results for an optE-generated weight set). However, using the protocol described in the next section, optE is fantastic at refitting reference energies.

4.5. A sequence-profile recovery protocol for fitting reference energies

Dissatisfyingly, the p_{NatAA} loss function produced weight sets that overdesigned the common amino acids (e.g., leucine) and underdesigned the rare amino acids (e.g., tryptophan). To address this shortcoming, we created an alternative protocol within optE for fitting only the amino acid reference energies, keeping all other weights fixed. This protocol does not use any of the loss functions described above; instead, it adjusts the reference energies directly based on the results of complete protein redesign.

The protocol iteratively performs complete protein redesign on a set of input protein structures and adjusts the reference energies upwards for amino acids it over designs and downwards for those it under designs, where the target frequencies are taken from the input set. After each round, optE computes both the sequence-recovery rate and the Kullback–Leibler (KL) divergence of the designed sequence profile against the observed sequence profile, given by $-\sum_{aa} \ln(p_{aa}/q_{aa})$, where p_{aa} is the naturally occurring frequency of amino acid aa in the test set and q_{aa} is the frequency of amino acid aa in the redesigned structures. The final reference energies chosen are those that maximize the sum -0.1 KL-divergence + seq. rec. rate. This protocol is used to refit reference energies after each of the three energy function changes described in Section 6.

The training set we use, which we call the "HiQ54," is a new collection of 54 nonredundant, monomeric proteins from the PDB through 2010 that have 60–200 residues (avg. = 134) and no tightly bound or large ligands. All were required to have both resolution and MolProbity score (Chen et al., 2010) at or below 1.4, very few bond length or angle outliers, and deposited structure-factor data. The HiQ54 set is available in the online appendix and at http://kinemage.biochem.duke.edu/.

5. LARGE-SCALE BENCHMARKS

Scientific benchmarking allows energy function comparison. The tests most pertinent to the Rosetta community often aim toward recapitulating observations from crystal structures. In this section, we describe a curated set of previously published benchmarks, which together provide a comprehensive view of an energy function's strengths and weaknesses. We continually test the benchmarks on the RosettaTests server to allow us to immediately detect changes to Rosetta that degrades its overall performance (Lyskov & Gray, 2012).

5.1. Rotamer recovery

One of the most direct tests for an energy function is its ability to correctly identify the observed rotamers in a crystal structure against all other possible rotamers while keeping the backbone fixed. Variants of the rotamer recovery test have long been used to evaluate molecular structure energy functions (Liang & Grishin, 2002; Petrella, Lazaridis, & Karplus, 1998), including extensive use to evaluate the Rosetta energy function (Dantas et al., 2007; Dobson, Dantas, Baker, & Varani, 2006; Jacak et al., 2012; Kortemme et al., 2003), the ORBIT energy function (Sharabi, Yanover, Dekel, & Shifman, 2010), and the SCWRL energy function (Shapovalov & Dunbrack, 2011).

Here, we test rotamer recovery in four tests combining two bifurcated approaches to the task: discrete versus continuous rotamer sampling and one-at-a-time versus full-protein rotamer optimization. The discrete, one-at-a-time rotamer optimization protocol is called *rotamer trials*. It builds rotamers, calculates their energies in the native context, and compares the lowest-energy rotamer against the observed crystal rotamer. The continuous, one-at-a-time rotamer optimization protocol is called *rt-min* (Wang, Schueler-Furman, & Baker, 2005). It similarly builds rotamers in the native context but minimizes each rotamer before comparing the lowest-energy rotamer against the crystal rotamer. The discrete, full-protein optimization protocol is called *pack rotamers*. This builds rotamers for all positions and then seeks to find the lowest-energy assignment of rotamers to the structure using a Monte Carlo with simulated annealing protocol (Kuhlman & Baker, 2000) where at a random position, a random rotamer is substituted into the current environment, its energy is calculated, and the substitution is accepted or rejected based on the Boltzmann criterion. The rotamers in the final assignment are compared against the crystal rotamers. The continuous, full-protein rotamer optimization task is called *min pack*. It is similar to the pack rotamers protocol except that each rotamer is minimized before the Boltzmann decision. A similar protocol has been described before (Ding & Dokholyan, 2006).

Recovery rates are measured on a set of 152 structures, each having between 50 and 200 residues and a resolution less than 1.2 Å. Rotamers are considered recovered if all their χ dihedrals are less than 20° from the crystal χ dihedrals, taking into account symmetry for the terminal dihedrals in PHE, TYR, ASP, and GLU. For the discrete optimization tests, rotamers are built at the center of the rotamer wells, with extra samples included at $\pm\sigma_i$ from $\bar{\chi}_i$ for χ_1 and χ_2. For the continuous optimization test, samples are only taken at $\bar{\chi}_i$ but can move away from the starting conformation through minimization.

5.2. Sequence recovery

In the sequence-recovery benchmark, we perform complete-protein fixed-backbone redesigns on a set of crystal structures, looking to recapitulate the native amino acid at each position. For this chapter, we used the test set of 38 large proteins from Ding and Dokholyan (2006). Sequence recovery was performed with the discrete, full-protein rotamer-and-sequence optimization protocol called `PackRotamers`, described above. Rotamer samples were taken from the given rotamer library (either the 2002 or the 2010 library), and extra samples were chosen at $\pm\sigma_i$ for χ_1 and χ_2. The multi-cool annealer-simulated annealing protocol (Leaver-Fay, Jacak, Stranges, & Kuhlman, 2011) was employed instead of Rosetta's standard simulated annealing protocol. We measured the sequence-recovery rate and the KL-divergence of the designed amino acid profile from the native amino acid profile: the sequence-recovery rate should be high, and the KL-divergence should be low.

5.3. $\Delta\Delta G$ prediction

The $\Delta\Delta G$ benchmark consists of running the high-resolution protocol described in Kellogg et al. (2011), on a curated set of 1210 point mutations for which crystal structures of the wild-type protein are available. The protocol predicts a change in stability induced by the mutation by comparing the Rosetta energy for the wild-type and mutant sequences after applying the same relaxation protocol to each of them. The Pearson correlation coefficient of the measured versus predicted $\Delta\Delta G$s is used to assess Rosetta's performance.

5.4. High-resolution protein refinement

Rosetta's protocol to predict protein structures from their sequence alone runs in two phases: a low-resolution phase (*ab initio*) relying on fragment insertion (Simons, Kooperberg, Huang, & Baker, 1997) to search through a relatively smooth energy landscape (Simons et al., 1999), and a high-resolution phase (*relax*) employing sidechain optimization and gradient-based minimization. This *abrelax* protocol offers the greatest ability to broadly sample conformation space in an attempt to find nonnative conformations that Rosetta prefers to near-native conformations.

Unfortunately, the *abrelax* protocol requires significantly more sampling than could be readily performed to benchmark a change to the energy function. The problem is that most low-resolution structures produced by the first *ab initio* phase do not yield low-energy structures in the second *relax* phase, so finding low-energy decoy conformations requires hundreds of thousands of trajectories. To reduce the required amount of sampling, Tyka et al. (2010) curated four sets of low-resolution decoys for 114 proteins by taking the low-resolution structures that generated low-energy structures after being put through high-resolution refinement. The benchmark is then to perform high-resolution refinement on these low-resolution structures with the new energy function. Each of the low-resolution sets was generated from different sets of fragments; some sets included fragments from homologs, while others included fragments from the crystal structure of the protein itself. This gave a spectrum of structures at varying distances from native structures.

These sets are used as input to the relax protocol. The decoy energies and their RMSDs from the crystal structure are used to assess the ability of Rosetta to discriminate natives from low-energy decoys. This is reported in two metrics by the benchmark: the number of proteins for which the probability of the native structure given by the Boltzmann distribution exceeds 80% ($\text{pNat} = \exp(E(\text{nat}))/(\exp(E(\text{nat})) + \Sigma\exp(E(d)))$ with $E(\text{nat})$ representing the best energy of any structure under 2 Å RMSD from the crystal structure, and *d* representing any structure greater than 2 Å RMSD. pNat should not be confused with the $p_{NatStruct}$ loss function in optE, and the number of proteins for which the lowest-energy near-native conformation has a lower energy than the lowest-energy decoy conformation. For the benchmarks reported here, we relaxed 6000 decoys randomly sampled with replacement from each of the four sets, resulting in 24,000 decoys per protein. The statistical significance of the difference between two runs can be assessed with a paired *t*-test, comparing pNat for each of the 114 targets.

The proteins included in this test set include many where the crystal structure of the native includes artifacts (e.g., loop rearrangements to form crystal contacts), where the protein coordinates metal ions or ligands, or where the protein forms an obligate multimer in solution. For this reason, perfect performance at this benchmark is not expected, and interpreting the results is somewhat complicated; an improvement in the benchmark is easier to interpret than a degradation.

5.5. Loop prediction

The loop-prediction benchmark aims to test Rosetta's accuracy at *de novo* protein loop reconstruction. For this, we used the kinematic closure (KIC) protocol, which samples mechanically accessible conformations of a given loop by analytically determining six dihedral angles while sampling the remaining loop torsions probabilistically from Ramachandran space (Mandell, Coutsias, & Kortemme, 2009).

We used the benchmark set of 45 12-residue loops as described in Mandell et al. (2009). For each loop, we generated 8000 structures and calculated their Cα loop RMSD to the native.

In some cases, KIC generates multiple clusters of low-energy conformations of which one cluster is close to the native structure, while the other(s) can be several Ångstroms away. Because the Rosetta energy function does not robustly distinguish between these multiple clusters, we considered not just a single structure, but the five lowest-energy structures produced. Of these five, we used the lowest-RMSD structure when calculating benchmark performance. Overall loop reconstruction accuracy is taken as the median Cα loop RMSD of all best structures across the entire 45-loop dataset. The first and third quartile, though of lesser importance than the median, should be examined as well, as they offer a picture of the rest of the distribution.

6. THREE PROPOSED CHANGES TO THE ROSETTA ENERGY FUNCTION

In this final section, we describe three changes to Rosetta's energy function. After describing each change and its rationale, we present the results of the benchmarks described above.

6.1. Score12′

We used the sequence-profile-recovery protocol in optE to fit the reference energies for *Score12* to generate a new energy function that we called *Score12′*. Refitting the *Score12* reference energies in Rosetta3 (Leaver-Fay, Tyka, et al., 2011) was necessary because the components of *Score12* in Rosetta3 differed in several ways from those in Rosetta2. First, in Rosetta2, there was a disagreement between the energy function used during sequence optimization and the one used throughout the rest of Rosetta. In the sequence optimization module ("the packer"), the knowledge-based Ramachandran potential was disabled, and the weight on the $P(\mathrm{aa}\,|\,\varphi, \psi)$ was doubled from 0.32 to 0.64. In Rosetta3, there is no schism between the energy functions used in the packer and elsewhere. Second, the Lennard–Jones and implicit solvation (Lazaridis & Karplus, 1999) terms were extended from 5.5 to 6 Å and spline-smoothed to be zero and flat at 6 Å. Previously, they ramped linearly from 5 to 5.5 Å, causing discontinuities in the derivatives, which—when combined with gradient-based minimization—created peaks in atom-pair radial distributions (W. Sheffler & D. Baker, unpublished observations). Also, the Lennard–Jones potential now starts counting the contribution of the repulsive component at the bottom of the well, instead of at the x-intercept. This change eliminates a derivative discontinuity at the x-intercept that forms if the attractive and repulsive weights differ (as they do in *Score12*). Third, in Rosetta2, interactions were inappropriately omitted because for certain types of residue pairs, the Cβ–Cβ interaction threshold was too short.

Table 6.2 gives the sequence-recovery rates for Rosetta2 and Rosetta3 using *Score12* and using reference weights trained with the p_{NatAA} loss function (Rosetta3-p_{NatAA}) and with the sequence-profile recovery protocol (Rosetta3-sc12′). Though the p_{NatAA} objective function achieves satisfactory sequence-recovery rates, it overdesigns leucine and lysine and never designs tryptophan. *Score12′*, on the other hand, does an excellent job recapitulating the sequence profile in the testing set while also outperforming the Rosetta3-

Table 6.2 Sequence recovery rates (% Rec.), KL-divergence of the designed sequence profiles from the native sequence profile (KL-div.), and amino acid profiles measured on the Ding & Dokholyan-38 set

	% Rec.	KL-div.	A	C	D	E	F	G	H	I	K	L	M	N	P	Q	G	S	T	V	W	Y
Test set	–	–	8.9	1.2	6.6	6.7	4.3	8.3	2.2	5.2	6.3	7.9	2.4	4.5	4.5	3.4	4.7	5.2	5.7	6.8	1.5	3.6
Rosetta2-sc12	38.0	0.019	6.3	1.2	6.2	6.3	5.5	7.6	1.9	6.3	5.2	10.3	1.9	4.1	6.0	4.2	5.3	5.2	5.0	5.6	2.0	3.9
Rosetta3-sc12	32.6	0.141	6.4	0.0	7.7	8.7	5.4	7.4	6.3	4.7	6.8	8.8	1.7	3.2	1.3	3.5	6.3	7.2	4.2	4.2	2.5	4.0
Rosetta3-p_{NatAA}	36.7	0.391	9.8	0.0	6.4	6.2	0.7	8.3	0.8	5.7	11.8	13.5	0.2	2.7	7.6	0.2	3.1	4.7	6.3	10.6	0.0	1.5
Rosetta3-sc12′	37.0	0.008	8.4	0.7	6.6	7.5	3.9	8.2	2.1	5.0	6.4	8.0	2.2	3.5	3.7	3.3	6.1	6.4	5.8	6.4	1.8	4.1

The Rosetta3-sc12 energy function represents the *Score12* reference energies taken directly from Rosetta2. The Rosetta3-p_{NatAA} weight set keeps the same weights as *Score12*, except that the reference energies were fit by optimizing the p_{NatAA} loss function; the Rosetta3-sc12′ (*Score12′*) reference energies were generated using the sequence-profile optimization protocol.

p_{NatAA} energy function at sequence recovery. The rotamer-recovery, high-resolution refinement, and loop-prediction benchmarks were not run for this proposed change to the energy function as fixed-sequence tasks are unaffected by changes to the reference energies.

6.2. Interpolating knowledge-based potentials with bicubic splines

Three knowledge-based potentials in Rosetta are defined on the ϕ, ψ map: the Ramachandran term which gives $E_{\text{rama}}(\phi, \psi \mid \text{aa}) = -\ln p(\phi, \psi \mid \text{aa})$, the `p_aa_pp` term (sometimes called the design term) which gives an energy from the log of the probability of observing a particular amino acid at a given ϕ, ψ, and the rotamer term (almost always called the Dunbrack term, or `fa_dun`), which gives $E_{\text{dun}}(\chi \mid \phi, \psi, \text{aa})$. These terms use bins on the ϕ, ψ map to collect the data that define these potentials and use bilinear interpolation between the bins to define a continuous function.

The `p_aa_pp` term in *Score12* is given by

$$E_{\text{paapp}}(\text{aa}|\phi,\psi) = -\ln\frac{p(\text{aa}|\phi,\psi)}{p(\text{aa})}. \qquad [6.12]$$

For any particular ϕ and ψ, the energy is given as the negative log of the bilinearly interpolated $p(\text{aa} \mid \phi, \psi)$ divided by $p(\text{aa})$. The Ramachandran term similarly defines the energy for the off-grid-point ϕ and ψ values as the negative log of the bilinearly interpolated $p(\phi, \psi \mid \text{aa})$. Both the `p_aa_pp` and the Ramachandran term place their bin centers every 10° starting from 5°.

The Dunbrack term in *Score12* is given by

$$E_{\text{dun}}(\chi|\phi,\psi,\text{aa}) = -\ln(p(\text{rot}|\phi,\psi,\text{aa})) + \sum_i \left(\frac{\chi_i - \bar{\chi}_i(\phi,\psi|\text{aa,rot})}{\sigma_i(\phi,\psi|\text{aa,rot})}\right)^2 \qquad [6.13]$$

where the rotamer bin, rot, which gives the probability of the rotamer, $p(\text{rot} \mid \phi, \psi, \text{aa})$, is computed from the assigned χ dihedrals, and both $\bar{\chi}_i(\phi,\psi|\text{aa,rot})$ and $\sigma_i(\phi, \psi \mid \text{aa, rot})$ are the measured mean χ values and standard deviations for the rotamer. This effectively models the probability for the sidechain conformation as the product of the rotamer probability and several (height-unnormalized) Gaussians. The 2002 library gives the $p(\text{rot} \mid \phi, \psi, \text{aa})$, $\bar{\chi}_i(\phi,\psi|\text{aa,rot})$, and $\sigma_i(\phi, \psi \mid \text{aa, rot})$ every 10°. Given a particular assignment of ϕ and ψ, the values for $p(\text{rot} \mid \phi, \psi, \text{aa})$, $\bar{\chi}_i(\phi,\psi|\text{aa,rot})$, and $\sigma_i(\phi, \psi \mid \text{aa,}$

rot) are bilinearly interpolated from the four surrounding bin centers. The Dunbrack term divides the ϕ, ψ plane into 10° bins, starting from 0°.

Bilinear interpolation leaves derivative discontinuities every 5° in the ϕ, ψ plane. These discontinuities frustrate the minimizer causing pileups at the bin boundaries. Looking at the ϕ, ψ distribution for nonhelical residues, the grid boundaries are unmistakable (Fig. 6.4B). Indeed, *Score12* predicts 23% of all ϕ, ψ pairs of lying within 0.05° of a grid boundary.

We propose to fix this problem by using bicubic splines to interpolate between the grid points. We fit bicubic splines with periodic boundary conditions for both the Ramachandran and p_aa_pp terms on the energies, interpolating in energy space. For the Dunbrack energy, we fit bicubic splines for the $-\ln(p(\text{rot} \mid \phi, \psi, \text{aa}))$ portion, but, to avoid increasing our memory footprint too much, continued to use bilinear interpolation for the χ-mean and χ-standard deviations. We refit the reference energies using the

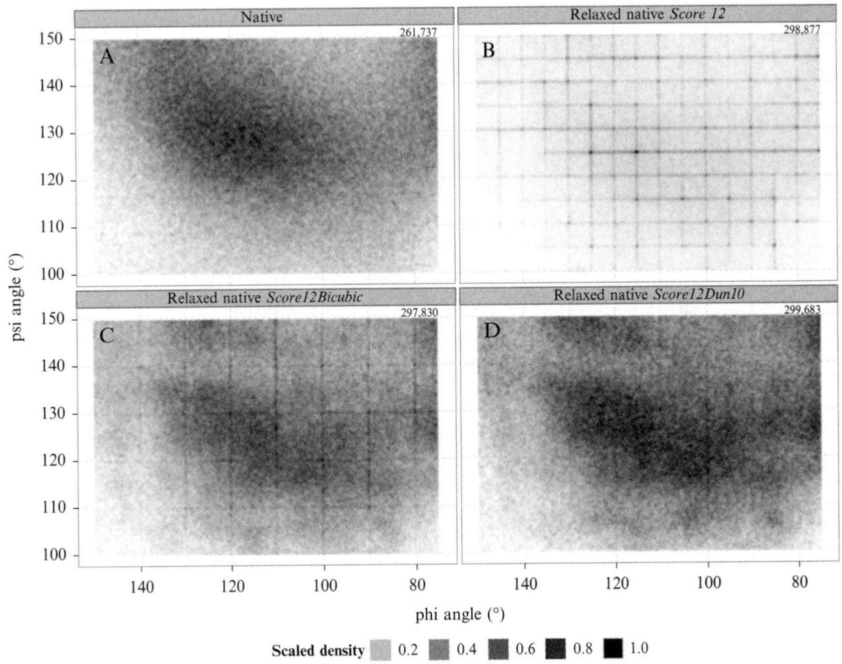

Figure 6.4 Backbone torsion angles in the beta-region with B-factors less than 30. (A) The distribution for the top 8000; counts in upper right. (B) In *Score12*, density accumulates on the 5° bins due to derivative discontinuities caused by bilinear interpolation. (C) *Score12Bicubic* has only a few remaining artifacts on the 10° bin boundaries due to the continued use of bilinear interpolation for parts of the Dunbrack energy. (D) Score12Dun10 has very few remaining artifacts.

sequence-profile recovery protocol to create a new energy function that, for this chapter, we refer to as *Score12Bicubic*; Fig. 6.4C shows that bicubic-spline interpolation dramatically reduces the pileups. Accumulation on the 10°-grid boundaries starting at 5° produced by the Ramachandran and `p_aa_pp` terms is completely gone. Modest accumulation on the 10°-grid boundaries starting at 0° persists because bicubic splines were not used to interpolate the χ-mean and χ-standard deviations.

6.3. Replacing the 2002 rotamer library with the extended 2010 rotamer library

In 2010, Shapovalov and Dunbrack (Shapovalov & Dunbrack, 2011) defined a new rotamer library that differs significantly from the 2002 library in the way the terminal χ is handled for eight amino acids: ASP, ASN, GLU, GLN, HIS, PHE, TYR, and TRP. Because these terminal χ dihedrals are about bonds between sp3-hybridized atoms and sp2-hybridized atoms, there are no well-defined staggered conformations. Instead of modeling the probability landscape for the terminal χ within a particular bin as a Guassian, the new library instead provides a continuous probability distribution over all the bins. This last χ is in effect nonrotameric, though the rest of the χ in the sidechain are still rotameric; these eight amino acids can be called *semi-rotameric*. In the 2002 library, there were discontinuities in both the energy function and the derivatives when crossing over these grid boundaries. Once a rotamer boundary is crossed, an entirely different set of $\bar{\chi}$ and σ_i are used to evaluate the rotamer energy. Further, Rosetta's treatment of the terminal χ probability distributions as Gaussians means that Rosetta structures display Gaussian distributions (the gray lines in Fig. 6.5) that do not resemble the native distributions (the black lines in Fig. 6.5).

With the 2010 library, the energy for a rotameric residue is computed in the same way as for the 2002 library, and the energy for one of the semi-rotameric amino acids is computed as

$$E_{\text{dun}}(\chi|\phi,\psi,\text{aa}) = -\ln(p(\text{rot}|\phi,\psi,\text{aa})p(\chi_T|\text{rot},\phi,\psi,\text{aa}))\sum_{i<T}\left(\frac{\chi_i-\bar{\chi}_i}{\sigma_i}\right)^2 \quad [6.14]$$

where T denotes the terminal χ, and where $\bar{\chi}_i(\phi,\psi|\text{aa,rot})$ and $\sigma_i(\phi,\psi\,|\,\text{aa, rot})$ from Eq. (6.13) have been abbreviated as $\bar{\chi}_i$ and σ_i, though they retain their dependence on the rotamer and amino acid and are a function of ϕ and ψ. The 2010 library provides data for $p(\chi_T|\,\text{rot},\phi,\psi,\text{aa})$ every 10° for ϕ and ψ, and

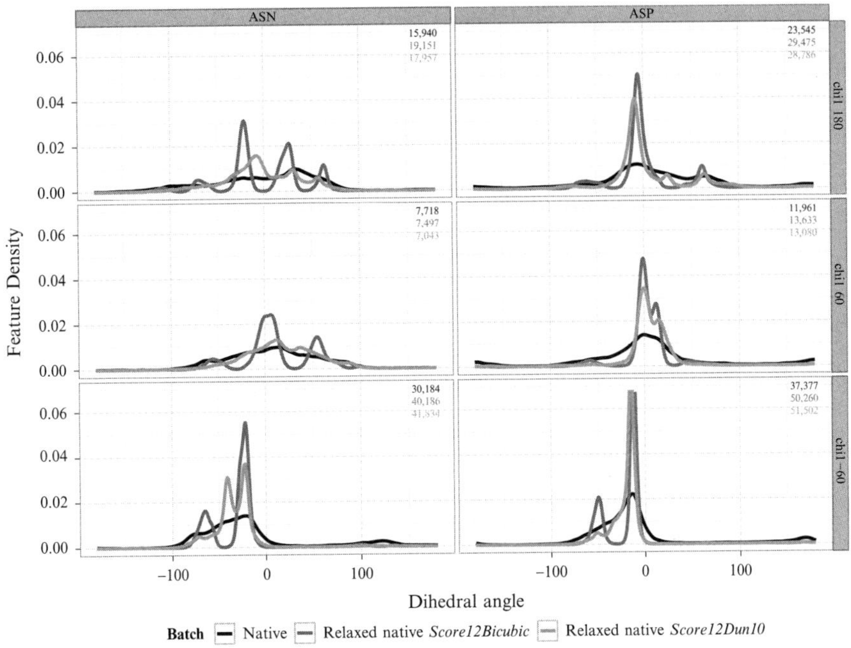

Figure 6.5 ASN/ASP χ_2 distribution by χ_1 bin. Comparison of χ_2 distributions for the semirotameric amino acids ASN and ASP, broken down by χ_1 rotamer; counts in upper right. Rosetta's implementation of the 2002 library produces Gaussian-like distributions for χ_2 in relaxed natives (gray), though the native distributions do not resemble Gaussians (black). Using the 2010 library (light gray), the distributions improve considerably though they remain too peaky in places.

every 5° or every 10° for χ_T depending on whether the amino acid, aa, is symmetric about χ_T. We fit tricubic splines to interpolate $-\ln(p(\text{rot}\,|\,\phi,\psi,\text{aa})\; p(\chi_T|\,\text{rot},\phi,\psi,\text{aa}))$ in ϕ,ψ and χ_T. As in the 2002 library, $\bar{\chi}_i$ and σ_i are interpolated bilinearly from the four surrounding grid points. We refit the reference energies using the sequence-profile recovery protocol to create a new energy function that, for this chapter, we refer to as *Score12Dun10*. *Score12Dun10* builds on top of *Score12Bicubic*.

6.4. Benchmark results

The results of the benchmarks, given in Table 6.3, show that *Score12′* is a clear improvement over *Score12*, substantially improving sequence recovery, and that *Score12Bicubic* is a clear improvement over *Score12′*, behaving as well as *Score12′* on most benchmarks and giving a slight, but statistically insignificant improvement at the high-resolution refinement benchmark ($p = 0.07$).

Score12Dun10 shows mixed results: at the rotamer-recovery benchmarks, it shows a clear improvement over *Score12Bicubic*; the improvement can be most clearly seen in the rotamer-recovery rates for the semirotameric amino acids (Table 6.4).

Score12Dun10 performed worse at the high-resolution refinement benchmark than *Score12Bicubic*. The two principal metrics for this benchmark are slightly worse: the number of proteins where the lowest energy near-native structure has a lower energy than the lowest energy decoy (#(eNat < eDec); 104 vs. 105), and the number of proteins where the probability of the native structure calculated by the Boltzmann distribution is greater than 80% (#(pNat > 0.8); 67 vs. 60). To estimate the significance of these results, we compared the distribution of ($pNat_{sc12Dun10} - pNat_{sc12Bicubic}$) for each of the 114 targets against the null hypothesis that this distribution had a mean of 0, using a two-tailed *t*-test. This gave a *p*-value for the difference of 0.01. Because this benchmark includes proteins whose accurate prediction is unlikely given protocol limitations (disulfides are not predicted), and crystal artifacts (loops which adopt conformations supported only by crystal contacts), its results are more difficult to interpret. We therefore restricted our focus to 29 proteins in the set which are absent of these issues (these are listed in the online appendix) and repeated the comparison of ($pNat_{sc12Dun10} - pNat_{sc12Bicubic}$). Here too, *Score12Dun10* showed a statistically significant degradation relative to *Score12Bicubic*, with a *p*-value of 0.04. The mean difference of the pNat statistic for this subset (0.06) was similar to the mean difference over the entire set (0.05).

For the $\Delta\Delta G$ benchmark, the differences between the four tested energy functions were very slight. *Score12′*'s performance was somewhat degraded relative to *Score12*, though this should be weighed against the dramatic improvement *Score12′* showed at sequence recovery. The other two energy functions performed in the same range as *Score12′*.

At the loop-modeling benchmark, the differences between the three methods were slight. To estimate the significance of the differences, we took 100 bootstrap samples and measured the three RMSD quartiles. The differences in median RMSDs between *Score12* and *Score12Bicubic* ($p < 0.74$), and *Score12Bicubic* and *Score12Dun10* ($p < 0.11$) were not statistically significant. However, the third quartile improved for both *Score12Bicubic* and *Score12Dun10*. In two cases (see the online appendix), KIC simulations using *Score12Dun10* correctly identified near-native structures by their lowest energy, whereas the *Score12* simulations did not.

Table 6.3 Benchmark results for score12 and for the three proposed energy function modifications

	Rotamer recovery benchmark				Seq. rec. bench		$\Delta\Delta G$ bench	High-res. refinement benchmark			Loop-modeling benchmark		
Energy function	Pack rots (%)	Min pack (%)	Rot. trials (%)	Rt-min (%)	% Rec.	KL-div.	*R*-value	# (pNat > 0.8)	ΣpNat	# (eNat < eDec)	First quart. (Å)	Med. (Å)	Third quart. (Å)
Score12	66.19	69.07	71.49	73.12	32.6	0.019	0.69	67	74.6	104	0.468	0.637	1.839
Score12′	–	–	–	–	37.0	0.008	0.67	–	–		–	–	–
Score12Bicubic	66.24	67.51	71.52	73.15	37.6	0.010	0.68	68	77.9	105	0.499	0.644	1.636
Score12Dun10	67.82	70.50	72.60	74.23	37.6	0.009	0.67	60	72.0	104	0.461	0.677	1.463

Score12′ differs from *Score12* only in its reference energies, which have no effect on rotamer-recovery, high-resolution refinement, or loop modeling, and so data for these benchmarks are not given.

Table 6.4 Percentage rotamer recovery by amino acid

	R	K	M	I	L	S	T	V	N	D	Q	E	H	W	F	Y
Score12	24.7	31.1	51.3	84.0	86.7	71.8	92.9	94.4	55.2	59.2	21.8	28.5	51.8	78.7	84.5	79.9
Score12Bicubic	25.7	31.8	51.1	85.2	86.8	71.5	92.9	94.4	54.8	58.7	20.5	28.7	52.0	80.1	83.3	79.9
Score12Dun10	26.7	31.7	49.6	85.4	87.5	72.5	92.6	94.3	56.8	60.4	30.7	33.6	55.0	85.0	85.4	82.9

Rotameric amino acids are listed on the left; semirotameric amino acids on the right.

It is not immediately clear why the 2010 rotamer library causes a degradation in Rosetta's ability to discriminate native structures from decoys. A possible reason for this might be pointed to in the distribution of off-center χ angles. Structures refined with the new library have χ-angles more tightly distributed around the reported $\bar{\chi}_i$. The 2010 library, which was generated using more stringent data filters than the 2002 library, reports smaller standard deviations on average: for example, the mean σ_1 for leucine for rotamers in all ϕ, ψ bins with probability >5% is 10.8° (with a median of 13.9°) in the 2002 library, but is down to 7.9° (with a median of 7.2°) in the 2010 library. (For all leucine rotamers, the 2002 library reports a 13.1° mean, and a 9.9° median; the 2010 library reports a 9.2° mean, and a 9.9° median.) By decreasing the weight on the `fa_dun` term or by merely weakening the "off-rotamer penalty" (the $\sum_i (\chi - \bar{\chi}/\sigma)^2$ component of Eq. 6.14), the distributions may broaden and performance at the high-resolution refinement benchmark might improve. Encouragingly, decreasing the `fa_dun` weight down to one-half of its *Score12* weight does not substantially worsen rotamer recovery for the 2010 library (see the online appendix). There is still significant work, however, before we are ready to conclude that the new library should be adopted for general use in Rosetta.

7. CONCLUSION

We have described three tools that can be used to evaluate and improve macromolecular energy functions. Inaccuracies in the energy function can be identified by comparing features from crystal structures and computationally generated structures. New or reparameterized energy terms can be rapidly tested with optE to determine if the change improves structure prediction and sequence design. When a new term is ready to be rigorously tested, we can test for unintended changes to feature distributions by relying upon the existing set of feature analysis scripts, refit reference energies for protein design using the sequence-profile recovery protocol in optE, and measure the impact of the new term on a wide array of modeling problems by running the benchmarks curated here. Of the three changes we benchmarked in this paper, we recommend that the first two should be adopted. In the context of Rosetta, this means using *Score12Bicubic* rather than the current *Score12*.

ACKNOWLEDGMENTS

Support for A. L. F., M. J. O., and B. K. came from GM073151 and GM073960. Support for J. S. R. came from NIH R01 GM073930. Thanks to Steven Combs for bringing the bicubic-spline implementation to Rosetta.

REFERENCES

Bower, M. J., Cohen, F. E., & Dunbrack, R. L., Jr. (1997). Prediction of protein side-chain rotamers from a backbone-dependent rotamer library: A new homology modeling tool. *Journal of Molecular Biology*, *267*, 1268–1282.

Chen, V. B., Arendall, W. B., Headd, J. J., Keedy, D. A., Immormino, R. M., Kapral, G. J., et al. (2010). MolProbity: All-atom structure validation for macromolecular crystallography. *Acta Crystallographica. Section D: Biological Crystallography*, *66*, 12–21.

Chen, H.-M., Liu, B.-F., Huang, H.-L., Hwang, S.-F., & Ho, S.-Y. (2007). Sodock: Swarm optimization for highly flexible protein-ligand docking. *Journal of Computational Chemistry*, *28*, 612–623.

Dantas, G., Corrent, C., Reichow, S. L., Havranek, J. J., Eletr, Z. M., Isern, N. G., et al. (2007). High-resolution structural and thermodynamic analysis of extreme stabilization of human procarboxypeptidase by computational protein design. *Journal of Molecular Biology*, *366*, 1209–1221.

Das, R., Karanicolas, J., & Baker, D. (2010). Atomic accuracy in predicting and designing noncanonical RNA structure. *Nature Methods*, 7, 291–294.

Ding, F., & Dokholyan, N. V. (2006). Emergence of protein fold families through rational design. *PLoS Computational Biology*, *2*, e85.

Dobson, N., Dantas, G., Baker, D., & Varani, G. (2006). High-resolution structural validation of the computational redesign of human U1A protein. *Structure*, *14*, 847–856.

Dunbrack, R. L., Jr. (2002). Rotamer libraries in the 21st century. *Current Opinion in Structural Biology*, *12*, 431–440.

Dunbrack, R. L., Jr., & Karplus, M. (1993). Backbone dependent rotamer library for proteins: Application to side chain prediction. *Journal of Molecular Biology*, *230*, 543–574.

Fabiola, G. F., Krishnaswarmy, S., Nagarajan, V., & Pattabhi, V. (1997). C-H···O hydrogen bonds in β-sheets. *Acta Crystallographica. Section D: Biological Crystallography*, *53*, 316–320.

Fleishman, S. J., Leaver-Fay, A., Corn, J. E., Strauch, E.-M., Khare, S. D., Koga, N., et al. (2011). RosettaScripts: A scripting language interface to the Rosetta macromolecular modeling suite. *PLoS One*, *6*, e20161.

Gilis, D., & Rooman, M. (1997). Predicting protein stability changes upon mutation using database-derived potentials: Solvent accessibility determines the importance of local versus non-local interactions along the sequence. *Journal of Molecular Biology*, *272*, 276–290.

Guerois, R., Nielsen, J. E., & Serrano, L. (2002). Predicting changes in the stability of proteins and protein complexes: A study of more than 1000 mutations. *Journal of Molecular Biology*, *320*, 369–387.

Hamelryck, T., Borg, M., Paluszewski, M., Paulsen, J., Frellsen, J., Andreetta, C., et al. (2010). Potentials of mean force for protein structure prediction vindicated, formalized and generalized. *PLoS One*, *5*, e13714.

Jacak, R., Leaver-Fay, A., & Kuhlman, B. (2012). Computational protein design with explicit consideration of surface hydrophobic patches. *Proteins*, *80*, 825–838.

Jacobson, M. P., Kaminski, G. A., Friesner, R. A., & Rapp, C. S. (2002). Force field validation using protein side chain prediction. *The Journal of Physical Chemistry B*, *106*, 11673–11680.

Jorgensen, W. L., Maxwell, D. S., & Tirado-Rives, J. (1996). Development and testing of the OPLS all-atom force field on conformational energetics and properties of organic liquids. *Journal of the American Chemical Society*, *118*, 11225–11236.

Keedy, D. A., Arendall III, W. B., Chen, V. B., Williams, C. J., Headd, J. J., Echols, N., et al. (2012). 8000 Filtered Structures, 2012 http://kinemage.biochem.duke.edu/databases/top8000.php.

Kellogg, E. H., Leaver-Fay, A., & Baker, D. (2011). Role of conformational sampling in computing mutation-induced changes in protein structure and stability. *Proteins: Structure, Function, and Bioinformatics*, *79*, 830–838.

Khatib, F., Cooper, S., Tyka, M., Xu, K., Makedon, I., Popovic, Z., et al. (2011). Algorithm discovery by protein folding game players. *Proceedings of the National Academy of Sciences of the United States of America, 109*, 5277–5282.

Kortemme, T., Morozov, A. V., & Baker, D. (2003). An orientation-dependent hydrogen bonding potential improves prediction of specificity and structure for proteins and protein-protein complexes. *Journal of Molecular Biology, 326*, 1239–1259.

Kuhlman, B., & Baker, D. (2000). Native protein sequences are close to optimal for their structures. *Proceedings of the National Academy of Sciences of the United States of America, 97*, 10383–10388.

Kuhlman, B., Dantas, G., Ireton, G., Varani, G., Stoddard, B., & Baker, D. (2003). Design of a novel globular protein fold with atomic-level accuracy. *Science, 302*, 1364–1368.

Lazaridis, T., & Karplus, M. (1999). Effective energy function for proteins in solution. *Proteins: Structure, Function, and Genetics, 35*, 133–152.

Leaver-Fay, A., Jacak, R., Stranges, P. B., & Kuhlman, B. (2011). A generic program for multistate protein design. *PLoS One, 6*, e20937.

Leaver-Fay, A., Tyka, M., Lewis, S. M., Lange, O. F., Thompson, J., Jacak, R., et al. (2011). ROSETTA3: An object-oriented software suite for the simulation and design of macromolecules. *Methods in Enzymology, 487*, 545–574.

LeCun, Y., & Jie, F. (2005). Loss functions for discriminative training of energy-based models. In: *Proceedings of the 10th international workshop on artificial intelligence and statistics, Society for AI and Statistics (AISTATS'05).*

Liang, S., & Grishin, N. (2002). Side-chain modeling with an optimized scoring function. *Protein Science, 11*, 322–331.

Lyskov, S., & Gray, J. J. (2012). RosettaTests. http://rosettatests.graylab.jhu.edu.

Mandell, D. J., Coutsias, E. A., & Kortemme, T. (2009). Sub-angstrom accuracy in protein loop reconstruction by robotics-inspired conformational sampling. *Nature Methods, 6*, 551–552.

Meiler, J., & Baker, D. (2003). Rapid protein fold determination using unassigned NMR data. *Proceedings of the National Academy of Sciences of the United States of America, 100*, 15404–15409.

Miyazawa, S., & Jernigan, R. L. (1985). Estimation of effective interresidue contact energies from protein crystal structures: Quasi-chemical approximation. *Macromolecules, 18*, 534–552.

Morozov, A. V., & Kortemme, T. (2005). Potential functions for hydrogen bonds in protein structure prediction and design. *Advances in Protein Chemistry, 72*, 1–38.

Murphy, P. M., Bolduc, J. M., Gallaher, J. L., Stoddard, B. L., & Baker, D. (2009). Alteration of enzyme specificity by computational loop remodeling and design. *Proceedings of the National Academy of Sciences of the United States of America, 106*, 9215–9220.

Novotný, J., Bruccoleri, R., & Karplus, M. (1984). An analysis of incorrectly folded protein models. Implications for structure predictions. *Journal of Molecular Biology, 177*, 787–818.

Park, B., & Levitt, M. (1996). Energy functions that discriminate X-ray and near native folds from well-constructed decoys. *Journal of Molecular Biology, 258*, 367–392.

Petrella, R. J., Lazaridis, T., & Karplus, M. (1998). Protein sidechain conformer prediction: A test of the energy function. *Folding and Design, 3*, 353–377.

Pierce, N., & Winfree, E. (2002). Protein design is NP-hard. *Protein Engineering, 15*, 779–782.

Ponder, J. W., & Richards, F. M. (1987). Tertiary templates for proteins. Use of packing criteria in the enumeration of allowed sequences for different structural classes. *Journal of Molecular Biology, 193*, 775–791.

Potapov, V., Cohen, M., & Schreiber, G. (2009). Assessing computational methods for predicting protein stability upon mutation: Good on average but not in the details. *Protein Engineering, Design & Selection, 22*, 553–560.

Richardson, J. S., Keedy, D. A., & Richardson, D. C. (2013). The Plot Thickens: More Data, More Dimensions, More Uses. In M. Bansai & N. Srinivasan (Eds.), *Biomolecular Forms and Functions: A Celebration of 50 Years of the Ramachandran Map* (pp. 46–61). World Scientific.

Rohl, C. A., Strauss, C. E. M., Misura, K. M. S., & Baker, D. (2004). Protein structure prediction using Rosetta. *Methods in Enzymology, 383*, 66–93.

Shapovalov, M. V., & Dunbrack, R. L. (2011). A smoothed backbone-dependent rotamer library for proteins derived from adaptive kernel density estimates and regressions. *Structure, 19*, 844–858.

Sharabi, O. Z., Yanover, C., Dekel, A., & Shifman, J. M. (2010). Optimizing energy functions for protein-protein interface design. *Journal of Computational Chemistry, 32*, 23–32.

Sheffler, W., & Baker, D. (2009). RosettaHoles: Rapid assessment of protein core packing for structure prediction, refinement, design and validation. *Protein Science, 18*, 229–239.

Simons, K., Kooperberg, C., Huang, E., & Baker, D. (1997). Assembly of protein tertiary structures from fragments with similar local sequences using simulated annealing and bayesian scoring functions. *Journal of Molecular Biology, 268*, 209–225.

Simons, K. T., Ruczinski, I., Kooperberg, C., Fox, B. A., Bystroff, C., & Baker, D. (1999). Improved recognition of native-like protein structures using a combination of sequence-dependent and sequence-independent features of proteins. *Proteins, 34*, 82–95.

Sippl, M. J. (1990). Calculation of conformational ensembles from potentials of mean force: An approach to the knowledge-based prediction of local structures in globular proteins. *Journal of Molecular Biology, 213*, 859–883.

Song, Y., Tyka, M., Leaver-Fay, A., Thompson, J., & Baker, D. (2011). Structure guided forcefield optimization. *Proteins: Structure, Function, and Bioinformatics, 79*, 1898–1909.

Taylor, R., & Kennard, O. (1982). Crystallographic evidence for the existence of the C-H···O, C-H···N and C-H···Cl hydrogen bonds. *Journal of the American Chemical Society, 104*, 5063–5070.

Tyka, M. D., Keedy, D. A., André, I., Dimaio, F., Song, Y., Richardson, D. C., et al. (2010). Alternate states of proteins revealed by detailed energy landscape mapping. *Journal of Molecular Biology, 405*, 607–618.

Wang, C., Bradley, P., & Baker, D. (2007). Protein-protein docking with backbone flexibility. *Journal of Molecular Biology, 373*, 503–519.

Wang, C., Schueler-Furman, O., & Baker, D. (2005). Improved side-chain modeling for protein-protein docking. *Protein Science, 14*, 1328–1339.

Weiner, S. J., Kollman, P. A., Case, D. A., Singh, U. C., Ghio, C., Alagona, G., et al. (1984). A new force field for molecular mechanical simulation of nucleic acids and proteins. *Journal of the American Chemical Society, 106*, 765–784.

Wickham, H. (2010). A layered grammar of graphics. *Journal of Computational and Graphical Statistics, 19*, 3–28.

Wickham, H. (2011). The split-apply-combine strategy for data analysis. *Journal of Statistical Software, 40*, 1–29.

Wilkinson, L. (1999). *The grammar of graphics*. New York: Springer.

Word, J. M., Lovell, S. C., Richardson, J. S., & Richardson, D. C. (1999). Asparagine and glutamine: Using hydrogen atom contacts in the choice of side-chain amide orientation. *Journal of Molecular Biology, 285*, 1735–1747.

Yanover, C., & Bradley, P. (2011). Extensive protein and DNA backbone sampling improves structure-based specificity prediction for C2H2 zinc fingers. *Nucleic Acids Research, 39*, 4564–4576.

CHAPTER SEVEN

Molecular Dynamics Simulations for the Ranking, Evaluation, and Refinement of Computationally Designed Proteins

Gert Kiss*, Vijay S. Pande*, K.N. Houk†,[1]

*Department of Chemistry, Stanford University, Stanford, California, USA
†Department of Chemistry and Biochemistry, University of California, Los Angeles, California, USA
[1]Corresponding author: e-mail address: houk@chem.ucla.edu

Contents

Abstract

Computational methods have been developed to redesign proteins so that they can perform novel functions such as the catalysis of nonnatural reactions. Active sites are constructed from the inside out by stochastically exploring mutations that favor the binding of transition states, small molecule binders, and protein surfaces—depending

Methods in Enzymology, Volume 523
ISSN 0076-6879
http://dx.doi.org/10.1016/B978-0-12-394292-0.00007-2

on the task at hand. The approach allows the use of many proteins for engineering scaffolds upon which to erect the necessary functionality. Beyond being of practical value for producing proteins with new applications, the approach tests our understanding of protein chemistry. The current success rate, however, is rather modest, and so the designers have become good only at making catalysts with low catalytic efficiencies. Directed evolution can be used to enhance function and stability, while more advanced computational techniques and physics-based simulations are useful at elucidating structural flaws and at guiding the design process. Here, we summarize work that focuses on the dynamic properties of computationally designed enzymes and their directed evolution variants. We utilized *in silico* methods to address three questions: (1) What are the shortcomings of these designs? (2) Can they be improved? (3) Can we screen out designs that are likely to be inactive?

1. INTRODUCTION

Enzymes are highly evolved catalysts that make metabolism and life possible. They are proteins, often functionalized with cofactors that catalyze a wide variety of reactions. Enzymes are so proficient that often the diffusion of substrate onto the protein in water becomes the slow step of the reaction. Because the expression of proteins in organisms such as *Escherichia coli* has become a routine activity in molecular biology, the idea of making protein catalysts for any desired reaction has gained traction. If methods to design proteins that could catalyze any desired reaction were perfected, enormous capabilities for synthesis, therapeutics, defense against biological threats, incorporation of new synthetic capabilities into organisms, and many other applications could be realized.

The factors that influence biological catalysis have been explored but not all aspects are yet understood in their entirety (Fersht, 1999). The design of new enzymes is built upon the premise that we do understand things well enough to create protein catalysts without the aid of evolution. Pauling (1948) proposed that enzymes complement transition states (TSs) similar to the way that antibodies complement and bind antigens. Over the decades, the view was augmented to include the idea of covalent catalysis and a modification of the reaction mechanism from what it is in solution to account for the up to 10^{23} acceleration that occurs with some of the most proficient enzyme catalysts (Zhang & Houk, 2005). It might be thought that screening of proteins could find potential catalysts that could then be subjected to directed evolution to produce efficient catalysts (Jäckel et al., 2008; Zhao, 2007). However, various estimates have been made which show that the probability of finding an active catalyst for an arbitrary reaction is vanishingly small (Axe, 2004; Taylor, Kast, & Hilvert, 2001). Catalytic antibodies have been created for a variety of reactions (Gouverneur et al., 1993; Janda, Shevlin, & Lerner, 1993; Lerner, Benkovic, & Schultz, 1991; Lewis, Kramer, Robinson, & Hilvert, 1991; Li, Janda, Ashley, & Lerner, 1994; Thorn, Daniels, Auditor, & Hilvert, 1995; Wagner, Lerner, & Barbas, 1995) and have given up to 10^6-fold rate accelerations (Kikuchi, Hannak, Guo, Kirby, & Hilvert, 2006; Müller et al., 2007), but the technology required is somewhat daunting for a typical lab. Many other avenues have been explored toward generating enzymes with altered properties such as substrate scope, enantioselectivity or preference, thermostability, pH profile, or solvent tolerance. For instance, bioinformatics tools play an important role in the discovery of new enzymes that are related to known ones; directed evolution techniques have been employed that range from random mutagenesis to targeted randomization, CASTing/ISM, ProSAR, and the 3DM consensus approach (Behrens, Hummel, Padhi, Schaetzle, & Bornscheuer, 2011); natural evolution-based redesign strategies have been used to exploit promiscuities toward related substrates and chemistries within enzyme superfamilies (Gerlt & Babbitt, 2009); and rational protein design has utilized chemical intuition to generate proteins with new substrate binding pockets (Sasaki & Kaiser, 1989), inverted coiled-coils (Harbury, Plecs, Tidor, Alber, & Kim, 1998), spatial clustering of charged residues (Johnsson, Allemann, Widmer, & Benner, 1993), and novel multihelical structures (Rossi, Tecilla, Baltzer, & Scrimin, 2004; Tommos, Skalicky, Pilloud, Wand, & Dutton, 1999). Highlights include the work of Benner (Johnsson et al., 1993), Mayo (Bolon, 2001), Scrimin (Rossi et al., 2004), and deGrado (Kaplan & DeGrado, 2004). Recently, the development of new computational tools has allowed for an *in silico* approach to generate biocatalysts for nonnatural reactions (Dahiyat, 1996; Zanghellini et al., 2006).

Protein design is an evolving and challenging field. While some advances are only incremental, the results are encouraging. This chapter provides a short overview of the technology behind the inside-out approach to computational enzyme design (Jiang et al., 2008; Röthlisberger et al., 2008; Siegel et al., 2010) and focuses on exploring the dynamic and structural properties of designed enzymes using physics-based simulations. Examples are highlighted in which molecular dynamics (MD) were used to gain insights about the inactivity of certain designs and the limited activity of others, and cases in which MD was utilized to guide the computational design process are discussed.

2. INSIDE-OUT COMPUTATIONAL ENZYME DESIGN

The inside-out approach (Fig. 7.1) is guided by the natural precedence of catalytic groups that have evolved to be effective. At the core of the computational design protocol is a theoretical active site (theozyme) with the appropriate functionality for catalysis. Quantum mechanical (QM) calculations are employed to determine the catalytic units that will be most effective at stabilizing the TS. Protein scaffolds are selected from the PDB and are used as templates into which the QM transition-state geometry is grafted (using RosettaMatch). Amino acid residues in the first and second shell surrounding the QM-theozyme are mutated and optimized to complement the geometric and electronic features of the TS using RosettaDesign (Dantas,

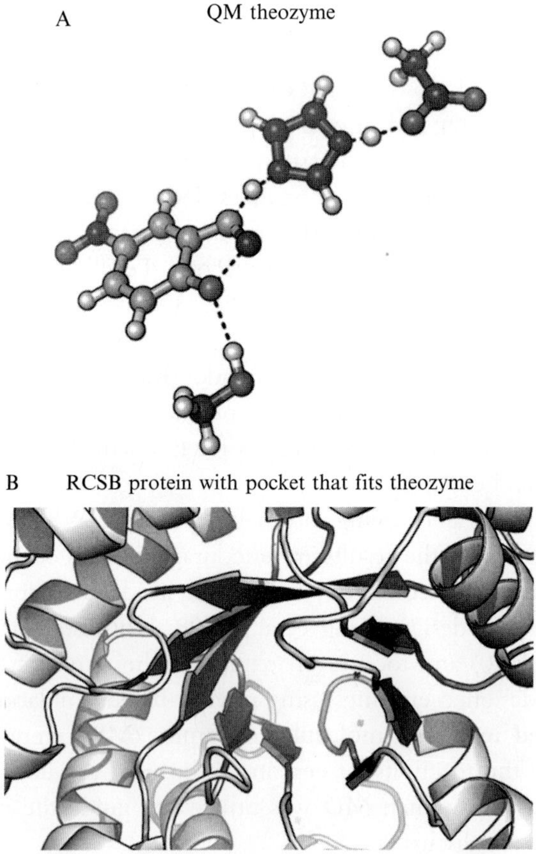

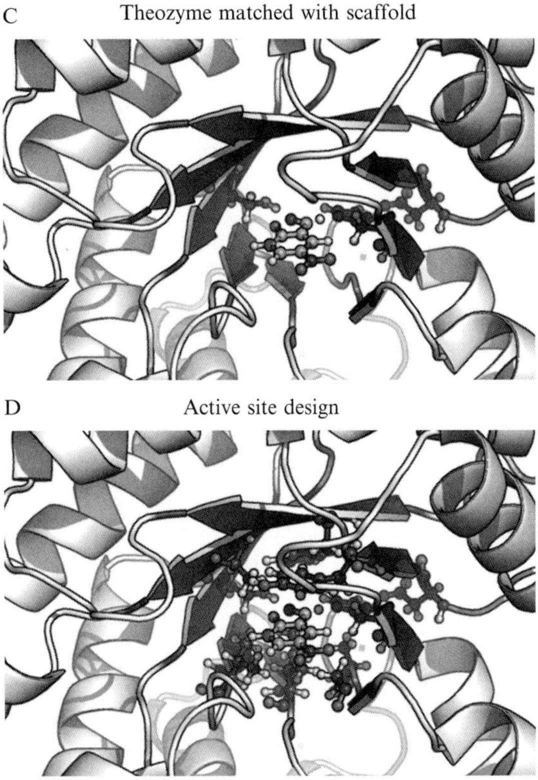

Figure 7.1 Key steps in the inside-out enzyme design protocol for the Kemp elimination reaction. *Reprint from Kiss et al. (2011).* (See Color Insert.)

Kuhlman, Callender, Wong, & Baker, 2003; Kiss et al., 2011; Richter, Leaver-Fay, Khare, Bjelic, & Baker, 2011).

3. FILTERING, RANKING, AND EVALUATION OF FINAL DESIGNS

The current collection of Rosetta modules (Rosetta3) (Richter et al., 2011) allows the use of any structure from the PDB as a template protein, and it extends the matching algorithm of previous versions by a nondiscretized component; this combination can result in a large number of new sequences. Thus, final designs (Fig. 7.1D) are assessed for their capability to stabilize the key catalytic residues prior to experimental workup. Designs are ranked based

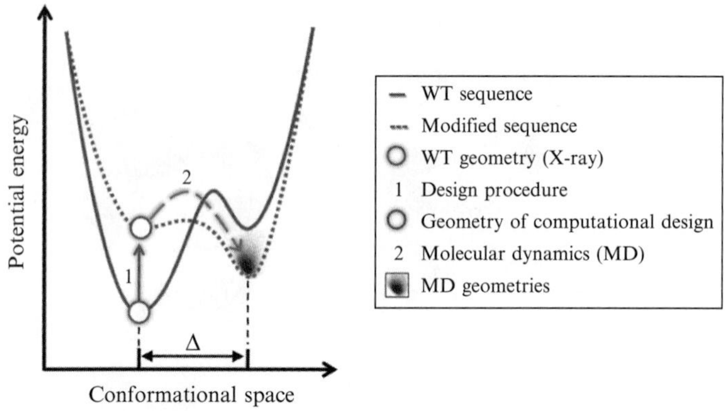

Figure 7.2 Schematic representation of the enzyme design process in terms of potential energy and conformational space. *Reprint from Kiss et al. (2010).* (For the color version of this figure, the reader is referred to the online version of this chapter.)

on empirical criteria such as Rosetta energy, ligand-binding scores, and packing scores. Manual inspection is still a large component in this process, but tools such as FoldIt, EDGE, and more rigorous computational tests that probe the dynamics of the systems are continuously being developed and refined with the goal of maximizing the hit rate, particularly, as more challenging reactions are pursued.

MD simulations were found to be particularly useful for assessing the quality of a newly designed active site by testing its structural integrity and by mapping out competing interactions, in essence exposing design flaws that are inaccessible to static evaluations (Kiss, Röthlisberger, Baker, & Houk, 2010). Briefly, the sequence of a final computational design generally differs by 10% or more from that of the wild-type protein that is used as the template. Consequently, the potential energy surface of the new sequence is in general significantly perturbed (Fig. 7.2, step 1). Cycling through repacking and geometry optimization during the design process ensures that the overall conformation of the new protein is at a potential energy minimum. But in response to the new sequence and the resulting perturbations, neighboring minima (corresponding to alternative conformations of side chains and loops) can become thermodynamically more favorable than the native conformation. The actual active site geometry of the new protein sequence in solution might thus significantly differ from that of the computational geometry (large Δ values in Fig. 7.2). Assessing whether a design is susceptible to such change was shown to be tractable with nanosecond-scale MD simulations (step 2).

MD simulation-based evaluations are now performed on a routine basis for selected designs as a means to pinpoint structural weaknesses and to guide adjustments. We find that a dynamic treatment of the systems in the presence of explicitly modeled water molecules is helpful at identifying structural flaws, competing hydrogen-bonding contacts, water influx, and other sources for inactivity. MD-derived geometric descriptors are instrumental for this. A direct comparison of the reaction profiles predicted for different designs is desirable, but the high computational cost of obtaining activation barriers with *ab initio* or density functional QM limits such high-level approaches to a static treatment of the systems. Based on experimental results and MD evaluations, we further note that even the most active designs appear to have considerable shortcomings, underlining the notion that there is much room for improvement.

4. DISCERNING ACTIVE FROM INACTIVE DESIGNS WITH MD

4.1. Hydrogen bond distance and directionality

We studied the dynamics of 23 designed Kemp eliminases (14 actives, 9 inactives) and concluded that the failed computational designs are unable to maintain essential active-site hydrogen bonds (Kiss et al., 2010). This becomes particularly clear at the example of the inactive design KE38 (Fig. 7.3C). Compared to the catalytic His-Asn contact in the naturally evolved cathepsin K (Fig. 7.3A) and the catalytic His-Asp dyad in the active KE70 (Fig. 7.3B), there is no significant population in which the KE38

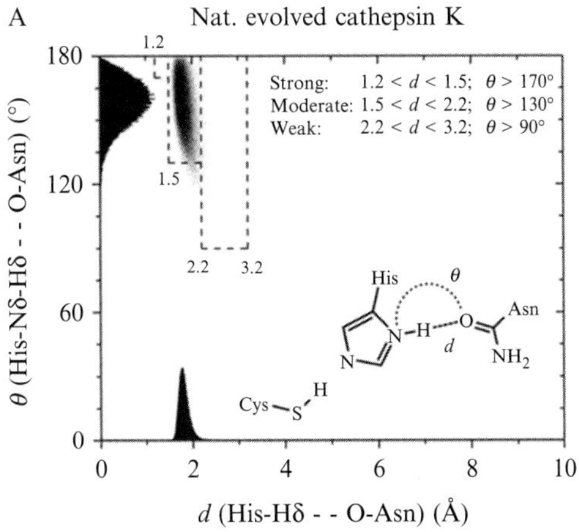

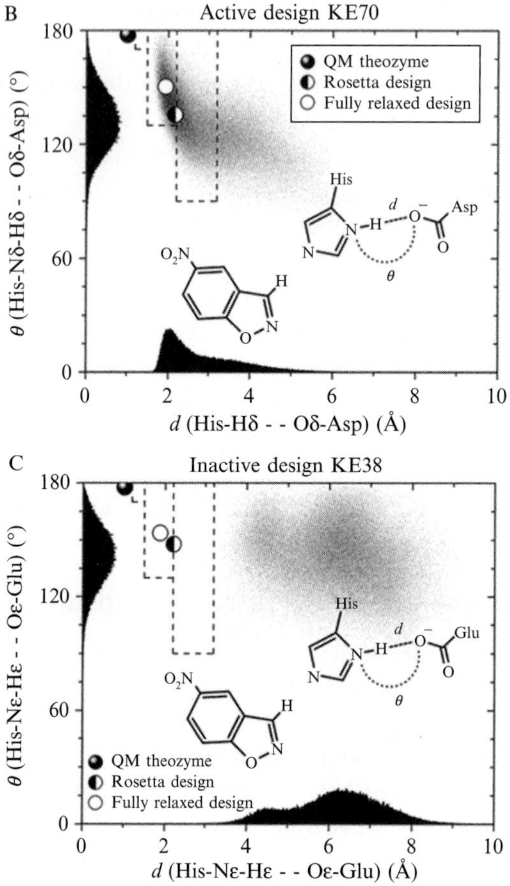

Figure 7.3 Angle vs. distance scatter plots. (A) His-Asn contact of the naturally evolved cathepsin K catalytic triad; (B) His-Asp contact of the active design KE70; (C) His-Glu contact of the inactive design KE38. Data points are from 20-ns MD. Three hydrogen bond categories (Steiner, 2002) are outlined with dotted lines. The individual distributions are projected onto the axes. The progression of the catalytic contact from QM theozyme, to final design, and the fully relaxed MD starting geometry is plotted with filled, half-filled, and empty discs, respectively. *Modified reprint from Kiss et al. (2010).*

His-Glu dyad is intact, and His alone is too weak a base to deprotonate the substrate on its own.

Overall, it was observed that the designed catalytic contacts can disassemble because of competing interactions (with solvent molecules and/or non-catalytic polar amino acids), but also because of alternative side-chain packing modes.

4.2. Structural integrity of the active site

Figure 7.4 puts the distributions of Fig. 7.3 into geometric context. KE70/1jcl is an active Kemp elimination design and employs a His-Asp dyad as the catalytic base. Ser137 was designed to serve as the TS-stabilizing hydrogen-bond contact and Tyr47 as the π-stacking residue (Fig. 7.4A). KE38/1lbm is an inactive design. Its binding site consists of a Glu-His dyad (as the base), a Trp residue (as the phenoxy-hydrogen bond), and a Tyr/Trp π-stacking arrangement (Fig. 7.4C).

4.3. Water accessibility and coordination

Solvent molecules that come into direct contact with the carboxylate oxygens can significantly reduce their base strength and catalytic power (up to 10^6 in terms of k_{cat}). Figure 7.5A shows this trend for a cross section of the dataset. On average, the active sites of functional designs are less hydrated than those of inactive designs (Fig. 7.5B), but even the microenvironments of the most

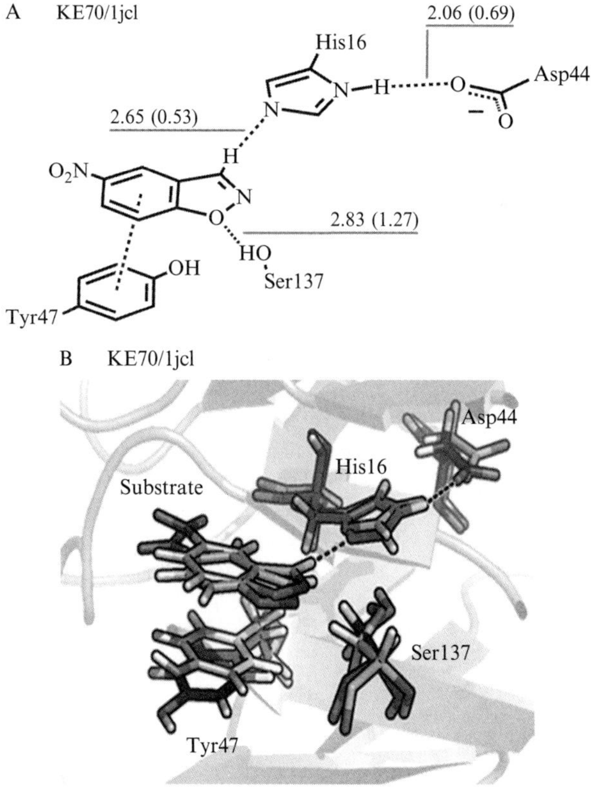

C KE38/1lbm

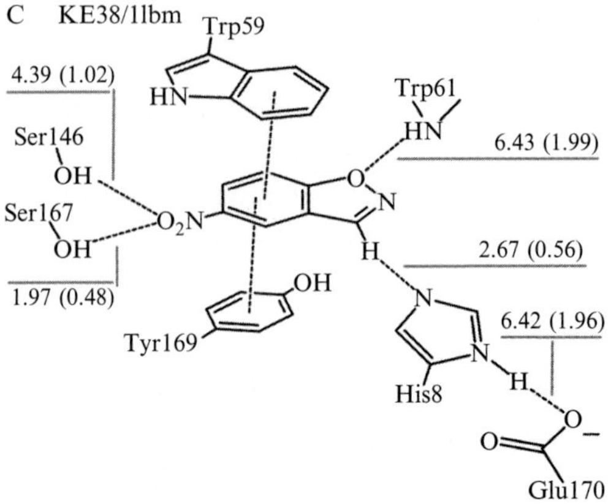

D KE38/1lbm

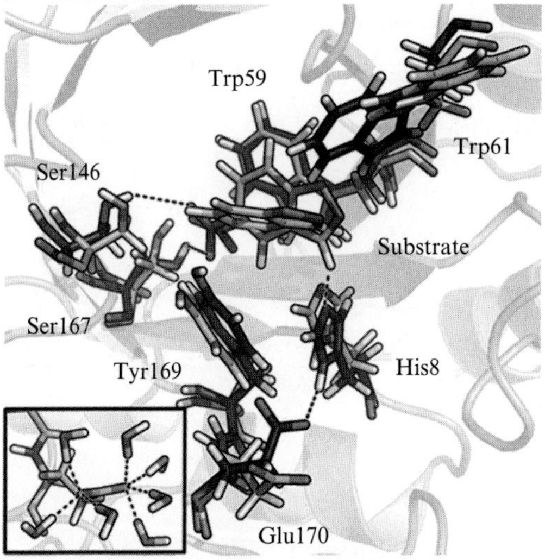

Figure 7.4 Design vs. MD. Schematic representation of the catalytic unit (A, C) and representative MD geometry (light blue) over Rosetta design geometry (black with orange substrate) (B, D). Bond labels in (A) and (C) are maxima of distance distributions with full widths at half-maximum in parentheses. All values are in Å. The backbone RMSD of the catalytic unit of KE70 and KE38 is 0.57 and 0.76 Å; the side chain RMSD is 0.95 and 2.24 Å, respectively. The inset in (D) shows Glu170 in direct contact with seven water molecules on average. *Reprint from Kiss et al. (2010).* (See Color Insert.)

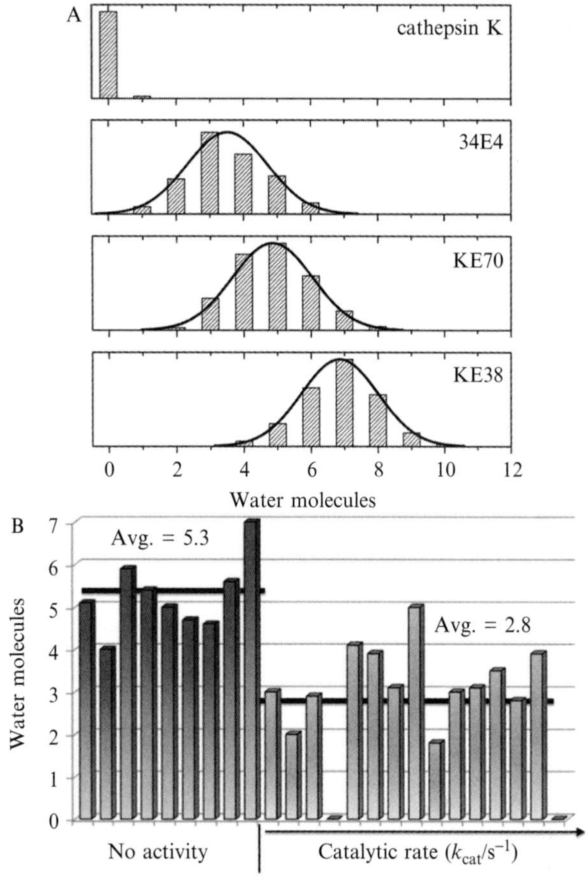

Figure 7.5 Water coordination distributions from MD at $d < 3.2$ Å to the catalytic carboxylate oxygen atoms. (A) Asn182 in the naturally evolved cathepsin K ($k_{cat} = 4.2\ s^{-1}$), GluH50 in the catalytic antibody 34E4 ($k_{cat} = 0.7\ s^{-1}$), Asp44 in the active KE70 ($k_{cat} = 0.2\ s^{-1}$), and Glu170 in the inactive KE38. (B) Maxima of the water coordination distributions 23 distinct Kemp eliminases, including antibody 34E4, and cathepsin K. *Reprint from Kiss et al. (2010).* (For the color version of this figure, the reader is referred to the online version of this chapter.)

active designs are still far from ideal, such as that around the acid/base component of natural enzymes like cathepsin K. The observation is relevant to computational enzyme design in general, but in particular for the catalysis of reactions that depend on a carboxylate base as they are highly sensitive to polar protic solvents such as water.

4.4. Summary

Taken together, active designs can be discerned from inactive ones by reducing a multidimensional problem to a simplified two-dimensional model. By querying the dynamics of a protein–substrate complex in the presence of explicitly represented solvent molecules, one can gather a wealth of information about the system at hand and relate it to experimental observables. On this basis, it has become a useful approach to combine MD-based analyses with the design and refinement of new enzymes, but also to use MD in the interpretation of results from directed evolution experiments.

5. MD EVALUATION EXAMPLES

5.1. Computationally designed Kemp eliminases

5.1.1 KE59 and KE59/G130S

KE59 is the most active of the Kemp elimination designs. Glu230 functions as the catalytic base and Trp109 is the π-stacking residue (SI Fig. 1 of Kiss et al., 2010). The catalytic machinery of this design does not include a TS-stabilizing hydrogen-bond residue, but the G130S variant was subsequently made for this purpose. Surprisingly, the kinetic characterization of the variant showed a ninefold reduction of k_{cat}/K_M. MD simulations suggest that position 130 is not suitable for hydrogen-bond stabilization of the developing TS oxyanion, and show that the substrate-binding mode in MD differs from that in the Rosetta design, such that the aryl ring rather than the heterocycle comes closest to position 130 (SI Fig. 1a of Kiss et al., 2010).

Instead of stabilizing the developing negative charge in the TS then, the increased steric demand of the G130S mutation causes the side chain of the catalytic base to point towards the bulk solvent when the substrate is bound (SI Fig. 1b and c of Kiss et al., 2010). As a result, Glu230 is roughly 30% more solvent-exposed when position 130 holds a Ser instead of a Gly (SI Fig. 1d of Kiss et al., 2010). PROPKA (Li et al., 2005; Bas et al., 2008) calculations are in line with this observation and predict post-MD $pK_{a(Glu230)}$ values of 7.8 and 6.5 for KE59 and KE59/G130S, respectively. The ninefold decrease in activity can thus be attributed to a decrease in the base strength of the catalytic Glu230.

5.1.2 KE07

The catalytic base of KE07 is a glutamate at position 101. Lys222 and Ser48 were both placed close to the substrate heterocycle-binding site to stabilize the developing negative charge in the TS. Kinetic characterization showed a relatively weak activity for KE07, which was attributed to the possibility that Lys222 forms a salt-bridge with the catalytic Glu101, thus reducing its activity. MD simulations support this and show a strong contact between the two residues (SI Fig. 2a of Kiss et al., 2010). They further highlight that Glu101 is in direct contact with Thr78 and two water molecules (avg. 2.0 waters with σ of 0.4) (SI Fig. 2b of Kiss et al., 2010). The waters form hydrogen bonds with both ionic residues (Glu101, Lys222) and effectively weaken the salt-bridge that would otherwise render this design inactive. Water molecules are thus beneficial in the case of KE07, a conclusion that is also supported by the heightened activity of the Lys222Ala variant, and subsequent directed evolution experiments by the Tawfik lab (Khersonsky et al., 2010). Variants of KE07 that resulted from seven rounds of directed evolution give a 70-fold increase in k_{cat} over that of the original design. All these variants have a characteristic Ile7Asp mutation, the side chain of which extends into the water-accessible cavity behind the active site. In MD simulations, Asp7 is salt-bridged to Arg5, which is positioned opposite to Lys222 (SI Fig. 2a of Kiss et al., 2010). The placement of a charged aspartate at this buried site facilitates the recruitment of solvent molecules to the Glu-Lys dyad. As a consequence, the catalytic Glu101 carboxylate oxygens are coordinated in average by four (previously two) water molecules. (SI Fig. 2b of Kiss et al., 2010). The presence of a negatively charged residue at position 7 then disrupts the Glu101-Lys222 salt-bridge and effectively increases the base-strength of the catalytic Glu101. Several well-ordered water molecules mediate this effect.

5.2. Directed evolution of computationally designed Kemp eliminases

5.2.1 Directed evolution of KE70

5.2.1.1 Active site preorganization

KE70 was subjected to nine rounds of directed evolution, giving rise to a >400-fold increase in k_{cat}/K_M. MD simulations on the series show that backbone RMSD values and atomic fluctuations of active site residues decrease as KE70 is evolved. Each round of evolution rigidified the active site and gave rise to an increased degree of preorganization. By focusing on the catalytic contacts, for instance, one can query the relationship between $(\Delta r)^2$ and $-\ln(k_{cat})$. Δr denotes the deviation of the His17-Asp45 distance (His-NεH–Oδ-Asp) when compared to the unconstrained hydrogen bond distance of 1.8 Å between these functional groups. Assuming a simple harmonic model, the energetic cost of deviation from ideality is proportional to $(\Delta r)^2$, and assuming a simple TS model, $\ln(k_{cat})$ is proportional to the activation free energy. Analysis of MD results from seven KE70 variants shows a linear relationship between $(\Delta r)^2$ and $\ln(k_{cat})$. This suggests that the increased k_{cat} of evolved variants results to a significant degree from a tightening of the hydrogen bond between the catalytic dyad members, as the active site residues of the KE70 variants become more optimally placed and less mobile.

5.2.1.2 The dynamic profile of the full sequence

Figure 7.6 summarizes the atomic fluctuation profiles of the evolved variants (stacked) and contrasts them to the original computational KE70 design (front). The peaks correspond to loops with elevated flexibility. Two trends are of particular interest: the D loop (residues 20 through 25, peaking at 22) rigidifies, while the loop that spans residues 170 through 179 (peaking at 176) becomes more flexible. MD shows that an Asn at position 29 (Lys29Asn) allows a tight set of hydrogen bonds to form with the carboxylate of Asp25,[1] effectively stabilizing the loop and thus reducing random fluctuations within the nearby catalytic dyad. This puts the linear relationship between $(\Delta r)^2$ and $-\ln(k_{cat})$ into perspective and shows how a decrease in active site mobility can be at the core of the observed increase of catalysis.

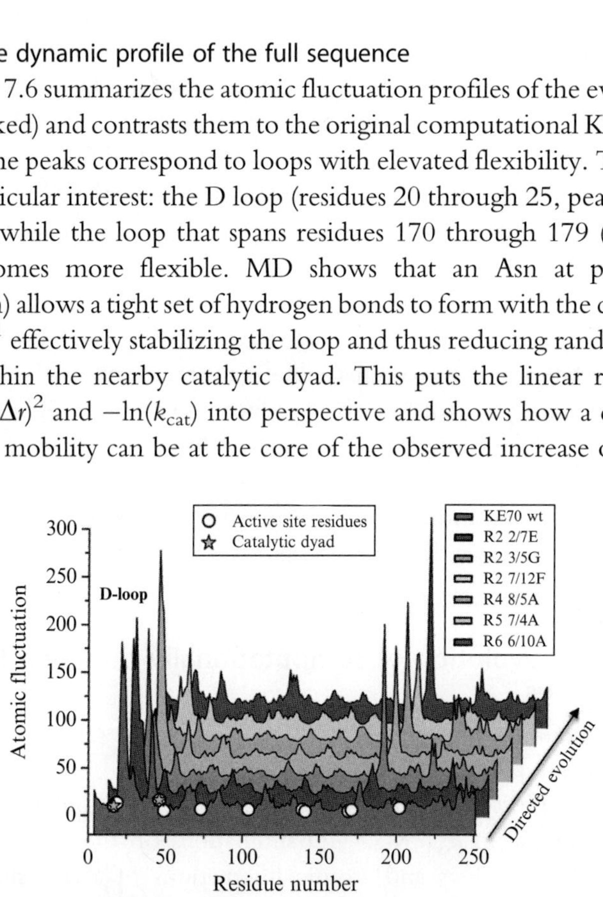

Figure 7.6 Atomic fluctuation profiles from MD. The computational design KE70 (front) is compared to directed evolution variants (stacked). Active site residues and catalytic dyad are labeled. *Reprint from Khersonsky et al. (2011).* (For the color version of this figure, the reader is referred to the online version of this chapter.)

[1] Lys29 and Asp25 are in close proximity and the side chain of Lys is too long for an ideal contact with Asp25.

5.2.2 Directed evolution of KE59

5.2.2.1 Arrangement of E230, S210, and W109 in the active site

Sixteen rounds of directed evolution on the computational design KE59 resulted in a >2000-fold increase in k_{cat}/K_M. The catalytic base, Glu230, adopts similar rotamers in the designed model and in the crystal structures of the evolved variants (Fig. 7.7A). MD simulations expand on this and show that when the proteins are fully solvated, the catalytic Glu230 has a strong preference for alternative conformations that differ from the designed one: rather than extending into the active site, the glutamate side chain extends toward bulk solvent such that the carboxylate forms hydrogen bonds with the backbone NH of Ser210 and with solvent molecules (Fig. 7.7B). The alternative conformations of Glu230 are still catalytically competent, but to a much lesser extent than expected from the design.

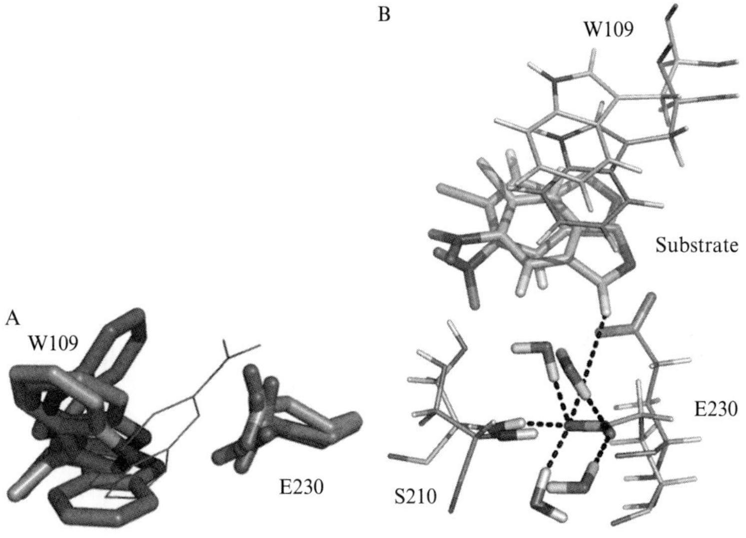

Figure 7.7 Trp109 and Glu230 rotamers in the KE59 structures. (A) Trp109 and Glu230 rotamers in the KE59 model (magenta) and in the apo structures of R1-7/10H (pink), R5-11/5F (green), and R8-2/7A (violet). The 5-nitrobenzisoxazole substrate (magenta lines) is overlayed from the KE59 model. (B) The predominant conformation of Glu230 observed in MD simulations (blue) vs. the conformation observed in the crystal structures (green), shown here for R13-3/11H variant. *Reprint from Khersonsky et al. (2012).* (See Color Insert.)

5.2.2.2 Desolvation and base-positioning effects

The increase in catalytic efficiency of the evolved KE59 variants, and specifically, in the catalytic power of Glu230, can arise from two sources: activating medium effects and improved positioning of the catalytic base. Both factors play a role in catalysis of the Kemp elimination, but their relative contribution is not easily quantifiable by kinetic measurements (Hollfelder, Kirby, & Tawfik, 2001; Hu, Houk, Kikuchi, Hotta, & Hilvert, 2004; Thorn et al., 1995).

The MD data show that fewer water molecules are in direct contact with the catalytic oxygens as the KE59 variants become more active (Fig. 7.8). Improved base desolvation is consistent with the higher pK_a values for the catalytic base of the more active KE59 variants. In addition, substituents at the 5 and 7 benzisoxazole positions are also found to better shield Glu230

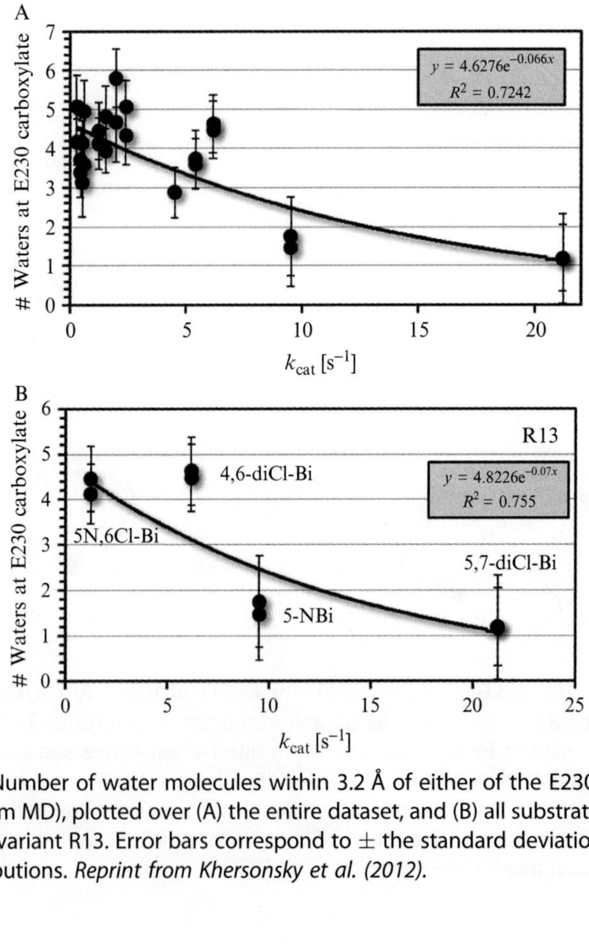

Figure 7.8 Number of water molecules within 3.2 Å of either of the E230 carboxylate oxygens (from MD), plotted over (A) the entire dataset, and (B) all substrates with available k_{cat} for variant R13. Error bars correspond to $\pm$ the standard deviation of the MD-based distributions. *Reprint from Khersonsky et al. (2012).*

from bulk solvent than substituents at positions 4 and 6 (Fig. 7.8B), thus accounting for the higher k_{cat} values of the former.

No correlation was detected between k_{cat} and the relative positioning of Glu230. Although improved geometries can certainly result in higher rates, as was observed for the catalytic His-Asp dyad in the KE70 series (no polarizing contact is needed to activate the catalytic carboxylate in the case of KE59: the catalytic base strength of Glu230 is drawn from its embedment into a hydrophobic pocket with no contacts that could favor a particular side chain conformation).

The active sites of the evolved KE59 variants differ from that of the designed model in one other aspect, however. The apo structures of KE59 variants show the stacking Trp109 in rotamers that differ from that in the design, and these rotamers are incompatible with substrate binding. Similar "flipping" of the stacking Trp in the absence of substrate is also observed in the cases of catalytic antibody 34E4 (Debler, Müller, Hilvert, & Wilson, 2008) and of KE design HG-1 (Fig. 7.9C) (Privett et al., 2012). MD simulations support a notion of induced fit, but they also show a direct competition between Trp109 and substrate, particularly for monosubstituted benzisoxazoles.

6. MD REFINEMENT EXAMPLES

6.1. Computational design of Diels-Alderases

Siegel et al. (2010) describe the computational design and experimental characterization of enzymes catalyzing a bimolecular Diels-Alder reaction with high stereoselectivity and substrate specificity. No naturally occurring enzymes are known that can do this. TS coordinates were obtained using QM calculations in the presence of specific hydrogen bond donors and acceptors. A diverse ensemble of 1.3×10^{11} distinct minimal active sites was then generated by systematically varying the identity and rotameric state of each catalytic side chain, the hydrogen bonding geometry between these residues and the TS, and the internal degrees of freedom of the TS. Two of the designs displayed measurable activity over background: DA_20_00 and DA_42_00.

To further improve their catalytic activity, residues that are in direct contact with the TS in each designed enzyme were mutated individually to sets of residues that were predicted to retain or improve TS binding and bolster the two catalytic residues. The predictions were based on the structural and dynamic model of the enzymes generated by Rosetta and MD simulations.

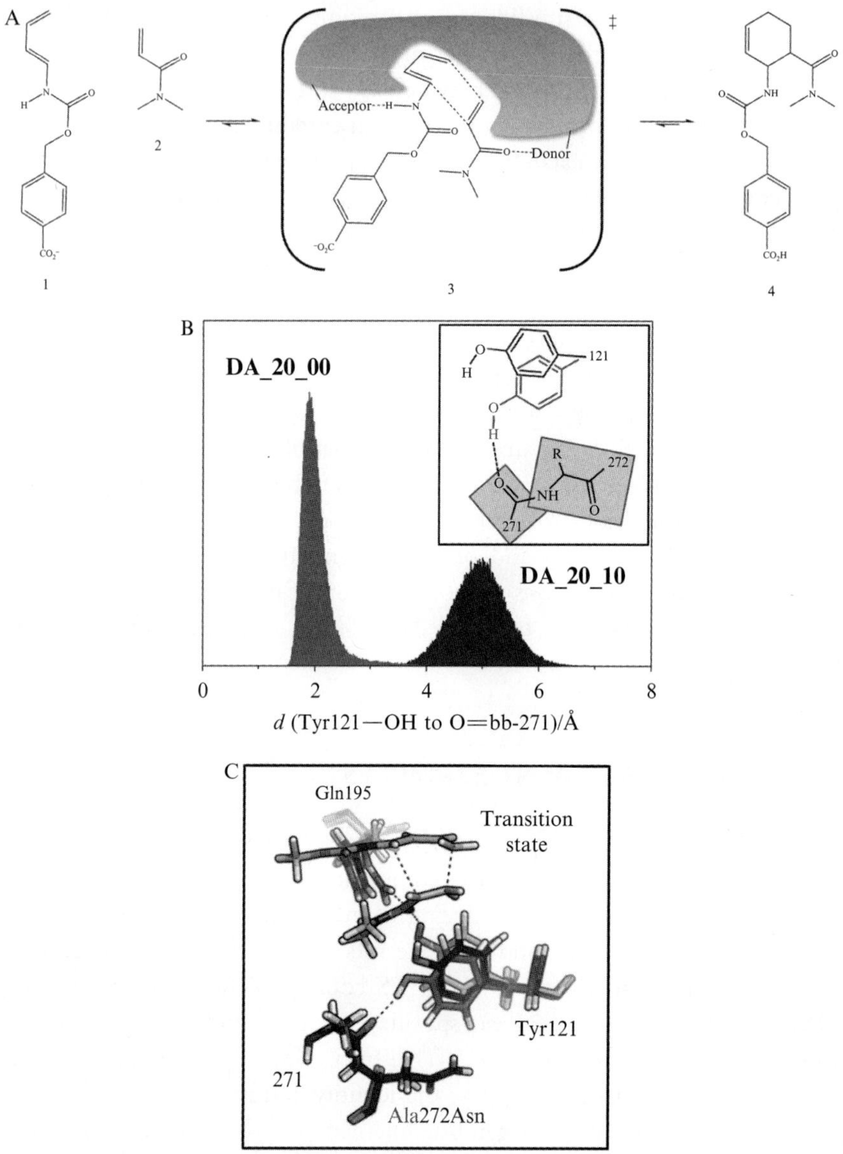

Figure 7.9 (A) The Diels–Alder reaction, showing the diene (1) and dienophile (2) for which active sites were designed to form a chiral cyclohexene ring (4). (B, C) Example of MD-guided mutation. (b) Tyr121-to-backbone271 distance distribution of DA_20_00 (red) and DA_20_10 (blue). (c) QM transition-state geometry in orange, and the equilibrated representative geometry from MD on DA_20_00 in red and that from MD on DA_20_10 in blue. *Modified reprint from Siegel et al. (2010).* (For interpretation of the references to color in this figure legend, the reader is referred to the online version of this chapter.)

A set of six mutations (A21T, A74I, Q149R, A173C, S271A, and A272N) was found to increase the overall catalytic efficiency of DA_20_00 by over 100-fold, resulting in variant DA_20_10.

The Ala272Asn mutation serves as a notable example of the interplay between sequence design and dynamics assessment. MD simulations of DA_20_00 show that the catalytic tyrosine can access a noncatalytic conformation in which its hydroxyl establishes a hydrogen bond with the backbone carbonyl of residue 271 (Fig. 7.9B, red histogram). The increased steric bulk at position 272 was proposed to interfere with this interaction (Fig. 7.9B, blue histogram), allowing tyrosine 121 to assume the conformation required for binding and catalysis (Fig. 7.9C).

6.2. Iterative design of a new Kemp eliminase

The Mayo and Houk labs together explored an iterative variation (Privett et al., 2012) to the Baker/Houk inside-out approach (Röthlisberger et al., 2008). Rather than expressing and characterizing a large number of computationally designed proteins, the researchers focused their efforts on a single template protein (PDB-ID 1gor; Lo Leggio et al., 2001). HG-1, the resulting inactive first-generation design, differs from the wild-type 1gor by seven mutations. Analysis of the structure and dynamics of HG-1 highlighted a number of problems. The side chains of the active site residues are highly mobile and their computationally designed orientations are lost when subjected to explicit solvent dynamics. The active site opens up and allows an influx of water molecules (Fig. 7.10B vs. A). The catalytic base Glu237 is readily coordinated by multiple water molecules that

A

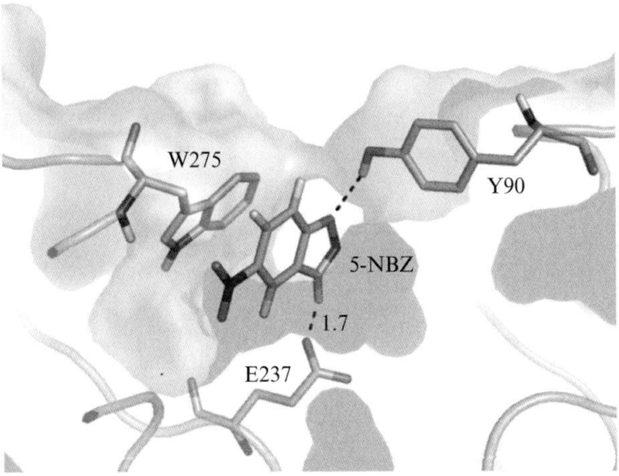

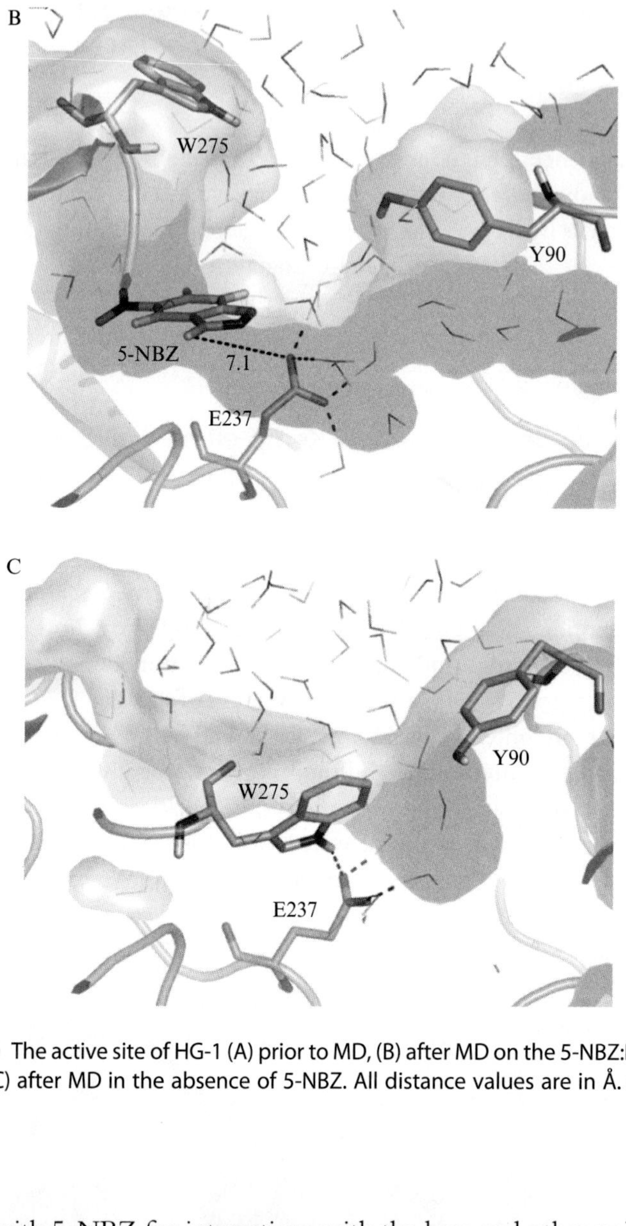

Figure 7.10 The active site of HG-1 (A) prior to MD, (B) after MD on the 5-NBZ:HG-1 complex, and (C) after MD in the absence of 5-NBZ. All distance values are in Å. (See Color Insert.)

compete with 5-NBZ for interactions with the base and other polar active site residues. MD of apo HG-1 further underlines the flexibility of the loop spanning residues 273 through 276 (Fig. 7.10C). Here, Trp275 can rotate into the active site where it hydrogen bonds to Glu237 while assuming a conformation that directly competes with the catalytically competent binding pose of 5-NBZ (compare with Fig. 7.10A).

Based on structural and dynamics studies of HG-1, a second round of design focused on the native Asp127 as the general base instead of Glu237. Compared to the HG-1 calculation, the active site search for this design was shifted 7 Å further into the barrel of the scaffold. The final catalytic configuration consisted of Asp127 as the general base, Trp44 as the π-stacking residue, and Ser265 as the hydrogen bond donor. Active site repacking produced the second-generation design HG-2, which was first subjected to MD simulations to test for structural and dynamic flaws, then expressed, and characterized. HG-2 showed a k_{cat}/K_M of 123 M^{-1} s^{-1} and an additional variation (Ser265Thr) gave HG-3 with a k_{cat}/K_M of 430 M^{-1} s^{-1}. The study highlights the value of MD simulations as informative and predictive tools in the endeavor to engineer new active sites. Similar to lessons learned from Ruscio, Kohn, Ball, and Head-Gordon (2009), Privett et al. (2012) make note of the importance of considering possible alternative substrate orientations when designing active sites from the inside out (Fig. 7.11).

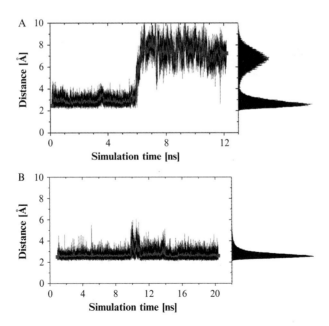

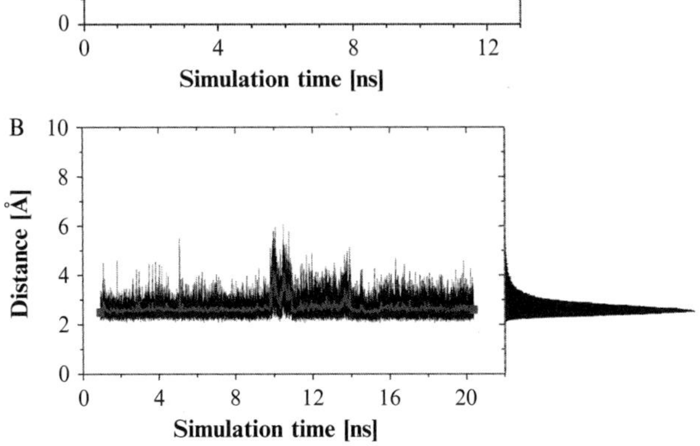

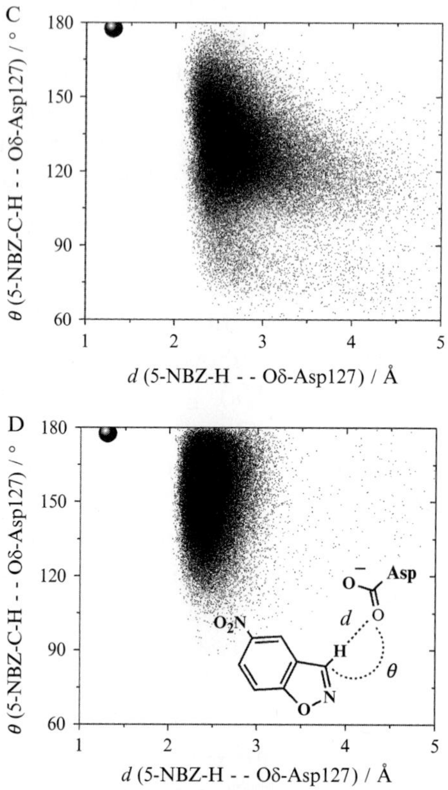

Figure 7.11 MD-assisted design refinement: from HG-1 to HG-2. (A, B) MD base–substrate distance vs. time plots (Asp OD to acidic hydrogen of 5-NBZ) for HG-1 (A) and HG-2 (B). (C, D) Angle vs. distance scatter plots of the catalytic contact (as displayed in inset). Data points were taken from 20-ns MD trajectories of HG-2 with 5-NBZ bound in orientation O1 (C) and O2 (=O1, with 5-NBZ flipped by 180°) (D). The coordinates of the transition-state geometry serve as a reference and are displayed in filled discs. *Reprint from Privett et al. (2012).* (For the color version of this figure, the reader is referred to the online version of this chapter.)

7. MOLECULAR DYNAMICS SIMULATIONS: PREPARATION AND SETUP

The structural models predicted by the computational design algorithm were used as starting structures where no crystallographic data was available. Substrate parameters were generated within the Antechamber module of AMBER 10 (Case et al., 2008) using the general AMBER force field, with partial charges set to fit the electrostatic potential generated at

HF/6-31G* by RESP (Bayly, Cieplak, Cornell, & Kollman, 1993). The charges were calculated according to the Merz–Singh–Kollman scheme (Besler, Merz, & Kollman, 1990; Wang, Wolf, Caldwell, Kollman, & Case, 2004) using Gaussian 03 (Frisch et al., 2004). Depending on the shape of the protein, each system was solvated with an orthorhombic or octahedral box of TIP3P water molecules (Jorgensen, Chandrasekhar, Madura, Impey, & Klein, 1983) (depending on which gave the smaller number of particles). The resulting water boxes generated a solvent layer of at least 10 Å from any point on the protein surface. Charged groups were neutralized by addition of explicit counter ions. Periodic boundary conditions were employed. Each protein–substrate complex was geometry-optimized and then heated from 0 to 300 K in six 50-ps steps for a total of 300 ps at NVT conditions. Weak harmonic restraints of 10 kcal/mol were applied to the solute, and the Langevin equilibration scheme was used to control and equalize the temperature. The time step was kept at 1 fs during the heating stages, allowing potential inhomogeneities to self-adjust. Each system was then equilibrated for 2 ns with a 2-fs time step at a constant pressure of 1 atm. Water molecules were triangulated with the SHAKE algorithm such that the angle between the hydrogen atoms is kept fixed. 20-ns NPT production MD simulations were carried out for each system (with and without substrate(s) bound to the active site) at 1 bar and 300 K. Geometries and velocities were saved every 100 steps (0.2 ps), which resulted in a total of 10,000 and 100,000 frames from each production run. Long-range electrostatic effects were modeled using the particle-mesh-Ewald method (Darden, York, & Pedersen, 1993).

8. CONCLUSIONS

Nature's enzymes achieve proficiency by precisely placing functional groups into active sites, by minimizing their conformational motions, and through a balanced management of active-site waters. They create finely tuned microenvironments in which catalytic residues display significantly shifted pK_a values and in which activated water molecules can become extensions of existing catalytic machineries. Computationally designed enzymes lag far behind in each of these categories. And so the future challenges for the inside-out approach go beyond preventative measures that are concerned with fortifying a final active site of a design against unproductive arrangements. Building on this, MD-based assessments are now routinely employed for uncovering inadequacies of designed active sites, and the findings are fed back into the continuously evolving design protocol. Active designs have well-defined catalytic contacts and active sites that closely resemble the QM-theozyme geometries that they are based on. Inactive proteins show poorly behaved catalytic contacts that deviate significantly from the designed active-site arrangements. To a large degree, this can be traced back to competing polar contacts of the catalytic unit with other residues within the active site and with solvent molecules. RMSD distributions are helpful at pinpointing structural inadequacies, pairwise contact distributions paint a more detailed picture of the active site arrangements, and the average number of water molecules around a given functional group can give a crude sense of the degree to which the pK_a of the microenvironment is modulated. More sophisticated statistical analyses of the MD data and alternative MD protocols are currently being explored.

ACKNOWLEDGMENTS

This work was supported by the National Institute of General Medical Sciences, National Institutes of Health, grant GM075962. G. K. is grateful for support from the LLNL Lawrence Scholars Program and from an NIH Simbios postdoctoral fellowship. Computational resources at ERDC, AFRL, LLNL, and IDRE were essential to this work and are gratefully acknowledged.

REFERENCES

Axe, D. D. (2004). Estimating the prevalence of protein sequences adopting functional enzyme folds. *Journal of Molecular Biology*, *341*, 1295–1315.

Bas, D. C., Rogers, D. M., & Jensen, J. H. (2008). Very fast prediction and rationalization of pKa values for protein–ligand complexes. *Proteins*, *73*, 765–783.

Bayly, C. I., Cieplak, P., Cornell, W., & Kollman, P. A. (1993). A well-behaved electrostatic potential based method using charge restraints for deriving atomic charges: The RESP model. *The Journal of Physical Chemistry*, *97*, 10269–10280.

Behrens, G. A., Hummel, A., Padhi, S. K., Schaetzle, S., & Bornscheuer, U. T. (2011). Discovery and protein engineering of biocatalysts for organic synthesis. *Advanced Synthesis and Catalysis*, *353*, 2191–2215.

Besler, B. H., Merz, K. M., Jr., & Kollman, P. A. (1990). Atomic charges derived from semiempirical methods. *Journal of Computational Chemistry*, *11*, 431–439.

Bolon, D. N. (2001). From the Cover: Enzyme-like proteins by computational design. *Proceedings of the National Academy of Sciences*, *98*, 14274–14279.

Case, D. A., Darden, T. A., Cheatham, T. E., Simmerling, C. L., Wang, J., Duke, R. E., et al. (2008). *Amber 10, a publication of the American Chemical Society*. San Francisco: University of California.

Dahiyat, B. I., & Mayo, S. L. (1996). Protein design automation. *Protein Science*, *5*, 895–903.

Dantas, G. G., Kuhlman, B. B., Callender, D. D., Wong, M. M., & Baker, D. D. (2003). A large scale test of computational protein design: Folding and stability of nine completely redesigned globular proteins. *Journal of Molecular Biology*, *332*, 449–460.

Darden, T., York, D., & Pedersen, L. (1993). Particle mesh Ewald: An N. log(N) method for Ewald sums in large systems. *The Journal of Chemical Physics*, *98*, 10089–10092.
Debler, E. W., Müller, R., Hilvert, D., & Wilson, I. A. (2008). Conformational isomerism can limit antibody catalysis. *The Journal of Biological Chemistry*, *283*, 16554–16560.
Fersht, A. (1999). *Structure and mechanism in protein science: A guide to enzyme catalysis and protein folding*. New York: W.H. Freeman.
Frisch, M. J., Trucks, G. W., Schlegel, H. B., Scuseria, G. E., Robb, M. A., Cheeseman, J. R., et al. (2004). *Gaussian 03, Revision C.02*. Wallingford, CT: Gaussian, Inc.
Gerlt, J. A., & Babbitt, P. C. (2009). Enzyme (re)design: Lessons from natural evolution and computation. *Current Opinion in Chemical Biology*, *13*, 10–18.
Gouverneur, V. E., Houk, K. N., de Pascual-Teresa, B., Beno, B., Janda, K. D., & Lerner, R. A. (1993). Control of the exo and endo pathways of the Diels-Alder reaction by antibody catalysis. *Science*, *262*, 204–208.
Harbury, P. B., Plecs, J. J., Tidor, B., Alber, T., & Kim, P. S. (1998). High-resolution protein design with backbone freedom. *Science*, *282*, 1462–1467.
Hollfelder, F., Kirby, A. J., & Tawfik, D. S. (2001). On the magnitude and specificity of medium effects in enzyme-like catalysts for proton transfer. *The Journal of Organic Chemistry*, *66*, 5866–5874.
Hu, Y., Houk, K. N., Kikuchi, K., Hotta, K., & Hilvert, D. (2004). Nonspecific medium effects versus specific group positioning in the antibody and albumin catalysis of the base-promoted ring-opening reactions of benzisoxazoles. *Journal of the American Chemical Society*, *126*, 8197–8205.
Jäckel, C., Kast, P., & Hilvert, D. (2008). Protein design by directed evolution. *Annual Review of Biophysics*, *37*, 153–173.
Janda, K. D., Shevlin, C. G., & Lerner, R. A. (1993). Antibody catalysis of a disfavored chemical transformation. *Science*, *259*, 490–493.
Jiang, L., Althoff, E. A., Clemente, F. R., Doyle, L., Röthlisberger, D., Zanghellini, A., et al. (2008). De novo computational design of retro-aldol enzymes. *Science*, *319*, 1387–1391.
Johnsson, K., Allemann, R. K., Widmer, H., & Benner, S. A. (1993). Synthesis, structure and activity of artificial, rationally designed catalytic polypeptides. *Nature*, *365*, 530–532.
Jorgensen, W. L., Chandrasekhar, J., Madura, J. D., Impey, R. W., & Klein, M. L. (1983). Comparison of simple potential functions for simulating liquid water. *The Journal of Chemical Physics*, *79*, 926–935.
Kaplan, J., & DeGrado, W. F. (2004). De novo design of catalytic proteins. *Proceedings of the National Academy of Sciences of the United States of America*, *101*, 11566–11570.
Khersonsky, O., Kiss, G., Röthlisberger, D., Dym, O., Albeck, S., Houk, K. N., et al. (2012). Bridging the gaps in design methodologies by evolutionary optimization of the stability and proficiency of designed Kemp eliminase KE59. *Proceedings of the National Academy of Sciences*, *109*, 10358–10363.
Khersonsky, O., Röthlisberger, D., Dym, O., Albeck, S., Jackson, C. J., Baker, D., et al. (2010). Evolutionary optimization of computationally designed enzymes: Kemp eliminases of the KE07 series. *Journal of Molecular Biology*, *396*, 1025–1042.
Khersonsky, O., Röthlisberger, D., Wollacott, A. M., Murphy, P., Dym, O., Albeck, S., et al. (2011). Optimization of the in-silico-designed kemp eliminase KE70 by computational design and directed evolution. *Journal of Molecular Biology*, *407*, 391–412.
Kikuchi, K., Hannak, R., Guo, M., Kirby, A., & Hilvert, D. (2006). Toward bifunctional antibody catalysis. *Bioorganic & Medicinal Chemistry*, *14*, 6189–6196.
Kiss, G., Johnson, S. A., Nosrati, G., Çelebi-Ölçüm, N., Kim, S., Paton, R., et al. (2011). Computational Design of New Protein Catalysts. In P. Comba (Ed.), *Modeling of molecular properties*. Weinheim, Germany: Wiley-VCH Verlag GmbH & Co. KGaA.
Kiss, G., Röthlisberger, D., Baker, D., & Houk, K. N. (2010). Evaluation and ranking of enzyme designs. *Protein Science*, *19*, 1760–1773.

Lerner, R. A., Benkovic, S. J., & Schultz, P. G. (1991). At the crossroads of chemistry and immunology: Catalytic antibodies. *Science, 252*, 659–667.
Lewis, C., Kramer, T., Robinson, S., & Hilvert, D. (1991). Medium effects in antibody-catalyzed reactions. *Science, 253*, 1019–1022.
Li, T., Janda, K. D., Ashley, J. A., & Lerner, R. A. (1994). Antibody catalyzed cationic cyclization. *Science, 264*, 1289–1293.
Li, H., Robertson, A. D., & Jensen, J. H. (2005). Very fast empirical prediction and rationalization of protein pKa values. *Proteins, 61*, 704–721.
Lo Leggio, L., Kalogiannis, S., Eckert, K., Teixeira, S. C., Bhat, M. K., Andrei, C., et al. (2001). Substrate specificity and subsite mobility in T. aurantiacus xylanase 10A. *FEBS Letters, 509*, 303–308.
Müller, R., Debler, E. W., Steinmann, M., Seebeck, F. P., Wilson, I. A., & Hilvert, D. (2007). Bifunctional catalysis of proton transfer at an antibody active site. *Journal of the American Chemical Society, 129*, 460–461.
Pauling, L. (1948). Nature of forces between large molecules of biological interest. *Nature, 161*, 707–709.
Privett, H. K., Kiss, G., Lee, T. M., Blomberg, R., Chica, R. A., Thomas, L. M., et al. (2012). Iterative approach to computational enzyme design. *PNAS, 109*, 3790–3795.
Richter, F., Leaver-Fay, A., Khare, S. D., Bjelic, S., & Baker, D. (2011). De novo enzyme design using Rosetta3. *PLoS One, 6*, e19230.
Rossi, P., Tecilla, P., Baltzer, L., & Scrimin, P. (2004). De novo metallonucleases based on helix-loop-helix motifs. *Chemistry, 10*, 4163–4170.
Röthlisberger, D., Khersonsky, O., Wollacott, A. M., Jiang, L., DeChancie, J., Betker, J., et al. (2008). Kemp elimination catalysts by computational enzyme design. *Nature, 453*, 190–195.
Ruscio, J. Z., Kohn, J. E., Ball, K. A., & Head-Gordon, T. (2009). The influence of protein dynamics on the success of computational enzyme design. *Journal of the American Chemical Society, 131*, 14111–14115.
Sasaki, T., & Kaiser, E. T. (1989). Helichrome: Synthesis and enzymic activity of a designed hemeprotein. *Journal of the American Chemical Society, 111*, 380–381.
Siegel, J. B., Zanghellini, A., Lovick, H. M., Kiss, G., Lambert, A. R., St Clair, J. L., et al. (2010). Computational design of an enzyme catalyst for a stereoselective bimolecular Diels-Alder reaction. *Science, 329*, 309–313.
Steiner, T. (2002). The hydrogen bond in the solid state. *Angewandte Chemie (International Ed. in English), 41*, 48–76.
Taylor, S. V., Kast, P., & Hilvert, D. (2001). Investigating and engineering enzymes by genetic selection. *Angewandte Chemie (International Ed. in English), 40*, 3310–3335.
Thorn, S. N., Daniels, R. G., Auditor, M. T., & Hilvert, D. (1995). Large rate accelerations in antibody catalysis by strategic use of haptenic charge. *Nature, 373*, 228–230.
Tommos, C., Skalicky, J. J., Pilloud, D. L., Wand, A. J., & Dutton, P. L. (1999). De novo proteins as models of radical enzymes. *Biochemistry, 38*, 9495–9507.
Wagner, J., Lerner, R. A., & Barbas, C. F. (1995). Efficient aldolase catalytic antibodies that use the enamine mechanism of natural enzymes. *Science, 270*, 1797–1800.
Wang, J., Wolf, R. M., Caldwell, J. W., Kollman, P. A., & Case, D. A. (2004). Development and testing of a general amber force field. *Journal of Computational Chemistry, 25*, 1157–1174.
Zanghellini, A., Jiang, L., Wollacott, A. M., Cheng, G., Meiler, J., Althoff, E. A., et al. (2006). New algorithms and an in silico benchmark for computational enzyme design. *Protein Science, 15*, 2785–2794.
Zhang, X., & Houk, K. N. (2005). Why enzymes are proficient catalysts: Beyond the Pauling paradigm. *Accounts of Chemical Research, 38*, 379–385.
Zhao, H. (2007). Directed evolution of novel protein functions. *Biotechnology and Bioengineering, 98*, 313–317.

CHAPTER EIGHT

Multistate Protein Design Using CLEVER and CLASSY

Christopher Negron*, **Amy E. Keating***,†,1
*Computational and Systems Biology Program, Massachusetts Institute of Technology, Cambridge, Massachusetts, USA
†Department of Biology, Massachusetts Institute of Technology, Cambridge, Massachusetts, USA
[1]Corresponding author: e-mail address: keating@mit.edu

Contents

Abstract

Structure-based protein design is a powerful technique with great potential. Challenges in two areas limit performance: structure scoring and sequence-structure searching. Many of the functions used to describe the relationship between protein sequence and energy are computationally expensive to evaluate, and the spaces that must be searched in protein design are enormous. Here, we describe computational tools that can be used in certain situations to provide enormous accelerations in protein design. Cluster expansion is a technique that maps a complex function of three-dimensional atomic coordinates to a simple function of sequence. This is done by expanding the sequence-energy relation as a linear function of sequence variables, which are fit using training examples. Generating a simpler function speeds up scoring dramatically, relative to all-atom methods, and facilitates the use of new types of search strategies. The application of cluster expansion in protein modeling is new but has shown utility for design problems that require simultaneous consideration of multiple states. In this chapter, we describe cases where cluster expansion can be useful, outline how to

Methods in Enzymology, Volume 523
ISSN 0076-6879
http://dx.doi.org/10.1016/B978-0-12-394292-0.00008-4

generate a cluster-expanded version of any existing scoring procedure using the software CLEVER, and describe how to apply a cluster-expanded potential to multistate protein design using the CLASSY method.

1. INTRODUCTION: ACCOMPLISHMENTS AND LIMITATIONS OF STRUCTURE-BASED DESIGN

Nearly 30 years ago, Drexler suggested that proteins had the potential to be manipulated to create molecular machines with predefined functions (Drexler, 1981). At that time, a realistic strategy for designing proteins rationally could not be envisioned in detail. Several groups subsequently tackled protein design using computational structure-based methods, culminating in the first fully automated design of a folding protein sequence in 1997 (Dahiyat, Gordon, & Mayo, 1997). Many researchers have now demonstrated impressive accomplishments in this area, including the engineering of protein inhibitors of therapeutically relevant targets (Fleishman et al., 2011), the creation of novel enzymes (Jiang et al., 2008; Röthlisberger et al., 2008), and the assembly of molecular structures that incorporate proteins and other materials (Grigoryan et al., 2011).

Modern structure-based design requires two things: an energy function for evaluating candidate sequences and an algorithm that can search the enormous space of sequence-structure possibilities. The requirements for each are linked because the nature of the scoring function dictates what kinds of searches are possible. One of the many limitations of commonly used scoring functions is that they can be costly to calculate. For example, all-atom scoring functions must, at a minimum, evaluate interactions between all pairs of atoms that lie within a prescribed distance. Computing electrostatic interactions can be particularly expensive, depending on the method used. Several techniques have been developed to increase the speed of energy evaluation. For example, Leaver-Fay et al. implemented a tree data structure in RosettaDesign to eliminate redundant calculation of atom–atom interactions, and this gave a fourfold speedup in the calculation of pairwise energy terms (Leaver-Fay, Kuhlman, & Snoeyink, 2005). Many groups have also worked on speeding up the search component of design. Early recognition that optimal search strategies using algorithms such as dead-end elimination (DEE) are often too slow for real design problems led to widespread adoption of stochastic sampling methods such as Monte Carlo optimization with simulated annealing and genetic algorithms (Havranek & Harbury, 2003; Kuhlman et al., 2003; Voigt, Gordon, & Mayo, 2000). FASTER is a particularly noteworthy stochastic sampling method that was shown to be 100–1000 times faster than DEE (Desmet, Spriet, & Lasters, 2002) and subsequently improved further (Allen & Mayo, 2006).

Despite these innovations, design problems involving large proteins, extensive structural sampling, or a large number of states can still be computationally intractable. In this chapter, we focus particularly on challenges posed by multistate design. In a multistate design problem, the designer is concerned not just with a single desired structure or function of interest but with numerous states either desired or undesired. For example, when designing dominant-negative inhibitors, it is important to avoid self-interaction or interactions with other proteins in the cell (Chen, Reinke, & Keating, 2011). Harbury was among the first to treat multistate design, designing topologically specific coiled-coil structures using multiple backbone templates (Harbury, Plecs, Tidor, Alber, & Kim, 1998). Since then, several groups have proposed different approaches (Allen & Mayo, 2010; Bolon, Grant, Baker, & Sauer, 2005; Havranek & Harbury, 2003; Humphris & Kortemme, 2007; Leaver-Fay, Jacak, Stranges, & Kuhlman, 2011; Yanover, Fromer, & Shifman, 2007; Kortemme et al., 2004; Sammond et al., 2010), and several excellent reviews cover this topic (Erijman, Aizner, & Shifman, 2011; Havranek, 2010; Karanicolas & Kuhlman, 2009).

In one example from our laboratory, Grigoryan et al. used multistate design to engineer specific binding partners for representative members of 20 human basic-region leucine zipper (bZIP) transcription factor families (Grigoryan, Reinke, & Keating, 2009). For this purpose, a novel computational solution provided both a dramatic acceleration of energy evaluation and an efficient way to search a complex, multistate design landscape. The approach used a method called cluster expansion (CE) to convert structure-based models of protein energetics into sequence-based models. Grigoryan et al. showed that CE can speed up energy calculations by seven orders of magnitude (Grigoryan et al., 2006). Furthermore, the use of cluster-expanded energy functions allowed a novel application of integer linear programming (ILP) to solve the multistate design problem. With the expectation that this approach can be applied to other problems in protein design, we illustrate the use of the open-source program CLEVER 1.0 to generate CE scoring functions, and discuss how such energy functions can be used in conjunction with ILP in the multistate design method CLASSY.

2. THEORY

CE is a general technique for deriving a simple linear function that approximates a complex mathematical expression. It involves fitting a set of coefficients to describe a space covered by a user-defined set of relevant variables. CE is used extensively in modeling alloys (de Fontaine, 1994; Sanchez, Ducastelle, & Gratias, 1984), and Zhou et al. demonstrated how to use CE to score the fitness of a protein sequence for a given protein structure (Zhou et al., 2005). That is, these authors showed how to apply CE when the complex mathematical expression to be described is the energy of a protein sequence adopting a particular three-dimensional structure. In this application, the energy of a protein sequence is written as an expansion around a reference sequence, with energies from specific amino acids and groups of amino acids contributing to the expansion. Two key assumptions are that lower order terms such as interactions between pairs of amino acids at pairs of sites contribute more to the energy than higher order clusters involving many residues, and that, consistent with this, a limited number of residue interaction terms are sufficient to approximate the energy of a protein. These assumptions are well aligned with the physical intuition of structural biologists, who expect short-range pairwise interactions to dominate an energy expression.

A brief description of the theory of CE as used for protein energetics is presented here. For a more detailed description, see Grigoryan et al. (2006). Let the variable σ^i index the amino acid at site i. If there are M allowed amino acids at site i then σ^i can take the values from 0 to $(M-1)$. For a protein of length L amino acids, values of i range from 1 to L, and an amino acid sequence is represented by the vector $\boldsymbol{\sigma} = [\sigma^1 \ldots \sigma^L]$. The energy of a sequence, $E(\boldsymbol{\sigma})$, based on an expansion around a reference sequence, is expressed as:

$$E(\boldsymbol{\sigma}) = J_0 + \sum_{\substack{i \\ \sigma^i \neq 0}} J^i_{\sigma^i} + \sum_{\substack{i \\ \sigma^i \neq 0}} \sum_{\substack{j>i \\ \sigma^j \neq 0}} J^{ij}_{\sigma^i \sigma^j} + \cdots \qquad [8.1]$$

The J parameters are *effective cluster interaction* (ECI) values. J_0 is a constant term that reflects the energy of the reference sequence, and the other J values give the contributions of amino acids and amino acid combinations relative to this reference. $J^i_{\sigma^i}$ represents the energetic contribution of a single amino acid σ^i at site i, and $J^{ij}_{\sigma^i \sigma^j}$ represents the energetic contribution from a pair of amino acids σ^i and σ^j at sites i and j, etc. All higher order interactions, up to L-body terms, would be needed to obtain an exact expansion. The goal in deriving a CE is to find a minimal set of ECI values that provide an accurate estimate of the energy. ECI values are eliminated by truncating the expansion so that it does not include high-order terms, and by testing ECI values that capture low-order terms to confirm that they make important contributions (and, if they do not, removing them as described below).

ECI values are determined by fitting, using a training set of sequences for which the correct function value according to some model or experiment is known. In our application, this is the protein energy $E(\boldsymbol{\sigma})$ computed using an all-atom model. Based on Eq. (8.1), for any training set with N sequences we can write

$$\boldsymbol{E} = \boldsymbol{XJ} \quad [8.2]$$

where $\boldsymbol{E}$ is an N-dimensional column vector of energies for the training sequences, $\boldsymbol{J}$ is a P-dimensional column vector of ECI values, and $\boldsymbol{X}$ is an $N \times P$ binary matrix indicating which residues and combinations of residues contribute to the energy in each training sequence. The presence/absence of different sets of residues is stored in "cluster functions" (or CFs) that compose matrix $\boldsymbol{X}$. For example, a CF indicating the presence of alanine at site 1 and leucine at site 2 in a protein would evaluate to 1 for a given sequence only if that sequence had that combination of amino acids and to 0 otherwise. Each CF has an associated ECI. For a given set of CFs, the training-set sequences and their energies define matrix $\boldsymbol{X}$ and vector $\boldsymbol{E}$ in Eq. (8.2). When there are more training sequences than ECI values, the system is over-determined, which allows techniques such as least-squares fitting to be used to find the optimal values for the unknown parameters $\boldsymbol{J}$.

The procedure used by CLEVER 1.0 to select CFs/ECIs to be included is described in Fig. 8.1 and more details can be found in Hahn, Ashenberg, Grigoryan, and Keating (2010). At the outset, the user defines a set of candidate CFs (candidate amino acid combinations) likely to contribute significantly to the total energy. An iterative procedure is then used to determine which of these should be included in the final expansion, using a leave-sequence-out cross-validation procedure to avoid overfitting. After fitting, the accuracy of the expansion can be evaluated by scoring another set of sequences, known as the test set, with both the original energy function and the cluster-expanded version of it.

3. BENEFITS OFFERED BY CLUSTER EXPANSION IN PROTEIN MODELING AND DESIGN

There are several reasons a protein modeler or designer might want to develop a cluster-expanded version of their energy function of interest. To understand these, it is necessary to focus on what CE delivers, which is an approximate version of a scoring procedure that is extremely rapid and convenient to evaluate. The cost of obtaining this benefit is the diminished

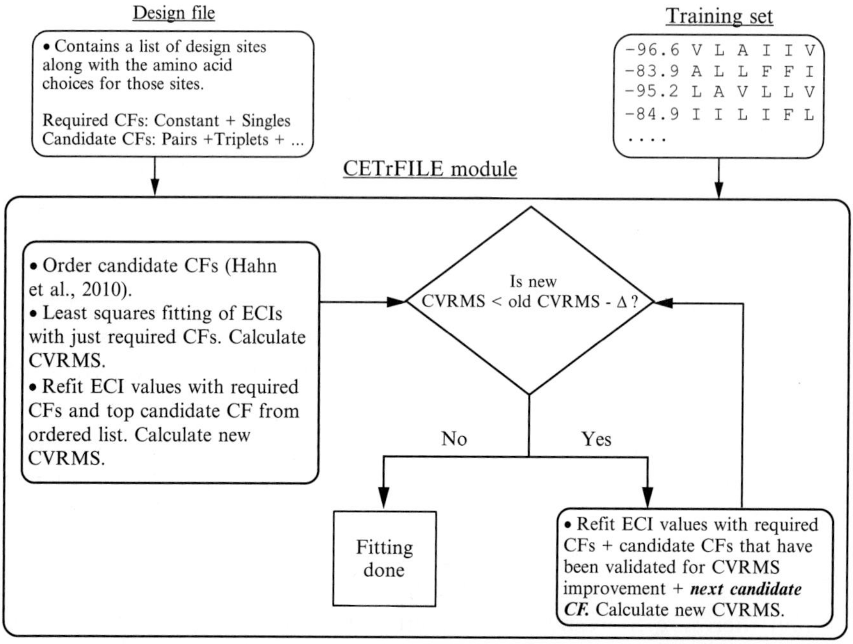

Figure 8.1 Procedure for fitting a cluster expansion. Two inputs are required to train a cluster expansion for protein design using the CLEVER package: the design file (see Fig. 8.2) and the training-set file. The design file lists a set of candidate cluster functions (CFs) and defines the sequence space the user is interested in describing. The training-set file provides a set of sequences with associated energies. The fitting procedure fits a subset of the variables listed in the design file to reproduce the training-set data. In our implementation, the constant and point CFs are always included in the fitting process. To avoid overfitting, pair and higher order CFs are incrementally added, following an order that is predetermined at an early stage of the fitting routine. The progress of the fitting is monitored using the cross-validated root mean square (CVRMS) error. When a new CF is added, all terms are refit, and a new CVRMS is calculated. If the CVRMS score improves by at least Δ, the new CF is accepted and used in the final expansion. If the CVRMS does not improve, the fitting process ends. The goal is to find the smallest number of CFs that must be included to give a good model.

accuracy of the CE function and the time required to develop it. Also, it should be emphasized that in all cases tested so far, the relationships derived between sequence and energy using CE have assumed conservation of the underlying protein backbone structure for all sequences. That is, CE delivers a structure-specific scoring function. Apgar et al. observed good performance when cluster expanding structure-based models that included some treatment of backbone flexibility, but these structural changes were very small (Apgar, Hahn, Grigoryan, & Keating, 2009).

CE provides a significant speedup to energy evaluation and thus will be of greatest benefit when the energy evaluation confronted in design is especially challenging. For example, electrostatic energies are often treated in a very crude way in protein design, in order to make the resulting functions expressible as a sum over residue pairs. More accurate functions can be much more costly to compute (Lippow, Wittrup, & Tidor, 2007). CE provides an attractive solution in such cases, and Grigoryan et al. explored the expansion of various scoring methods including a generalized Born treatment of electrostatics (Grigoryan et al., 2006). In another, more extreme example, CE can be used when protein energies are determined experimentally for a training set of interest. The time and expense of experimental protein characterization means that only a part of sequence space will ever be covered this way. But if sufficient examples are available, it may be possible to train a CE expression that can be used to guide protein design. Hahn et al. presented an example of using experimental data to train a CE for SH3 domain protein–peptide interactions (Hahn et al., 2010).

Importantly, CE energy expressions also help address the search problem in design. Beyond just speeding up standard Monte Carlo searches, CE energy functions can be used to formulate protein design as an integer linear program in sequence space. The ILP provides provably optimal solutions and flexibility in optimization (see below). Another advantage stems from the fact that in multistate design, the best approach for combining the scores of many states into one objective may not be clear. Deriving CE functions for all of the states allows facile searching and researching using a variety of different objectives and also allows tradeoffs to be rigorously explored, for example, between optimizing stability and specificity. This is discussed further below, where we illustrate one way to do this.

Overall, the suitability of CE for a particular problem will depend on many things, including the accuracy with which a desired scoring method can be approximated by its expansion. Previous work has shown that this varies considerably for different scoring functions and different structures. When a cluster-expanded scoring method has low error, it provides a tremendous advantage to the search part of the problem and is worth the cost of fitting. When the error is moderate, CE may still provide a useful filter to help identify promising parts of the sequence space that can then be examined in more detail with more expensive calculations. As more work is done, it will become easier to judge those problems for which CE will provide the greatest benefit.

4. HOW TO RUN A CLUSTER EXPANSION WITH CLEVER 1.0

Hahn et al. created the open-source package CLEVER 1.0, available at http://web.mit.edu/biology/keating/software/, to aid users in developing their own CE models (Hahn et al., 2010). Here, we provide an overview of how to use CLEVER to cluster expand an arbitrary scoring method provided by the user. The discussion is geared toward a new user of the program. More details can be found in the original papers. Instructions for installing CLEVER 1.0 can be found in the clever1.0 manual at http://web.mit.edu/biology/keating/software/, or in the docs subdirectory that is created when unzipping clever1.0-package.zip.

There are three executable modules in the CLEVER 1.0 package. First is the GenSeqs module, which can help the user generate sequences for both training and testing a CE. The second executable is the CETrFILE module. This program executes the crucial step of fitting the ECI values, as described in the theory section. The third module is the CEEnergy module, which uses the CE trained by CETrFILE to score sequences. This module can be combined with other data to assess CE performance.

5. GenSeqs

The GenSeqs, or Generate Sequences, module helps with the preparation of unique training and test-set sequences. It requires two inputs. The first is the desired number of training sequences the user would like GenSeqs to return. A rule of thumb is to use at least 2.5 times the number of training sequences as the number of CFs, although a larger number can reduce the error, as discussed in some detail in Hahn et al. (2010). The other input for the module is the design file, which is crucial to many aspects of the CLEVER package. An example design file is shown in Fig. 8.2. The design file states which amino acids are allowed at each of the design sites. It is not necessary to include any site where residues will not be changed. Given the appropriate input, GenSeqs generates random sequences by selecting amino acids uniformly from the allowed amino acids at each of the sites. Each sequence generated is checked for uniqueness such that the training set contains no repeated sequences. In addition, the -o flag combined with the training-set file can be used to generate test sequences not present in the training set. It should be noted that sampling uniformly from the set of amino acids at each of the positions may not provide a good description

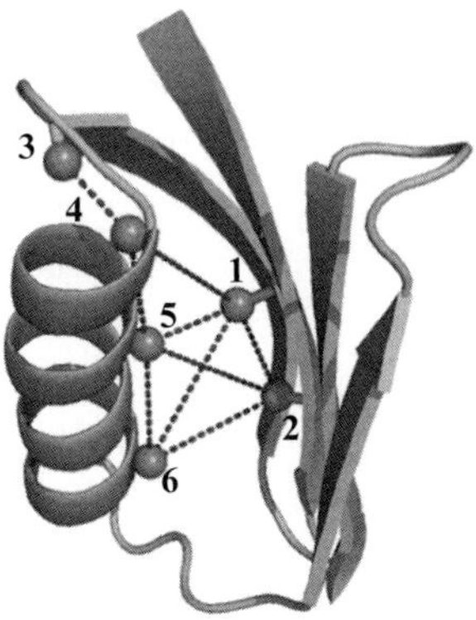

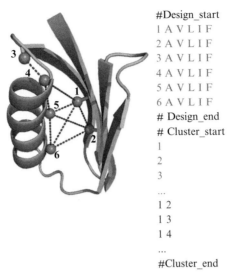

```
#Design_start
1 A V L I F
2 A V L I F
3 A V L I F
4 A V L I F
5 A V L I F
6 A V L I F
# Design_end
# Cluster_start
1
2
3
...
1 2
1 3
1 4
...
#Cluster_end
```

Figure 8.2 Example design file. On the left is a structure of streptococcal protein Gβ1 from crystal structure 1PGA (Gallager, Alexander, Bryan, & Gillard, 1994). The Cβ atoms of the six sites chosen for variation in design are shown as spheres. For clarity, only subsets of the possible pair interactions between the sites are shown as dashed lines. On the right is a sample design file. It is composed of two parts. The first half lists sites where the sequence will vary, and these sites are known as design sites. Each design site line lists the amino acids that will be allowed at that site. In the second half of the design file, the user lists the cluster functions to consider for inclusion in the expansion. A user must list all single sites. Only a subset of the cluster functions used for fitting in this example is shown. (For color version of this figure, the reader is referred to the online version of this chapter.)

of the sequence space the user wants the CE to cover. For example, some amino acid substitutions at certain sites may be considered much more common or important, and a user might want to include these with higher frequency. In such a case, the user should generate sequences using their own distribution for the amino acids at each of the sites. An example command line for GenSeqs that would generate 3000 random sequences based on information in design.file is

GenSeqs -n 3000 -d design.file

6. CETrFILE

The CETrFILE, or CE Training File, module uses the procedure in Fig. 8.1 to fit ECI values. This is the heart of the CE method. The module requires two inputs. The CETrFILE module, like GenSeqs, requires a design file. As mentioned earlier, the design file states which amino acids are allowed at each of the design sites. The first amino acid in each design site position is taken as the reference amino acid for that position. The design file also lists which single and higher order interactions among the design sites should be considered during fitting. This is an important choice, made by the user, which can strongly influence performance. There must be a single body term for each of the design sites in order for the code to run; inclusion of pair or higher terms is optional. The second required input is the training-set file, which includes a list of energies paired with sequences. An example of the format for this file can be found in the clever 1.0 manual. Briefly, this file should have a column of energies that can come from any source. Each energy is followed by the sequence of residues at the design site positions. All sequences should be the same length because all sequences should have the same number of design sites. CETrFILE outputs several things to "standard out" such as a table containing all of the ECI values for all of the CFs. CETrFILE also outputs a binary file containing ECI values trained from the input data. An example command line for CETrFILE is

CETrFile -d design.file -s sequence.file –r training.result

7. CEEnergy

CEEnergy scores sequences with the derived CE. This module uses the binary output of CETrFILE, for example, *training.result*, and a sequence file with a list of test sequences. The format of the test-sequence file is the same as the training-sequence file, except the energy column is not used. To get a good idea of expected performance on new problems, test sequences should not overlap with sequences that the CE was trained on. Test sequences are specified using only the design site residues. An example command line for CEEnergy is

CEEnergy -r training.result -s sequence.file

8. CLUSTER EXPANSION CASE STUDY

In this section, we provide a simple illustrative example of how the RosettaDesign conformational energy of selected sites in streptococcal protein Gβ1 can be cluster expanded using CLEVER 1.0. The Gβ1 structure is composed of an alpha helix lying across a beta sheet made up of two beta hairpins. Dahiyat et al. choose the Gβ1 domain as one of the first targets for redesign using automated software (Dahiyat & Mayo, 1997). Unlike coiled coils, for which CE has been used in numerous published examples (Apgar et al., 2009; Grigoryan et al., 2006; Hahn et al., 2010; Zhou et al., 2005), Gβ1 lacks any structural or sequence symmetry and thus represents a generic globular fold. Here, we select only a few residues for modeling, to keep the example very simple.

The first step is to create a design file as shown in Fig. 8.2. We selected six positions in and around the core of the Gβ1 structure to vary and designated these as design sites, labeled 1 through 6, in the design file. Our choices correspond to residue positions 5, 7, 20, 26, 30, and 34 in PDB structure 1PGA (Gallager et al., 1994). We allowed the same small set of hydrophobic residues (A, V, L, I, and F) at each of these mostly buried positions. Alanine is listed first for each design position and serves as the reference at each site. In the bottom half of the design file, we specified the interactions between the design sites that should be considered for CE. Only single and pair interactions between residues were considered, for simplicity, though it is possible to include triplets of amino acids or even higher order terms. Higher order terms may improve the accuracy of a CE, and suggested techniques for choosing them can be found in (Grigoryan et al., 2006). In this case study, we considered 15 pair clusters and 25 possible amino acid combinations for each pair, resulting in 375 ECI values to be fit. To fit these terms, 2000 random training sequences, drawn from the possible design space of $5^6 = 15{,}625$ sequences, were generated using the GenSeqs module.

The training sequences were modeled on the 1PGA structure and scored using RosettaDesign with the "soft-potential" and "minimize side-chain" flags. For simplicity, only the side chains of the design site residues were optimized. All other side chains were fixed in their crystal-structure coordinates. The corresponding RosettaDesign energy was then paired with each of the 2000 training sequences to make the training-set file, which spanned an energy range over 70 Rosetta energy units (also referred to as kcal/mol). The training-set file and the design file were used to train a CE using the CETrFILE module.

Once a CE is trained, it is crucial to evaluate its accuracy, as this can vary widely based on the type of problem, the selection of candidate CFs, and the underlying scoring method being approximated (Apgar et al., 2009; Grigoryan et al., 2006; Hahn et al., 2010). The CETrFILE module reports the cross-validated root mean square (CVRMS) error. This is a metric for assessing how well the results of a predictive model generalize to an independent data set. In this specific example, it is a measure of how well the CE would be expected to predict the RosettaDesign energy of a sequence threaded onto the Gβ1 structure. CE of the RosettaDesign energy function on the 1PGA structure gave a CVRMS score of 1.5 kcal/mol, as shown in Fig. 8.3A. An additional test of the error is to generate a test set of sequences, independent of the training set, and score them using both the structure-based method and the newly derived CE. For our example, the GenSeqs module was used to generate 2000 test-set sequences nonoverlapping with the training set that spanned an energy range of nearly 80 kcal/mol. The root mean square deviation (RMSD) between the scores for the test-set sequences derived from the CE model and the structure-based model was 1.6 kcal/mol, which is shown in Fig. 8.3B. Overall, this CE performed very well at approximating the structure-based method it was derived from.

The example shown here is very simple. Often, good performance requires refinement of the expansion protocol. There are several techniques for reducing error discussed by Hahn et al. (2010). For example, introducing higher order CFs such as triplets, increasing training-set size, or decreasing the number of variable amino acids are approaches for reducing error. Additionally, Hahn et al. presented techniques for identifying and removing from the design space CFs that are particularly poorly fit.

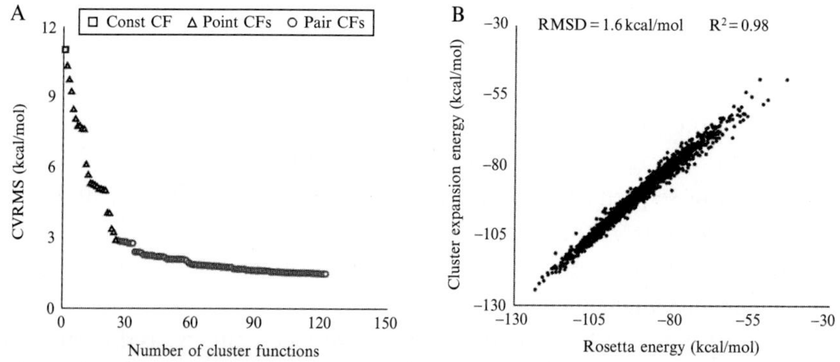

Figure 8.3 Cluster expansion error in the Gβ1 example. (A) Evolution of the CVRMS as CFs were added to the model. The type of CF added in each iteration is indicated by shape, as shown in the legend. A total of 122 CFs were added, giving a CVRMS of 1.5 Rosetta energy units (also referred to as kcal/mol). (B) The performance of the cluster expansion on 2000 randomly generated sequences not included in the training set. The RMSD between the CE test-set energies and the RosettaDesign test-set energies was 1.6 kcal/mol. (For color version of this figure, the reader is referred to the online version of this chapter.)

9. USING CLUSTER EXPANSION WITH INTEGER LINEAR PROGRAMMING

As mentioned earlier, CE energy functions are amenable to linear optimization techniques. Grigoryan et al. combined CE with ILP, resulting in the computational protocol called CLASSY (Chen et al., 2011; Grigoryan et al., 2009). The use of ILP for protein design was described previously by Kingsford et al. and is most easily explained for single-state design using the graph shown in Fig. 8.4. Here, each cluster of nodes represents a design site with the associated nodes corresponding to design choices (Kingsford, Chazelle, & Singh, 2005). For Kingsford et al., these choices represented different conformations of one or more residues; in our case, they represent different residues because we are designing at the sequence level. A design solution corresponds to selecting one node at each site and connecting the selected nodes into a fully connected graph. A brief description of the ILP protocol for protein design based on this graph now follows.

Using notation similar to Kingsford et al., the sequence space for designing a protein L amino acids long can be represented with a node set $V = V_1$ U. . .U V_L. Each subset, V_i, contains a set of nodes that represent the possible amino acids at site i. Nodes (u) of V_i have a weight (E^T_{uu}) representing the

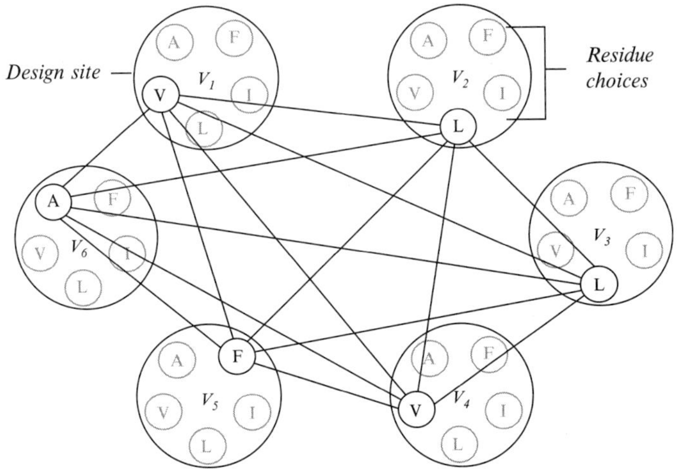

Figure 8.4 Integer linear programming (ILP) formulation for protein design. Each design site consists of nodes representing the allowed residue choices at that position. Edges between nodes represent interactions between those nodes. Our Gβ1 example had six design positions with five choices at each position. One set of nodes and the corresponding edges are highlighted to show one possible design solution.

energetic contribution of that node to the target structure, T. Edges between nodes of the graph $D=\{(u, v): u \in V_i \text{ and } v \in V_j, i \neq j\}$ also have a weight corresponding to the energetic contribution of that edge (E^T_{uv}) to the target structure. E^T represents the energy of a sequence as evaluated by the CE model. E^T is obtained by summing over the node energies (E^T_{uu}) and edge energies (E^T_{uv}). When using a cluster-expanded scoring method, both types of contributions can readily be written as sums of linear terms in sequence variables. Thus, minimization of the total energy (Eq. 8.3) can be done in sequence space using ILP, the only requirement being addition of constraints that enforce a unique and consistent choice of amino acid at each site; this can be done using Eqs. (8.4)–(8.6). Equation (8.4) forces the design solution to have only one amino acid at each design site. Equation (8.5) can then be used such that only the edges from the amino acid being chosen at each design site are used for edge energies. In Eq. (8.6), the terms x_{uu} and x_{uv} are the optimization variables and can have values of 0 or 1 corresponding to the absence or presence of a node or an edge, respectively.

Minimize:

$$E^T = \sum_{u \in V} E^T_{uu} x_{uu} + \sum_{\{u,v\} \in D} E^T_{uv} x_{uv} \qquad [8.3]$$

subject to:

$$\sum_{u \in V_i} x_{uu} = 1 \text{ for } i = 1, \ldots, L \qquad [8.4]$$

$$\sum_{u \in V_i} x_{uv} = x_{vv} \text{ for } i = 1, \ldots, L \text{ and } v \in V \setminus V_i \qquad [8.5]$$

$$x_{uu}, x_{uv} \in \{0, 1\} \qquad [8.6]$$

ILP has several attractive features for protein design. First, it is an optimal search technique and thus ensures, if any solution is returned, that it will be the global minimum energy according to the cluster-expanded scoring method. Second, we have found in practice that for protein design problems of the type described here, the ILP optimization converges reliably and quickly. Further, ILP readily accommodates the addition of arbitrary constraints that are linear in the optimization variables. Such constraints can include limits on the sequence composition or total charge or helical propensity. Grigoryan et al. constrained designed sequences to have at least a minimum score based on a position-specific scoring matrix; this was used to ensure that designed sequences resembled natural sequences in their overall characteristics (Grigoryan et al., 2009).

In our Gβ1 example, ILP can be used to find the lowest energy sequence on the Gβ1 template. In this case, the values E_{uu}^{T} correspond to point ECI values, while the values for E_{uv}^{T} come from the ECI values for pair interactions. For a multicriterion problem, a user can add constraints, for example, restricting solutions that are similar to the wild-type sequence. An example of such a constraint can be seen in Eq. (8.7). Here, WT_u takes on the value: 0.16 (one out of six sites) for wild-type residues at their respective positions and 0 for all other residues at those positions. A user can then define the maximum allowed sequence identity between the design solution and the wild-type sequence of Gβ1.

$$\sum_{u \in V} \mathrm{WT}_u x_{uu} < \mathrm{Allowed\_SeqID} \qquad [8.7]$$

An open-source tool kit for solving ILP problems can be found at http://www.gnu.org/software/glpk/ and can be used with any CE of the type described here.

10. CLASSY APPLIED TO MULTISTATE DESIGN

As mentioned in Section 1, Grigoryan et al. used CE to design specific peptide inhibitors for human bZIP proteins (Grigoryan et al., 2009). The bZIPs are transcription factors that can homo- and/or heterodimerize by forming a parallel coiled coil. They provide an interesting design challenge because, due to the extensive sequence similarity between different bZIPs, designing a peptide to specifically interact with one bZIP but not others is challenging (Mason, Muller, & Arndt, 2007). Grigoryan et al. selected one member of each of the 20 human bZIP families as a target for design and used members of the remaining families as examples of off-targets, to which binding of the design was not desirable. This was accomplished by using CLASSY.

The advantage of using an ILP framework for this design problem is that it enabled optimization of the design for interaction with the target with the addition of linear constraints enforcing simultaneous consideration of the additional competing states. This is because both the objective function, E^T, and the energies of the undesired states, E^i, were written as linear functions of design variables x_{uu} and x_{uv}. Thus, the difference in energy between E^T and each E^i could also be written this way. A series of equations of the form $E^i - E^T > \Delta$ was constructed and used as constraints to enforce an energy gap between the designed target and undesired competitors.

Figure 8.5 shows how the constrained ILP optimization was used in a protocol known as a "specificity sweep." In the first step of a specificity sweep, the design is optimized for binding to a target with no constraints on interaction energies with off-target partners. Due to the high sequence similarity between bZIP families, this can often lead to sequences predicted to interact more favorably with off-target sequences than with the target. Therefore, in a subsequent round of optimization, a constraint can be imposed with a given value of Δ, such that the design is required to bind

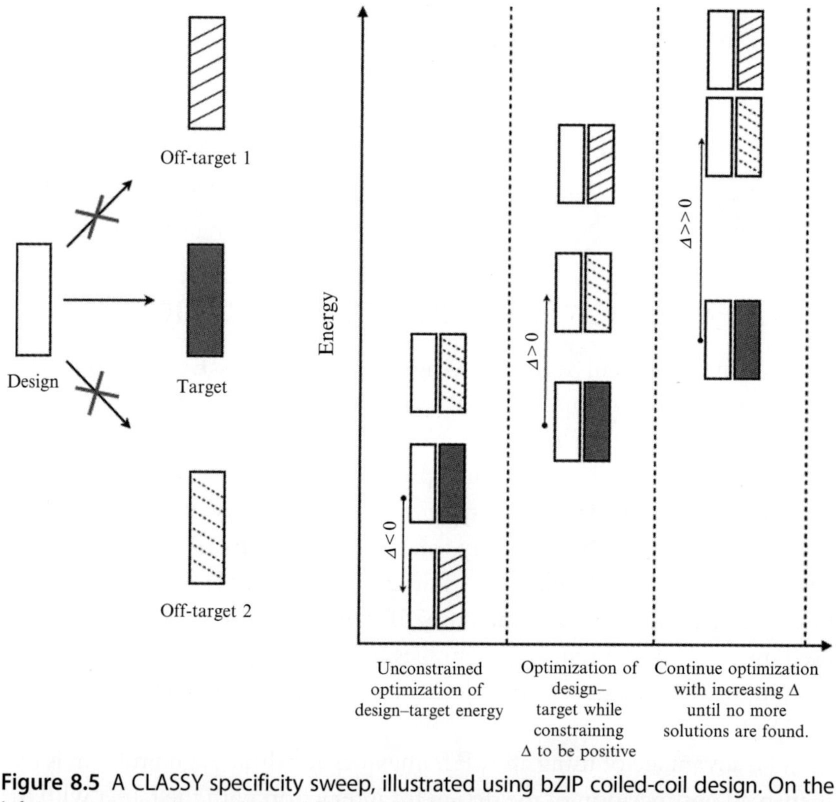

Figure 8.5 A CLASSY specificity sweep, illustrated using bZIP coiled-coil design. On the left is a cartoon representation of the bZIP multistate design problem. The goal in this problem is to design a sequence (white rectangle) that will interact with a target (gray rectangle), yet avoid interactions with off-target sequences (striped rectangles). On the right is a plot of the energies of the various states. Initially, the design–target interaction is predicted not to be the most stable state for the design (far left). Constraints are added in subsequent rounds of design (moving to the right) that impose specificity for the target at the price of the stability of the design–target complex. The constraint on specificity can be increased until the most specific sequence in the space defined for the search is found.

the target at least this much better than the off-targets. In subsequent rounds, Δ can be systematically increased until no more solutions can be found. The complete specificity sweep protocol generates an extensive and systematic search of sequences with different predicted stabilities and target versus off-target specificities. A user can then select any sequence or set of sequences along the specificity sweep for experimental testing. This protocol was successful in generating experimentally validated specific binders, and the interested reader is referred to the original work for details (Grigoryan et al., 2009).

A multistate design criterion can be introduced to our example of redesigning Gβ1. For example, a quadruple mutant of Gβ1 has been shown to form a domain-swapped homodimer (Byeon, Louis, & Gronenborn, 2003). Three of the design sites chosen for our CE case study overlap with the four positions that can bring about the domain swap. To disfavor sequences likely to adopt this domain-swap dimer in design, the domain-swap dimer can be introduced as an explicit undesired state. This requires rescoring the training-set sequences with the RosettaDesign energy function on the domain-swapped homodimer backbone 1Q10 (Byeon et al., 2003). These energies can then be used to derive a new CE that describes the undesired dimer state. An expression just like Eq. (8.3) can be written for this state, E^i, and a linear constraint like that in Eq. (8.8) can be imposed to require that the energy gap between the undesired state and target state be greater than Δ. Similar to Grigoryan et al., Δ could be varied and a specificity sweep conducted to give multiple solutions with different values of Δ.

$$E^i - E^T > \Delta, \text{ where } E^i = \sum_{u \in V} E^i_{uu} x_{uu} + \sum_{\{u,v\} \in D} E^i_{uv} x_{uv} \qquad [8.8]$$

11. CONCLUSION

CE provides a way of converting a complex nonlinear function into a simple approximation of that function that has a linear dependence on sequence variables. This not only allows protein engineers to convert structure-based models into sequence-based models but also opens the door to new search protocols that operate in sequence space for protein design. In particular, CE combined with ILP promises to be a powerful tool for multistate design, and previous analyses have suggested that even single-state problems can benefit (Grigoryan et al., 2006). With the rapid acceleration of sequence-based energy evaluation, and the flexibility that it affords in searching sequence space, designing protein–protein interaction networks using protocols where numerous possible interactions are considered may soon be possible. Excitingly, CE is not limited to expanding stabilities or binding energies resulting from theoretical structure-based models. Hahn et al. demonstrated that it is possible to cluster-expand experimental data directly (Hahn et al., 2010), and it may also be possible to cluster expand other protein properties like association and dissociation rates. As larger data sets emerge from high-throughput experiments linking sequences to protein properties, methods like CE will prove to be increasingly powerful tools.

ACKNOWLEDGMENTS

We thank members of the Grigoryan and Keating labs, especially G. Grigoryan, J. D. Curuksu, and O. Ashenberg, for helpful comments. This work was funded by the National Science Foundation Graduate Research Fellowship Program to C. N., and by NSF CAREER award MCB-0347203 and NIH award GM67681 to A. K. We used computer resources provided by National Science Foundation award 0821391.

REFERENCES

Allen, B. D., & Mayo, S. L. (2006). Dramatic performance enhancements for the FASTER optimization algorithm. *Journal of Computational Chemistry*, *27*, 1071–1075.

Allen, B. D., & Mayo, S. L. (2010). An efficient algorithm for multistate protein design based on FASTER. *Journal of Computational Chemistry*, *31*, 904–916.

Apgar, J., Hahn, S., Grigoryan, G., & Keating, A. E. (2009). Cluster expansion models for flexible-backbone protein energetics. *Journal of Computational Chemistry*, *30*, 2401–2413.

Bolon, D. N., Grant, R. A., Baker, T. A., & Sauer, R. T. (2005). Specificity versus stability in computational protein design. *Proceedings of the National Academy of Sciences*, *102*, 12724–12729.

Byeon, I. J., Louis, J. M., & Gronenborn, A. M. (2003). A protein contortionist: Core mutations of GB1 that induce dimerization and domain swapping. *Journal of Molecular Biology*, *333*, 141–152.

Chen, T. S., Reinke, A. W., & Keating, A. E. (2011). Design of peptide inhibitors that bind the bZIP domain of Epstein–Barr virus protein BZLF1. *Journal of Molecular Biology*, *408*, 304–320.

Dahiyat, B. I., Gordon, B., & Mayo, S. L. (1997). Automated design of the surface positions of protein helices. *Protein Science*, *6*, 1333–1337.

Dahiyat, B. I., & Mayo, S. L. (1997). Probing the role of packing specificity in protein design. *Proceedings of the National Academy of Sciences*, *94*, 10172–10177.

de Fontaine, D. (1994). Cluster approach to order-disorder transformations in alloys. In H. Ehrenreich & D. Turnbull (Eds.), *Solid state physics: advances in research and applications*. *47*. (pp. 33–176). New York: Academic Press.

Desmet, J., Spriet, J., & Lasters, I. (2002). Fast and accurate side-chain topology and energy refinement (FASTER) as a new method for protein structure optimization. *Proteins: Structure, Function, and Bioinformatics*, *48*, 31–43.

Drexler, K. E. (1981). Molecular engineering an approach to the development of general capabilities for molecular manipulation. *Proceedings of the National Academy of Sciences*, *78*, 5275–5278.

Erijman, A., Aizner, Y., & Shifman, J. M. (2011). Multispecific recognition: Mechanism, evolution, and design. *Biochemistry*, *50*, 602–611.

Fleishman, S. J., Whitehead, T. A., Ekiert, D. C., Dreyfus, C., Corn, J. E., Strauch, E. M., et al. (2011). Computational design of proteins targeting the conserved stem region of influenza hemagglutinin. *Science*, *332*, 816–821.
Gallager, T., Alexander, P., Bryan, P., & Gillard, G. L. (1994). Two crystal structures of the B1 immunoglobulin binding domain of streptococcal protein G and comparison with NMR. *Biochemistry*, *33*, 4721–4729.
Grigoryan, G., Kim, Y. H., Acharya, R., Axelrod, K., Jain, R. M., Willis, L., et al. (2011). Computational design of virus-like protein assemblies on carbon nanotube surfaces. *Science*, *332*, 1071–1076.
Grigoryan, G., Reinke, A. W., & Keating, A. E. (2009). Design of protein-interaction specificity gives selective bZIP-binding peptides. *Nature*, *458*, 859–864.
Grigoryan, G., Zhou, F., Lustig, S. R., Ceder, G., Morgan, D., & Keating, A. E. (2006). Ultra-fast evaluation of protein energies directly from sequence. *PLoS Computational Biology*, *2*, 551–563.
Hahn, S., Ashenberg, O. A., Grigoryan, G., & Keating, A. E. (2010). Identifying and reducing error in cluster-expansion approximations of protein energies. *Journal of Computational Chemistry*, *31*, 2900–2914.
Harbury, P. B., Plecs, J. J., Tidor, B., Alber, T., & Kim, P. S. (1998). High-resolution protein design with backbone freedom. *Science*, *282*, 1462–1467.
Havranek, J. J. (2010). Specificity in computational protein design. *The Journal of Biological Chemistry*, *285*, 31095–31099.
Havranek, J. J., & Harbury, P. B. (2003). Automated design of specificity in molecular recognition. *Nature Structural Biology*, *10*, 45–52.
Humphris, E. L., & Kortemme, T. (2007). Design of multi-specificity in protein interfaces. *PLoS Computational Biology*, *3*, 1591–1604.
Jiang, L., Althoff, E. A., Clemente, F. R., Doyle, L., Rothlisberger, D., Zanghellini, A., et al. (2008). De novo computational design of retro-aldol enzymes. *Science*, *319*, 1387–1391.
Karanicolas, J., & Kuhlman, B. (2009). Computational design of affinity and specificity at protein–protein interfaces. *Current Opinion in Structural Biology*, *19*, 458–463.
Kingsford, C. L., Chazelle, B., & Singh, M. (2005). Solving and analyzing side-chain positioning problems using linear and integer programming. *Bioinformatics*, *21*, 1028–1036.
Kortemme, T., Joachimiak, L. A., Bullock, A. N., Schuler, A. D., Stoddard, B. L., & Baker, D. (2004). Computational redesign of protein-protein interaction specificity. *Nature Structural & Molecular Biology*, *11*, 371–379.
Kuhlman, B., Dantas, G., Ireton, G. C., Varani, G., Stoddard, B. L., & Baker, D. (2003). Design of a novel globular protein fold with atomic-level accuracy. *Science*, *302*, 1364–1368.
Leaver-Fay, A., Jacak, R., Stranges, P. B., & Kuhlman, B. (2011). A generic program for multistate protein design. *PLoS One*, *6*, 1–17.
Leaver-Fay, A., Kuhlman, B., & Snoeyink, J. (2005). Rotamer-pair energy calculations using a trie data structure. *Lecture Notes in Computer Science*, *3692*, 389–400.
Lippow, S. M., Wittrup, K. D., & Tidor, B. (2007). Computational design of antibody-affinity improvement beyond in vivo maturation. *Nature Biotechnology*, *25*, 1171–1176.
Mason, J. M., Muller, K. M., & Arndt, K. M. (2007). Positive aspects of negative design: Simultaneous selection of specificity and interaction stability. *Biochemistry*, *46*, 4804–4814.
Röthlisberger, D., Khersonsky, O., Wollacott, A. M., Jiang, L., Dechancie, J., Betker, J., et al. (2008). Kemp elimination catalysts by computational enzyme design. *Nature*, *453*, 190–195.
Sammond, D., Eletr, Z. M., Purbeck, C., & Kuhlman, B. (2010). Computational design of second-site suppressor mutations at protein–protein interfaces. *Proteins: Structure, Function, and Bioinformatics*, *78*, 1055–1065.
Sanchez, J. M., Ducastelle, F., & Gratias, D. (1984). Generalized cluster description of multicomponent systems. *Physica A: Statistical Mechanics and its Applications*, *128*, 334–350.

Voigt, C. A., Gordon, B., & Mayo, S. L. (2000). Trading accuracy for speed: A quantitative comparison of search algorithms in protein sequence design. *Journal of Molecular Biology*, *299*, 789–803.

Yanover, C., Fromer, M., & Shifman, J. M. (2007). Dead-end elimination for multistate protein design. *Journal of Computational Chemistry*, *28*, 2122–2129.

Zhou, F., Grigoryan, G., Lustig, S. R., Keating, A. E., Ceder, G., & Morgan, D. (2005). Coarse-graining protein energetics in sequence variables. *Physical Review Letters*, *95* (148103), 1–4.

CHAPTER NINE

Using Analyses of Amino Acid Coevolution to Understand Protein Structure and Function

Orr Ashenberg[*,†], **Michael T. Laub**[*,†,‡,1]

[*]Department of Biology, Massachusetts Institute of Technology, Cambridge, Massachusetts, USA
[†]Computational & Systems Biology Initiative, Massachusetts Institute of Technology, Cambridge, Massachusetts, USA
[‡]Howard Hughes Medical Institute, Massachusetts Institute of Technology, Cambridge, Massachusetts, USA
[1]Corresponding author: e-mail address: laub@mit.edu

Contents

Abstract

Determining which residues of a protein contribute to a specific function is a difficult problem. Analyses of amino acid covariation within a protein family can serve as a useful guide by identifying residues that are functionally coupled. Covariation analyses have been successfully used on several different protein families to identify residues that work together to promote folding, enable protein–protein interactions, or contribute to an enzymatic activity. Covariation is a statistical signal that can be measured in a multiple sequence alignment of homologous proteins. As sequence databases have expanded dramatically, covariation analyses have become easier and more powerful. In this chapter, we describe how functional covariation arises during the evolution of proteins and how this signal can be distinguished from various background signals. We discuss the basic methodology for performing amino acid covariation analysis, using bacterial two-component signal transduction proteins as an example. We provide practical

Methods in Enzymology, Volume 523
ISSN 0076-6879
http://dx.doi.org/10.1016/B978-0-12-394292-0.00009-6

suggestions for each step of the process including assembly of protein sequences, construction of a multiple sequence alignment, measurement of covariation, and analysis of results.

1. INTRODUCTION

Proteins are central to most biological functions. They catalyze enzymatic reactions, relay cellular signals, and build cellular structures, sometimes acting alone but often in concert with other proteins. Determining which residues within a protein control an activity of interest often requires laborious structural and functional experiments. However, rapidly expanding protein sequence databases and analyses of a protein's evolutionary record can serve as powerful complements to experimental approaches. Patterns of amino acid conservation across a large set of homologs can, of course, identify important residues. However, patterns of correlated substitutions, or amino acid covariation, can also provide insight into functionally important and related sets of residues. Here, we briefly review recent studies of amino acid coevolution and then provide a practical guide to performing such analyses.

For most proteins, several functions are usually under strong purifying selection, including folding and maintenance of a particular structure, the ability to carry out an enzymatic activity, and the ability to interact with various protein partners. Amino acids critical for these functions are often invariant in homologous proteins as substitutions at these positions will often disrupt function and be eliminated by natural selection. However, in many cases, substitutions can be tolerated, provided they are accompanied by a compensatory substitution, a form of epistasis at the amino acid level (Fitch & Markowitz, 1970; Korber, Farber, Wolpert, & Lapedes, 1993). In other words, functionally related residues in a protein will often coevolve. The extent of this coevolution can be quantified from protein sequence alignments.

The most strongly covarying pairs of residues within a protein, or in a pair of interacting proteins, are often in physical contact (Morcos et al., 2011). Consider the interaction between two proteins A and B involving a glycine at position 2 and a phenylalanine at position 5 (Fig. 9.1). A mutation of the glycine in protein A to phenylalanine would result in a steric clash, thereby weakening the interaction; however, a subsequent mutation

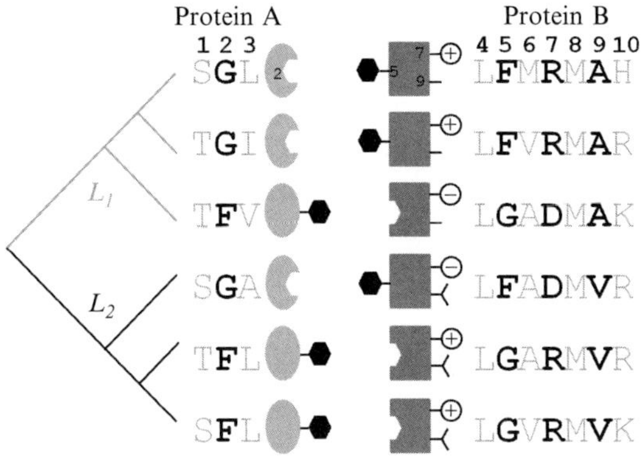

Figure 9.1 Covariation between interacting proteins. Protein A interacts with Protein B. Each row of the multiple sequence alignment consists of an interacting pair of protein A and B homologs from a single species. The phylogenetic tree designates that the protein sequences come from species in two different lineages: L_1 (gray branches) and L_2 (black branches). Positions 2 and 5 covary, thereby preserving a hydrophobic packing interaction at the protein–protein interface, while positions 2 and 7 are independent of one another. Positions 5 and 9 also covary, but this signal comes from shared phylogeny in protein B homologs.

in protein B from a phenylalanine to a glycine may compensate and restore the interaction. Covariation between these positions can be detected by examining the identity of residues at these positions in homologs of proteins A and B. This type of coevolution, in which a deleterious mutation leads to selection for compensatory mutations at other positions that restore protein function, is often referred to as second-site suppression (Kulathinal, Bettencourt, & Hartl, 2004; Yanofsky, Horn, & Thorpe, 1964). In such cases, the compensatory change on its own is also often deleterious. Scenarios in which an initial mutation is neutral, rather than deleterious, but creates a favorable context for an otherwise deleterious mutation have also been suggested (Ortlund, Bridgham, Redinbo, & Thornton, 2007).

1.1. Covariation applications

Amino acid covariation analysis has been used to predict residue pairs critical to a wide range of protein functions, and follow-up experiments have validated many of these predictions. Covariation analysis can provide functional insight into both individual and interacting proteins. Many different scoring metrics for covariation exist, with two of the most widely used being mutual information (MI) and statistical coupling analysis (SCA; Korber et al., 1993; Lockless & Ranganathan, 1999). One extensively studied application of covariation has been the prediction of amino acid contacts within a protein (Dunn, Wahl, & Gloor, 2008), which are often under selective pressure because a protein must stably fold to perform its biological function. Direct coupling analysis (DCA), which uses a metric based on MI, has been somewhat successful in predicting contacts. When tested on a set of 131 domain families with crystal structures, an average of 84% of the most significant residue pairs within a family were found to be in direct contact (Morcos et al., 2011). Further, DCA-based scores have successfully been used as distance restraints in NMR-like structure determination calculations to predict the tertiary structure of proteins (Hopf et al., 2012; Marks et al., 2011). In another example of contact prediction, highly destabilizing mutations in the hydrophobic core of an SH3 domain were compensated by second and third site substitutions that covaried with and contacted the destabilizing site (Larson, Di Nardo, & Davidson, 2000). Mutations in networks of covarying contacts were also shown to affect enzymatic activity for phosphoglycerate kinase and transketolase (Gloor et al., 2010; Strafford et al., 2012). More generally, finding a position in an enzyme that is physically surrounded by covarying residues turns out to be a strong predictor for that position being a catalytic residue (Buslje, Teppa, Di Domenico, Delfino, & Nielsen, 2010).

Although covarying residues often make direct contact, many pairs do not. Such residue pairs are not necessarily false positives. These distant covarying pairs may come in direct contact during the folding of a protein, during protein oligomerization, or in a relevant conformation not captured in the experimental structure (Dago et al., 2012; Marks et al., 2011). Covariation signal may also result from residues that affect the same function but at different sites within a protein (Chakrabarti & Panchenko, 2009; Lunzer, Golding, & Dean, 2010). For instance, residues that promote a particular protein conformation may be distributed throughout a protein, as in allosterically regulated proteins. SCA has been used to probe allostery in several protein families, including PDZ domains and serine proteases (Halabi, Rivoire, Leibler, & Ranganathan, 2009; Lockless & Ranganathan, 1999).

Beyond individual proteins, covariation also has many exciting applications in understanding and engineering protein–protein interactions (Skerker et al., 2008; Thattai, Burak, & Shraiman, 2007). For many protein complexes, no structures exist, and even in cases where a structure of the complex has been solved, the residues critical to the interaction are often unclear. The identification of residues that coevolve in two interacting proteins can highlight residues important for complex formation; if a mutation in one of the proteins destabilizes the complex, an interaction can often be restored through a compensatory mutation in the other protein. Such an approach has recently been applied to two-component signaling proteins, which will be used as an example through the rest of this chapter.

Two-component signaling pathways are a predominant means by which bacteria sense and respond to changes in their environments. The prototypical system consists of an input sensor histidine kinase that autophosphorylates and then transfers its phosphoryl group to a response regulator which can effect a cellular output (Stock, Robinson, & Goudreau, 2000). Most species encode dozens of histidine kinases and response regulators, but individual pathways are usually insulated from one another, with a given kinase interacting only with a single, cognate regulator. Analyses of amino acid coevolution in large sets of cognate kinase–regulator pairs led to the identification of a small set of strongly covarying residues (Capra et al., 2010; Skerker et al., 2008). The majority of these residue pairs were in direct contact at the interaction interface formed by the two proteins. The coevolving residues were further validated as critical determinants of specificity in two-component signaling pathways. Substituting these residues in one kinase with those found in another kinase was sufficient to completely reprogram phosphotransfer specificity (Capra et al., 2010; Skerker et al., 2008). A similar covariation analysis was applied to understand the specificity of histidine kinase homodimerization (Ashenberg, Rozen-Gagnon, Laub, & Keating, 2011). Using covariation to identify functionally important residues, such as those directing protein–protein interactions, can complement protein design efforts.

1.2. Covariation algorithms

Performing covariation analysis requires two basic steps: (1) construction of a multiple sequence alignment and (2) an assessment of amino acid covariation between pairs of positions in the multiple sequence alignment. Many different algorithms for detecting covariation have been proposed, and they can be divided into two general classes: algorithms such as MI or SCA that score covariation between all pairs of columns in a sequence alignment, and global probabilistic models such as DCA that assess the likelihood of covariation between sites. We focus on the former, which are conceptually straightforward, technically simpler to implement, and often sufficiently powerful to provide useful insights into the proteins of interest. The latter class of algorithms can, in some cases, separate direct physical interactions from indirect effects and may prove particularly useful in guiding protein structure predictions (Burger & van Nimwegen, 2010; Thomas, Ramakrishnan, & Bailey-Kellogg, 2009; Weigt, White, Szurmant, Hoch, & Hwa, 2009). However, it is important to note that coevolving residues need not be in direct contact to be functionally coupled.

To calculate MI between columns i and j in a multiple sequence alignment, both single and joint probability distributions of amino acids are used. Let p_x^i and p_y^j be the probabilities (or frequencies) of amino acids x in column i and y in column j, respectively, and let $p_{xy}^{i,j}$ be the frequency of simultaneously finding residues x in column i and y in column j of an alignment. Given n different types of amino acids in column i and m different types of amino acids in column j,

$$\mathrm{MI}(i,j) = \sum_{x=1}^{n}\sum_{y=1}^{m} p_{xy}^{i,j} \log \frac{p_{xy}^{i,j}}{p_x^i p_y^j} \qquad [9.1]$$

MI measures how much the uncertainty at one position is reduced given information about another position. For example, consider the alignment in Fig. 9.1 where the amino acid distribution is $p_{\mathrm{G}}^2 = 0.5$ and $p_{\mathrm{F}}^2 = 0.5$ at position 2; $p_{\mathrm{G}}^5 = 0.5$ and $p_{\mathrm{F}}^5 = 0.5$ at position 5; and $p_{\mathrm{R}}^7 = 0.67$ and $p_{\mathrm{D}}^7 = 0.33$ at position 7. If position 2 is a G, position 5 is always an F whereas if position 2 is an F, position 5 is always a G. This covariation is reflected in the joint probability distribution term $p_{xy}^{i,j}$, where $p_{\mathrm{GF}}^{2,5} = 0.5$, $p_{\mathrm{FG}}^{2,5} = 0.5$, $p_{\mathrm{GG}}^{2,5} = 0$, and $p_{\mathrm{FF}}^{2,5} = 0$. Knowing the residue at position 2 completely removes all uncertainty at position 5, resulting in a large MI score for this pair. In contrast, consider the MI score for positions 2 and 7. Knowing whether position 2 is an F or G does not reduce the uncertainty as to whether R or D occupies position 7. Consequently, the MI score is zero as the joint probability distribution $p_{xy}^{2,7}$ is simply the product of probability distributions p_x^2 and p_y^7. Assessing the statistical significance of MI scores is discussed below.

In many sequence alignments, there will often be significant covariation signal even when two columns in a sequence alignment are functionally independent of one another. This background signal can arise from statistical noise and, sometimes, from the shared ancestry of the sequences (Wollenberg & Atchley, 2000). The statistical noise, a form of sampling bias, stems from the fact that columns in an alignment are being used to estimate an amino acid probability distribution; hence, too few sequences in the alignment can produce inaccurate estimates of the true probability distribution. The appropriate number of sequences for covariation analysis is discussed in the next section. The shared ancestry, or phylogenetic signal, arises because no pair of positions in a protein can ever be truly independent as they are ultimately derived from a common ancestor and have been inherited as a unit. As an example, consider position 9 in the alignment of Fig. 9.1, which is not functionally coupled to the other positions, but due to drift is an A in lineage L_1 and a V in lineage L_2. This phylogenetic pattern results in a relatively high MI score between positions 5 and 9, even though no functional coupling exists between this pair of positions.

Many methods have been proposed to correct for the signals arising from statistical and phylogenetic noise. Tree-aware methods calculate the likelihood of observing a pair of sites in a dependent versus independent model of evolution on a phylogenetic tree (Yeang & Haussler, 2007). However, in at least one comparison, covariation methods that do not explicitly model the phylogeny, like MI, performed as well as or better than tree-aware methods (Caporaso et al., 2008). Simpler, tree-independent extensions of MI, including MIp, ZNMI, and adjusted MI (MI_{adj}), have also been proposed to help correct for background signal, (Brown & Brown, 2010; Capra et al., 2010; Dunn et al., 2008) and are each discussed briefly below.

For MIp, the MI between two positions is corrected by subtracting the product of the mean MI, $\mu_{MI(i)}$, at each position normalized by the mean MI of the alignment, μ_{MI}. The underlying hypothesis for this approach is that a large majority of residue pairs are not functionally linked; hence, the average MI between a column and all other columns in the alignment reflects background signals from statistical noise and phylogeny. This correction greatly improves contact prediction in many sequence alignments (Dunn et al., 2008). Using the same variables and definitions as in Eq. (9.1) above, MIp is defined as follows:

$$\mathrm{MIp}(i,j) = \mathrm{MI}(i,j) - \frac{\mu_{\mathrm{MI}(i)}\mu_{\mathrm{MI}(j)}}{\mu_{\mathrm{MI}}} \qquad [9.2]$$

where $\mu_{\mathrm{MI}(i)} = (1/c-1)\sum_{x=1,x\neq i}^{c}\mathrm{MI}(i,x)$, $\mu_{\mathrm{MI}} = (2/c(c-1))\sum_{i=1}^{c-1}\sum_{j=i+1}^{c}$ $\mathrm{MI}(i,j)$, and c is the number of columns in the multiple sequence alignment.

ZNMI extends the approach of MIp by modeling the background MI distribution using both the mean, $\mu_{\mathrm{MI}(i)}$, and standard deviation, $\sigma_{\mathrm{MI}(i)}$, of column MI scores (Brown & Brown, 2010).

Another simple extension of MI, adjusted MI, corrects for the fact that positions with higher sequence diversity, or entropy, tend to exhibit higher MI (Martin, Gloor, Dunn, & Wahl, 2005). The higher the entropy, the smaller the values of p_x^i and p_x^j (see Eq. 9.1) and, consequently, the higher the MI score. In adjusted MI, the raw MI score for a given pair of columns is converted into an average of Z-scores for each of the two columns.

$$\mathrm{MI}_{\mathrm{adj}}(i,j) = \frac{\left(\left(\mathrm{MI}(i,j) - \mu_{\mathrm{MI}(i)}\right)/\sigma_{\mathrm{MI}(i)}\right) + \left(\left(\mathrm{MI}(i,j) - \mu_{\mathrm{MI}(j)}\right)/\sigma_{\mathrm{MI}(j)}\right)}{2} \qquad [9.3]$$

where $\sigma_{\mathrm{MI}(i)} = \sqrt{(1/c-2)\sum_{x=1,x\neq i}^{c}\left(\mathrm{MI}(i,x) - \mu_{\mathrm{MI}(i)}\right)^2}$.

This corrects for the underlying distribution of MI at each column and significantly lessens the dependence on entropy. MIp and ZNMI also correct for the dependence on entropy in different ways, and all three metrics are significant improvements over MI.

While these various corrections to raw MI scores should help distinguish functional coupling of residues from statistical and phylogenetic noise, judging each method's value is complicated since there is no "gold standard" alignment where the complete set of functionally relevant covarying residues are known (Brown & Brown, 2010; Caporaso et al., 2008; Dickson, Wahl, Fernandes, & Gloor, 2010). The performance of covariation methods is often assessed based on the ability to predict direct contacts. However, as discussed above, covarying residue pairs may not always be physically adjacent.

1.3. Choosing sequences for covariation analyses

Two important considerations when performing covariation analyses are (1) the number of sequences needed and (2) the phylogenetic distribution, or diversity, of those sequences. Calculating MI requires a reliable estimate of the joint probability distribution for a pair of columns. As each position in the alignment may sample 20 amino acids, at least a few hundred sequences are necessary for MI-based algorithms, for DCA, and for SCA (Halabi et al., 2009; Morcos et al., 2011; White, Szurmant, Hoch, & Hwa, 2007), although as few as 150 sequences have given meaningful signal (Martin et al., 2005).

Along with the number of available sequences, sequence diversity can determine the success of covariation analysis. The primary requirement for sequence diversity is sufficient variation within columns of a multiple sequence alignment, meaning that homologs must be phylogenetically diverse. As a corollary, positions that never vary such as conserved active site residues of an enzyme will be refractory to covariation analyses. Additionally, orthologous sequences taken from closely related species may have very few changes, even at less critical positions. Highly similar sequences can severely disrupt covariation analysis, by reducing the overall diversity of the sequences and, in some cases, by increasing the phylogenetic noise as discussed above. Setting a threshold for the elimination of closely related sequences is inherently arbitrary, but several laboratories have used 90% identity (Ashkenazy, Unger, & Kliger, 2009; Fodor & Aldrich, 2004; Skerker et al., 2008).

Although some sequence diversity is necessary, sequences should also not be so different that confidence is lost as to their functional homology. A typical lower limit for homology is 20–30% identity. Including paralogs of a protein of interest can provide helpful sequence diversity and improve covariation analyses, provided the paralogs have similar functional constraints to the orthologs (Ashkenazy et al., 2009).

Whatever the overall sequence similarity, multiple sequence alignments typically exhibit different levels of conservation at different positions. Importantly, the ability of different algorithms to detect covariation depends on these levels of conservation. As previously noted, MI-based algorithms tend to produce higher scores for columns with higher entropy. Conversely, the SCA algorithm tends to produce higher scores for columns with lower entropy as the algorithm essentially weights scores from such columns more heavily (Halabi et al., 2009). This fundamental difference between MI and SCA ultimately impacts the types of functional covariation that can be detected, and means that choosing an algorithm depends on the biological problem of interest (L. Colwell, personal communication). For relatively well-conserved positions in which substitutions are evolutionarily rare, such as those sites important for the folding of a protein, SCA will generally outperform MI. In contrast, for sites with more frequent substitutions, MI will typically outperform SCA. For example, MI will often be the appropriate algorithm when identifying covarying residues involved in transient protein–protein interfaces; such residues have been shown to evolve more quickly than those in obligate protein–protein interfaces (Mintseris & Weng, 2005; Teichmann, 2002).

2. PREDICTING SPECIFICITY DETERMINING RESIDUES USING MI

In this section, we detail the steps necessary to carry out covariation analysis, using two-component signal transduction proteins from bacteria as an example for discussing practical issues that arise during coevolution analyses. These signaling pathways involve phosphotransfer from an autophosphorylated sensor histidine kinase to a cognate response regulator. The specificity of the phosphotransfer reaction relies predominantly on molecular recognition (Skerker, Prasol, Perchuk, Biondi, & Laub, 2005); consequently, the residues dictating specificity are under strong selective pressure to remain constant or coevolve to maintain the kinase–regulator interaction.

The phosphotransfer reaction primarily involves the dimerization and histidine phosphotransfer (DHp) domain of the histidine kinase and the receiver domain of the response regulator. In the following analysis, we show how to build an alignment of concatenated pairs of cognate DHp and receiver domains, and then measure MI between all pairs of positions between the two domains. This process is broken into four stages: (1) sequence retrieval, (2) sequence alignment, (3) covariation measurement, and (4) analysis of covariation results (Fig. 9.2A).

2.1. Sequence retrieval

To perform reliable covariation analysis, a sequence alignment comprising hundreds to thousands of homologous sequences is typically needed. Such sequence alignments can be obtained from an online database or built manually (Table 9.1). An excellent starting point is the Pfam database, which contains, as of July 2012, multiple sequence alignments for 13,672 protein families (Punta et al., 2012). For each protein family in Pfam, a profile hidden Markov model (HMM) was built from a curated seed alignment. This HMM was then searched against the most recent release of UniProtKB, allowing identification and alignment of all protein sequences belonging to a given family. Users can conveniently search Pfam by protein sequence and get both comprehensive multiple sequence alignments along with the profile HMMs needed to align additional sequences. At the time of writing, Pfam contained an alignment of 55,506 DHp domains built using the HMM model HisKA, and an alignment of 103,232 receiver domains built using the HMM model Response_reg (Punta et al., 2012).

If a protein is not already represented by a family in a database, programs such as PSI-BLAST or HMMER can be used to identify homologous sequences (Altschul et al., 1997; Eddy, 2011). PSI-BLAST uses protein–protein BLAST to search a database for hits to a given protein sequence. Sequence hits with significance scores below a threshold are then used to derive a position-specific scoring matrix, and new sequence hits are identified by searching the database again, but with this new scoring matrix.

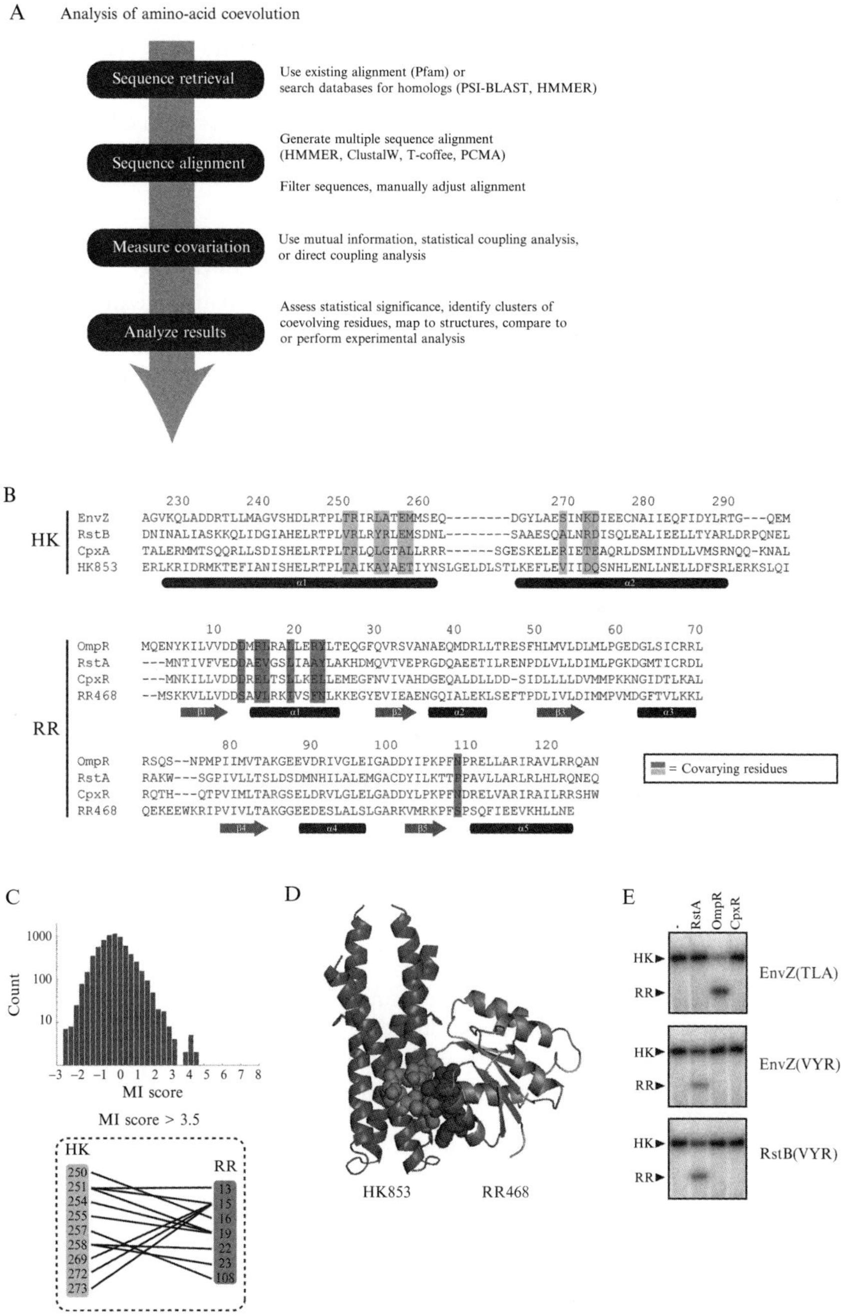
A Analysis of amino-acid coevolution
Sequence retrieval
Use existing alignment (Pfam) or search databases for homologs (PSI-BLAST, HMMER)
Sequence alignment
Generate multiple sequence alignment (HMMER, ClustalW, T-coffee, PCMA)
Filter sequences, manually adjust alignment
Measure covariation
Use mutual information, statistical coupling analysis, or direct coupling analysis
Analyze results
Assess statistical significance, identify clusters of coevolving residues, map to structures, compare to or perform experimental analysis
B
HK
230 240 250 260 270 280 290
EnvZ AGVKQLADDRTLLMAGVSHDLRTPLTRIRLATEMMSEQ--------DGYLAESINKDIEECNAIIEQFIDYLRTG---QEM
RstB DNINALIASKKQLIDGIAHELRTPLVRLRYRLEMSDNL--------SAAESQALNRDISQLEALIEELLTYARLDRPQNEL
CpxA TALERMMTSQQRLLSDISHELRTPLTRLQLGTALLRRR------SGESKELERIETEAQRLDSMINDLLVMSRNQQ-KNAL
HK853 ERLKRIDRMKTEFIANISHELRTPLTAIKAYAETIYNSLGELDLSTLKEFLEVIIDQSNHLENLLNELLDFSRLERKSLQI
α1 α2
RR
10 20 30 40 50 60 70
OmpR MQENYKILVVDDDMRLRALLERYLTEQGFQVRSVANAEQMDRLLTRESFHLMVLDLMLPGEDGLSICRRL
RstA ---MNTIVFVEDDAEVGSLIAAYLAKHDMQVTVEPRGDQAEETILRENPDLVLLDIMLPGKDGMTICRDL
CpxR ---MNKILLVDDDRELTSLLKELLEMEGFNVIVAHDGEQALDLLDD-SIDLLLLDVMMPKKNGIDTLKAL
RR468 --MSKKVLLVDDSAVLRKIVSFNLKKEGYEVIEAENGQIALEKLSEFTPDLIVLDIMMPVMDGFTVLKKL
β1 α1 β2 α2 β3 α3
80 90 100 110 120
OmpR RSQS--NPMPIIMVTAKGEEVDRIVGLEIGADDYIPKPFNPRELLARIRAVLRRQAN
RstA RAKW---SGPIVLLTSLDSDMNHILALEMGACDYILKTTPPAVLLARLRLHLRQNEQ
CpxR RQTH---QTPVIMLTARGSELDRVLGLELGADDYLPKPFNDRELVARIRAILRRSHW
RR468 QEKEEWKRIPVIVLTAKGGEEDESLALSLGARKVMRKPFSPSQFIEEVKHLLNE
β4 α4 β5 α5
= Covarying residues
C
1000
100
10
Count
-3 -2 -1 0 1 2 3 4 5 6 7 8
MI score
MI score > 3.5
HK
250 251 254 255 257 258 269 272 273
RR
13 15 16 19 22 23 108
D
HK853
RR468
E
- RstA OmpR CpxR
HK►
RR►
EnvZ(TLA)
EnvZ(VYR)
RstB(VYR)

New hits are incorporated into the next scoring matrix. This iterative process enables the identification of remote homologs; however, it also often identifies false positives and results should be curated. In contrast to PSI-BLAST, HMMER uses an initial profile HMM, built manually or taken from Pfam, to search for homologs and generate an expanded sequence alignment.

For our example of two-component signaling proteins, we used the *hmmsearch* program from the HMMER package. HMMER programs (Eddy, 2011) are easily run from a web server or a command line in a Linux, MacOSX, or Windows environment:

hmmsearch -E 0.01 HisKA.hmm genome_proteins.fasta

hmmsearch -E 0.01 Response_reg.hmm genome_proteins.fasta

These commands queried the desired profile HMMs, HisKA.hmm and Response_reg.hmm, against a sequence database, genome_proteins.fasta, keeping those sequences below an E-value cutoff of 0.01. In this case, genome_proteins.fasta was a FASTA-formatted database containing the entire set of predicted protein sequences from approximately 1000 prokaryotic genomes in GenBank (ftp://ftp.ncbi.nih.gov/genomes/Bacteria).

Figure 9.2 Amino acid coevolution analysis. (A) Overview of steps for performing amino acid coevolution analysis. (B)–(E) Coevolution analysis applied to the problem of identifying phosphotransfer specificity determinants in two-component signal transduction proteins. (B) A set of cognate histidine kinases (HK) and response regulators (RR) was retrieved by searching across all available prokaryotic genomes with HMMER. These sequences were aligned using HMMER, and the resulting alignment was filtered to eliminate pairwise sequence redundancy and gapped columns. Covariation was then measured using adjusted mutual information. The multiple sequence alignment shown includes a subset of cognate kinase–regulator pairs and highlights significantly coevolving residues in the DHp domains of histidine kinases and the receiver domains of response regulators. Secondary structure elements of the kinases and regulators are labeled underneath the sequences and assigned using the *T. maritima* kinase–regulator cocrystal structure. Significant covarying pairs are identified in the tail of the covariation score distribution and then grouped into clusters (C). Pairs are mapped onto the kinase–regulator cocrystal structure (D) with covarying residues highlighted as in (B) and shown in space-filling form. (E) Rewiring phosphotransfer specificity through rational mutagenesis. Phosphotransfer specificity was measured by autophosphorylating each kinase with [γ^{32}P]-ATP and then incubating the phosphorylated kinase with a response regulator (RstA, OmpR, or CpxR). The histidine kinase EnvZ, labeled TLA based on the identities of the three specificity residues T250, L254, and A255, specifically phosphotransfers to OmpR (top panel) whereas RstB, labeled VYR based on the identities of its three specificity residues, specifically phosphotransfers to RstA (bottom panel). Mutating the specificity residues in EnvZ to match those found in RstB rewires phosphotransfer specificity from OmpR to RstA (middle panel).

Table 9.1 Sequence alignment and structure prediction programs

Program	URL	Reference
Sequence alignment programs		
ClustalW	www.clustal.org/clustal2	Larkin et al. (2007)
MUSCLE	www.drive5.com/muscle	Edgar (2004)
MAFFT	www.mafft.cbrc.jp/alignment/server	Katoh and Toh (2008)
T-Coffee	www.tcoffee.crg.cat	Di Tommaso et al. (2011)
PCMA	www.prodata.swmed.edu/pcma/pcma.php	Pei, Sadreyev, and Grishin (2003)
PSI-BLAST	www.blast.ncbi.nlm.nih.gov/Blast.cgi	Altschul et al. (1997)
HMMER	www.hmmer.org	Finn, Clements, and Eddy (2011)
Jalview	www.jalview.org	Waterhouse, Procter, Martin, Clamp, and Barton (2009)
Sequence alignment databases		
Pfam	www.pfam.sanger.ac.uk	Punta et al. (2012)
EMBL SMART	www.smart.embl-heidelberg.de	Letunic, Doerks, and Bork (2012)
NCBI CDD	www.ncbi.nlm.nih.gov/cdd	Marchler-Bauer et al. (2012)
Structure prediction programs		
PSIPRED	www.bioinf.cs.ucl.ac.uk/web_servers	Bryson et al. (2005)
Phyre2	www.sbg.bio.ic.ac.uk/phyre2	Kelley and Sternberg (2009)
Robetta	www.robetta.bakerlab.org	Kim, Chivian, and Baker (2004)

When building sequence alignments to examine covariation between interacting proteins, sequences for both proteins in the interaction must be collected, aligned, concatenated, and then effectively treated as a single sequence. To identify interacting DHp and receiver domains, we applied the fact that histidine kinases and their cognate response regulators are often encoded in the same operon. Kinases and regulators were considered co-operonic if their Genbank ID numbers or organism-specific ID numbers differed by one, or if their genes were less than 500 bases apart and both genes were encoded on the same strand. These properties were assessed using a custom-designed script that accesses the GenBank annotation file for each sequenced bacterial genome.

From ~1000 bacterial genomes, we extracted ~7000 co-operonic pairs. This set included a diverse set of orthologous and paralogous signaling proteins. Importantly, as we were interested in identifying residues that determine the interaction specificity of two-component signaling proteins, we assumed that the manner in which DHp and receiver domains dock was conserved and stereotypical across this entire set of proteins. A similar assumption is necessary in most cases of covariation analysis.

2.2. Sequence alignment

Choosing an appropriate sequence alignment algorithm can be difficult as there are dozens of methods available, each with different strengths and weaknesses (Kemena & Notredame, 2009). One of the most widely used programs is ClustalW (Larkin et al., 2007), a progressive alignment algorithm in which all possible pairwise alignments are first used to construct a guide tree. Sequences are then aligned again and built into a profile, using the guide tree to determine the order in which sequences are added to the profile, eventually producing a complete alignment. Although fast, even with thousands of sequences, and often reliable, the drawback to ClustalW and other progressive alignment algorithms is that errors made early on become "frozen" in the growing profile and can never be corrected. Two popular alignment algorithms, MUSCLE and MAFFT, deal with this issue through an iterative strategy and can produce more accurate alignments (Edgar, 2004; Katoh & Toh, 2008). Consistency-based alignment methods, like T-Coffee, can be even more accurate, but their computational run times scale cubically with the number of sequences, making them impractical for most alignments needed in covariation analysis (Di Tommaso et al., 2011). However, algorithms like PCMA combine heuristic approaches with these consistency-based methods and can align thousands of sequences in a reasonable amount of time (Pei et al., 2003).

In contrast to the above methods, sequence alignments can also be built using preexisting information encoded in a profile HMM (Eddy, 2011). The model can either come directly from the Pfam database, or it may be built from an existing sequence alignment using the HMMER program *hmmbuild*. Alignments in HMMER are built with *hmmalign*, which is fast and scales linearly with the number of sequences to be aligned. The same HMMs we used to identify kinase and regulator domains, HisKA.hmm and Response_reg.hmm, were used to build an alignment:

hmmalign −o HK_align.txt −−trim −−outformat PSIBLAST HisKA.hmm HK_seq.fasta

hmmalign −o RR_align.txt −−trim −−outformat PSIBLAST Response_reg.hmm RR_seq.fasta

These commands aligned sequences, previously identified by *hmmsearch* (HK_seq.fasta or RR_seq.fasta), using the appropriate HMM and placed the output in HK_align.txt or RR_align.txt. The PSI-BLAST flag generated the sequence alignment in a convenient format, and the trim flag removed nonaligned residues from the alignment. We then concatenated co-operonic DHp and receiver domain sequences and treated them as a single sequence thereafter.

Using custom Perl scripts, the resulting sequence alignment was filtered to remove columns with more than 10% gaps as the alignment accuracy for such columns is low. We also filtered the alignment such that no pair of sequences was greater than 90% identical, helping to reduce the background phylogeny signal in the subsequent calculations of covariation. The final sequence alignment contained 4375 pairs of interacting histidine kinases and response regulators (Fig. 9.2B). The alignment included the sequences corresponding to the solved structure of a kinase in complex with its regulator (Casino, Rubio, & Marina, 2009) to guide later interpretation of covariation results. Final sequence alignments were visually examined in the alignment editor Jalview (Waterhouse et al., 2009) to verify that kinase and regulator sequence motifs of functional importance were properly aligned. Having high confidence in a sequence alignment is critical to any comparative study, and the detrimental effect of sequence misalignment on covariation studies is well documented (Brown & Brown, 2010; Dickson et al., 2010).

2.3. Measuring covariation

For the histidine kinase-response regulator sequence alignment, we measured adjusted MI for every pair of positions between the kinase and the regulator. As discussed earlier, the choice of algorithm depends on the sequence variability of the alignment and, in particular, the variability of the residues of interest. In this particular case, we were interested in the specificity-determining residues in a set of paralogous proteins; such residues

Table 9.2 Covariation algorithms available by software or Web server

Description	URL	Reference
Several methods, web server	www.coevolution.gersteinlab.org/coevolution	Yip et al. (2008)
Several methods, software	www.afodor.net	Fodor and Aldrich (2004)
SCA, software	www.ais.swmed.edu/rrlabs/register.htm	Halabi et al. (2009)
ZNMI, software	Available from authors upon request	Brown and Brown (2010)
EVFold	www.evfold.org	Marks et al. (2011)

are likely highly variable, so MI is an appropriate choice. We implemented adjusted MI using simple custom Perl scripts (Fig. 9.2C). As an alternative, web servers that measure covariation and facilitate some limited pre- and postprocessing tasks are available (Table 9.2).

2.4. Analyzing covariation

The final step of covariation analysis is to assess the significance of covarying pairs and interpret their possible functional significance. One common way to set a significance threshold is to compare the MI scores derived from the sequence alignment to the MI scores found in a randomized sequence alignment (Skerker et al., 2008). Within each column of the alignment, the amino acids can be randomly permuted so as to maintain overall column conservation while scrambling any functional coupling, and hence covariation, with other columns. Such randomization will typically eliminate most covariation signal resulting from functional coupling or shared ancestry, with the remaining covariation representing statistical noise. In principle, randomization should be performed independently at least 100 times with the covariation scores for each pair averaged across all randomizations, thereby producing a reliable estimate of the statistical noise. The maximum covariation score for all residue pairs from these randomized alignments can then be used as a threshold for assessing the significance of scores obtained with the original, nonrandomized alignment. Any covariation score from the original alignment above this threshold can be considered significant. Assessing significance through column randomization is especially useful when an alignment contains only a few hundred sequences as statistical noise signal can be quite high in such cases.

For the kinase–regulator analysis, we performed a randomization analysis and set a score threshold of MI $\geq$3.5, as the randomization procedure produced no column pairs with scores above this threshold (Capra et al., 2010). This process resulted in 12 significant covarying pairs, involving nine positions in the DHp domain of the kinases and seven positions in the receiver domain of the regulator (Fig. 9.2C).

In addition to randomizing the sequence alignment, significance can also be assessed by measuring how sensitive covariation scores are to the sequences chosen for the sequence alignment. In one proposed cross-validation scheme, 50% of the aligned sequences are randomly chosen, and the covariation scores recalculated for this subset of sequences. This process is repeated many times, and the more often a given pair appears, the higher the confidence the covariation is significant (Brown & Brown, 2010).

A second useful step in assessing the significance of covariation scores is to consider higher order couplings; that is, to determine whether there are larger sets of covarying positions, beyond the pairs identified based on scores and the randomization procedures described earlier (Fig. 9.2C). One simple clustering approach is to apply a transitive rule in which if position 1 covaries with 2 and position 2 covaries with 3, then 1, 2, and 3 are assigned to the same cluster. For example, position 258 in the kinase forms a cluster with positions 22 and 23 in the regulator (Fig. 9.2C). Another clustering approach is to apply principal component analysis to a matrix of all covariation scores from a sequence alignment. This method, previously applied to a covariation score matrix for serine proteases, identified collective modes that represent three functionally different clusters of covarying residues (Halabi et al., 2009).

The final, and perhaps most important, step in assessing the output of a covariation analysis is to determine whether any experimental information exists for the residues that comprise the most significant covarying pairs. Existing structural information can help assess whether covarying pairs are in physical contact and whether covarying pairs are clustered or dispersed throughout the structure. For our example, we examined a crystal structure of a *Thermotoga maritima* histidine kinase, HK853, in complex with its cognate response regulator, RR468 (Casino et al., 2009). This immediately suggested that the identified pairs may be phosphotransfer specificity determinants as the covarying pairs map to the primary protein–protein interface of the complex (Fig. 9.2D). Covarying positions on the kinase are found at the base of the α-helix bundle in the DHp domain, while the covarying partners on the regulator are found on the face of an α-helix and on a nearby loop. Many of these residue pairs form direct contacts; however, not all interfacial contacts are covarying.

Even in the absence of a solved structure, structural bioinformatic approaches can be explored. For instance, in the kinase–regulator analysis, many of the significant covarying positions in both proteins are separated by three or four residues. This (i, $i+3$) or (i, $i+4$) pattern is a signature of an α-helix and could have, even in the absence of a solved structure, indicated that the residues of interest occur on the same face of an α-helix (Caporaso et al., 2008). Secondary structure prediction programs can lend further support to such patterns. Indeed, the PSIPRED server accurately predicts that the covarying residues we identified occur within α helices (Jones, 1999), suggesting that their docking is governed by α-helix packing. There are many excellent structure prediction servers useful for dissecting the output of covariation analyses (Table 9.1).

Functional hypotheses for covarying pairs can also be tested through mutagenesis studies (Ashenberg et al., 2011; Skerker et al., 2008). To test whether covarying residues in the kinase–regulator analysis were in fact phosphotransfer specificity determinants, we attempted to rationally rewire the specificity of model two-component signaling proteins (Capra et al., 2010; Skerker et al., 2008). For instance, we mutated the strongest covarying positions in the kinase EnvZ to match the residues found at the equivalent positions in the kinase RstB. Strikingly, the triple mutant of EnvZ (T250V, L254Y, A255R) specifically phosphorylated RstA, the cognate substrate of RstB, rather than OmpR, the cognate substrate of wild-type EnvZ, indicating that phosphotransfer specificity had been completely rewired (Fig. 9.2E). Similar rewiring experiments demonstrated that substituting the highly covarying residues in OmpR with those in RstA produced a regulator that was preferentially phosphorylated by RstB rather than EnvZ (Capra et al., 2010). These mutagenesis experiments clearly demonstrated that the covarying pairs function as phosphotransfer specificity determinants. Although directed mutagenesis experiments are highly informative, they are often time-consuming and labor intensive. Recent approaches combining functional selections with high-throughput sequencing may soon allow more rapid, comprehensive analyses (Fowler et al., 2010).

An important consideration when making mutants to test the function of covarying pairs is that the sequence context of a mutation can have an enormous effect. In fact, the underpinning idea of covariation analyses is that the permissible mutations at a given position can depend on the residue identities at other positions. Hence, a mutation at a given site could be deleterious in one sequence context but advantageous in another context. Such epistasis at the amino acid level has been observed for both histidine kinases and response regulators (Ashenberg et al., 2011; Capra et al., 2010) and in other protein families (Gloor et al., 2010; Lunzer et al., 2010; Ortlund et al., 2007; Weinreich, Delaney, Depristo, & Hartl, 2006).

3. CONCLUDING REMARKS

Amino acid covariation analysis is a powerful approach that leverages the increase in size of sequence databases to predict residue pairs contributing to protein function. Over the past 2 decades, these approaches have been applied to a diverse set of questions in protein function. In this review, we have used the example of two-component signal transduction to illustrate the many practical considerations in implementing covariation analysis. We emphasize, however, that there is no single optimal approach and performing covariation analysis requires a careful choice of protein sequences, sequence alignment algorithms, and covariation scoring metrics. Moreover, because multiple functions, including folding, enzymatic activity, and protein–protein interaction, are often relevant to, and under selection in, a given protein, each of these functions can produce covariation signals. Determining which residues contribute to a particular function ultimately requires experimental testing, but we anticipate that analyses of amino acid coevolution will become increasingly important guides for such studies.

ACKNOWLEDGMENTS

We thank members of the Laub laboratory (C.D. Aakre and A.I. Podgornaia) and Keating laboratory (T.S. Chen, J.B. Kaplan, and C. Negron) for manuscript comments. This work is supported by a National Science Foundation CAREER Award to M.T. Laub and a National Science Foundation GRFP fellowship to O. Ashenberg and M.T. Laub is an Early Career Scientist of the Howard Hughes Medical Institute. We used computer resources provided by National Science Foundation award 0821391.

REFERENCES

Altschul, S. F., Madden, T. L., Schaffer, A. A., Zhang, J., Zhang, Z., Miller, W., et al. (1997). Gapped BLAST and PSI-BLAST: A new generation of protein database search programs. *Nucleic Acids Research*, *25*, 3389–3402.

Ashenberg, O., Rozen-Gagnon, K., Laub, M. T., & Keating, A. E. (2011). Determinants of homodimerization specificity in histidine kinases. *Journal of Molecular Biology*, *413*, 222–235.

Ashkenazy, H., Unger, R., & Kliger, Y. (2009). Optimal data collection for correlated mutation analysis. *Proteins*, *74*, 545–555.

Brown, C. A., & Brown, K. S. (2010). Validation of coevolving residue algorithms via pipeline sensitivity analysis: ELSC and OMES and ZNMI, oh my! *PloS One*, *5*, e10779.

Bryson, K., McGuffin, L. J., Marsden, R. L., Ward, J. J., Sodhi, J. S., & Jones, D. T. (2005). Protein structure prediction servers at University College London. *Nucleic Acids Research*, *33*, W36–W38.

Burger, L., & van Nimwegen, E. (2010). Disentangling direct from indirect co-evolution of residues in protein alignments. *PLoS Computational Biology*, *6*, e1000633.

Buslje, C. M., Teppa, E., Di Domenico, T., Delfino, J. M., & Nielsen, M. (2010). Networks of high mutual information define the structural proximity of catalytic sites: Implications for catalytic residue identification. *PLoS Computational Biology*, *6*, e1000978.

Caporaso, J. G., Smit, S., Easton, B. C., Hunter, L., Huttley, G. A., & Knight, R. (2008). Detecting coevolution without phylogenetic trees? Tree-ignorant metrics of coevolution perform as well as tree-aware metrics. *BMC Evolutionary Biology*, *8*, 327.

Capra, E. J., Perchuk, B. S., Lubin, E. A., Ashenberg, O., Skerker, J. M., & Laub, M. T. (2010). Systematic dissection and trajectory-scanning mutagenesis of the molecular interface that ensures specificity of two-component signaling pathways. *PLoS Genetics*, *6*, e1001220.

Casino, P., Rubio, V., & Marina, A. (2009). Structural insight into partner specificity and phosphoryl transfer in two-component signal transduction. *Cell*, *139*, 325–336.

Chakrabarti, S., & Panchenko, A. R. (2009). Coevolution in defining the functional specificity. *Proteins*, *75*, 231–240.

Dago, A. E., Schug, A., Procaccini, A., Hoch, J. A., Weigt, M., & Szurmant, H. (2012). Structural basis of histidine kinase autophosphorylation deduced by integrating genomics, molecular dynamics, and mutagenesis. *Proceedings of the National Academy of Sciences of the United States of America*, *109*, E1733–E1742.

Dickson, R. J., Wahl, L. M., Fernandes, A. D., & Gloor, G. B. (2010). Identifying and seeing beyond multiple sequence alignment errors using intra-molecular protein covariation. *PloS One*, *5*, e11082.

Di Tommaso, P., Moretti, S., Xenarios, I., Orobitg, M., Montanyola, A., Chang, J. M., et al. (2011). T-Coffee: A web server for the multiple sequence alignment of protein and RNA sequences using structural information and homology extension. *Nucleic Acids Research*, *39*, W13–W17.

Dunn, S. D., Wahl, L. M., & Gloor, G. B. (2008). Mutual information without the influence of phylogeny or entropy dramatically improves residue contact prediction. *Bioinformatics*, *24*, 333–340.

Eddy, S. R. (2011). Accelerated profile HMM searches. *PLoS Computational Biology*, *7*, e1002195.

Edgar, R. C. (2004). MUSCLE: Multiple sequence alignment with high accuracy and high throughput. *Nucleic Acids Research*, *32*, 1792–1797.

Finn, R. D., Clements, J., & Eddy, S. R. (2011). HMMER web server: Interactive sequence similarity searching. *Nucleic Acids Research*, *39*, W29–W37.

Fitch, W. M., & Markowitz, E. (1970). An improved method for determining codon variability in a gene and its application to the rate of fixation of mutations in evolution. *Biochemical Genetics*, *4*, 579–593.

Fodor, A. A., & Aldrich, R. W. (2004). Influence of conservation on calculations of amino acid covariance in multiple sequence alignments. *Proteins*, *56*, 211–221.

Fowler, D. M., Araya, C. L., Fleishman, S. J., Kellogg, E. H., Stephany, J. J., Baker, D., et al. (2010). High-resolution mapping of protein sequence-function relationships. *Nature Methods*, *7*, 741–746.

Gloor, G. B., Tyagi, G., Abrassart, D. M., Kingston, A. J., Fernandes, A. D., Dunn, S. D., et al. (2010). Functionally compensating coevolving positions are neither homoplasic nor conserved in clades. *Molecular Biology and Evolution*, *27*, 1181–1191.

Halabi, N., Rivoire, O., Leibler, S., & Ranganathan, R. (2009). Protein sectors: Evolutionary units of three-dimensional structure. *Cell, 138*, 774–786.

Hopf, T. A., Colwell, L. J., Sheridan, R., Rost, B., Sander, C., & Marks, D. S. (2012). Three-dimensional structures of membrane proteins from genomic sequencing. *Cell, 149*, 1607–1621.

Jones, D. T. (1999). Protein secondary structure prediction based on position-specific scoring matrices. *Journal of Molecular Biology, 292*, 195–202.

Katoh, K., & Toh, H. (2008). Recent developments in the MAFFT multiple sequence alignment program. *Briefings in Bioinformatics, 9*, 286–298.

Kelley, L. A., & Sternberg, M. J. (2009). Protein structure prediction on the Web: A case study using the Phyre server. *Nature Protocols, 4*, 363–371.

Kemena, C., & Notredame, C. (2009). Upcoming challenges for multiple sequence alignment methods in the high-throughput era. *Bioinformatics, 25*, 2455–2465.

Kim, D. E., Chivian, D., & Baker, D. (2004). Protein structure prediction and analysis using the Robetta server. *Nucleic Acids Research, 32*, W526–W531.

Korber, B. T., Farber, R. M., Wolpert, D. H., & Lapedes, A. S. (1993). Covariation of mutations in the V3 loop of human immunodeficiency virus type 1 envelope protein: An information theoretic analysis. *Proceedings of the National Academy of Sciences of the United States of America, 90*, 7176–7180.

Kulathinal, R. J., Bettencourt, B. R., & Hartl, D. L. (2004). Compensated deleterious mutations in insect genomes. *Science, 306*, 1553–1554.

Larkin, M. A., Blackshields, G., Brown, N. P., Chenna, R., McGettigan, P. A., McWilliam, H., et al. (2007). Clustal W and Clustal X version 2.0. *Bioinformatics, 23*, 2947–2948.

Larson, S. M., Di Nardo, A. A., & Davidson, A. R. (2000). Analysis of covariation in an SH3 domain sequence alignment: Applications in tertiary contact prediction and the design of compensating hydrophobic core substitutions. *Journal of Molecular Biology, 303*, 433–446.

Letunic, I., Doerks, T., & Bork, P. (2012). SMART 7: Recent updates to the protein domain annotation resource. *Nucleic Acids Research, 40*, D302–D305.

Lockless, S. W., & Ranganathan, R. (1999). Evolutionarily conserved pathways of energetic connectivity in protein families. *Science, 286*, 295–299.

Lunzer, M., Golding, G. B., & Dean, A. M. (2010). Pervasive cryptic epistasis in molecular evolution. *PLoS Genetics, 6*, e1001162.

Marchler-Bauer, A., Lu, S., Anderson, J. B., Chitsaz, F., Derbyshire, M. K., DeWeese-Scott, C., et al. (2012). CDD: A Conserved Domain Database for the functional annotation of proteins. *Nucleic Acids Research, 39*, D225–D229.

Marks, D. S., Colwell, L. J., Sheridan, R., Hopf, T. A., Pagnani, A., Zecchina, R., et al. (2011). Protein 3D structure computed from evolutionary sequence variation. *PloS One, 6*, e28766.

Martin, L. C., Gloor, G. B., Dunn, S. D., & Wahl, L. M. (2005). Using information theory to search for co-evolving residues in proteins. *Bioinformatics, 21*, 4116–4124.

Mintseris, J., & Weng, Z. (2005). Structure, function, and evolution of transient and obligate protein-protein interactions. *Proceedings of the National Academy of Sciences of the United States of America, 102*, 10930–10935.

Morcos, F., Pagnani, A., Lunt, B., Bertolino, A., Marks, D. S., Sander, C., et al. (2011). Direct-coupling analysis of residue coevolution captures native contacts across many protein families. *Proceedings of the National Academy of Sciences of the United States of America, 108*, E1293–E1301.

Ortlund, E. A., Bridgham, J. T., Redinbo, M. R., & Thornton, J. W. (2007). Crystal structure of an ancient protein: Evolution by conformational epistasis. *Science, 317*, 1544–1548.

Pei, J., Sadreyev, R., & Grishin, N. V. (2003). PCMA: Fast and accurate multiple sequence alignment based on profile consistency. *Bioinformatics*, *19*, 427–428.

Punta, M., Coggill, P. C., Eberhardt, R. Y., Mistry, J., Tate, J., Boursnell, C., et al. (2012). The Pfam protein families database. *Nucleic Acids Research*, *40*, D290–D301.

Skerker, J. M., Perchuk, B. S., Siryaporn, A., Lubin, E. A., Ashenberg, O., Goulian, M., et al. (2008). Rewiring the specificity of two-component signal transduction systems. *Cell*, *133*, 1043–1054.

Skerker, J. M., Prasol, M. S., Perchuk, B. S., Biondi, E. G., & Laub, M. T. (2005). Two-component signal transduction pathways regulating growth and cell cycle progression in a bacterium: A system-level analysis. *PLoS Biology*, *3*, e334.

Stock, A. M., Robinson, V. L., & Goudreau, P. N. (2000). Two-component signal transduction. *Annual Review of Biochemistry*, *69*, 183–215.

Strafford, J., Payongsri, P., Hibbert, E. G., Morris, P., Batth, S. S., Steadman, D., et al. (2012). Directed evolution to re-adapt a co-evolved network within an enzyme. *Journal of Biotechnology*, *157*, 237–245.

Teichmann, S. A. (2002). The constraints protein-protein interactions place on sequence divergence. *Journal of Molecular Biology*, *324*, 399–407.

Thattai, M., Burak, Y., & Shraiman, B. I. (2007). The origins of specificity in polyketide synthase protein interactions. *PLoS Computational Biology*, *3*, 1827–1835.

Thomas, J., Ramakrishnan, N., & Bailey-Kellogg, C. (2009). Graphical models of protein-protein interaction specificity from correlated mutations and interaction data. *Proteins*, *76*, 911–929.

Waterhouse, A. M., Procter, J. B., Martin, D. M., Clamp, M., & Barton, G. J. (2009). Jalview Version 2—A multiple sequence alignment editor and analysis workbench. *Bioinformatics*, *25*, 1189–1191.

Weigt, M., White, R. A., Szurmant, H., Hoch, J. A., & Hwa, T. (2009). Identification of direct residue contacts in protein-protein interaction by message passing. *Proceedings of the National Academy of Sciences of the United States of America*, *106*, 67–72.

Weinreich, D. M., Delaney, N. F., Depristo, M. A., & Hartl, D. L. (2006). Darwinian evolution can follow only very few mutational paths to fitter proteins. *Science*, *312*, 111–114.

White, R. A., Szurmant, H., Hoch, J. A., & Hwa, T. (2007). Features of protein-protein interactions in two-component signaling deduced from genomic libraries. *Methods in Enzymology*, *422*, 75–101.

Wollenberg, K. R., & Atchley, W. R. (2000). Separation of phylogenetic and functional associations in biological sequences by using the parametric bootstrap. *Proceedings of the National Academy of Sciences of the United States of America*, *97*, 3288–3291.

Yanofsky, C., Horn, V., & Thorpe, D. (1964). Protein structure relationships revealed by mutational analysis. *Science*, *146*, 1593–1594.

Yeang, C. H., & Haussler, D. (2007). Detecting coevolution in and among protein domains. *PLoS Computational Biology*, *3*, e211.

Yip, K. Y., Patel, P., Kim, P. M., Engelman, D. M., McDermott, D., & Gerstein, M. (2008). An integrated system for studying residue coevolution in proteins. *Bioinformatics*, *24*, 290–292.

CHAPTER TEN

Evolution-Based Design of Proteins

Kimberly A. Reynolds, William P. Russ, Michael Socolich, Rama Ranganathan[1]

Green Center for Systems Biology, Department of Pharmacology, University of Texas Southwestern Medical Center, Dallas, Texas, USA

[1]Corresponding author: e-mail address: rama.ranganathan@utsouthwestern.edu

Contents

Abstract

Statistical analysis of protein sequences indicates an architecture for natural proteins in which amino acids are engaged in a sparse, hierarchical pattern of interactions in the tertiary structure. This architecture might be a key and distinguishing feature of evolved proteins—a design principle providing not only for foldability and high-performance function but also for robustness to perturbation and the capacity for rapid adaptation to new selection pressures. Here, we describe an approach for systematically testing this design principle for natural-like proteins by (1) computational design of synthetic sequences that gradually add or remove constraints along the hierarchy of interacting residues and (2) experimental testing of the designed sequences for folding and biochemical function. By this process, we hope to understand how the constraints on fold, function, and other aspects of fitness are organized within natural proteins, a first step in understanding the process of "design" by evolution.

Methods in Enzymology, Volume 523
ISSN 0076-6879
http://dx.doi.org/10.1016/B978-0-12-394292-0.00010-2

1. INTRODUCTION

Natural proteins can fold under physiological conditions into compact three-dimensional structures and are capable of remarkably complex and high-performance biochemical functions. Because these properties require great accuracy in the position and dynamics of certain amino acids, it is tempting to think of proteins as precisely engineered systems in which interactions between the components (amino acids) are finely tuned and exactly arranged throughout the structure. However, other aspects of natural proteins are inconsistent with this view and demand a deeper examination of the basic underlying design principles through the process of evolution. For example, proteins are typically robust to random mutation; that is, they tolerate perturbations in many amino acid positions without much alteration in function (Bowie, Reidhaar-Olson, Lim, & Sauer, 1990; McLaughlin, Poelwijk, Gosal, & Ranganathan, 2012; Reidhaar-Olson & Sauer, 1990). In addition, they are plastic; that is, they have the ability to adapt to changing selection pressures by allowing specific variation of a few residues to profoundly alter function (McLaughlin et al., 2012; Orencia, Yoon, Ness, Stemmer, & Stevens, 2001). This curious combination of robustness to random mutation and yet sensitivity to targeted perturbation is interesting because it suggests that despite the appearance of precise construction throughout, strong heterogeneity exists in the design of proteins such that some residues and interactions between residues (the "core" machinery) are much more important than others. A major current goal in protein biology is to define and then mechanistically understand this heterogeneous architecture of natural proteins.

What is a good approach for this problem? The first step is to systematically map the energetic value of amino acid interactions globally in proteins, a nontrivial task by computational or experimental approaches. The main reasons are well known: (1) the relationship between the observed structural features of amino acid interactions and their net energetic value is extremely subtle and (2) amino acids can interact through complex high-order cooperative groups that can induce functionally significant energetic couplings between noncontacting amino acids. In general, physics-based approaches to computing protein energetics are based on making simplifying approximations for these two issues that result in quasi-empirical potential functions that consider only local interactions in protein structure. These approximations are necessary to make the calculations feasible and have led to remarkable successes in the engineering of protein folds (Dahiyat & Mayo, 1997; Dantas, Kuhlman, Callender, Wong, & Baker, 2003; Harbury, Plecs, Tidor, Alber, & Kim, 1998; Kuhlman et al., 2003). Nevertheless, it is important to realize that the design principles imposed by the approximations made can, in principle, deviate significantly from the natural evolutionary design of proteins.

The remarkable recent advances in genome sequencing efforts suggest an alternative *statistical* approach to this problem. The basic idea is that extant sequences have been selected through a long process of random mutation and selection and that a protein family that shares an overall fold and basic aspects of function should reveal the architecture of key amino acid interactions in the pattern of statistical constraints on and between amino acid positions. This idea is nothing more than a quantitative generalization of the widely accepted principle of sequence conservation as a metric of structural or functional importance. The conjectures are twofold: (1) positions that are important should experience an evolutionary constraint and should show a degree of conservation that reflects this constraint and (2) positions that energetically interact (whether through direct structural interactions or through indirect pathways of amino acid interactions) should experience a joint evolutionary constraint and should show correlated conservation or coevolution. Below, we discuss primarily one quantitative approach based on these conjectures called the statistical coupling analysis (SCA). For a given protein family, this analysis yields a mechanistically unbiased global map of amino acid interactions that encapsulates evolutionary constraints over all biochemical and biophysical properties that contribute to fitness.

Application of SCA in many protein families has led to a general model for the architecture of natural proteins—protein structure and function are hierarchically encoded by a subset of residues (termed the sector) embedded within the protein. In this chapter, we first give a brief description of our mathematical approach to sequence analysis and summarize the basic findings of protein sectors. We then describe a design method for testing this model for natural proteins by the creation of synthetic proteins that explore the hierarchy of statistical constraints. It is our intent to provide a general recipe for experiments to investigate how the pattern of amino acid correlations specifies folding, stability, and function in natural proteins.

2. SCA: THE PATTERN OF EVOLUTIONARY CONSTRAINT IN PROTEINS

The details of SCA have been described elsewhere (Halabi, Rivoire, Leibler, & Ranganathan, 2009; Smock et al., 2010), but here we give an overview as a preliminary to describing our design methodology. Matlab scripts for performing these calculations in full as well as tutorials illustrating SCA for several representative protein families are provided on our website (http://systems.swmed.edu/rr_lab).

2.1. The basic calculations

The process of SCA begins with assembly of a large and diverse multiple sequence alignment (MSA) for a particular protein family. For example, consider an MSA comprising 240 sequences of the WW domain family of small protein interaction modules that bind to proline-containing target peptides (Fig. 10.1A). The suitability of an MSA for SCA depends on multiple factors, such as the number of sequences, the sampling of phylogenetic space, and the general quality of the alignment (lack of large gapped regions, correct alignment of key functional residues). However, in practice, a general (though not strict) guideline for alignment construction is the inclusion of more than 100 sequences with a mean sequence identity between sequence pairs in the range of 15–50%.

From the MSA, the first-order analysis is to compute the conservation of each amino acid a at position i considered independently of other positions. In SCA, conservation is measured by $D_i^{(a)}$, an information-theoretic quantity called the Kullback–Leibler (K–L) relative entropy. This quantity indicates the deviation in the frequency of amino acid a at position i $\left(f_i^{(a)}\right)$ from the background probability of amino acid a ($q^{(a)}$) estimated from the non-redundant protein database. In the limit of large sampling (number of sequences > 80; Halabi et al., 2009), this calculation reduces to

$$D_i^{(a)} = f_i^{(a)} \ln \frac{f_i^{(a)}}{q^{(a)}} + \left(1 - f_i^{(a)}\right) \ln \frac{1 - f_i^{(a)}}{1 - q^{(a)}}. \quad [10.1]$$

The K–L entropy basically describes how unexpected the observed frequency $f_i^{(a)}$ is, given an expected probability of $q^{(a)}$, and has the following two properties: (1) $D_i^{(a)} = 0$ if $f_i^{(a)} = q^{(a)}$ and (2) $D_i^{(a)}$ increases nonlinearly more and more steeply as f deviates from q. An overall positional K–L entropy D_i can also be computed that takes into account all the amino acids per position (Fig. 10.1B, bar graph):

$$D_i = \sum_{a=0}^{20} f_i^{(a)} \ln \left(\frac{f_i^{(a)}}{\bar{q}^{(a)}} \right), \quad [10.2]$$

where $\bar{q}^{(a)}$ represents the background frequencies including gaps (Halabi et al., 2009). For the WW domain, the conservation pattern is as expected;

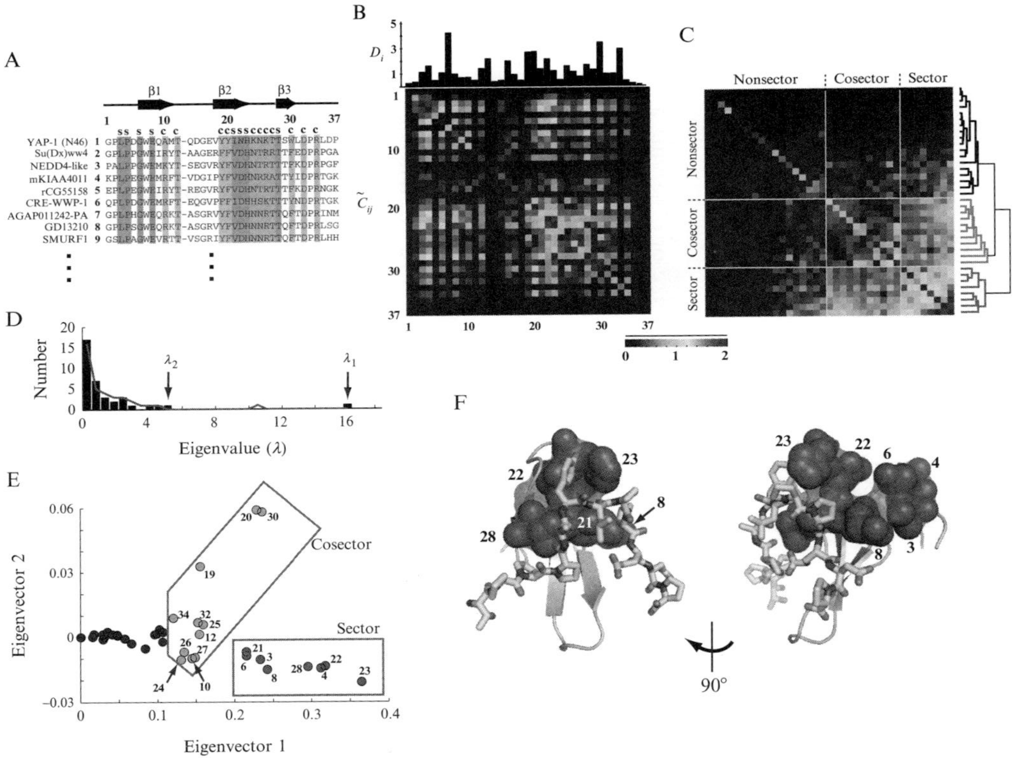

Figure 10.1 Statistical coupling analysis (SCA). (A) A portion of the alignment for the WW domain family. Sector (s), cosector (c), or nonsector (unmarked) positions (defined below) show no obvious arrangement in primary or secondary structure. (B) The site-independent conservation (D_i, bar graph) and the SCA matrix of coevolution between all pairs of amino acids $\left(\widetilde{C}_{ij}\right)$. Values in the matrix are as indicated by the color bar. (C) Clustering in the matrix reveals three main groups of residues: sector, cosector, and nonsector. (D) Comparison of the eigenspectrum of the SCA matrix generated from the natural alignment (bars) to eigenspectra for randomized versions of the alignment (line) indicates that just the top two eigenvalues, λ_1 and λ_2, are distinguished from noise. (E) The corresponding eigenvectors reveal the positions that contribute the most to the top eigenvalues and define the sector positions (red) and cosector positions (blue). (F) The sector shown as red space-filling spheres on a representative WW domain structure (PDB ID: 2LAW, gray cartoon) in complex with a peptide ligand (stick bonds). (See Color Insert.)

the two positions with the eponymous tryptophan residues (7 and 30) are the most conserved, together with a proline at position 33 that is a key part of the protein core. In general, prior work shows that the pattern of positional conservation is an effective predictor of residue burial in the tertiary structure (Halabi et al., 2009).

The second-order analysis is to compute a conservation-weighted correlation matrix $\left(\tilde{C}_{ij}\right)$ that represents the coevolution of each pair of positions in the MSA. To do this, we compute the weighted correlation tensor $\tilde{C}_{ij}^{(ab)}$:

$$\tilde{C}_{ij}^{(ab)} = \phi_i^{(a)} \phi_j^{(b)} C_{ij}^{(ab)}, \qquad [10.3]$$

where $C_{ij}^{(ab)} = f_{ij}^{(ab)} - f_i^{(a)} f_j^{(b)}$ represents the raw frequency-based correlations between each pair of amino acids (a,b) at each pair of positions (i, j), and ϕ represents a conservation-based weighting function. In the current implementation of SCA, $\phi_i^{(a)} = \left|\partial D_i^{(a)} / \partial f_i^{(a)}\right|$, the gradient of relative entropy. Just as $D_i^{(a)}$ represents positional conservation by the significance of observing a frequency $f_i^{(a)}$ given a background expectation, $\tilde{C}_{ij}^{(ab)}$ represents coevolution by the significance of observing a raw correlation in $C_{ij}^{(ab)}$ as judged by the weighting functions $\phi_i^{(a)}$ and $\phi_j^{(b)}$. Thus, $\tilde{C}_{ij}^{(ab)}$ up-weights correlations between conserved positions and damps correlations between less conserved positions. This conservation weighting serves to minimize the contribution of purely phylogenetic correlations between weakly conserved positions that are expected to emerge from small clades of sequences that have not had sufficient time to decorrelate unconstrained pairs of sequence positions. Other weighting functions are possible and are the subject of ongoing studies, and they will not be discussed further here. Regardless, the salient concept is that SCA considers conservation-weighted correlations between amino acids.

The result of this calculation is a four-dimensional tensor, $\tilde{C}_{ij}^{(ab)}$, that contains the correlation of every amino acid pair (a, b) for every position pair (i, j). For a protein with N positions, the dimensions of $\tilde{C}_{ij}^{(ab)}$ are $N \times N \times 20 \times 20$. We then reduce this tensor to an $N \times N$ matrix of positional correlations $\left(\tilde{C}_{ij}\right)$ by taking the Frobenius norm of each 20×20 amino acid correlation matrix for each amino acid pair:

$$\tilde{C}_{ij} = \sqrt{\sum_{(ab)} \left(\tilde{C}_{ij}^{(ab)}\right)^2}. \qquad [10.4]$$

This matrix norm gives the overall magnitude of correlation between each pair of positions (i, j) arising through all possible amino acid pairs. $\tilde{C}_{ij}$ is also referred to, in short, as the SCA matrix—a global examination of statistical coupling between all pairs of positions in the long-term evolutionary record of a protein family. Figure 10.1B shows $\tilde{C}_{ij}$ for the WW family; in this matrix, diagonal elements are related to the intrinsic conservation of each position, and each off-diagonal element indicates the coevolution between a pair of amino acid positions.

2.2. Analysis of the SCA positional coevolution matrix

How can we analyze the pattern(s) of amino acid coevolution in the SCA matrix? Visual examination of $\tilde{C}_{ij}$ for many different protein families leads to two main observations. First, the pattern of correlations is not obviously organized with respect to proximity in primary structure or to the pattern of contacts between secondary structure elements (Fig. 10.1A). Second, the matrix is remarkably sparse—the majority of amino acids appear to evolve relatively independently (as indicated by the large number of weak correlations in Fig. 10.1B), while a few show strong indications of coevolution.

Clustering of the SCA matrix makes this result more obvious—a subset of amino acids located in the bottom right corner are strongly coevolving while the majority of amino acids are more weakly coupled to one another (Fig. 10.1C). A closer inspection of the clustered matrix suggests a hierarchical organization of amino acid interactions. The bottom cluster (Fig. 10.1C) constitutes a small group of residues that collectively coevolve with one another and the group contains the majority of the signal in the matrix; we define such a group of residues as a "protein sector." The middle cluster (Fig. 10.1C) comprises residues that show little direct coupling to each other but that show systematic coevolution with sector positions (Fig. 10.1C). As these positions cluster by association to the sector, we call them the "cosector." The third and largest set (the nonsector, Fig. 10.1C) shows very little coupling at all.

Clustering provides one means of examining the pattern of evolutionary constraints within the matrix, but a more rigorous approach derives from the principles of spectral decomposition and random matrix theory (Halabi et al., 2009). The spectral decomposition mathematically transforms a correlation matrix between initial variables (e.g., the SCA matrix of amino acid correlations) into eigenmodes, which are described by a set of eigenvectors and eigenvalues. In this representation, each eigenvector contains the weights for linearly combining the initial variables (e.g., the amino acid positions) and each associated eigenvalue indicates the quantity of overall variance captured by that eigenmode. For the SCA matrix, each eigenmode represents a group of residues that share a similar pattern of coevolution, and the eigenvalue spectrum—the histogram of ($\lambda_1 > \lambda_2 > \lambda_3 > \ldots$)—reveals how the coevolutionary signal is quantitatively distributed among the eigenmodes. Comparison of this distribution with the eigenvalue spectra of correlation matrices derived from randomized alignments (shown as a line in Fig. 10.1D) shows that most of the lowest modes are indistinguishable from noise, while the top few modes capture statistically significant correlations.

For the WW domain, examination of the eigenvectors associated with the top two eigenmodes confirms the findings from clustering; a small set of amino acid positions contribute to the majority of the covariation in the matrix and emerge along the first eigenmode (the sector) (Fig. 10.1D). Consistent with coevolution with the sector, the cosector is evident as positions with weaker weights on the first eigenmode and projection along the second eigenmode (Fig. 10.1E). Indeed, the sector identified by eigendecomposition corresponds exactly, in this case, to the set of residues identified by clustering the matrix, though an exact match between the two methods need not always be true. In general, the spectral decomposition (rather than clustering) is a more quantitative approach for sector identification and is valuable in the process of SCA-based protein design, described later.

What is the structural interpretation of the sector? In the WW domain, the sector forms a sparse, distributed, and physically contiguous network that is distinct from known classifications of proteins based on primary, secondary, and tertiary structure (Russ, Lowery, Mishra, Yaffe, & Ranganathan, 2005; Fig 10.1F). Mutational studies show that sector residues, whether near or far from the ligand, contribute cooperatively to binding affinity—an extended network underlying the WW domain function (Russ et al., 2005). In the PDZ family of protein interaction modules, the sector connects the ligand-binding site to an allosteric site located on the opposite face (Fig. 10.2A; Halabi et al., 2009; Lockless & Ranganathan, 1999; McLaughlin et al., 2012). Sectors have been found in all protein families studied to date, and like in WW and PDZ domains, are empirically observed to share three properties: they are (1) sparse (comprising ~20% of the protein structure), (2) they are physically contiguous, and (3) they connect the active site or ligand-binding site to distant surfaces distributed throughout the structure (Ferguson et al., 2007; Halabi et al., 2009; Hatley, Lockless, Gibson, Gilman, & Ranganathan, 2003; Lee et al., 2008; Lockless & Ranganathan, 1999; Shulman, Larson, Mangelsdorf, & Ranganathan, 2004; Smock et al., 2010; Suel, Lockless, Wall, & Ranganathan, 2003).

Interestingly, the mapping of sectors to domains is not necessarily one to one. Using more advanced extensions of eigendecomposition (Smock et al.,

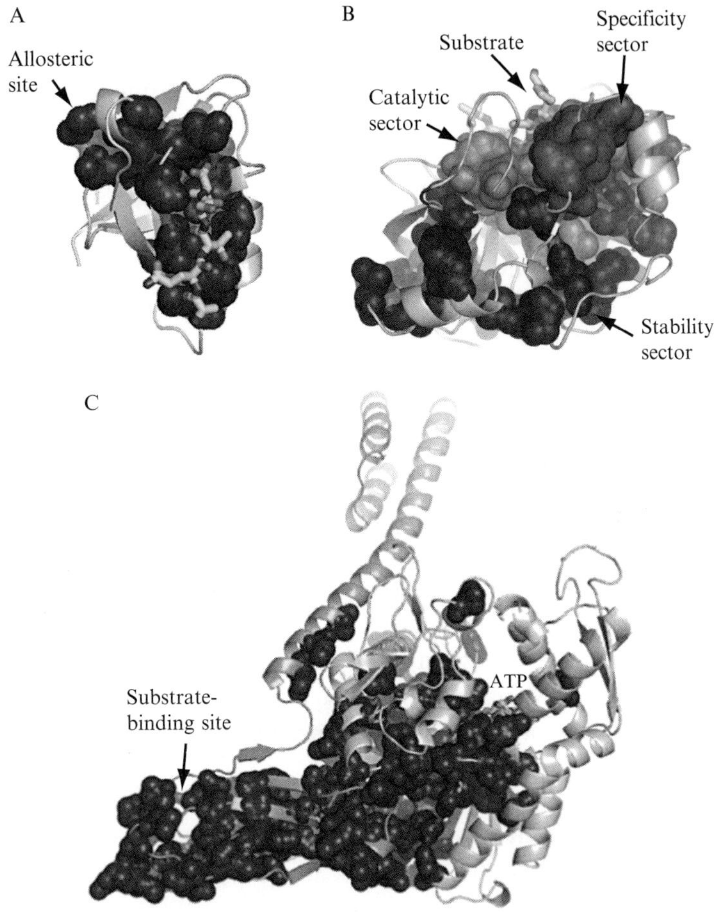

Figure 10.2 Sectors in three protein families. A single sector in the PDZ domain connects the ligand-binding pocket to a distant allosteric site (A), three quasi-independent sectors occur in the S1A family of serine proteases (B), and a single interdomain sector functionally connects the ATP-binding site in the nucleotide-binding domain (white) to the ligand-binding site in the substrate-binding domain (gold) in the Hsp70 family of molecular chaperones. (For interpretation of the references to color in this figure legend, the reader is referred to the online version of this chapter.)

2010), it is possible to find multiple independent sectors within a single-domain protein. For example, in the S1A serine proteases, three near-independent sectors are evident, each of which comprises a distinct but physically contiguous subnetwork within the tertiary structure (Fig. 10.2B). Studies in one member of the S1A family (rat trypsin) show that each sector corresponds to a distinct biochemical property—catalytic mechanism, substrate specificity, and stability—indicating that this decomposition of the protein by patterns of coevolution is functionally relevant (Halabi et al., 2009). Conversely, it is also possible to find a single sector spanning two distinct domains of a protein (e.g., the Hsp70 chaperone, Fig. 10.2C; Smock et al., 2010). In Hsp70, the single interdomain sector links the ATP-binding site in the nucleotide-binding domain with the ligand-binding site in the substrate-binding domain through the interdomain interface, a feature that reflects the fact that the conserved functional activity of Hsp70 proteins is allosteric communication between the two domains. Thus, sectors expose the architecture of fitness constraints on or between proteins, independent of structural basis or mechanistic detail.

3. SCA-BASED PROTEIN DESIGN

How can we test the sufficiency of the sector model for protein structure, function, stability, and adaptability—the basic features of natural proteins? Targeted mutational analyses provide a first-order test that sectors specify aspects of protein function. But a more global and complete test comes through synthetic protein design. The idea is to carry out computational simulations that start with random sequences and evolve (*in silico*) synthetic sequences that are constrained by the observed evolutionary statistics. Experimental study of libraries of the designed sequences represents a deep test of the sufficiency of the applied constraints for recapitulating the properties of natural proteins.

3.1. Defining an objective function

The approach in SCA-based protein design is to use the Metropolis Monte Carlo simulated annealing (MCSA) algorithm to explore the sequence space consistent with a set of applied constraints between amino acids. The MCSA algorithm is an iterative numerical method for searching for the global minimum energy configuration of a system starting from any arbitrary state and is especially useful when the number of possible states is very large and the energy landscape is rugged and characterized by many local minima (Kirkpatrick, Gelatt, & Vecchi, 1983; Metropolis, Rosenbluth, Rosenbluth, Teller, & Teller, 1953). The energy function (or "objective function") to be minimized can, in general, depend on many parameters of the system and represents the constraints that define the size and shape of the final solution space. In essence, the objective function can be thought of as the hypothesis being tested—the set of applied constraints that we wish to test for specifying folding, thermodynamic stability, function, and any other aspects of protein fitness.

For SCA-based protein design, the system under consideration is a MSA (rather than a single sequence), and the objective function (E) is the summed difference between the correlation tensor for a MSA of protein sequences during iterations of the design process and the target correlation matrix deduced from the natural MSA:

$$E = \sum_{ijab} \left| \tilde{C}^{(ab)}_{ij(\text{design})} - \tilde{C}^{(ab)}_{ij(\text{natural})} \right|. \qquad [10.5]$$

Thus, the lowest energy configuration for the designed MSA is the set of sequences that gives a pattern of correlations in the designed sequences $\left(\tilde{C}^{(ab)}_{ij(\text{design})}\right)$ that most closely reproduces that of the natural MSA $\left(\tilde{C}^{(ab)}_{ij(\text{natural})}\right)$. At the limit of large numbers of sequences, this result is tantamount to drawing sequences from a maximum entropy probability distribution consistent with the applied set of observed correlations (Bialek & Ranganathan, 2007).

What correlations should be included in the objective function? At the extreme limit, the objective function could involve the full correlation tensor in which $\tilde{C}^{(ab)}_{ij}$ has indices i and j that run over all positions in the MSA and a and b that run over all 20 amino acids; this is a trivial simulation because the only ensemble of sequences that lies at the global minimum of the objective function is the same sequences that comprise the natural MSA. However, a large number of (weak) correlations in the full $\tilde{C}^{(ab)}_{ij}$ are indistinguishable from noise due to finite sampling or phylogeny and are therefore proposed to be functionally insignificant. In addition, even the statistically significant correlations are not likely to be all equally important; indeed, there is a hierarchy of correlations within the sector such that some amino acids are more strongly coevolving and are surrounded by residues making lesser contributions (Fig. 10.1C). The key goal in SCA design is then is to find appropriate "reduced" objective functions that comprise a hypothesis for the "relevant" constraints and then to test these for sufficiency with regard to protein structure and function.

Reduced objective functions can be obtained by two general approaches: (1) elimination (or masking) of correlations in the $\tilde{C}^{(ab)}_{ij}$ tensor based on heuristic knowledge or on statistical cutoffs on the distribution of correlations and (2) partial convergence on the full correlation tensor. It is important to say that these techniques are not entirely different from each other; indeed, the collective group of correlations defining sectors tend to be larger in magnitude, and as we will describe below, partial convergence by the MCSA algorithm has the property of building in the collective modes defining sectors at the expense of the weaker, more idiosyncratic correlations. Nevertheless, these strategies represent different practical ways of posing hypotheses for testing the information content of protein sequences through design.

For example, in an initial study on SCA-based design of WW domains, the objective function involved a subset of $\tilde{C}_{ij}^{(ab)}$ corresponding to the correlations for only five sector residues—a test that the correlations between just these few specific positions and all other sites is sufficient for protein folding and function (Russ et al., 2005; Socolich et al., 2005). However, other objective functions might be imagined; for example, for multisector proteins such as the S1A serine proteases, one can envision designs that target only the correlations that define a single protein sector—a design that should cause variation in one functional property while leaving the roles of other sectors relatively unperturbed. More generally, it would be interesting to test objective functions that sample different eigenmodes of the SCA correlation matrix to experimentally understand if and how the spectral decomposition of information content in protein sequences corresponds to a decomposition of the different biochemical properties of proteins that contribute to fitness.

3.2. The simulated annealing algorithm

Given an objective function, the strategy of the MCSA algorithm underlying SCA-based design follows the standard simulated annealing process. We initiate the simulation with an alignment that has been randomized by a process we term "vertical shuffling"—randomly permuting each column of the MSA independently. Vertical shuffling removes all nonrandom correlations between positions, but by nature, preserves the frequency distribution (i.e., the conservation) of amino acids at positions exactly. The MCSA method then involves many iterations of a two-step process to converge on the set of constraints specified by the objective function. In step one, we choose one position and two sequences from the MSA at random and swap the corresponding amino acids (Fig. 10.3A). Since the swap is always done within one position, this perturbation never influences the independent conservation of positions, but it could introduce a change to the pattern of correlations. In step 2, we evaluate the objective function (E) to give the impact of the swap on the overall set of included correlations and compute $\Delta E = E_{\text{current}} - E_{\text{previous}}$, the change in the objective function from the previous iteration. If $\Delta E \leq 0$, the perturbation is favorable (i.e., the pattern of correlations has become more natural-like), and we always accept the swap.

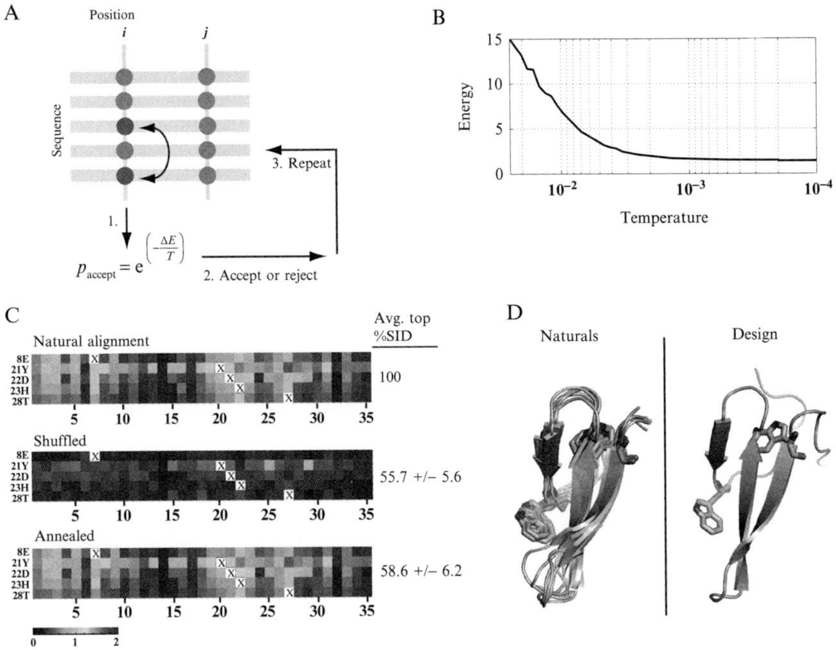

Figure 10.3 Evolution-based design of the WW domain. (A) The Monte Carlo design process. Each iteration involves two steps: (1) a swap of amino acids between two randomly chosen sequences at one randomly chosen position and (2) a decision to either accept or reject the swap based on the difference in the objective function (ΔE, see text) and a computational "temperature" (T). If $\Delta E > 0$, the swap is accepted with a probability determined by the Boltzmann distribution. (B) A MCSA trajectory for the WW domain family. The process starts with a high temperature and exponentially cools the MSA, converging toward a minimum for the objective function. (C) In Socolich et al. (2005), the objective function involved correlations for five sector positions. This portion of the SCA matrix (with self-correlations blanked) is shown for the natural WW alignment, the vertically shuffled (high temperature) alignment, and the annealed alignment. At right is indicated the average sequence identity (± SD) to the closest sequence in the natural WW domain MSA. (D) Comparison of experimental structures for a collection of natural WW domains and one of the synthetic WW domains. The eponymous tryptophan residues are indicated in stick bonds. (See Color Insert.)

If $\Delta E > 0$, the perturbation is unfavorable, and we accept the swap with a probability given by the Boltzmann distribution:

$$p = e^{\frac{-\Delta E}{T}}. \quad [10.6]$$

This property of MCSA—in which unfavorable swaps are probabilistically accepted—is a key feature that prevents the search algorithm from becoming trapped in local minima. The "temperature" factor T is a purely computational term that controls the probability of accepting unfavorable swaps. At high temperatures, swaps causing even significant perturbation to correlations are likely, while at low temperatures such swaps become exponentially less probable. The basic idea of the MCSA is to gradually cool the MSA from a high temperature along a near-equilibrium path until the simulation converges on a set of synthetic sequences that reproduces the correlations included in the objective function (Fig. 10.3B). It can be intuitively seen that in the path of an MCSA simulation, the strongest collectively evolving modes of the correlation matrix (that define sectors) will anneal first and cooperatively over a narrow range of temperatures, with the remainder of the weaker and less collective correlations converging gradually as the temperature cools further. The simulation exits when the temperature cools sufficiently that no further swaps are accepted.

3.3. SCA-based design of WW domains

The first application of protein design using evolutionary correlations was carried out for the WW family of protein interaction modules (Russ et al., 2005; Socolich et al., 2005). The WW domain adopts a curved, three-stranded antiparallel β-sheet configuration and binds to proline-rich peptide ligands along one face of the sheet (Fig. 10.1F). As mentioned earlier, the objective function in this initial design experiment involved a matrix comprising just the correlations between one dominant amino acid at five sector positions and all amino acids at all other positions—a heuristic choice based on the fact that these correlations capture much of the total information content in the SCA matrix for the WW family. Starting from the vertically shuffled MSA as the initial state (IC, or site-*i*ndependent *c*onservation sequences), the MCSA algorithm was used to converge on new MSA of sequences (CC, or "*c*oupled *c*onservation" sequences) that recapitulate the applied constraints (i.e., minimizing the objective function, Fig. 10.3B–C).

Four libraries of synthetic genes were constructed and analyzed to experimentally test the likelihood of native folding and function in SCA-based design: (1) 48 natural WW domains drawn randomly from the natural MSA (as a positive control), (2) 48 IC sequences that represent the input for MCSA, (3) 48 CC sequences that represent the output of MCSA, and (4) 23 completely random sequences (R) with amino acids at each position chosen from the mean frequency of each amino acid in the WW MSA (as a negative control). Statistically, the natural, IC, and CC sequences showed a mean amino acid identity to natural WW domains of ~36%, an expected result given the constraint to preserve the conservation of amino acids at sites. Also expected, the random sequences show a much lower mean identity to natural WW domains (~6%). However, the IC and CC sequences show a similar "top-hit" identity on average with natural sequences—the percent identity to their closest counterpart in the natural world (Fig. 10.3C). Thus, the CC sequences are statistically indistinguishable in sequence divergence from IC domains indicating that, by this measure, the number of extra constraints from correlations is small. In essence, the difference between IC and CC sequences is not the magnitude of similarity to natural sequences but the pattern by which they are similar.

Analysis of solubility, folding thermodynamics, and ligand-binding specificity for all soluble domains showed a clear result: no random or IC sequences were folded, but a significant fraction of CC sequences showed both native folding and biochemical function that quantitatively recapitulated the behavior of natural WW proteins (Socolich et al., 2005). In addition, structure determination of one synthetic CC domain demonstrated recapitulation of the characteristic tertiary structure of the WW domain, with an atomic-level accuracy that is within the variance of known natural WW structures (Fig. 10.3D). Thus, for this domain, fold and function can be recapitulated by the information contained in the portion of the SCA matrix that contains the pairwise correlations for a set of sector positions. This experiment provides a first test that the pattern of amino acid coevolution in the SCA matrix represents one solution for specifying natively folded and functional proteins. The small fraction of total correlations used in the objective function implies a surprising simplicity in the evolutionary design of this protein domain.

4. SCA-BASED PARSING OF PROTEIN STABILITY AND FUNCTION

The objective function used in the initial design of WW domains included the correlations for five sector positions over all other positions. The near-sufficiency of this design is interesting, but this experiment does not decompose the contribution of sector and nonsector positions. Indeed, does the hierarchy of correlations differentially encode properties of protein folding and function? To examine this, let us revisit the clustered matrix for the WW domain family (Fig. 10.1C). This matrix shows a hierarchical pattern of organization with three groups of residues identified by clustering:

sector, cosector, and nonsector. To understand how this organization encodes protein stability and function, we conducted a simple "positional shuffling" experiment on the WW domain sequence alignment (Fig. 10.4A–E). In this experiment, we designed synthetic variants of one WW domain sequence within the overall MSA (the second WW domain from the Yes-kinase associated protein 1 (YAP-1), referred to as N46; Socolich et al., 2005) in which statistical couplings are systematically eliminated for sector, cosector, and nonsector groups, either alone or in combination. To do this, we simply vertically shuffle the amino acids independently at each position comprising a group and select the N46 variant. As described in Section 3.2, this process removes all nonrandom correlations between positions comprising a selected group(s) while preserving the conservation of amino acids at individual sites (Fig. 10.4E).

In total, we generated four sets of 20 shuffled N46 sequences each (80 total proteins), which are named according to the residue clusters in which correlations were retained: N46(1)—sector intact, all other positions shuffled (Fig 10.4A); N46(1 + 2)—sector/cosector intact, nonsector positions shuffled (Fig 10.4B); N46(2 + 3)—cosector/nonsector intact, sector positions shuffled (Fig 10.4C); and N46(3)—nonsector intact, sector, and cosector positions shuffled (Fig 10.4D). All the synthetic shuffled sequences were characterized for three properties: solubility upon expression in *Escherichia coli*, presence of a cooperative unfolding transition by thermal denaturation (Socolich et al., 2005), and function as assessed by class-specific peptide binding (Russ et al., 2005). N46 is a class I WW domain, recognizing PPxY containing target peptides; accordingly, functional synthetic WW domains were scored as those that recapitulate this same binding specificity.

The results of these experiments are summarized in Fig. 10.4F and G. First, the data show that for all four sequence sets only a small number of domains are insoluble (gray wedges), leaving the majority for analysis of fold stability and function. Second, both the necessity and near-sufficiency of the sector/cosector for N46-like class I binding specificity are evident. Shuffling the sector positions (N46(2 + 3)) results in only one of 16 tested domains with class I specificity, and shuffling of both the sector and cosector positions (N46(3)) results in a total loss of function for all domains (Fig. 10.4G). In contrast, preserving only the eight sector residues and shuffling the remainder (N46(1)) results in 5 of 16 functional domains, and retaining both the sector and the cosector (N46(1 + 2)) results in 16 of 18 class I domains (Fig. 10.4G). Thus, protein function in the WW domain largely emerges from the sparse network of amino acid positions that define the sector.

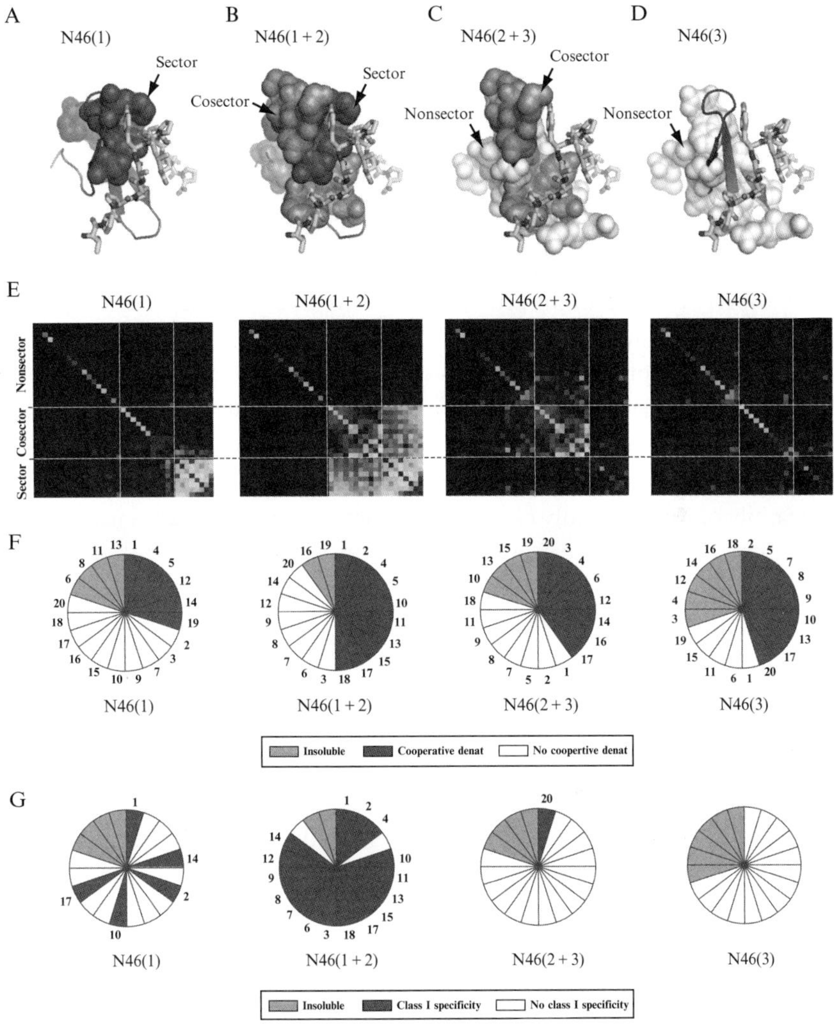

Figure 10.4 SCA-based design of shuffled WW domains. (A–D) In each panel, the group of residues for which couplings were preserved (nonshuffled positions) are shown in space-filling spheres on the structure of the "N46" WW domain in complex with a Class I peptide ligand (PDB 2LAW). Sector, cosector, and nonsector positions are labeled. (E) SCA matrices for each set of shuffled N46 sequences, visually illustrating the correlations that are scrambled. (F–G) Ordered pie charts showing the outcome of thermal denaturation (F) and Class I peptide binding specificity (G) experiments for each of the four groups of synthetic WW domains. The order of slices is the same in the two panels to facilitate comparison. (See Color Insert.)

Interestingly, fold stability seems to obey a different rule. Regardless of whether one shuffles only the sector positions (N46(2+3)), the sector and cosector positions (N46(3)), or nonsector positions (N46(1+2)), the results are the same: 30–50% of the resulting domains display a cooperative folding transition in the 4–90 °C temperature range (Fig. 10.4F). Taken together, these findings suggest that the capacity to fold and exhibit specific molecular recognition is localized to the subset of positions showing coevolution (the sector/cosector), while the stability of the fold is a more distributed property of the protein structure, even involving weakly correlated and less conserved positions.

From an evolutionary perspective, this suggests the possibility that the amino acid interactions underlying thermodynamic stability need not be deeply conserved in protein families, but rather can be easily varied. That is, the origin of thermodynamic stability in any particular member of a protein family may be a rapidly changing and perhaps even idiosyncratic feature emerging from small local groups of amino acids with many degenerate possible solutions. The generality and validity of this apparent parsing of fold stability and function should be more deeply examined in larger and more stable proteins and with greater sampling of synthetic designs.

5. FUTURE MONTE CARLO STRATEGIES FOR EXPLORING SEQUENCE SPACE

The experiments described above provide a simple coarse-grained preview of how properties of protein folding and function might be encoded in the hierarchy of positional correlations. But, to examine the pattern of residue couplings with greater resolution and less interpretational bias (i.e., without heuristically defined objective functions and parsing of correlations into discrete blocks such as sector and cosector), we need a way to systematically add or remove correlations along the hierarchy. Here, we describe a strategy to address this question. Implementing this method requires three things: (1) a computational method to design protein sequences that smoothly vary couplings along a hierarchy observed in the SCA matrix, (2) a way to synthesize genes corresponding to a large number of synthetic proteins along this trajectory cheaply and reliably, and (3) high-throughput methods to assess protein function for the libraries of synthetic designs. The methods for assessing protein function are specific for model systems and are not discussed here, but they should be generally possible for any protein in which cell growth rate or fitness can be coupled to protein activity. For example, proteins such as primary metabolic enzymes (Reynolds, McLaughlin, & Ranganathan, 2011; Taylor, Kast, & Hilvert, 2001) or enzymes that mediate antibiotic resistance (Weinreich, Delaney, Depristo, & Hartl, 2006) have obvious advantages in this respect. These experiments are likely to become feasible as advances in technologies for gene synthesis and automated screening for protein stability and function mature (Gerber, Maerkl, & Quake, 2009; Isom, Marguet, Oas, & Hellinga, 2011; Kosuri et al., 2010).

The computational approach is to design sequences as a function of temperature along a Monte Carlo trajectory (i.e., obeying Eq. 10.5), systematically testing for loss (in the case of heating) or gain (in the case of cooling) of protein properties of interest. The cooling trajectory is as described above in Section 3.2—we begin with a vertically shuffled alignment and lower the temperature along a near-equilibrium path to converge on our objective function. The result is an ensemble of sequences that can be characterized at each temperature. Constraints between residues anneal as a function of temperature according to their strength and collective character and thus this design has the property of building the top eigenmodes (defining sectors) first and then slowly annealing on the lower eigenmodes containing weaker and less collective correlations. A dense sampling of sequences along this trajectory would provide a rigorous test of how the statistical structure of correlations is related to various protein properties.

Monte Carlo simulated heating (MCSH) is conceptually similar to MCSA but differs in initial condition and direction of progress (Fig. 10.5A). In this experiment, we begin with a MSA of natural sequences (rather than a vertically shuffled MSA) and conduct a simulation while raising the temperature according to a specified heating schedule until the alignment is completely vertically shuffled. In effect, this is a strategy for computationally introducing mutations in natural sequences constrained by the positional conservation of amino acids and according to a pattern specified by the objective function and the heating protocol. For example, consider the heating trajectories in Fig. 10.5A for an MSA of 240 members of the PDZ family of protein interaction modules. The objective function is the full $\tilde{C}_{ij}^{(ab)}$ correlation tensor, and we show two trajectories differing by heating protocol starting from the MSA of natural sequences (marked N, Fig. 10.5A). Carrying out a simulation with the temperature set to infinity defines a path in which all positions mutate within their conservation pattern without regard to correlations (see Eq. 10.6), finally approaching the fully randomized, vertically shuffled limit (marked R, Fig. 10.5A). Accordingly, both global and sector identity

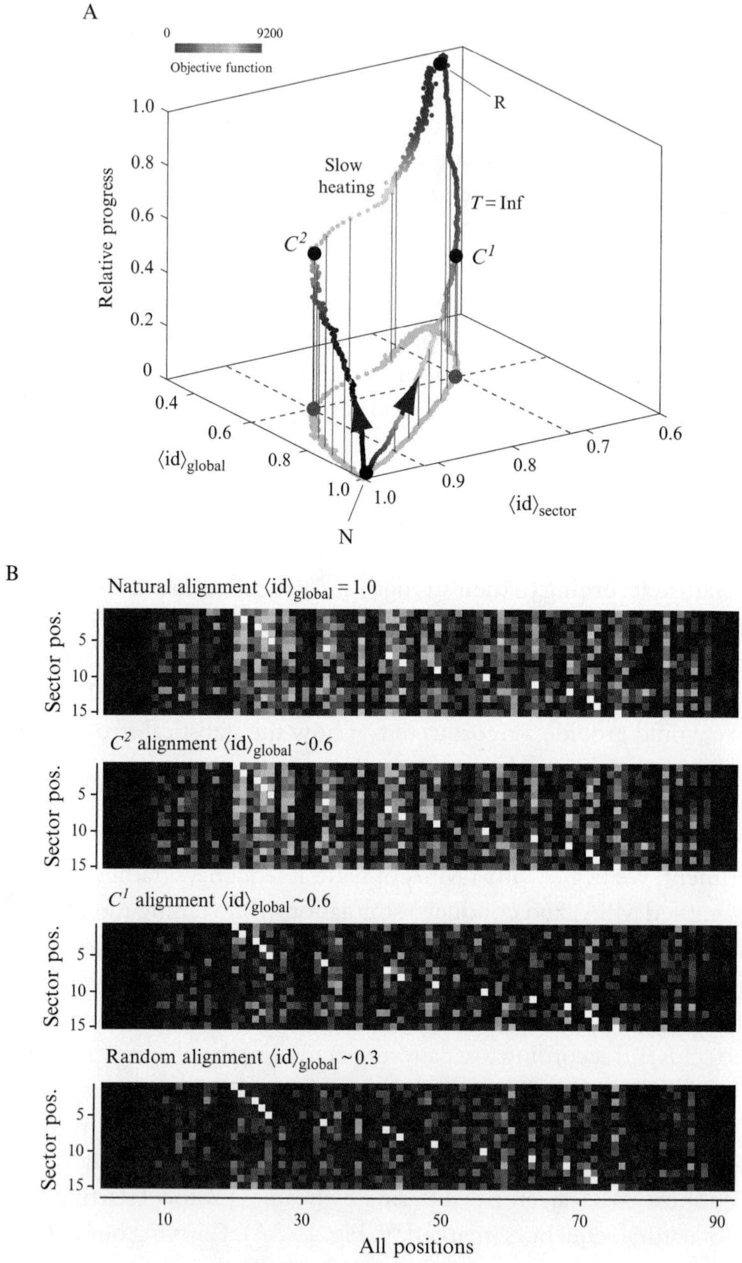
A
0
9200
Objective function
Relative progress
1.0
0.8
0.6
0.4
0.2
0
R
Slow heating
T = Inf
C²
C¹
0.4
0.6
⟨id⟩global
0.8
1.0
1.0
0.9
0.8
0.7
0.6
⟨id⟩sector
N
B
Natural alignment ⟨id⟩global = 1.0
Sector pos.
5
10
15
C² alignment ⟨id⟩global ~ 0.6
C¹ alignment ⟨id⟩global ~ 0.6
Random alignment ⟨id⟩global ~ 0.3
10
30
50
70
90
All positions

to the natural sequences are lost equally, and sequences at the position marked C^1 show a pattern of correlation for sector positions that is nearly randomized (Fig. 10.5B). In contrast, slow near-equilibrium heating produces a very different trajectory; in this process, global identity is initially lost without much loss in sector identity until a characteristic temperature at which the sector "melts" and the trajectory approaches the same fully vertically shuffled limit. Accordingly, sequences at the position marked C^2 have a pattern of correlations for sector positions that is nearly the same as for natural sequences, despite the same global divergence as C^1 sequences (Fig. 10.5B).

An interesting experiment is to choose one natural sequence within the MSA as a model system, carry out many trials of MCSH, and build versions of this protein at different temperatures along the trajectories. Experimental characterization of the natural sequence is a specific reference for folding, stability, and function in the synthetic "heated" sequences. Study of ensembles of synthetic variants sampled along the two different heating trajectories should provide a clear answer to how these properties of the selected natural sequence differentially diverge as a function of systematically removing the information contained in the SCA correlation tensor.

Both MCSA and MCSH methods provide a means to explore the mapping between statistical correlations and protein structure/function, a mapping that deserves study in several protein systems. Indeed, it will be important to see how the results compare for small versus large domains and for enzymes versus more simple binding proteins.

Figure 10.5 MCSH trajectories for the PDZ domain family. (A) A plot mapping the progress of two different heating trajectories against the average "top-hit" sequence identities of designed sequences calculated for the full-length sequence ($\langle id \rangle_{global}$) or for just sector positions ($\langle id \rangle_{sector}$). In the T=Inf trajectory, each position is allowed to mutate within its conservation pattern without regard to correlations, and global and sector identity drop together. In the slow heating trajectory, the temperature is gradually increased to "melt out" couplings between positions in an order that depends on the strength and collective nature of the correlations. The value of the objective function along the trajectories is indicated by the color bar, and four points are marked for reference with panel B: N, the natural MSA; R, the fully randomized, vertically shuffled MSA; and C^1 and C^2, two intermediary points that share the same global sequence divergence but differ significantly in sector divergence. (B) Subset of the SCA matrix $\tilde{C}_{ij}$ for 15 sector positions in the PDZ family to illustrate the property of the heating trajectories. Despite identical global sequence divergence, C^1 sequences show a pattern of correlations that nearly approaches the fully randomized case while C^2 sequences show correlations that are nearly the same as for the natural MSA. Experimental analysis of C^2 and C^1 sequences or more generally sequences drawn from both trajectories represent a systematic investigation of how properties of natural proteins are stored in the pattern of correlations. (See Color Insert.)

6. CONCLUSION

Protein design represents one approach for understanding how the pattern of pairwise residue couplings inferred from the statistics of natural protein sequences is related to the encoding of protein structure and function. Further, it may provide insight into decomposability of biochemical properties. In cases such as the S1A serine proteases, the finding of multiple statistically independent sectors offers the exciting possibility of orthogonal control of different biochemical properties of these enzymes. More generally, it may be that a broad study of the Monte Carlo-based design trajectories for proteins will reveal rules for tuning stability and function independently through targeted variation of protein sequences. The methods described here present one approach to distill the general principles, if any, for the design of natural proteins. Practically, such rules might permit the design of improved synthetic proteins that show natural-like properties of high catalytic efficiency, mutational robustness, and adaptability to new functional challenges.

ACKNOWLEDGMENTS

We thank Dr. Walraj Gosal for his contributions on Monte Carlo Simulated Heating for the PDZ domain. We acknowledge support from the NIH (R01EY018720-05, R. R.), The Robert A. Welch Foundation (I-1366, R. R.), and the Green Center for Systems Biology (R. R.).

REFERENCES

Bialek, W., & Ranganathan, R. (2007). *Rediscovering the power of pairwise interactions.* arXiv:0712.4397.

Bowie, J. U., Reidhaar-Olson, J. F., Lim, W. A., & Sauer, R. T. (1990). Deciphering the message in protein sequences: Tolerance to amino acid substitutions. *Science, 247*, 1306–1310.

Dahiyat, B. I., & Mayo, S. L. (1997). De novo protein design: Fully automated sequence selection. *Science, 278*, 82–87.

Dantas, G., Kuhlman, B., Callender, D., Wong, M., & Baker, D. (2003). A large scale test of computational protein design: Folding and stability of nine completely redesigned globular proteins. *Journal of Molecular Biology, 332*, 449–460.

Ferguson, A. D., Amezcua, C. A., Halabi, N. M., Chelliah, Y., Rosen, M. K., Ranganathan, R., et al. (2007). Signal transduction pathway of TonB-dependent transporters. *Proceedings of the National Academy of Sciences of the United States of America, 104*, 513–518.

Gerber, D., Maerkl, S. J., & Quake, S. R. (2009). An in vitro microfluidic approach to generating protein-interaction networks. *Nature Methods, 6*, 71–74.

Halabi, N., Rivoire, O., Leibler, S., & Ranganathan, R. (2009). Protein sectors: Evolutionary units of three-dimensional structure. *Cell, 138*, 774–786.

Harbury, P. B., Plecs, J. J., Tidor, B., Alber, T., & Kim, P. S. (1998). High-resolution protein design with backbone freedom. *Science, 282*, 1462–1467.

Hatley, M. E., Lockless, S. W., Gibson, S. K., Gilman, A. G., & Ranganathan, R. (2003). Allosteric determinants in guanine nucleotide-binding proteins. *Proceedings of the National Academy of Sciences of the United States of America, 100*, 14445–14450.

Isom, D. G., Marguet, P. R., Oas, T. G., & Hellinga, H. W. (2011). A miniaturized technique for assessing protein thermodynamics and function using fast determination of quantitative cysteine reactivity. *Proteins, 79*, 1034–1047.

Kirkpatrick, S., Gelatt, C. D., Jr., & Vecchi, M. P. (1983). Optimization by simulated annealing. *Science, 220*, 671–680.

Kosuri, S., Eroshenko, N., Leproust, E. M., Super, M., Way, J., Li, J. B., et al. (2010). Scalable gene synthesis by selective amplification of DNA pools from high-fidelity microchips. *Nature Biotechnology, 28*, 1295–1299.

Kuhlman, B., Dantas, G., Ireton, G. C., Varani, G., Stoddard, B. L., & Baker, D. (2003). Design of a novel globular protein fold with atomic-level accuracy. *Science, 302*, 1364–1368.

Lee, J., Natarajan, M., Nashine, V. C., Socolich, M., Vo, T., Russ, W. P., et al. (2008). Surface sites for engineering allosteric control in proteins. *Science, 322*, 438–442.

Lockless, S. W., & Ranganathan, R. (1999). Evolutionarily conserved pathways of energetic connectivity in protein families. *Science, 286*, 295–299.

McLaughlin, R. N., Jr, Poelwijk, F. J., Raman, A., Gosal, W. S., & Ranganathan, R. (2012). The spatial architecture of protein function and adaptation. *Nature, 491*, 138–142.

Metropolis, N., Rosenbluth, A. W., Rosenbluth, M. N., Teller, A. H., & Teller, E. (1953). Equation of state calculations by fast computing machines. *Journal of Chemical Physics, 21*, 1087–1092.

Orencia, M. C., Yoon, J. S., Ness, J. E., Stemmer, W. P. C., & Stevens, R. C. (2001). Predicting the emergence of antibiotic resistance by directed evolution and structural analysis. *Nature Structural and Molecular Biology, 8*, 238–242.

Reidhaar-Olson, J. F., & Sauer, R. T. (1990). Functionally acceptable substitutions in two alpha-helical regions of lambda repressor. *Proteins, 7*, 306–316.

Reynolds, K. A., McLaughlin, R. N., & Ranganathan, R. (2011). Hot spots for allosteric regulation on protein surfaces. *Cell, 147*, 1564–1575.

Russ, W. P., Lowery, D. M., Mishra, P., Yaffe, M. B., & Ranganathan, R. (2005). Natural-like function in artificial WW domains. *Nature, 437*, 579–583.

Shulman, A. I., Larson, C., Mangelsdorf, D. J., & Ranganathan, R. (2004). Structural determinants of allosteric ligand activation in RXR heterodimers. *Cell, 116*, 417–429.

Smock, R. G., Rivoire, O., Russ, W. P., Swain, J. F., Leibler, S., Ranganathan, R., et al. (2010). An interdomain sector mediating allostery in Hsp70 molecular chaperones. *Molecular Systems Biology, 6*, 414.

Socolich, M., Lockless, S. W., Russ, W. P., Lee, H., Gardner, K. H., & Ranganathan, R. (2005). Evolutionary information for specifying a protein fold. *Nature, 437*, 512–518.

Suel, G. M., Lockless, S. W., Wall, M. A., & Ranganathan, R. (2003). Evolutionarily conserved networks of residues mediate allosteric communication in proteins. *Nature Structural Biology, 10*, 59–69.

Taylor, S. V., Kast, P., & Hilvert, D. (2001). Investigating and engineering enzymes by genetic selection. *Angewandte Chemie (International Ed. in English), 40*, 3310–3335.

Weinreich, D. M., Delaney, N. F., Depristo, M. A., & Hartl, D. L. (2006). Darwinian evolution can follow only very few mutational paths to fitter proteins. *Science, 312*, 111–114.

CHAPTER ELEVEN

Protein Engineering and Stabilization from Sequence Statistics: Variation and Covariation Analysis

Venuka Durani*, Thomas J. Magliery*,†,1
*Department of Chemistry, The Ohio State University, Columbus, Ohio, USA
†Department of Biochemistry, The Ohio State University, Columbus, Ohio, USA
[1]Corresponding author: e-mail address: magliery.1@osu.edu

Contents

Abstract

The concepts of consensus and correlation in multiple sequence alignments (MSAs) have been used in the past to understand and engineer proteins. However, there are multiple ways of acquiring MSA databases and also numerous mathematical metrics that can be applied to calculate each of the parameters. This chapter describes an overall methodology that we have chosen to employ for acquiring and statistically analyzing MSAs. We have provided a step-by-step protocol for calculating relative entropy and mutual information metrics and describe how they can be used to predict mutations that have a high probability of stabilizing a protein. This protocol allows for flexibility for modification of formulae and parameters without using anything more complicated than Microsoft Excel.

Methods in Enzymology, Volume 523
ISSN 0076-6879
http://dx.doi.org/10.1016/B978-0-12-394292-0.00011-4

We have also demonstrated various aspects of data analysis by carrying out a sample analysis on the BPTI-Kunitz family of proteins and identified mutations that would be predicted to stabilize this protein based on consensus and correlation values.

1. INTRODUCTION

The information required to fold a protein into its native structure is encoded in its amino acid sequence (Anfinsen, 1973). Studying commonalities and differences between several sequences that fold into the same structure can provide information about the necessary and sufficient parameters of this code. Vast databases of sequences (Consortium, 2011; Finn et al., 2010) classified into structural folds based on sequence similarity (Chothia, 1992; Chothia & Lesk, 1986; Orengo, Jones, & Thornton, 1994) have become available with the advent of the genomic era. These databases have multiple sequence alignments (MSAs) of proteins with nearly the same structure and function and can be exploited to gain information about their folds. Upon comparison of sequences within an MSA, it is often observed that some amino acids occur again and again at a particular position. These might be important for the fitness of the protein in a structural, functional, or dynamic way and hence did not tolerate much mutation over the course of evolution. This concept of consensus has been used to design peptide motifs (Tripp & Barrick, 2003) such as tetratricopeptide repeats (TPRs) (Main, Xiong, Cocco, D'Andrea, & Regan, 2003) and ankyrin repeats (Binz, Stumpp, Forrer, Amstutz, & Pluckthun, 2003; Mosavi, Minor, & Peng, 2002) and even larger globular proteins such as fungal phytases (Lehmann et al., 2000). Recently, we demonstrated the consensus design of a triosephosphate isomerase with native-like activity and very high stability (Sullivan, Durani, & Magliery, 2011). Various methods have been used by researchers to estimate the degree of conservation of positions in an MSA. Some methods take into account the physical properties of amino acids such as hydrophobicity and charge (Mosavi et al., 2002), some others use global propensities (Main et al., 2003), some use statistical free energy-based metrics (Steipe, Schiller, Pluckthun, & Steinbacher, 1994), and yet others use information theory-based metrics (Magliery & Regan, 2005). Previous studies on antibodies, phytases, and thioredoxin suggest that consensus mutations—where a residue is mutated to the amino acid that is the most common at that position in the MSA—stabilize proteins about half the time (Godoy-Ruiz, Perez-Jimenez, Ibarra-Molero, & Sanchez-Ruiz, 2005; Knappik et al., 2000; Steipe et al., 1994).

Consensus design implicitly assumes that all amino acid positions function independently, but in reality, the amino acids in a protein interact with each other and work cooperatively to produce the optimum structure required for its function. These interactions of the amino acids can be studied using correlation analysis of MSA databases. Correlations can be calculated using various algorithms (Fodor & Aldrich, 2004) that employ approaches such as amino acid similarity matrices, χ^2-tests (Kass & Horovitz, 2002), perturbation-based approaches (Dekker, Fodor, Aldrich, & Yellen, 2004; Halabi, Rivoire, Leibler, & Ranganathan, 2009; Lockless & Ranganathan, 1999), and information theory (Ackerman & Gatti, 2011). Studies on WW domains have shown that capturing a structural fold is more successful when both consensus and correlation are used, as opposed to using consensus alone (Russ, Lowery, Mishra, Yaffe, & Ranganathan, 2005; Socolich et al., 2005). Studies on TPR motifs have shown that complex charge networks on the surface of proteins can skew results obtained by consensus, but correlation analysis can capture them, leading to better results in protein engineering (Magliery & Regan, 2004). The concept of correlation has also been used to predict structural contacts (Bartlett & Taylor, 2008; Miller & Eisenberg, 2008; Sullivan et al., 2012), probe allosteric communications (Kass & Horovitz, 2002; Suel, Lockless, Wall, & Ranganathan, 2003), and to identify sectors of interacting residues in proteins (Halabi et al., 2009; Lockless & Ranganathan, 1999). We wished to use consensus and correlation approaches to study proteins, particularly for protein engineering (Magliery, Lavinder, & Sullivan, 2011). We have also shown recently that a combination of consensus and correlation data can be used to predict stabilizing mutations to a protein with higher accuracy than using consensus information alone (Sullivan et al., 2012). However, there are multiple ways of acquiring MSA databases and also numerous mathematical metrics that can be applied to calculate each of the parameters. This chapter describes the overall methodology that we have employed to acquire and statistically analyze MSAs.

2. CASE STUDY: BPTI

Throughout this chapter, the workings of each method will be illustrated using examples from the bovine pancreatic trypsin inhibitor (BPTI) MSA. BPTI is a Kunitz domain, the active domains of proteins that inhibit the function of proteases and are called Kunitz-type protease inhibitors (Chen et al., 2001; Paesen et al., 2009; Schmidt, Chand, Cascio, Kisiel, & Bajaj, 2005). They are about 50–60 amino acids long with a molecular

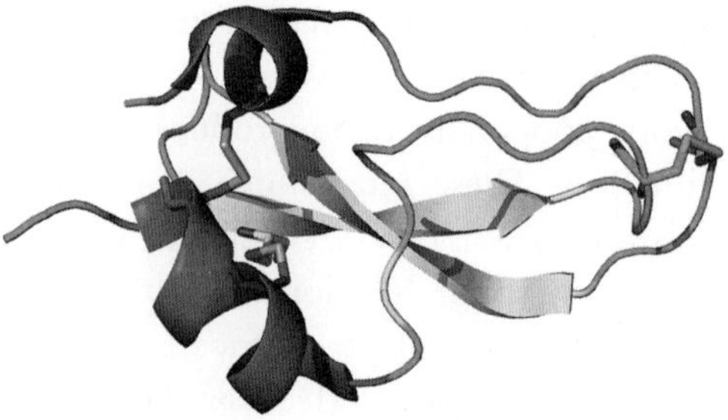

Figure 11.1 Structure of BPTI: In this structure (PDB ID: 1BPI) (Parkin et al., 1996), α-helices are shown in red, β-sheets in yellow, and loop regions in green. Three conserved disulfide bonds are shown as sticks. (For interpretation of the references to color in this figure legend, the reader is referred to the online version of this chapter.)

weight of about 6 kDa and fold into a disulfide-rich α/β structure (Fig. 11.1; Parkin, Rupp, & Hope, 1996). Standalone Kunitz domains have been used as a framework for the development of new pharmaceutical drugs. BPTI is an extensively studied model protein belonging to this family.

3. ACQUIRING AN MSA

This section discusses how MSAs can be acquired or built and curated based on the availability of information and the goal of the project. We will discuss two ways of acquiring MSAs. One is by accessing data already organized as MSAs in Pfam and the other by building MSAs using the MyHits Web site. Pfam (Punta et al., 2012) (http://pfam.janelia.org/) is an online database of MSAs and hidden Markov models (HMMs) (Eddy, 2004) powered by the HMMER (Finn, Clements, & Eddy, 2011) search tool. Each Pfam MSA represents a protein family or domain. Pfam has two categories, and Pfam-A families have better annotation and alignment quality than Pfam-B. For the purposes of this chapter, Pfam-B will be ignored and Pfam will refer to Pfam-A. Each protein family listed in Pfam has a seed alignment and a full alignment (Sonnhammer, Eddy, & Durbin, 1997). The seed alignment is a high quality, manually checked alignment of a few hand-picked sequences that represent the protein family in question. It is used to build an HMM profile which is used for automated searches of more sequences that belong to the protein family. The positions that occur in seed alignments are called canonical positions, and the alignment of these positions is usually of good quality for the full alignment too. Insertions and deletions with respect to the seed alignment are noncanonical positions, which tend to have low occupancies and poor alignment quality. Pfam has an option to download sequences in a format where canonical positions are represented with one-letter amino acid codes in capital letters (A, C, D, etc.) and gaps as hyphens (-), whereas the noncanonical positions are represented in small letters (a, c, d, etc.) with gaps as periods (.). This format allows for choosing only canonical positions for analysis if so required. To keep Pfam updated with new protein sequences that are discovered, the HMM-based search of available protein databases is carried out and released about once a year. While the full alignments change substantially between releases, the seed alignments remain almost invariant. If the quality of an alignment is not sufficient, it can be curated and realigned with more manual intervention using alignment tools such as ClustalX (Thompson, Gibson, & Higgins, 2002).

An ideal MSA for sequence statistics would have many unique sequences of equal length that align well with each other. However, typical MSA databases downloaded from Pfam are imperfect due to sequence repeats and truncated sequences. These imperfections must be accounted for, especially in correlation analysis. For example, an abundance of repeats may bias correlations between the pairs that occur in them, and small changes to consensus sequences from differently assembled databases can have large effects on protein properties (Magliery et al., 2011). To avoid such biases, the MSA used for correlation analysis can be curated by removing exact sequence repeats and short sequence fragments. This gives rise to a database that looks more like an ideal MSA (Fig. 11.2). Our lab tends to favor the use of Microsoft Excel for simple manipulations, sometimes with the addition of Visual Basic macros, and the use of Perl scripts for more complicated manipulations.

3.1. Protocol for acquiring an MSA from Pfam

Step 1. *Downloading sequences:* The "keyword search" tool on the home page of the Pfam Web site (http://pfam.janelia.org/) is used to search for the protein of interest. If the "description" for an entry matches the protein of interest, the entry for the protein is accessed by clicking on either the "accession" or the "ID." The MSA is accessed by going to the "sequences" tab. The following formatting options are typically used for the sequences: Alignment: Full; Format: Selex; Order: Tree; Sequence: All upper case; Gaps: Gaps as "-" (dashes); Download/view: Download. The "Generate" option generates a text file with all sequences and their names.

Figure 11.2 Curation of multiple sequence alignments: Ideally, an MSA would have many unique sequences of equal length (A). However, the MSAs available in Pfam often have sequence fragments and repeated sequences (B). This can cause biases in correlation analysis, and hence, the MSA should be curated by removing sequence fragments and exact repeats so that it resembles an ideal MSA (C). (For color version of this figure, the reader is referred to the online version of this chapter.)

Step 2. *Curating for length of sequences:* Sequences from the text file are imported into a Microsoft Excel file (office 2010) with the names in column A and sequences in column B. Column B is copied into column C. Using the "replace" function, all dashes (-) are removed from column C. The lengths of the sequences in column C are calculated in column D using the `LEN` function (`D2=LEN(C2)` and so on). Column D is selected and the filter function is used to filter the data for sequences that have a reasonable length. For example, in the BPTI database, the average length of a sequence was 53.5 amino acids and about half of the sequences in the database were 53 amino acids long. We kept all sequences that were 50–60 amino acids long. These filtered data were copied into another worksheet.

Step 3. *Removing sequence repeats:* Column C, that has all the sequences, is selected. Then the "Data" tab is accessed to select the "Advanced" filter option. The option for "Unique records only" is selected. This removes all the sequence repeats. Columns A and B of the filtered data are copied to another worksheet.

Step 4. *Occupancy of positions:* The curated database is first tabulated using the `MID` function that returns a specified number of characters from a specified position in a text string. Column A has the sequence names, column B has all the aligned sequences, and row 1 has position numbers. Since the BPTI alignment containing 880 sequences had 184 aligned positions, numbers 1–184 were written in row 1. The formula `C2=MID($B2,C$1,1)` was copied over the entire table `C2:GD881`. Then the occupancy for each column is calculated by counting the number of gaps (-) and subtracting it from the total number of sequences `(C882=100*(880-COUNTIF(C2:C881,"-")/880))`. Columns with high occupancy are chosen. A quick way to filter sequences based on occupancy is to transpose the table so that each position is in a row and use the "filter" function on the column containing occupancy values. The numerical value used for the occupancy cutoff can be chosen such that the length of the final sequence is close to the length of a wild-type protein, typically eliminating sites with less than about 65% occupancy.

3.2. Building an MSA using the MyHits Web site

Sometimes, a project may require an MSA with different specifications than are applied in making the Pfam MSAs. As an example, we generated the BPTI49 MSA of standalone structures (as opposed to Kunitz domains in larger proteins).

Step 1. *Seed alignment:* First, a seed alignment is required which can be constructed by compiling known sequences of the protein of interest and aligning them using a program such as ClustalX. Since the Kunitz_BPTI family has a Pfam MSA, its seed alignment was downloaded. The 151 sequences in this alignment were crosschecked for domain organization in the UniProt database and any sequence shown as a part of a multidomain construct was discarded. At the end of this process, only 30 sequences remained. These sequences were aligned using ClustalX and saved in the FASTA format. This MSA was used as the seed alignment.

Step 2. *Search based on HMM profile:* The seed alignment is used to create an HMM using HMMER3 on the MyHits (Pagni et al., 2007) Web site (http://myhits.isb-sib.ch/). The HMMER3 option under the "search" section of "Tools" is selected. The seed MSA from step 1 is copied into the space provided. For constructing the BPTI database, the Swiss-Prot option was selected under the "seq_source" section and this "search" provided about 200 hits.

Step 3. *Screening of search results:* The search results from step 2 can be curated based on parameters such as sequence identity and length of sequences. For the BPTI database, we screened for sequences that were annotated as standalone structures. In order to do this, the "entry" link was used to access the UniProt data for each of the hits from step 2. Sequences that were annotated as standalone sequences were selected. "Send checked matches" option was set to "MSA Hub" and the sequences were copied. The final alignment for the BPTI sequences selected using this process had 49 sequences.

4. RELATIVE ENTROPIES: QUANTIFYING THE DEGREE OF POSITIONAL VARIATION

We estimate the conservation of positions in the MSAs using the relative entropy (RE, sometimes symbolized as *D*) metric. This is an information theoretic metric that gives the "distance" of a distribution from a reference distribution. It is logarithmically related to the multinomial probability of observing a particular distribution if you expect the reference distribution (Magliery & Regan, 2005). If a position is highly conserved, its amino acid distribution would be very different from that expected at random, making its RE value very high relative to a neutral reference state (Fig. 11.3). The reference distribution chosen in this case was based on codon usages in the yeast proteome. This is a good estimate of the amino acid frequency expected in an average eukaryotic protein, accounting for factors such as codon usage and side chain chemistry. This reference state is conveniently invariant over time. RE (*D*) is calculated as

$$D = \sum_i p(x_i) \log \frac{p(x_i)}{q(x_i)}$$

where $p(x_i)$ is the probability of observing residue i at position x in the MSA and $q(x_i)$ is the probability of observing residue i in the yeast proteome. We used fraction of sequences in each category as approximation of probability.

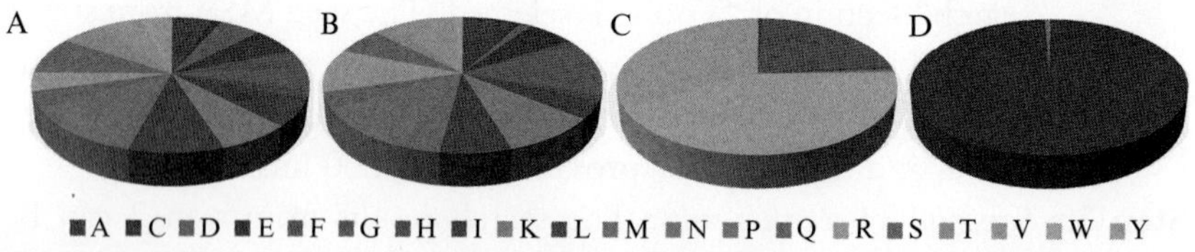

Figure 11.3 Amino acid distributions corresponding to various relative entropy values: Each color in the pie charts represents the frequency of occurrence of an amino acid residue. (A) Amino acid distribution expected at random in an average eukaryotic protein. (B) Amino acid distribution at a position with low a RE value of 0.2. (C) Amino acid distribution at a position with a high RE value of 2.6. (D) Amino acid distribution at a position with a very high RE value of 4.3. (For interpretation of the references to color in this figure legend, the reader is referred to the online version of this chapter.)

4.1. Protocol for calculating RE values

The formulae for calculating consensus values need to be set up so that disallowed functions like division by zero or taking log of a negative number do not crop up in any calculation. Keeping this in mind, the following protocol was established.

Step 1. *MSA format:* The tabulated form of the curated MSA is used (Section 3.1, step 4), and the first column of the table (column C) is assigned the name `pos1`. This worksheet is named "seq."

Step 2. *Counting amino acid frequencies:* On another worksheet named "RE," a table is made to record the number of times various amino acids occur at various positions. Position numbers are recorded in column A and one-letter amino acid codes in row 1. Then the formula `B2=COUNTIF(OFFSET(pos1,0,$A2-1),B$1)` is entered followed by *`Ctrl+Shift+Return`*. (This is how array formulae are entered in Excel.) This formula counts the number of occurrences of amino acids and is copied into the entire table. In order to calculate frequency of occurrence, each value is divided by the sum of the row it is contained in and the values are recorded in another table.

Step 3. *Calculating relative entropies:* The formula used to calculate relative entropies is `AB4=IF(B4=0,0,B4*LN(B4/B$2))` copied over table `AB4:AU56` (each row of the table is summed for RE corresponding to each position) where `B4:U56` is the table of amino acid frequencies in the MSA, and `B2:U2` has the amino acid frequencies in the yeast proteome (A, 0.056; C, 0.013; D, 0.058; E, 0.065; F, 0.045; G, 0.051; H, 0.021; I, 0.065; K, 0.073; L, 0.095; M, 0.021; N, 0.061; P, 0.044; Q, 0.039; R, 0.044; S, 0.089; T, 0.059; V, 0.057; W, 0.010; Y, 0.034).

4.2. Relative entropy in BPTI

The frequency of occurrence of all the amino acids, although similar, is not the same in the reference state (Fig. 11.3A). The RE values depend on both the frequency of an amino acid and its identity. While an average amino acid occurs in nature about 5% of the time, the Cys residue has only 1.3% frequency of occurrence in the reference set while the most common residue Leu occurs 9.5% of the time. When a rarer residue such as Cys dominates a position, then the RE becomes higher than when a more common amino acid such as Leu dominates the position. In the BPTI database, due to three

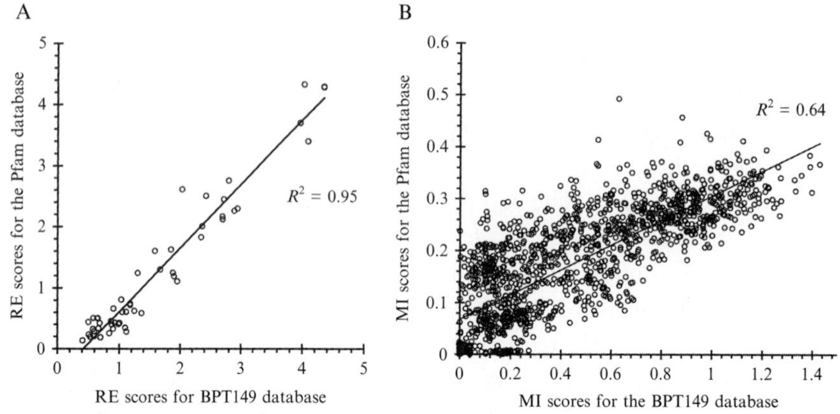

Figure 11.4 (A) A scatter plot of RE scores for the Pfam database versus RE scores for the BPTI49 database. (B) Comparison of MI scores from the two databases.

conserved disulfide bonds, six Cys residues were the most conserved. The extremely high RE values for these positions are in part because Cys residues are relatively less common in nature. The amino acid distributions for positions with varying RE scores are shown in Fig. 11.3. While the amino acid distribution at a position with a small RE value of 0.2 (Fig. 11.3B) looks very similar to that of the reference state, a position dominated by three residues Phe, Tyr, and Trp has an RE value of 2.6 (Fig. 11.3C) and a position dominated by Cys residue has an RE value of 4.3 (Fig. 11.3D).

Comparison of RE values for the two BPTI databases shows that while they agree with each other very well (Fig. 11.4A), the smaller database has slightly higher RE values. The average RE value for the smaller BPTI49 database is 1.7, while the average RE value for the significantly larger Pfam database is 1.3. The good agreement between data from both databases ($R^2 = 0.95$, Fig. 11.4A) is in part because of the dominant Cys residues that have very high RE values in both databases. At lower RE values, the databases are not as concordant.

5. MUTUAL INFORMATION: QUANTIFYING THE DEGREE OF COVARIATION

Positional correlations within the MSA can be calculated using the mutual information (MI) statistic from information theory (Applebaum, 1996; Shannon & Weaver, 1949). MI is the RE between the actual joint probability distribution for two sites, and the hypothetical independent joint probability distribution determined from the product of the marginal distributions for each site. For instance, if Ala occurs at position x 50% of the time and Leu occurs at position y 50% of the time, then if both positions are independent of each other, the Ala–Leu pair would be observed in 25% of the sequences. If this pair is observed much more (or much less) than 25% of the time, then the pair of positions is correlated (or anticorrelated). The mutual information MI_{xy} between sites x and y is calculated as,

$$MI_{xy} = \sum_i \sum_j p(x_i, y_i) \log \frac{p(x_i, y_i)}{p(x_i)p(y_i)}$$

where $p(x_i)$ is the probability of observing residue i at position x in the MSA, $p(y_i)$ is the probability of observing residue j at position y, and $p(x_i, y_j)$ is the joint probability of observing residue i at position x and residue j at position y. For each pair of positions, this value is summed over all the 400 possible pairs of amino acid residues. MI is symmetric (the extent of information position x contains about position y is the same as the extent of information position y contains about position x), and there is no meaning to the MI of a position with itself. Consequently, we need to calculate only $((a \times a) - a)/2$ values. For large MSAs, this can save a considerable amount of computational time. MI does not distinguish between correlations and anticorrelations, and it does not give details of correlations at the amino acid level, but it is easily calculated and visualized.

It is not simple to calculate a statistical significance for MI values. Instead, we calculate a threshold "noise level" from MI values for a scrambled dataset with the same amino acid distribution in each position. This takes into account any biases that may arise due to database size and correlation patterns. The maximum MI value over all pairs obtained from this scrambled data is set as the noise level (Fig. 11.5). For better statistical significance, this value can be calculated for multiple randomized datasets.

5.1. Protocol for calculating mutual information values

Step 1. *MSA and frequency table:* The MSA and amino acid frequency table can be copied from the RE calculation workbook (worksheets "seq" and "RE" from Section 4.1).

Step 2. *Reference distribution:* The reference distribution is calculated using the formula `B28=$C3*VLOOKUP(B$27,$B$3:$D$22,3,FALSE)` copied over all the cells in table `B28:U47`. Range `C3:C22` has the amino acid frequencies for position 1, and range `D3:D22` contains amino acid

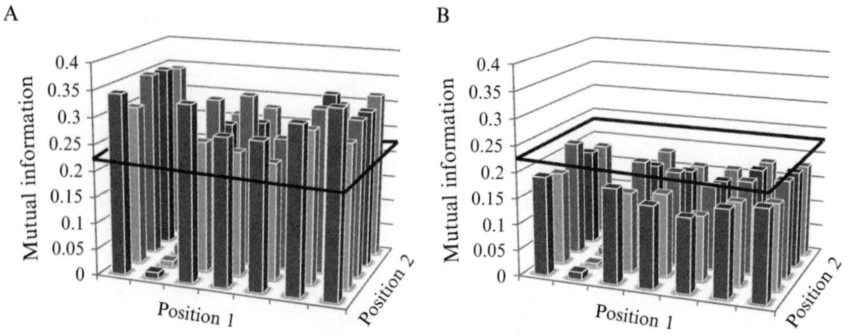

Figure 11.5 Noise level in correlation calculations: A sample correlation calculation of a real dataset (A) and column-randomized dataset (B) The black lines indicate the noise level estimated using the protocols described here.

frequencies for position 2. One-letter codes for amino acids are listed in B3:B22, B27:U27, and A28:A47.

Step 3. *Observed distribution:* The observed distribution is calculated using the formula B52 =SUM(IF(OFFSET(pos1,0,$C$1-4)=$A52,1,0)*IF(OFFSET (pos1,0,$D$1-4)=B$51,1,0))/$V$51 followed by *Ctrl+Shift+Return* and copied over all the cells of table B52:U71. Cell C1 contains the position number of one position, and D1 contains position number of the other position in question. Cell V51 contains the frequency of cooccurrence of amino acids in the two positions and is calculated using the formula V51=SUM((IF(OFFSET(pos1,0,$C$1-4)<>"-",1,0)*IF(OFFSET (pos1,0,$D$1-4)<>"-",1,0))) followed by *Ctrl+Shift+Return*.

Step 4. *MI calculation:* MI is calculated using the formula B75 = IF (B52=0,0,B52*LN(B52/(B28)) copied over all the cells in table B75:U94 and then summed over the whole table. In this worksheet, range B52:U71 is the table of observed frequencies of cooccurrence (from step 3) and range B28:U47 is the table of expected frequencies of cooccurrence assuming mutual independence (from step 2).

Step 5. *Repeating the calculation for all pairs of positions:* While calculating correlation values, the same set of calculations need to be repeated for each pair of positions. Once the excel spreadsheet is set up to calculate one iteration of the calculation, the following macro is used to iteratively calculate these values and tabulate them. Since MI is symmetric, calculating values on one side of the diagonal is sufficient.

```
Sub MI()
  '
  'This script calculates MI values for each pair and tabulates them
  'There are 53 positions in this alignment.
  'B96 is the cell that contains MI value calculated in each iteration
  'Row 100 and column A have position numbers
  'The MI values are tabulated in B101:BB153
  '
  Application.ScreenUpdating = False
  For ColumnCounter = 2 To 54
  For RowCounter = 100 + ColumnCounter To 153
  Worksheets("MI").Activate
  Range("C1") = Cells(RowCounter, 1)
  Range("D1") = Cells(100, ColumnCounter)
  Calculate
  Cells(RowCounter, ColumnCounter) = Range("B96")
  Next RowCounter
  Next ColumnCounter
  Application.ScreenUpdating = True
  ActiveWorkbook.Save
  End Sub
```

Step 6. *Converting a table into a list:* When correlation values are calculated, the output format is in a matrix or table form. In order to sort the values, it is more convenient to format them into a list. This Excel macro converts a 53 × 53 MI table into a list. This macro takes the values from only half of the matrix (below the diagonal). The matrix is located in a worksheet titled "matrix" and starts from cell A1. The first row and first column contain position numbers. Another blank worksheet called "list" needs to be created before the macro is run. The RowCounter and ColumnCounter values can be edited if the table in question is of a different size.

```
Sub matrix_list()
  '
  'This script converts a 53x53 table in "matrix" worksheet
  'into a list in "list" worksheet
  'In the "matrix" worksheet, Row 1 and column A have data labels
  '
  Application.ScreenUpdating = False
  Dim ColumnCounter As Integer
```

```
    Dim RowCounter As Integer
    Dim MyCounter As Long
    MyCounter = 2
    For RowCounter = 2 To 54
    For ColumnCounter = (RowCounter + 1) To 54
    Worksheets("list").Cells(MyCounter, 1).Value = Worksheets
("matrix").Cells(RowCounter, 1).Value
    Worksheets("list").Cells(MyCounter, 2).Value = Worksheets
("matrix").Cells(1, ColumnCounter).Value
    Worksheets("list").Cells(MyCounter, 3).Value = Worksheets
("matrix").Cells(RowCounter, ColumnCounter).Value
    MyCounter = MyCounter + 1
    Next ColumnCounter
    Next RowCounter
    Application.ScreenUpdating = True
    ActiveWorkbook.Save
    End Sub
```

Step 7. *Calculating noise level:* In order to calculate noise level for a correlation calculation, a randomized MSA is created where each column is scrambled. This keeps the consensus information the same while scrambling the correlations. In order to randomize the MSA, the `RAND`, `RANK`, and `INDEX` functions of Microsoft Excel are used. For the BPTI database, the worksheet containing the original MSA (table `C2:BC881`) was named "seq" and a table of the same size was created in another worksheet named "rand" where each cell of the table was `=RAND()` and hence contained a random number between 0 and 1. In a third worksheet named "scramble," another table of equal size was created where the cell `C2` was `=INDEX(seq!C$2:C$881,RANK(rand!C2,rand!C$2:C$881))`. This formula was copied over the whole table `C2:BC881`. Every time the worksheet was refreshed/recalculated (F9 on manual calculation mode), a new column-randomized MSA was generated. MI calculation was carried out for the randomized dataset as per the procedure described in steps 1–5, and the maximum MI value obtained from the column-randomized MSA was accepted as the noise level.

5.2. Mutual information in BPTI

Correlation values were calculated for each pair of positions using both the BPTI49 database and the curated Pfam database. The correlation data show a much stronger dependence on size of the database than the consensus data. As can be seen from the plot comparing correlation values between these datasets, the smaller dataset has significantly higher values for correlations (Fig. 11.4B). The disparity between ranking of pairs of positions in terms of their correlation values is also quite high between the databases ($R^2=0.64$, Fig. 11.4B). This shows that the database size has a greater effect on correlation values than on consensus values. For the larger Pfam database, the maximum MI value was 0.49 and that for the randomized dataset was 0.22 leaving 550 correlations above noise. On the other hand, for the smaller BPTI49 database, the maximum MI value was 1.4 and that for the randomized dataset was 1.3 leaving only nine correlations above noise. This shows another aspect of the finite-size effect and reflects the difficulty of achieving statistical significance for smaller MSAs. Previous studies have shown that to avoid "finite-size effects" it is preferable to use MSAs of at least a few hundred sequences for correlation analysis (Gloor, Martin, Wahl, & Dunn, 2005; Weil, Hoffgaard, & Hamacher, 2009). Based on these estimates, the BPTI49 database is too small for reliable correlation calculations. Henceforth, all the correlation calculations discussed were carried out on the curated Pfam database of BPTI.

Correlation calculations generate a lot of data and analyzing it can be challenging. A very visual and intuitive way of representing this information is via a heat map (Fig. 11.6). For most of our applications, we make heat maps by using conditional formatting options in Microsoft Excel to color-code the table of MI values. The MI values are binned into intervals based on the noise value, mean value, and standard deviation. In the heat map shown, all values below the noise level of 0.22 are colored blue, and values up the mean and one, two, or three, or more σ above the mean are green, yellow, orange, and red, respectively.

To evaluate how the extent of conservation of positions affects their correlation distribution, the correlation distribution corresponding to positions with below average conservation scores can be compared to that of positions with higher than average conservation scores (Fig. 11.7). Positions with low conservation have a greater number of correlations than positions with high conservation. This is a consequence of the meaning of "correlation" as defined by MI. If a site x is all Ala, then all x,y pairs are Ala and something, and no additional information about x can be given by the identity of y.

Although MI is a very useful metric for getting an overall picture of the level of correlation in the database, if we want to look at the details of the

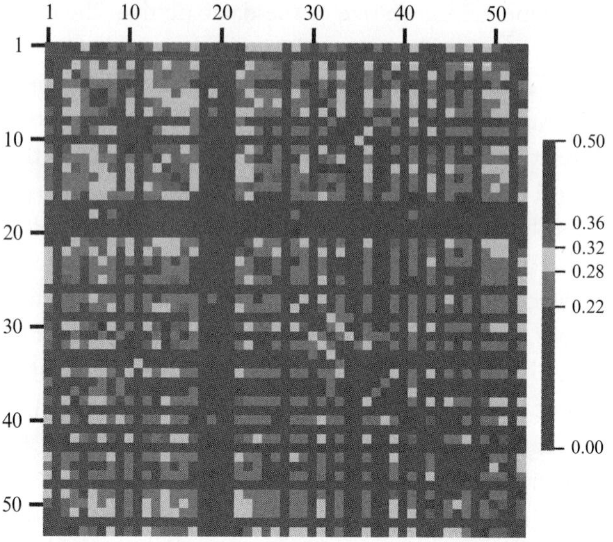

Figure 11.6 A heat map representing the MI data of BPTI: The position numbers as per the MSA are specified on the left and top of the heat map. The color scale is shown on the right side. (See Color Insert.)

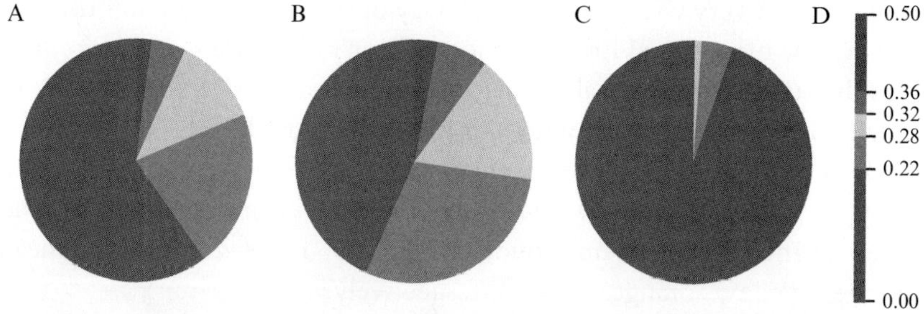

Figure 11.7 Effect of residue conservation on MI distributions: (A) MI distribution for all positions in curated BPTI_Kunitz Pfam alignment. (B) MI distribution for positions with below average conservation. (C) MI distribution for positions with above average conservation. (D) Color scale for binning the data. (For interpretation of the references to color in this figure legend, the reader is referred to the online version of this chapter.)

correlations at the amino acid level then other methods including perturbation-based approaches such as statistical coupling analysis (Lockless & Ranganathan, 1999) or correlation coefficients (Wang et al., 2009) are useful. However, while these methods furnish more information at the amino acid level, they are computationally more demanding than MI and require more sequences for statistical significance.

6. PROTOCOL FOR PREDICTING STABILIZING MUTATIONS

Making "consensus mutations" (i.e., mutating an amino acid to that observed most frequently at that position in an MSA of related proteins) stabilizes a protein about half the time. Recently, we demonstrated that avoiding positions with strong correlations improves the success of predicting stabilizing consensus mutations (Sullivan et al., 2012). This protocol for predicting stabilizing mutations has three steps.

Step 1. *Conservation filter:* All positions with above average RE values are chosen. Positions with low RE values do not contain much information at the level of consensus and are avoided. For the BPTI database, the average RE value was 1.3 and there were 18 positions that had RE values above that.

Step 2. *Correlation filter:* The positions are ranked based on the maximum MI value for any pair they are involved in and positions with correlations in the top 1% of all $(n^2 - n)/2$ values are discarded. Positions that have high correlation values may need compensatory mutations which are difficult to identify and hence they are avoided. Also, positions that are nearly invariant may have "hidden correlations" (i.e., may be physically coupled even though that cannot be mathematically distinguished from being independently required) and are also avoided. In the BPTI database, the positions with MI values above 0.36 and those with RE values above 3 were discarded thus leaving 11 positions.

Step 3. *Identifying mutation sites:* Several positions identified by applying the above filters may already be consensus residues in the protein of interest, so in the final step, we compare the consensus residue and wild-type residue at these sites; the sites where these residues are different are identified as mutation targets. If we wanted to make consensus mutations to the bovine version of the BPTI-Kunitz domain (PDB file 1BPI), then 7 out of the 11 sites would already have the consensus residue, leaving 4 sites as candidates for mutation. The mutations to BPTI suggested based on this protocol are Y21W, A40G, K41N, and F22Y.

7. SUMMARY

This chapter provides an overview of the methods that we use for consensus and correlation analyses and how they can be calculated and interpreted. Most of the calculations are carried out in Microsoft Excel using various formulae and macros (the details provided are for Microsoft Office 2010). The general trends for such calculations have been illustrated by using the Kunitz-BPTI domain for a sample analysis. We also demonstrated the protocol for predicting stabilizing mutations in a protein using the metrics we calculated.

REFERENCES

Ackerman, S. H., & Gatti, D. L. (2011). The contribution of coevolving residues to the stability of KDO8P synthase. *PloS One*, *6*, e17459.

Anfinsen, C. B. (1973). Principles that govern folding of protein chains. *Science*, *181*, 223–230.

Applebaum, D. (1996). *Probability and information: an integrated approach.* New York: Cambridge University Press.

Bartlett, G. J., & Taylor, W. R. (2008). Using scores derived from statistical coupling analysis to distinguish correct and incorrect folds in de-novo protein structure prediction. *Proteins-Structure Function and Bioinformatics*, *71*, 950–959.

Binz, H. K., Stumpp, M. T., Forrer, P., Amstutz, P., & Pluckthun, A. (2003). Designing repeat proteins: Well-expressed, soluble and stable proteins from combinatorial libraries of consensus ankyrin repeat proteins. *Journal of Molecular Biology*, *332*, 489–503.

Chen, C. P., Hsu, C. H., Su, N. Y., Lin, Y. C., Chiou, S. H., & Wu, S. H. (2001). Solution structure of a Kunitz-type chymotrypsin inhibitor isolated from the elapid snake Bungarus fasciatus. *The Journal of Biological Chemistry*, *276*, 45079–45087.

Chothia, C. (1992). One thousand families for the molecular biologist. *Nature*, *357*, 543–544.

Chothia, C., & Lesk, A. M. (1986). The relation between the divergence of sequence and structure in proteins. *The EMBO Journal*, *5*, 823–826.

Consortium, T. U. (2011). Ongoing and future developments at the Universal Protein Resource. *Nucleic Acids Research*, *39*, D214–D219.

Dekker, J. P., Fodor, A., Aldrich, R. W., & Yellen, G. (2004). A perturbation-based method for calculating explicit likelihood of evolutionary co-variance in multiple sequence alignments. *Bioinformatics*, *20*, 1565–1572.

Eddy, S. R. (2004). What is a hidden Markov model? *Nature Biotechnology*, *22*, 1315–1316.

Finn, R. D., Clements, J., & Eddy, S. R. (2011). HMMER web server: Interactive sequence similarity searching. *Nucleic Acids Research*, *39*, W29–W37.

Finn, R. D., Mistry, J., Tate, J., Coggill, P., Heger, A., Pollington, J. E., et al. (2010). The Pfam protein families database. *Nucleic Acids Research*, *38*, D211–D222.

Fodor, A. A., & Aldrich, R. W. (2004). Influence of conservation on calculations of amino acid covariance in multiple sequence alignments. *Proteins-Structure Function and Bioinformatics*, *56*, 211–221.

Gloor, G. B., Martin, L. C., Wahl, L. M., & Dunn, S. D. (2005). Mutual information in protein multiple sequence alignments reveals two classes of coevolving positions. *Biochemistry*, *44*, 7156–7165.

Godoy-Ruiz, R., Perez-Jimenez, R., Ibarra-Molero, B., & Sanchez-Ruiz, J. M. (2005). A stability pattern of protein hydrophobic mutations that reflects evolutionary structural optimization. *Biophysical Journal*, *89*, 3320–3331.

Halabi, N., Rivoire, O., Leibler, S., & Ranganathan, R. (2009). Protein sectors: Evolutionary units of three-dimensional structure. *Cell*, *138*, 774–786.

Kass, I., & Horovitz, A. (2002). Mapping pathways of allosteric communication in GroEL by analysis of correlated mutations. *Proteins-Structure Function and Genetics*, *48*, 611–617.

Knappik, A., Ge, L. M., Honegger, A., Pack, P., Fischer, M., Wellnhofer, G., et al. (2000). Fully synthetic human combinatorial antibody libraries (HuCAL) based on modular consensus frameworks and CDRs randomized with trinucleotides. *Journal of Molecular Biology, 296*, 57–86.

Lehmann, M., Kostrewa, D., Wyss, M., Brugger, R., D'Arcy, A., Pasamontes, L., et al. (2000). From DNA sequence to improved functionality: Using protein sequence comparisons to rapidly design a thermostable consensus phytase. *Protein Engineering, 13*, 49–57.

Lockless, S. W., & Ranganathan, R. (1999). Evolutionarily conserved pathways of energetic connectivity in protein families. *Science, 286*, 295–299.

Magliery, T. J., Lavinder, J. J., & Sullivan, B. J. (2011). Protein stability by number: High-throughput and statistical approaches to one of protein science's most difficult problems. *Current Opinion in Chemical Biology, 15*, 443–451.

Magliery, T. J., & Regan, L. (2004). Beyond consensus: Statistical free energies reveal hidden interactions in the design of a TPR motif. *Journal of Molecular Biology, 343*, 731–745.

Magliery, T. J., & Regan, L. (2005). Sequence variation in ligand binding sites in proteins. *BMC Bioinformatics, 6*, 240.

Main, E. R. G., Xiong, Y., Cocco, M. J., D'Andrea, L., & Regan, L. (2003). Design of stable alpha-helical arrays from an idealized TPR motif. *Structure, 11*, 497–508.

Miller, C. S., & Eisenberg, D. (2008). Using inferred residue contacts to distinguish between correct and incorrect protein models. *Bioinformatics, 24*, 1575–1582.

Mosavi, L. K., Minor, D. L., & Peng, Z. Y. (2002). Consensus-derived structural determinants of the ankyrin repeat motif. *Proceedings of the National Academy of Sciences of the United States of America, 99*, 16029–16034.

Orengo, C. A., Jones, D. T., & Thornton, J. M. (1994). Protein superfamilles and domain superfolds. *Nature, 372*, 631–634.

Paesen, G. C., Siebold, C., Dallas, M. L., Peers, C., Harlos, K., Nuttall, P. A., et al. (2009). An ion-channel modulator from the saliva of the brown ear tick has a highly modified Kunitz/BPTI structure. *Journal of Molecular Biology, 389*, 734–747.

Pagni, M., Ioannidis, V., Cerutti, L., Zahn-Zabal, M., Jongeneel, C. V., Hau, J., et al. (2007). MyHits: Improvements to an interactive resource for analyzing protein sequences. *Nucleic Acids Research, 35*, W433–W437.

Parkin, S., Rupp, B., & Hope, H. (1996). Structure of bovine pancreatic trypsin inhibitor at 125 K: Definition of carboxyl-terminal residues Gly57 and Ala58. *Acta Crystallographica. Section D, Biological Crystallography, 52*, 18–29.

Punta, M., Coggill, P. C., Eberhardt, R. Y., Mistry, J., Tate, J., Boursnell, C., et al. (2012). The Pfam protein families database. *Nucleic Acids Research, 40*, D290–D301.

Russ, W. P., Lowery, D. M., Mishra, P., Yaffe, M. B., & Ranganathan, R. (2005). Natural-like function in artificial WW domains. *Nature, 437*, 579–583.

Schmidt, A. E., Chand, H. S., Cascio, D., Kisiel, W., & Bajaj, S. P. (2005). Crystal structure of Kunitz domain 1 (KD1) of tissue factor pathway inhibitor-2 in complex with trypsin—Implications for KD1 specificity of inhibition. *The Journal of Biological Chemistry, 280*, 27832–27838.

Shannon, C., & Weaver, W. (1949). *The mathematical theory of information*. Urbana, IL: University of Illinois Press.

Socolich, M., Lockless, S. W., Russ, W. P., Lee, H., Gardner, K. H., & Ranganathan, R. (2005). Evolutionary information for specifying a protein fold. *Nature, 437*, 512–518.

Sonnhammer, E. L. L., Eddy, S. R., & Durbin, R. (1997). Pfam: A comprehensive database of protein domain families based on seed alignments. *Proteins-Structure Function and Genetics, 28*, 405–420.

Steipe, B., Schiller, B., Pluckthun, A., & Steinbacher, S. (1994). Sequence statistics reliably predict stabilizing mutations in a protein domain. *Journal of Molecular Biology, 240*, 188–192.

Suel, G. M., Lockless, S. W., Wall, M. A., & Ranganathan, R. (2003). Evolutionarily conserved networks of residues mediate allosteric communication in proteins. *Nature Structural Biology*, *10*, 232. *Vol. 10*, p. 59.

Sullivan, B. J., Durani, V., & Magliery, T. J. (2011). Triosephosphate isomerase by consensus design: Dramatic differences in physical properties and activity of related variants. *Journal of Molecular Biology*, *413*, 195–218.

Sullivan, B. J., Nguyen, T., Durani, V., Mathur, D., Rojas, S., Thomas, M., et al. (2012). Stabilizing proteins from sequence statistics: The interplay of conservation and correlation in triosephosphate isomerase stability. *Journal of Molecular Biology*, *420*, 384–399.

Thompson, J. D., Gibson, T. J., & Higgins, D. G. (2002). Multiple sequence alignment using ClustalW and ClustalX. *Current protocols in bioinformatics/editoral board, Andreas D. Baxevanis ... [et al.]* Chapter 2, Unit 2.3. See: http://www.ncbi.nlm.nih.gov/pubmed/18792934

Tripp, K. W., & Barrick, D. (2003). Folding by consensus. *Structure*, *11*, 486–487.

Wang, N., Smith, W. E., Miller, B. R., Aivazian, D., Lugovskoy, A. A., Reff, M. E., et al. (2009). Conserved amino acid networks involved in antibody variable domain interactions. *Proteins-Structure Function and Bioinformatics*, *76*, 99–114.

Weil, P., Hoffgaard, F., & Hamacher, K. (2009). Estimating sufficient statistics in co-evolutionary analysis by mutual information. *Computational Biology and Chemistry*, *33*, 440–444.

CHAPTER TWELVE

Enzyme Engineering by Targeted Libraries

Moshe Goldsmith, Dan S. Tawfik[1]
Department of Biological chemistry, Weizmann Institute of Science, Rehovot, Israel
[1]Corresponding author: e-mail address: dan.tawfik@weizmann.ac.il

Contents

Abstract

This review outlines the strategies we apply for directed enzyme evolution using targeted libraries, namely, libraries that diversify specific residues with predefined mutational compositions. The theoretical grounds underlining the design of such libraries are described, including the mutational load, the ratio of beneficial versus deleterious mutations, and screening capacity. We point out the advantage of using mutational spiking strategies for "hedging the bets," exploring a large number of potentially beneficial mutations, and tuning the library's mutational load. Also highlighted are the merits of low-throughput screens that measure multiple parameters at high accuracy, and of using the desired substrate and reaction conditions rather than surrogates. We subsequently describe library construction strategies (rational and analytical) based on structure and sequence analyses, including ancestral libraries, which are particularly suitable for low-throughput screens. We also discuss the critical role of including compensatory, stabilizing mutations during library construction. Finally, the design efficiency and the optimal mutational loads of libraries are assessed by comparing targeted mutational libraries versus libraries of random mutations.

Methods in Enzymology, Volume 523
ISSN 0076-6879
http://dx.doi.org/10.1016/B978-0-12-394292-0.00012-6

1. INTRODUCTION

The most challenging test of our understanding of enzyme catalysis, and the "holy grail" of protein engineers, is to be able to predict a single amino acid sequence that would catalyze a given target reaction with catalytic efficiency and specificity that match those of natural enzymes.[1] However, this challenge, best put in Thomas Edison's words: "Until man duplicates a blade of grass, nature can laugh at his so-called scientific knowledge," is far from being met. As a result, the evolutionary strategy of trial and error has become an inevitable choice. From an enzyme engineering point of view, the challenge is therefore to narrow the exploration of sequence space, from the impractical size of all possible sequence permutations (20^n, n being the number of amino acids) to a manageable number of variants. Thus, the gap in our understanding and prediction capabilities can be partly bridged (Khersonsky et al., 2012).

How many trials may be needed depends on how wide is the knowledge gap with respect to two factors: (i) How well can we focus our exploration of sequence space and thereby maximize the frequency of mutations that confer the desired function; and (ii) How well can we minimize the error, namely, reduce the frequency of deleterious mutations. When exploring randomly mutated enzymes, the frequency of beneficial mutations is very low (in the order of 10^{-3}) and the frequency of deleterious mutations is high ($\geq$30% of all mutations are deleterious). The accumulation of random mutations, therefore, rapidly renders a protein completely inactive and the likelihood of finding an improved variant is accordingly reduced (Bershtein & Tawfik, 2008; Fig. 12.1). This is the reason why in general, both in randomly mutated and designed libraries, the mutational load (average number of mutations per library variant) must be kept low (as discussed in Sections 4 and 5.4).

To increase the rate of success, researchers generate libraries that target specific regions of the target protein, or even individual positions, and use specific mutations within these regions or positions. In addition, different strategies are employed for filtering or compensating for deleterious

[1] Biochemistry textbooks highlight the "superstars" of the enzyme world, enzymes that exhibit impressively high turnover rates (e.g., carbonic anhydrase with a $k_{cat} > 10^4\ s^{-1}$) and catalytic efficiencies that approach the diffusion rate limit (e.g., TIM, triose phosphate isomerase, k_{cat}/K_M approaching $10^9\ M^{-1}\ s^{-1}$). It is worth noting, however, that the "average" enzyme is far from being so efficient, with the average k_{cat} value being $\sim 14\ s^{-1}$, and the average $k_{cat}/K_M \sim 10^5\ M^{-1}\ s^{-1}$ (Bar-Even et al., 2011).

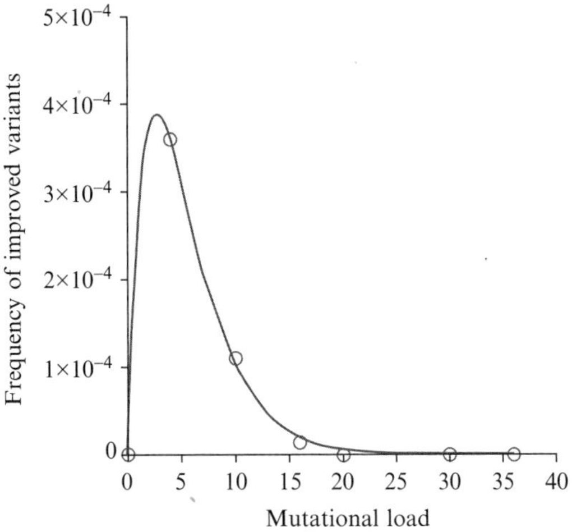

Figure 12.1 The frequency of improved variants in random mutagenesis libraries as a function of the mutational load. The experimental data were obtained from a selection of TEM-1 β-lactamase for variants that confer cefotaxime (CTX) resistance at 0.5 μg/ml (Bershtein & Tawfik, 2008). The curve fit follows the equation: $P_{new} = nf^{+}e^{-\alpha n}$; where P_{new} is the frequency of library variants with improved CTX resistance; n is the mutational load—the average number of mutations per library variant; f^{+} is the fraction of beneficial mutations in the library, and α is the fraction of deleterious mutations. As can be seen, P_{new} reaches a maximum at ~3 mutations per gene and rapidly declines as the mutational load increases. This pattern is the outcome of the frequency of deleterious mutations being ~0.36, and of beneficial ones ~4×10^{-4}.

mutations in libraries. Designed libraries therefore aim at increasing the ratio of advantageous to deleterious mutations beyond the daunting ratio of $\leq 1/300$ that underlines randomly incorporated mutations. The higher this ratio is, the better or "smarter" is the library, and the smaller is the number of sequence permutations that need to be explored to obtain the desired activity.

2. SCREENING VERSUS SELECTION[2]

The chosen engineering strategy is dictated foremost by what comprises a manageable a library. Larger gaps in our prediction abilities demand larger libraries and thereby the application of a high, or even ultrahigh-throughput selection methodology (10^6 to $>10^{12}$ variants processed in

[2] Parts of this section and of the next ones were adopted from Goldsmith et al. (2012).

parallel). But, such throughputs are often inapplicable. A successfully designed library can, however, be explored by a low-to-medium throughput screen (10^2–10^4 variants tested individually). Thus, high/ultrahigh-throughput screens or selections become advantageous with libraries of variants generated by random mutagenesis, or by simultaneous saturation mutagenesis (diversification to all 20 amino acids) of many positions. As powerful as they may be, selection methodologies have inherent disadvantages. They need to be tailored for a given enzyme and activity, and thus, developing an ultrahigh-throughput screen is usually a project in itself. Unlike screens, where the activity of each and every library variant becomes known, selections only lead to the isolation of the most improved variants. They are therefore harder to control and more prone to artifacts. For example, the most improved variant might be the one showing the highest expression level in a particular selection system, and not necessarily the most enzymatically active one. In addition, selections usually make use of surrogate substrates (e.g., fluorogenic analogues) and/or modified reaction conditions (e.g., single turnover), which may lead to the evolution of enzymes that have only minor improvements in activity with the target substrate.

Within academic laboratories, screens are mostly applicable at low-to-medium throughputs (10^2–10^4 clones/round). They can, however, utilize almost any biochemical or biophysical detection method including HPLC or NMR. Despite their lack of elegance, screens often "deliver the goods" and may actually save time and effort (see "the Okazaki maneuver" in Kornberg, 1989). In addition, screens also have several unique advantages unavailable for selections: (i) Rapid assessment of the diversification strategy—for example, if >90% of variants show no activity with the enzyme's original substrate—change strategy. (ii) The ability to monitor multiple parameters, for example, monitoring the original as well as the evolved activity. This enables the isolation of variants with higher expression levels versus variants with improved enzymatic activity (Gupta et al., 2011), targeting of higher activity as well as selectivity, screening for "generalist" variants that process more than one substrate (Goldsmith et al., 2012), and so on. (iii) Precision of activity measurements, which enables the identification of marginally improved variants, or even neutral ones, that can later yield higher improvements in combination with additional mutations (Gupta et al., 2011). (iv) Nearly unlimited dynamic range—the evolved variants only need to be further diluted as they become more improved. (v) The actual substrate and desired reaction conditions can be applied, thus, "you get what you select for" (Romero & Arnold, 2009).

It should be noted that screens do not necessarily impose low throughput. The use of FACS (fluorescence-activated cell sorting), for example, for screening enzyme libraries (using cell-entrapped substrates (Liu, Li, Liotta, & Lutz, 2009; Yang et al., 2010), by compartmentalization in emulsion droplets (Bawazer et al., 2012; Gupta et al., 2011; Hardiman, Gibbs, Reeves, & Bergquist, 2010; Stapleton & Swartz, 2010), or by cell-surface display (Chen, Dorr, & Liu, 2011; Varadarajan, Cantor, Georgiou, & Iverson, 2009; Varadarajan, Pogson, Georgiou, & Verson, 2009)) affords a screen with high throughput. Albeit, these high-throughput screens usually demand the use of surrogate substrates and/or specialized expression systems, and their accuracy is rather limited. Technological advents such as microfluidics may afford high throughput while maintaining refined sample control, high accuracy, and the ability to monitor in parallel multiple parameters (Agresti et al., 2010; Granieri, Baret, Griffiths, & Merten, 2010; Kintses, van Vliet, Devenish, & Hollfelder, 2010; Kintses et al., 2012; Shim et al., 2009). While these technologies are still under development, they may eventually lead to versatile, bench-top instruments that can be used in almost any laboratory, including laboratories that do not specialize in the development of screening technologies.

3. THE MERITS OF A DIRECT SCREEN: THE NERVE AGENT-DETOXIFYING ENZYMES

Screening directly for the target activity is a crucial factor, particularly in the engineering of enzymes for practical applications. Described below is an example taken from our own work.

When evolving enzymes for nerve agent detoxification (specifically, for the hydrolysis of G-type nerve agents), we began with ultrahigh-throughput FACS screens using an assay that has become the standard in this area (Briseno-Roa et al., 2006; Khare et al., 2012; Tsai et al., 2010), namely, monitoring the hydrolysis of fluorogenic substrate analogues of the phosphonate nerve agents (Amitai et al., 2007; Briseno-Roa et al., 2006; Hill, Li, Thoden, Holden, & Raushel, 2003; Tsai et al., 2012; Fig. 12.2A). By combining FACS and 96-well plate screens, we were able to increase the activity of a recombinant form of the enzyme serum paraoxonase-1 (rePON1) by >160-fold toward a racemic mixture of a cyclosarin surrogate and by $>6.3\times10^5$-fold toward its toxic isomer (Gupta et al., 2011). The surrogates, however, differ from the actual target substrates in three aspects. First, the nerve agent's leaving group size is much smaller than the size of the coumarin leaving group of the surrogate (fluoride

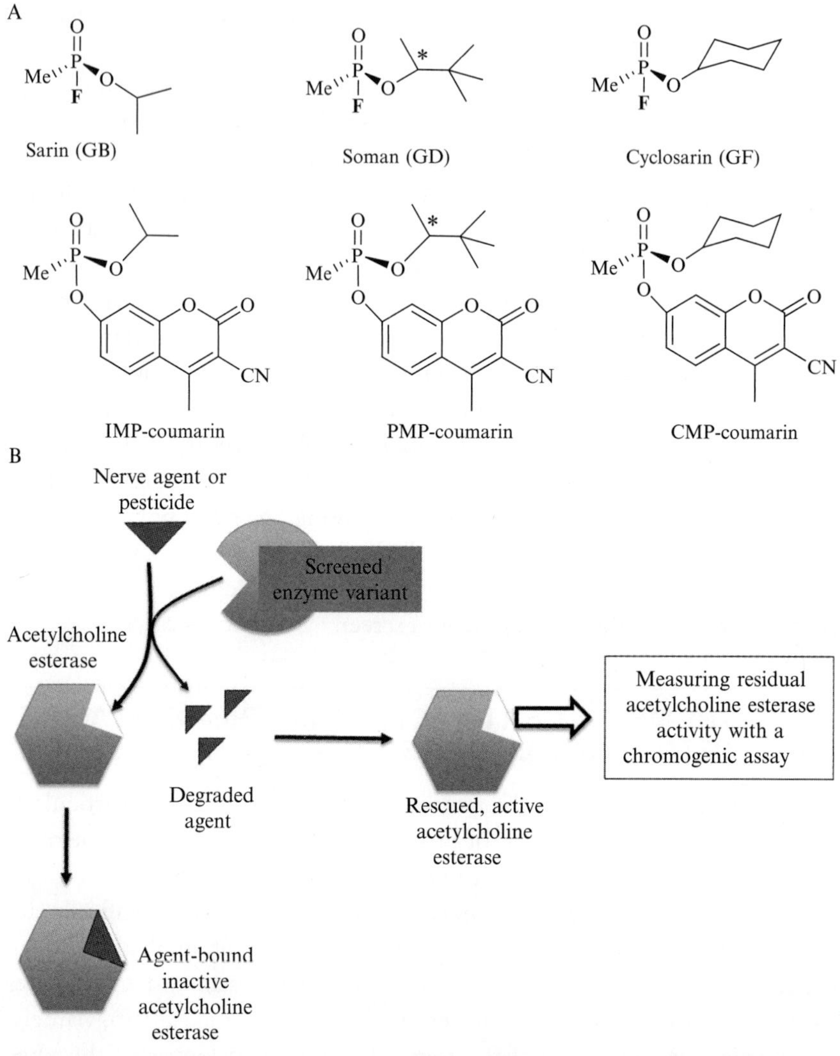

Figure 12.2 (A) The G-type nerve agents GB, GD, GF (the toxic S_P stereoisomer), and their coumarin surrogates. (* - Denotes a chiral carbon) (B) A schematic representation of the acetylcholinesterase protection screen for the identification of detoxifying enzyme variants. (For color version of this figure, the reader is referred to the online version of this chapter.)

in sarin, soman and cyclosarin or cyanide in tabun). Second, they differ in chemistry—the pK_a of the leaving group is 3.1 for fluoride and 6.5 for coumarin. Finally, they differ in acetylcholinesterase (AChE) inhibition constants and therefore in their toxicity (the coumarin surrogates are ≥10-fold less potent inhibitors than their nerve agent counterparts; Amitai et al., 2007; Briseno-Roa et al., 2006). AChE toxicity also influences another aspect of our ability to screen for improved variants: Nerve agents, and their coumarin analogues, are chiral with respect to the phosphorus atom (R_p, S_p; Fig. 12.2A). The S_p isomers of nerve agents inhibit AChE $\geq$ 1000-fold faster than their respective R_p isomers and therefore comprise the primary target for detoxification. A screen that monitors the release of the fluorescent coumarin, or even of the fluoride leaving group (Ben-David, Shoham, & Shoham, 2008), would not distinguish between the two isomers, and is most likely to drive the process of directed evolution toward efficient R_p hydrolysis, in particular as all known enzymes that hydrolyze nerve agents show R_p preference (diTargiani, Chandrasekaran, Belinskaya, & Saxena, 2010; Theriot & Grunden, 2011).

A direct screen comprised an optimal solution. Nerve agents inhibit the enzyme AChE thus causing a cholinergic crisis that may result in respiratory failure and death. The detoxifying enzyme must hydrolyze the agents at a rate that is high enough to avoid the irreversible inhibition of AChE. Further, the agents must be hydrolyzed at their physiologically relevant concentrations. Thus, in the direct screen we developed (Fig. 12.2B), AChE is added to the assayed library variants, and the reaction mixture is exposed to *in situ* prepared nerve agents at increasing concentrations up to the physiologically intoxicating concentrations in the micromolar range ($\leq$2 μ*M*). Variants capable of hydrolyzing the agent are detected by measuring the residual levels of AChE activity.

This assay demanded the development of suitable substrates, namely, the *in situ* synthesis of nerve agents at very low, nonhazardous concentrations. Nonetheless, it has proven optimal. Foremost, it enabled the isolation of variants that hydrolyze the toxic component of these agents only (the S_p isomer), at physiologically relevant concentrations, and fast enough to provide prophylactic protection (according to our model) (Goldsmith et al., 2012; Gupta et al., 2011). The assay also offers high sensitivity. AChE is a highly efficient and sensitive enzyme that is inhibited by nerve agents at near-stoichiometric ratios. Thus, the interception of nerve agents at $\geq$5 n*M* concentrations can be detected in cell lysates. In the initial rounds of directed evolution, variants with relatively slow hydrolysis rates could be detected by preincubation of the nerve agents with cell lysates of bacteria that expressed library variants (*Escherichia coli (E.coli)* lysates not expressing PON1 show no background hydrolysis), and addition of AChE to the pretreated agent. In this manner, library variants exhibiting k_{cat}/K_M values of $\geq 2\ M^{-1}\ S^{-1}$ for hydrolysis of the S_p isomers of the nerve agents could be readily identified using nerve agent concentrations as low as 5 n*M* (Goldsmith et al., 2012). In later rounds, the selection pressure was augmented by increasing the nerve agent's concentration up to 2.75 μ*M* in the last round (Goldsmith et al., 2012). The selection pressure was also augmented by simultaneously mixing the nerve agent with the intercepting enzyme and AChE, rather then preincubating the agent and the intercepting enzyme. Thus, by directly screening for the protection of AChE from inhibition by *in situ* generated nerve agents, we were able to isolate variants that were improved by up to 3400-fold with respect to wild-type PON1, at hydrolyzing the toxic isomers of the three most toxic G-agents (GB, GD, and GF) (Goldsmith et al., 2012). The assay we developed is also versatile, as it is suitable for detection of intercepting enzymes for any AChE-inhibiting pesticide or nerve agent irrespective of their chemistry.

4. HEDGING THE BETS: MUTATIONAL SPIKING APPROACHES

Almost regardless of the applied screen and library design strategies, the theoretical number of sequence permutations that can be explored is likely to exceed the screen's capacity. A typical active site is comprised of over 30 first- and second-shell residues. A 12-Å distance from the key active-site residue or element (e.g., the catalytic metal in metalloenzymes or the nucleophilic serine in serine hydrolases) is what we usually consider first *plus* second shell. However, this distance typically earmarks >90 residues (or ca. 25% of the enzyme's length) for diversification (Alcolombri, Elias, & Tawfik, 2011). Of these, as many as half of the positions may be totally conserved within the enzyme's family, and even within related families, and such positions are usually excluded from the library. But using even a conservative estimate of 20 positions, each diversified with five different amino acids, gives rise to $\sim 10^{14}$ variants. Alternatively, saturating (mutating to all 20 amino acids) as few as five active-site positions yields a library of $>10^{6}$ possible variants.

The library size hurdle can be overcome in three ways:

i. Exploring only a small fraction of the generated library diversity. This approach assumes that most of the mutations are neutral. Hence, advantageous mutations, or even combinations of advantageous mutations, are redundantly represented, and can be identified. The validity of this assumption depends on the starting point enzyme, and primarily on how effective the library design is, namely, what fraction of the mutations is beneficial for the target function, and how large is the fraction of neutral mutations. It is, however, likely that designed libraries that mostly focus on active-site residues will carry a relatively high fraction of deleterious mutations. In this case, the fraction of active variants drops exponentially with the number of mutations per variant (as in random libraries; see Fig. 12.1), and the low sampling approach is likely to fail.

ii. Targeted mutagenesis can be performed in an iterative manner (Reetz, 2011; Reetz, Prasad, Carballeira, Gumulya, & Bocola, 2010). In this way, the mutated positions are explored one by one in an order dictated by the experimentalist. The best replacement in the position explored first serves as the starting point for exploring diversity in the second position, and so on. This approach is highly economical in screening capacity (library diversity is $\leq$20 per round). However, it will fail to access non-additive, epistatic combinations of mutations—that is, when the effect of one mutation depends on whether another mutation is present or not. Additivity has been shown to dominate at least one such iterative search, suggesting that the optimal combination could be identified almost regardless of the order by which the active-site positions are explored (Reetz & Sanchis, 2008). However, the effects of mutations are often nonadditive. In particular, many mutations can be neutral, or even deleterious, but become advantageous in combination with other mutations (Guthrie, Allen, Camps, & Karchin, 2011; Parera, Perez-Alvarez, Clotet, & Martinez, 2009; Toth-Petroczy & Tawfik, 2011; Weinreich, Watson, & Chao, 2005). In many other cases, beneficial mutations at different positions interact with negative epistasis, namely, when combined, they result in lowering activity beyond the single mutants (Salverda et al., 2011) (see H115W and V346A in serum paraoxonase, in which the initial advantage gained by the double mutant was rapidly lost in later rounds, where additional mutations excluded V346A in favor of the dominant H115W mutation; Gupta et al., 2011). Nonadditivity greatly complicates the search for the optimal set of mutations that will eventually confer the highest activity. It often makes the order of exploration a critical factor, and impossible to know in advance.

iii. The third way is by spiking different combinations of oligonucleotides encoding the diversified positions, thereby hedging the bets, or bridging the above two strategies. In this way, a large number of positions can be explored, but individual library variants contain only a limited number of mutations as dictated by the screen's capacity. For example, as noted above, the parallel diversification of 20 positions, each to five different amino acids, gives rise to a library of $\sim 10^{15}$ variants (n^k, where n is the number of different amino acids per position, including the wild-type option, and k is the number of diversified positions). However, if the very same diversified positions are spiked into the library, with an average, say, of two mutated positions per library variant, the theoretical diversity becomes <7000 variants. The library size in this case is given by $n^l k!/(k-l)!l!$, where l is the average number of mutated positions per variant. In this manner, combinations of mutations can be identified even with a modest screening capacity, and the number of inactive variants is also minimized (Fig. 12.1).

To apply the spiking strategy, we developed a facile and versatile protocol for the combinatorial diversification of large sets of residues. This protocol uses diversifying oligonucleotides, each diversifying one position, and DNA shuffling (Fig. 12.3) (Herman & Tawfik, 2007). Examples for the application of this, or similar protocols include Alcolombri et al. (2011), Berger, Guttman, Amar, Zarivach, and Aharoni (2011), Gupta et al. (2011), Khersonsky et al. (2011), and Scanlon et al. (2010).

5. RATIONAL AND ANALYTICAL LIBRARY DESIGNS

5.1. Libraries by rational design

Despite the advent of computational approaches, and of other analytical approaches for designing libraries (e.g. see Chen, Snow, Vizcarra, Mayo, and Arnold, 2012; Chica, Moore, Allen, and Mayo, 2010; Privett et al., 2012), the researcher's intuition is still an often-used approach. However, with few exceptions, rational protein design, namely, engineering new functions through point mutations based on common sense and structural rationale, has failed to deliver consistent results. The classical example is the attempt to increase the specificity of chymotrypsin toward a trypsin substrate (Brannigan & Wilkinson, 2002). Despite the high structural homology between these two enzymes, and the clearly defined differences in their substrate-binding pockets, the rational design of a specificity switch from substrates with positively charged side chains (Arg, Lys at S1′ position of the cleaved polypeptide substrate) to aromatic/hydrophobic side chains has proven elusive. Contrary examples do exist. For example, by comparing structures and sequences of two related plant enzyme families, the specificity of tyrosine ammonia-lyase (TAL) was readily switched to phenylalanine (PAL) with a single mutation (His to Phe) (Watts, Mijts, Lee, Manning, &

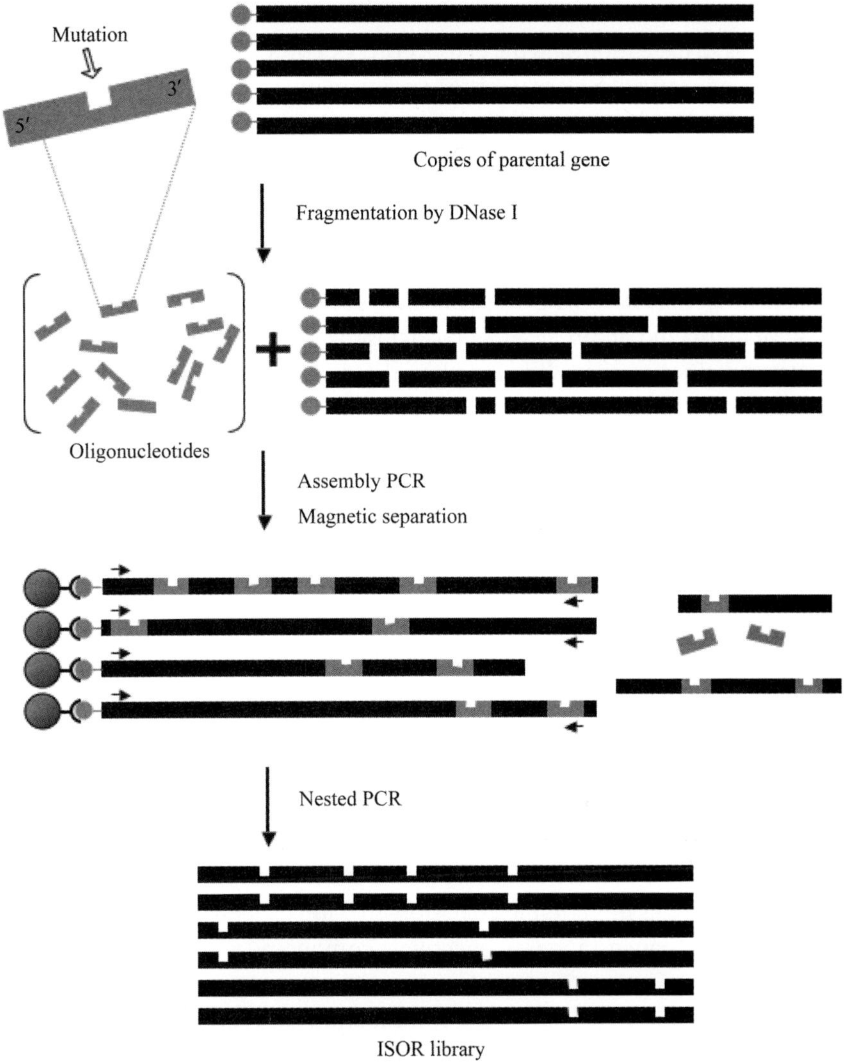

Figure 12.3 A schematic representation of ISOR (Incorporation of Synthetic Oligonucleotides) in the process of gene Reassembly protocol (Herman & Tawfik, 2007). The target gene is fragmented by digestion with DNaseI. The DNaseI fragments are mixed with a set of synthetic oligonucleotides in which each oligo diversifies one residue (to all 20 amino acids or fewer). The gene is assembled as in DNA shuffling by a process of self-primed extension using a thermophilic polymerase. The resulting libraries have a tunable average number of mutated positions that constitute a random subset of the entire set of diversified residues.

Schmidt-Dannert, 2006). However, the swift transition observed in the TAL to PAL case is an exception, and the engineering of a new activity, for example, an activity not observed in closely related family members of the target enzyme, is far more challenging. Overall, it seems that the term "rationalized design" applies to many cases where the amino acid exchanges that successfully triggered the desired activity could be retrospectively rationalized.

The rational design approach can become effective when given a wide margin of error. Namely, rational design can direct the choice of active-site positions that may modulate function, and of second-shell residues that interact with them, and of subsets of amino acids to be explored at these positions. Rational design applies primarily when the desired activity already exists in the starting point enzyme as a promiscuous one, and when detailed structural information is available. In two recent examples, this approach was used to engineer a P450 monooxygenase and a nucleotidyltransferase. By analyzing all residues in the active site within 5–8 Å of the bound substrate analogue, a small number of residues were targeted for saturation mutagenesis (Hoffmann, Bonsch, Greiner-Stoffele, & Ballschmiter, 2011; Moretti et al., 2011). Targeted mutagenesis (of selected positions, and amino acids within these positions) was also applied toward the directed evolution of an organophosphate hydrolase (Goldsmith et al., 2012; Gupta et al., 2011), a lipase (Reetz, 2011; Reetz, Prasad, et al., 2010), and for the introduction of allostery and an increase in the substrate specificity of a Bayer–Villiger monooxygenase (Wu, Acevedo, & Reetz, 2010).

Directed evolution via rationally designed libraries is an ongoing process whereby success or failure in one round dictates library design for the next round. This results in a complex decision-making process that is usually hard to describe in a concise and retrospectively convincing manner. Often, more than one design strategy is applied within different rounds, or even the same round. For example, we have used combinations of rationally designed and phylogenetic mutations (Goldsmith et al., 2012), or of the latter two with computationally designed as well as randomly incorporated mutations (Khersonsky et al., 2011, 2012). Nonetheless, this approach makes use of the researcher's knowledge of the enzyme in hand, and her/his intuition as to which positions may trigger change at given stage.

In our experience, detailed structural information is highly valuable. The structures of complexes with relevant substrate analogues, which are not always available, can be supplemented by docking models, provided that the latter are critically assessed and examined in view of the mutagenesis results (e.g., see Ben-David et al., 2012). The obvious targets for exploration are positions that are in direct contact with the substrate (the original substrate and the targeted one, although the latter is rarely available), except for those that play a key role in catalysis and whose mutagenesis may render the enzyme inactive. Beyond the obvious, there are a few strategies that are worth highlighting:

i. *Neighbor joining.* If a certain mutation in an active-site residue was found to trigger a significant improvement in activity (in the range of an order of magnitude or more), it is worth exploring the nearby residues. The neighboring positions to be explored should include not only first-shell positions (i.e., positions in direct contact with the substrate and/or a known catalytic function) but also second-shell sites—that is, neighboring residues that are not in direct contact but may influence the first shell mutation. For example, the first position explored in the directed evolution of serum paraoxonase (PON1) toward nerve agent hydrolysis was His115, as its mutation to Trp triggered a leap in the desired activity. In the next rounds, we explored position 69 (first-shell) but also 134 (second-shell; Fig. 12.4) (Gupta et al., 2011). Likewise, a highly beneficial mutation at position 222 led us to explore the adjacent positions 196, 291, and 292 (Goldsmith et al., 2012). In this way, a network of new interactions was created that not only reshaped the active site (e.g., see Fasan, Meharenna, Snow, Poulos, & Arnold, 2008) but also redirected the catalytic mechanism (e.g. see Tokuriki et al., 2012).

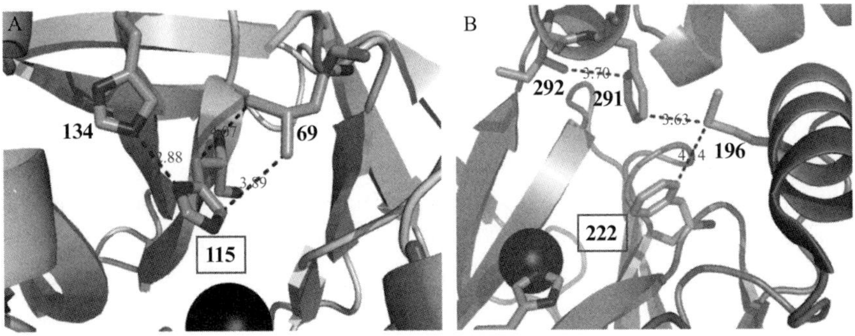

Figure 12.4 Neighbor joining and revisiting key positions in library design. Example taken from the directed evolution of serum paraoxonase-1 (Goldsmith et al., 2012; Gupta et al., 2011). Two key active-site positions that are close to PON1's catalytic calcium (gray sphere) and are in direct contact with substrates were explored: His115 (panel A) and Phe222 (panel B). The selection of advantageous mutations in these two first-shell positions led us to explore adjacent first- and second-shell residues (e.g., 134, 69, 292, 291, and 196) in subsequent rounds. Positions 115, 69, and 222 were then reexplored for their optimal compositions. (See Color Insert.)

ii. *Revisiting key positions.* It is often worth exploring the possibility that changes at previously explored positions may become beneficial at later stages, namely, after other positions have been modified. This relates to the epistatic nature of the evolutionary processes—what's at position *i* is often influenced by position *j*, thus resulting in cycles of sequence refinements. For example, following fixation of the first mutation that increased the activity of PON1 toward nerve agent hydrolysis, H115W, several other active-site positions were also fixed (e.g., F222S), including mutations in positions that are in direct contact with position 115 (e.g., L69G and H134R) (Gupta et al., 2011). Positions 115, 69, and 222 were reexplored in later rounds and their optimal composition was refined (Fig. 12.4). These cycles of optimization support the view that sequence changes are generally context dependent, and that revisiting key active-site positions that were mutated in early rounds is a valuable strategy.

5.2. Libraries by analytical design

The researcher's intuition can be supplemented by various analytical methods that direct library design based on structural and/or sequence-related information. We dub these approaches analytical, in contrast to computational approaches, as they use simple parameters such as family consensus to guide library design. Thus, analytical approaches do not aim at redesigning the active site toward a particular substrate, namely, at computing a single sequence that will confer the desired activity. Rather, these analyses direct library design in terms of which positions should be diversified, and with what compositions.

Several computer algorithms have been developed that aim at making more informed decisions regarding library composition. Some algorithms identify mutational hot-spots and combinations of mutations based on sequence and structure databases and functional information (Jochens & Bornscheuer, 2010; Kuipers et al., 2010 Ma & Berezovsky, 2010; Pavelka, Chovancova, & Damborsky, 2009; van Leeuwen, Wijma, Floor, van der Laan, & Janssen, 2012). Other algorithms have been developed to identify combinations of beneficial mutations that originally appeared in separate variants (Brouk, Nov, & Fishman, 2010; Fox et al., 2007; Hokanson et al., 2011), or variants that are simultaneously optimized for more than one trait (He, Friedman, & Bailey-Kellogg, 2012).

Phylogenetic analyses are also becoming instrumental in guiding the construction of small libraries enriched with active and stable variants. Comparing the sequence of the target enzyme to sequences of evolutionarily related enzymes provides useful information regarding positions for diversification and the kind of amino acids that could be accommodated at these positions (Jochens & Bornscheuer, 2010). A potentially more powerful approach is to explore the ancestral sequence of potential function-modifying positions. This approach is derived from the assumption that a direct transition between paralogs is a trajectory that has never been taken by nature. Rather, paralogs, for example, related enzymes with different enzymatic specificities, have diverged from a common ancestor. Ancestral libraries are therefore comprised of active-site substitutions predicted to have existed at various nodes of the evolutionary trajectories that gave raise to contemporary enzyme families, each exhibiting a different specificity. These ancestral forms are long gone, but their sequences can be reconstructed from the phylogenetic tree.

Typically, a Phylogenetic tree that includes the target enzyme, its family members (orthologs), and related family members (paralogs), is constructed, and the sequences of various nodes that connect the paralogous families are predicted (Fig. 12.5). The target enzyme is then diversified by combinatorially incorporating back-to-ancestor mutations at positions within or near the active-site region. In many positions, the ancestral amino acid corresponds to one of the extant family members. At some positions, however, the computationally predicted ancestral amino acid is different from all the amino acids found at that same position in extant family members, and comprises a unique sequence composition that cannot be obtained by a simple alignment analysis, or by family shuffling (e.g., the predicted residue in position 193 of the vertebrate ancestor of PON1 is Met, while all extent family members include other amino acids such as Leu, Phe, Ser, and others; Alcolombri et al., 2011). The exploration is systematic and includes every ancestral substitution in all positions that meet a certain criterion such as distance from the active site, or any other structural measure. This approach is therefore analytical, and relies to a much lesser degree on the researcher's intuition.

Several recent examples indicate that ancestral libraries can afford a large variety of improved variants by screening only few hundred examples. Although these approaches vary in the way by which the ancestral mutations are identified, and in the strategy for library construction, their outcome is similar (Alcolombri et al., 2011; Chen et al., 2010; Cole & Gaucher, 2011; see also Section 5.4). As detailed below (Section 5.4), by virtue of exhibiting an unusually high ratio of beneficial over deleterious mutations, ancestral libraries can be constructed with high mutational loads yet contain a remarkably high fraction of improved variants.

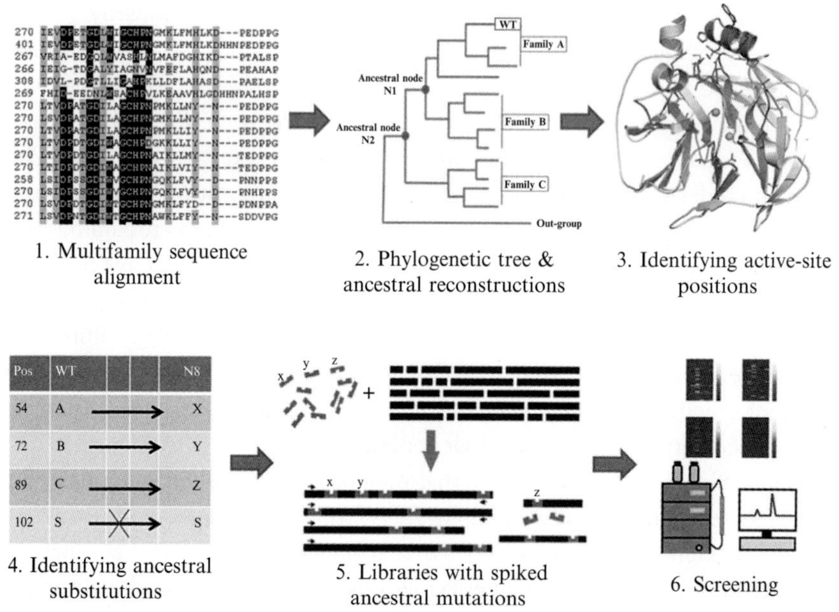

Figure 12.5 A schematic representation of the ancestral library methodology. 1. Orthologs and paralogs of the target enzyme are identified and their sequences are aligned. 2. A phylogenetic tree is generated, the ancestral nodes are assigned, and their sequences are predicted. The reference node(s) typically encompass sequences of several paralogous families that exhibit $\geq$50% sequence identity. 3. The positions targeted for diversification are determined based on the enzyme's 3D structure (e.g., active site, first- or second-shell positions). 4. The sequence of the chosen ancestral node (e.g., N8) is compared to that of the starting enzyme (WT) at the chosen positions. 5. Ancestral substitutions that differ from the starting enzyme are incorporated in a combinatorial manner into it by assembly PCR using oligonucleotides. 6. The resulting gene library is cloned, transformed to *E. coli*, and screened to identify variants with the desired specificity. See also Alcolombri et al. (2011).

5.3. Prediction of stabilizing compensatory mutations

Although not directly involved in modulating function, compensatory, stabilizing mutations are also a key component of successful library design as they dramatically increase the frequency of variants that are folded and active. The theoretical grounds of this phenomena are discussed in detail elsewhere (Soskine & Tawfik, 2010). In essence, most mutations are destabilizing, and function-modifying mutations tend to be more destabilizing than neutral ones (Tokuriki, Stricher, Serrano, & Tawfik, 2008; Wang, Minasov, & Shoichet, 2002). Proteins with marginal stability show low tolerance to mutations and hence limited evolvability (Bershtein, Segal, Bekerman,

Tokuriki, & Tawfik, 2006; Bloom, Labthavikul, Otey, & Arnold, 2006). One means of overcoming this barrier is by chaperonin overexpression (Tokuriki & Tawfik, 2009). Here, we outline analytical approaches that enable the identification of stabilizing mutations—approaches that are simple and easy to apply.

Ancestral mutations within or near the active site, but also in other parts of the protein, also act as stabilizing mutations that facilitate the acquisition of new functions by compensating for the destabilizing effects of active-site mutations. Indeed, ancestral mutations often overlap with "consensus" mutations that can be readily identified from sequence alignments without the need for an exact phylogenetic tree and ancestral reconstruction (Fig. 12.6). Restoring positions at which the target enzyme carries a different amino acid than most other family members to the consensus amino acid usually increases the protein's stability, and thereby boosts its evolvability (Bershtein et al., 2008). Stabilizing ancestral/consensus mutations have been applied to improve heterologous expression in *E. coli* (Khersonsky et al., 2009) and to increase thermostability (Jochens, Aerts, & Bornscheuer, 2010; Zhang, Yi, Pei, & Wu, 2010), and also to promote functional changes in evolving enzymes (Bershtein et al., 2008; Goldsmith et al., 2012; Khersonsky et al., 2012). Ancestor/consensus mutations can be included in rationally designed libraries (Goldsmith et al., 2012), or random mutation libraries (Khersonsky et al., 2012), and thereby boost the emergence of new variants. The critical role of stabilizing mutations has been demonstrated in the optimization by directed evolution of the computationally designed enzyme Kemp eliminase-dubbed KE59.

KE59 is the most catalytically efficient enzyme from a series of about 60 enzymes that were *de novo* designed to catalyze a reaction known as the Kemp elimination (Rothlisberger et al., 2008). The three most active designs were optimized by directed evolution (Khersonsky et al., 2010, 2011, 2012). Of these, KE59 exhibited the highest catalytic activity. However, KE59's stability was severely impaired—its soluble expression in *E. coli* was very poor and it could not be purified to homogeneity. The low stability of KE59 was somewhat surprising as it was generated by the redesign of a hyperthermophilic enzyme, and the design process aimed to minimize the protein's free energy. Indeed, other designed enzymes from the same series were found to be extremely stable (Khersonsky et al., 2010). Screening of random mutation libraries of KE59 (and of other unstable designs) gave no improved variants.

To enable KE59 to evolve, we included consensus mutations in the random mutation libraries. The sequence of the template enzyme from which KE59 was derived (indole-3-glycerolphophate synthase (IGPS) from

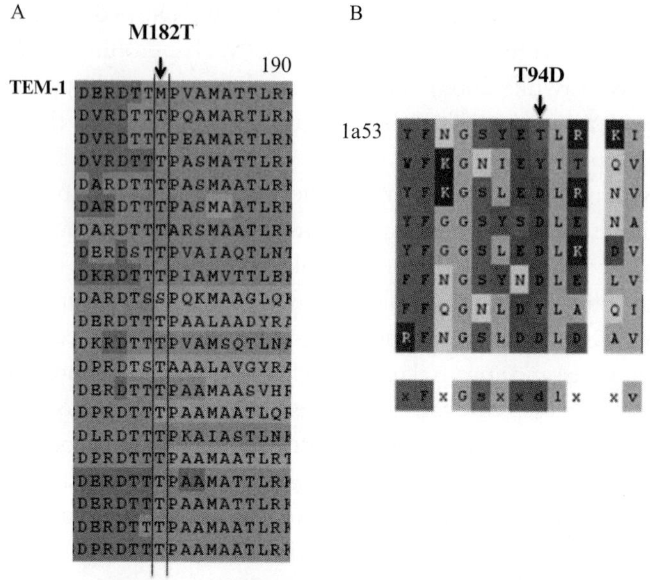

Figure 12.6 The identification of ancestral/consensus mutations. (A) A segment from the alignment of the TEM-1 β-lactamase family is displayed (for complete alignment and characterization see Bershtein, Goldin, & Tawfik, 2008). Highlighted is position 182 that differs in the starting point gene (Met, top sequence) from nearly all other family members. The mutation M182T restores the wild-type sequence both to its family consensus and to its predicted family ancestor sequence at this position. The incorporation of this mutation, and of other consensus mutations, into wild-type TEM-1 increases its stability and its evolvability (Bershtein et al., 2008). (B) A segment from the alignment of IGPS from *S. Solfataricus* used for the computational design of enzyme KE59 (Khersonsky et al., 2012) and of its related family members. Highlighted is position 94 (Thr, top sequence) in which the template and the designed KE59 carry a Thr, whereas the consensus is Asp. The mutation Thr94Asp was spiked into random mutation libraries of KE59 and was identified in every active variant isolated from these libraries, thus indicating its essential compensatory role. (See Color Insert.)

Sulfolobus Solfataricus) was aligned to orthologous sequences. In this way, several amino acids that deviate in KE59 from the family consensus were identified, as shown in Fig. 12.6. The consensus mutations were spiked into the libraries (see Section 4) to explore various combinations of stabilizing mutations and of other mutations from the random repertoire that improved the enzymatic function. The importance of this combinatorial incorporation approach was indicated by the fact that 10 out of 23 consensus mutations that were included in the library were incorporated into the evolved variants. The consensus mutations were mostly incorporated in the first two rounds of directed evolution. They boosted KE59's stability as indicated by >10-fold higher levels of soluble expression of the KE59 variants, and by higher T_m values relative to the designed KE59. The consensus mutations also enabled the accumulation of other mutations within and near KE59's active site, thus yielding >100-fold higher catalytic rates. Consensus, and other stabilizing mutations, were also included in the later rounds, altogether enabling >2000-fold increase in catalytic efficiency, with the most optimized KE59 variant exhibiting a k_{cat}/K_m value of $0.6 \times 10^6 M^{-1} s^{-1}$ (Khersonsky et al., 2012).

Stabilizing mutations can also be identified on the basis of high B-values in crystal structures. Highly mobile residues or segments (loops, etc.) on the enzyme's surface and away from the active site often contribute to instability (Giver, Gershenson, Freskgard, & Arnold, 1998). Mutating flexible residues has led to higher thermostability and resistance to organic solvents (Gumulya & Reetz, 2011; Reetz, Son, Fernandez, Gumulya, & Carballeira, 2010), and can also be applied to identify compensatory mutations that can be spiked into libraries. Another option is to initially select for higher stability, by screening variants that maintain activity at higher temperatures, or by identifying mutations that restore function following a strongly destabilizing mutation (Speck et al., 2012). The latter is a better option as this approach promotes the identification of compensatory mutations.

5.4. Optimizing efficiency and mutational loads of designed libraries

How effective are rational or analytical design strategies? And given their efficiency, what would be the optimal mutational loads? A simple yet informative way of assessing design effectiveness is to compare the fraction of beneficial mutations (i.e., mutations that, after screening, have proven advantageous for the target activity), and of deleterious mutations in the designed library to the corresponding fractions in randomly mutated gene libraries. Provided here are examples for such analyses of several designed libraries from our own work (Table 12.1). The fraction of deleterious mutations in these designed libraries has not been measured. However, by the most conservative approach, any mutation that did not appear in the most active variants isolated from a given round is considered deleterious. Beneficial mutations are mutations that were found in all improved variants in a given round and were also retained in subsequent rounds. The remaining fraction is dubbed neutral, and corresponds to mutations that sporadically appeared in active variants but were not fixed in subsequent rounds.

Table 12.1 Designed libraries: Design efficiency and mutational loads

	Library	Number of positions mutated[a]	Average number of amino acids per position	Total number of new mutations	Mutational load (average number of mutations per gene)[b]	Fraction of neutral mutations	Fraction of deleterious mutations (α)	Fraction of beneficial mutations (f^{+})
1	PON-GF-Round 5[c]	7	20	140	4	0.057	0.914	0.029
2	PON G-type Round 1[d]	8	4	29	4.2 ± 1.3	0.345	0.379	0.276
3	PON G-type Round 3[e]	12	4	46	2.4 ± 0.9	0.087	0.739	0.174
4	PON G-type Round 3[f]	8	5	44	0.6 ± 0.5	0.045	0.995	0
5	PON—Ancestral N6[g]	27	1	27	4 ± 2	0.277	0.259	0.444
6	PON—Ancestral N8[g]	21	1	21	4 ± 2	0.418	0.286	0.296
7	SULT—Ancestral N6[g]	24	1	24	5.1 ± 2.4	0.542	0.208	0.25

[a]The total number of diversified positions per gene.
[b]All libraries were constructed by mutational spiking using the ISOR protocol (Herman & Tawfik, 2007).
[c]Round 5 of directed evolution of PON1 (Gupta et al., 2011).
[d]Round 1 of directed evolution of PON1 (Goldsmith et al., 2012).
[e]Data form library 3.1 used in Round 3 of Goldsmith et al. (2012).
[f]Data from library 3.2 used in Round 3 of Goldsmith et al. (2012).
[g]Ancestral libraries from Alcolombri et al. (2011). Repetitive mutations in active variants were considered beneficial, nonrepetitive mutations were considered neutral, and mutations not observed in active variants were considered deleterious.

Our analysis indicates that the fraction of beneficial mutations (f^+; see model in Fig. 12.1) can be orders of magnitude higher in designed libraries (f^+ up to ~0.4 in ancestral libraries; Table 12.1) than in random libraries ($f^+ \approx 10^{-3}$; Fig. 12.1). However, the fraction of deleterious mutations (α) can be also much higher (up to 0.96, relative to $\leq$0.36 in random libraries). The high fraction of deleterious mutations is particularly dominant in "brute force libraries"—libraries constructed by NNS diversification to all 20 amino acids (Table 12.1, entry #1), and also "bad luck libraries", namely, libraries that failed to target the right positions (Table 12.1, entry #4). Thus, restricting the mutational load in designed libraries is as crucial in designed libraries as in random libraries (Fig. 12.7). When the fraction of deleterious mutations is high, libraries are easily "intoxicated" such that active variants become very scarce. High-throughput screening can solve this problem (e.g., the library described in entry #1, Table 12.1, was sorted by FACS). However, the inevitable conclusion from this analysis is that lower mutational loads are generally preferable. In fact, the simulations presented in Fig. 12.7 indicate that in the library described in entry #1, a mutational load of $\leq$2 mutations per gene is likely to have yielded a better result. This conclusion is also supported by the fact that the most active variants pooled out

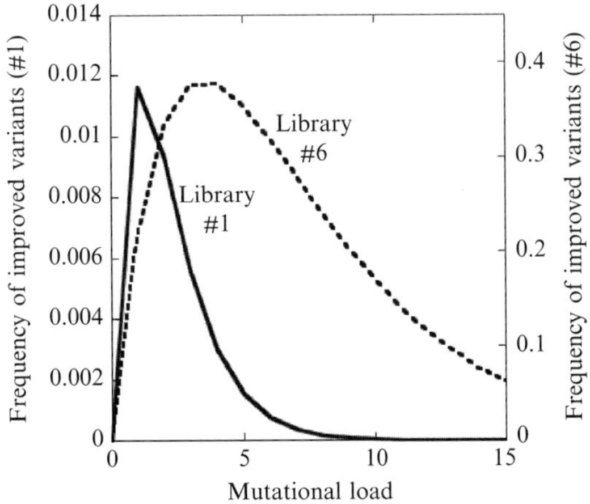

Figure 12.7 Optimizing the library's mutational load. The model described in Fig. 12.1 was applied to calculate the frequency of variants with improved activity in the designed libraries described in Table 12.1 at different mutational loads, based on their fractions of deleterious and beneficial mutations. The solid black line simulates the rationally designed substitution library in entry #1 (Fraction of beneficial mutations is 0.029 and of deleterious one, 0.914). The dashed line represents the ancestral library in entry #6 (Fraction of beneficial mutations is 0.296 and of deleterious one, 0.286). The optimal mutational load for Library #1 is ~2 mutations per gene versus ~5 mutations per gene for Library #6. Note the different scales for the frequency of improved variants.

of these libraries (entries #1–4) carried on average two beneficial mutations per round (Goldsmith et al., 2012) in spite of different mutational loads. Overall, given the limited diversities that can be screened, and the ratio of beneficial to deleterious mutations, beneficial combinations are not likely to exceed two mutated positions.

The ancestral libraries represent a clear exception to the low mutational loads rule (entries #5–7). The fraction of beneficial mutation is remarkably high, and that of deleterious mutations is smaller than in random libraries. Active variants isolated from these libraries carried 4–7 mutations (Alcolombri et al., 2011), and thus, the relatively high mutational load (~5 mutations per gene) was perfectly adequate (Fig. 12.7).

6. SUMMARY

The success of directed evolution experiments aimed at increasing a desired enzymatic activity can be greatly improved by screening small and targeted libraries of mutants with moderate mutational loads. The gradual increase in catalytic efficiency by iterative rounds of directed evolution is driven by two factors: the applied screening or selection method, and the library design. The former relates to the stringency, accuracy, and relevance of the screening or selection assay, and the latter requires a combination of knowledge, experience, and intuition. We have attempted to formulate some guidelines for screening assays and library design based on our own successes and failures. We advocate for medium-throughput screens that are direct (screening for the targeted activity and substrate) and accurate. For library design, we show strategies by which catalytic residues, and their first- and second-shell neighbors, are optimized. We specifically outline the principles of "neighbor joining" and "revisiting key positions" in library designs. Ancestral and consensus mutations are also included in our library designs as they provide both increased stability and activity. Finally, we emphasize the importance of tuning the library's mutational load, and outline the merits of combinatorial spiking approaches that control the mutational load while affording diverse libraries enriched with active variants.

REFERENCES

Agresti, J. J., Antipov, E., Abate, A. R., Ahn, K., Rowat, A. C., Baret, J. C., et al. (2010). Ultrahigh-throughput screening in drop-based microfluidics for directed evolution. *Proceedings of the National Academy of Sciences of the United States of America*, *107*, 4004–4009.

Alcolombri, U., Elias, M., & Tawfik, D. S. (2011). Directed evolution of sulfotransferases and paraoxonases by ancestral libraries. *Journal of Molecular Biology, 411*, 837–853.

Amitai, G., Adani, R., Yacov, G., Yishay, S., Teitlboim, S., Tveria, L., et al. (2007). Asymmetric fluorogenic organophosphates for the development of active organophosphate hydrolases with reversed stereoselectivity. *Toxicology, 233*, 187–198.

Bar-Even, A., Noor, E., Savir, Y., Liebermeister, W., Davidi, D., Tawfik, D. S., et al. (2011). The moderately efficient enzyme: Evolutionary and physicochemical trends shaping enzyme parameters. *Biochemistry, 50*, 4402–4410.

Bawazer, L. A., et al. (2012). Evolutionary selection of enzymatically synthesized semiconductors from biomimetic mineralization vesicles. *Proceedings of the National Academy of Sciences of the United States of America, 109*, E1705–1714. http://dx.doi.org/10.1073/pnas.1116958109.

Ben-David, M., Elias, M., Filippi, J. J., Dunach, E., Silman, I., Sussman, J. L., et al. (2012). Catalytic versatility and backups in enzyme active sites: The case of serum paraoxonase 1. *Journal of Molecular Biology, 418*, 181–196.

Ben-David, A., Shoham, G., & Shoham, Y. (2008). A universal screening assay for glycosynthases: Directed evolution of glycosynthase XynB2(E335G) suggests a general path to enhance activity. *Chemistry & Biology, 15*, 546–551.

Berger, I., Guttman, C., Amar, D., Zarivach, R., & Aharoni, A. (2011). The molecular basis for the broad substrate specificity of human sulfotransferase 1A1. *PloS One, 6*, e26794.

Bershtein, S., Goldin, K., & Tawfik, D. S. (2008). Intense neutral drifts yield robust and evolvable consensus proteins. *Journal of Molecular Biology, 379*, 1029–1044.

Bershtein, S., Segal, M., Bekerman, R., Tokuriki, N., & Tawfik, D. S. (2006). Robustness-epistasis link shapes the fitness landscape of a randomly drifting protein. *Nature, 444*, 929–932.

Bershtein, S., & Tawfik, D. S. (2008). Ohno's model revisited: Measuring the frequency of potentially adaptive mutations under various mutational drifts. *Molecular Biology and Evolution, 25*, 2311–2318.

Bloom, J. D., Labthavikul, S. T., Otey, C. R., & Arnold, F. H. (2006). Protein stability promotes evolvability. *Proceedings of the National Academy of Sciences of the United States of America, 103*, 5869–5874.

Brannigan, J. A., & Wilkinson, A. J. (2002). Protein engineering 20 years on. *Nature Reviews. Molecular Cell Biology, 3*, 964–970.

Briseno-Roa, L., Hill, J., Notman, S., Sellers, D., Smith, A. P., Timperley, C. M., et al. (2006). Analogues with fluorescent leaving groups for screening and selection of enzymes that efficiently hydrolyze organophosphorus nerve agents. *Journal of Medicinal Chemistry, 49*, 246–255.

Brouk, M., Nov, Y., & Fishman, A. (2010). Improving biocatalyst performance by integrating statistical methods into protein engineering. *Applied and Environmental Microbiology, 76*, 6397–6403.

Chen, I., Dorr, B. M., & Liu, D. R. (2011). A general strategy for the evolution of bond-forming enzymes using yeast display. *Proceedings of the National Academy of Sciences of the United States of America, 108*, 11399–11404.

Chen, F., Gaucher, E. A., Leal, N. A., Hutter, D., Havemann, S. A., Govindarajan, S., et al. (2010). Reconstructed evolutionary adaptive paths give polymerases accepting reversible terminators for sequencing and SNP detection. *Proceedings of the National Academy of Sciences of the United States of America, 107*, 1948–1953.

Chen, M. M., Snow, C. D., Vizcarra, C. L., Mayo, S. L., & Arnold, F. H. (2012). Comparison of random mutagenesis and semi-rational designed libraries for improved cytochrome P450 BM3-catalyzed hydroxylation of small alkanes. *Protein Engineering, Design & Selection, 25*, 171–178.

Chica, R. A., Moore, M. M., Allen, B. D., & Mayo, S. L. (2010). Generation of longer emission wavelength red fluorescent proteins using computationally designed libraries. *Proceedings of the National Academy of Sciences of the United States of America, 107*, 20257–20262.

Cole, M. F., & Gaucher, E. A. (2011). Exploiting models of molecular evolution to efficiently direct protein engineering. *Journal of Molecular Evolution, 72*, 193–203.

diTargiani, R. C., Chandrasekaran, L., Belinskaya, T., & Saxena, A. (2010). In search of a catalytic bioscavenger for the prophylaxis of nerve agent toxicity. *Chemico-Biological Interactions, 187*, 349–354.

Fasan, R., Meharenna, Y. T., Snow, C. D., Poulos, T. L., & Arnold, F. H. (2008). Evolutionary history of a specialized P450 propane monooxygenase. *Journal of Molecular Biology, 383*, 1069–1080.

Fox, R. J., Davis, S. C., Mundorff, E. C., Newman, L. M., Gavrilovic, V., Ma, S. K., et al. (2007). Improving catalytic function by ProSAR-driven enzyme evolution. *Nature Biotechnology, 25*, 338–344.

Giver, L., Gershenson, A., Freskgard, P. O., & Arnold, F. H. (1998). Directed evolution of a thermostable esterase. *Proceedings of the National Academy of Sciences of the United States of America, 95*, 12809–12813.

Goldsmith, M., Ashani, Y., Simo, Y., Ben-David, M., Leader, H., Silman, I., et al. (2012). Evolved stereoselective hydrolases for broad-spectrum g-type nerve agent detoxification. *Chemistry & Biology, 19*, 456–466.

Granieri, L., Baret, J. C., Griffiths, A. D., & Merten, C. A. (2010). High-throughput screening of enzymes by retroviral display using droplet-based microfluidics. *Chemistry & Biology, 17*, 229–235.

Gumulya, Y., & Reetz, M. T. (2011). Enhancing the thermal robustness of an enzyme by directed evolution: Least favorable starting points and inferior mutants can map superior evolutionary pathways. *Chembiochem, 12*, 2502–2510.

Gupta, R. D., Goldsmith, M., Ashani, Y., Simo, Y., Mullokandov, G., Bar, H., et al. (2011). Directed evolution of hydrolases for prevention of G-type nerve agent intoxication. *Nature Chemical Biology*, 7, 120–125.

Guthrie, V. B., Allen, J., Camps, M., & Karchin, R. (2011). Network models of TEM beta-lactamase mutations coevolving under antibiotic selection show modular structure and anticipate evolutionary trajectories. *PLoS Computational Biology*, 7, e1002184.

Hardiman, E., Gibbs, M., Reeves, R., & Bergquist, P. (2010). Directed evolution of a thermophilic beta-glucosidase for cellulosic bioethanol production. *Applied Biochemistry and Biotechnology, 161*, 301–312.

He, L., Friedman, A. M., & Bailey-Kellogg, C. (2012). A divide-and-conquer approach to determine the Pareto frontier for optimization of protein engineering experiments. *Proteins, 80*, 790–806.

Herman, A., & Tawfik, D. S. (2007). Incorporating synthetic oligonucleotides via gene reassembly (ISOR): A versatile tool for generating targeted libraries. *Protein Engineering, Design & Selection, 20*, 219–226.

Hill, C. M., Li, W. S., Thoden, J. B., Holden, H. M., & Raushel, F. M. (2003). Enhanced degradation of chemical warfare agents through molecular engineering of the phosphotriesterase active site. *Journal of the American Chemical Society, 125*, 8990–8991.

Hoffmann, G., Bonsch, K., Greiner-Stoffele, T., & Ballschmiter, M. (2011). Changing the substrate specificity of P450cam towards diphenylmethane by semi-rational enzyme engineering. *Protein Engineering, Design & Selection, 24*, 439–446.

Hokanson, C. A., Cappuccilli, G., Odineca, T., Bozic, M., Behnke, C. A., Mendez, M., et al. (2011). Engineering highly thermostable xylanase variants using an enhanced combinatorial library method. *Protein Engineering, Design & Selection, 24*, 597–605.

Jochens, H., Aerts, D., & Bornscheuer, U. T. (2010). Thermostabilization of an esterase by alignment-guided focussed directed evolution. *Protein Engineering, Design & Selection, 23*, 903–909.

Jochens, H., & Bornscheuer, U. T. (2010). Natural diversity to guide focused directed evolution. *Chembiochem, 11*, 1861–1866.

Khare, S. D., Kipnis, Y., Greisen, P., Jr., Takeuchi, R.,. Ashani, Y., Goldsmith, M., et al. (2012). Computational redesign of a mononuclear zinc metalloenzyme for organophosphate hydrolysis. *Nature Chemical Biology, 8*, 294–300.

Khersonsky, O., Kiss, G., Rothlisberger, D., Dym, O., Albeck, S., Houk, K. N., et al. (2012). Bridging the gaps in design methodologies by evolutionary optimization of the stability and proficiency of designed Kemp eliminase KE59. *Proceedings of the National Academy of Sciences of the United States of America, 109*, 10358–10363.

Khersonsky, O., Rosenblat, M., Toker, L., Yacobson, S., Hugenmatter, A., Silman, I., et al. (2009). Directed evolution of serum paraoxonase PON3 by family shuffling and ancestor/consensus mutagenesis, and its biochemical characterization. *Biochemistry, 48*, 6644–6654.

Khersonsky, O., Rothlisberger, D., Dym, O., Albeck, S., Jackson, C. J., Baker, D., et al. (2010). Evolutionary optimization of computationally designed enzymes: Kemp eliminases of the KE07 series. *Journal of Molecular Biology, 396*, 1025–1042.

Khersonsky, O., Rothlisberger, D., Wollacott, A. M., Murphy, P., Dym, O., Albeck, S., et al. (2011). Optimization of the in-silico-designed kemp eliminase KE70 by computational design and directed evolution. *Journal of Molecular Biology, 407*, 391–412.

Kintses, B., van Vliet, L. D., Devenish, S. R. A., & Hollfelder, F. (2010). Microfluidic droplets: New integrated workflows for biological experiments. *Current Opinion in Chemical Biology, 14*, 548–555.

Kintses, B., et al. (2012). Picoliter cell lysate assays in microfluidic droplet compartments for directed enzyme evolution. *Chemistry & Biology, 19*, 1001–1009. http://dx.doi.org/10.1016/j.chembiol.2012.06.009.

Kornberg, A. (1989). *For the love of enzymes*. Cambridge, London: Harvard University Press.

Kuipers, R. K., Joosten, H. J., van Berkel, W. J., Leferink, N. G., Rooijen, E., Ittmann, E., et al. (2010). 3DM: Systematic analysis of heterogeneous superfamily data to discover protein functionalities. *Proteins, 78*, 2101–2113.

Liu, L., Li, Y., Liotta, D., & Lutz, S. (2009). Directed evolution of an orthogonal nucleoside analog kinase via fluorescence-activated cell sorting. *Nucleic Acids Research, 37*, 4472–4481.

Ma, B. G., & Berezovsky, I. N. (2010). The MBLOSUM: A server for deriving mutation targets and position-specific substitution rates. *Journal of Biomolecular Structure & Dynamics, 28*, 415–419.

Moretti, R., Chang, A., Peltier-Pain, P., Bingman, C. A., Phillips, G. N., Jr., & Thorson, J. S. (2011). Expanding the nucleotide and sugar 1-phosphate promiscuity of nucleotidyltransferase RmlA via directed evolution. *The Journal of Biological Chemistry, 286*, 13235–13243.

Parera, M., Perez-Alvarez, N., Clotet, B., & Martinez, M. A. (2009). Epistasis among deleterious mutations in the HIV-1 protease. *Journal of Molecular Biology, 392*, 243–250.

Pavelka, A., Chovancova, E., & Damborsky, J. (2009). HotSpot wizard: A web server for identification of hot spots in protein engineering. *Nucleic Acids Research, 37*, W376–W383.

Privett, H. K., et al. (2012). Iterative approach to computational enzyme design. *Proceedings of the National Academy of Sciences of the United States of America, 109*, 3790–3795.

Reetz, M. T. (2011). Laboratory evolution of stereoselective enzymes: A prolific source of catalysts for asymmetric reactions. *Angewandte Chemie, 50*, 138–174.

Reetz, M. T., Prasad, S., Carballeira, J. D., Gumulya, Y., & Bocola, M. (2010). Iterative saturation mutagenesis accelerates laboratory evolution of enzyme stereoselectivity:

Rigorous comparison with traditional methods. *Journal of the American Chemical Society*, *132*, 9144–9152.

Reetz, M. T., & Sanchis, J. (2008). Constructing and analyzing the fitness landscape of an experimental evolutionary process. *Chembiochem*, *9*, 2260–2267.

Reetz, M. T., Soni, P., Fernandez, L., Gumulya, Y., & Carballeira, J. D. (2010). Increasing the stability of an enzyme toward hostile organic solvents by directed evolution based on iterative saturation mutagenesis using the B-FIT method. *Chemical Communications*, *46*, 8657–8658.

Romero, P. A., & Arnold, F. H. (2009). Exploring protein fitness landscapes by directed evolution. *Nature Reviews. Molecular Cell Biology*, *10*, 866–876.

Rothlisberger, D., Khersonsky, O., Wollacott, A. M., Jiang, L., DeChancie, J., Betker, J., et al. (2008). Kemp elimination catalysts by computational enzyme design. *Nature*, *453*, 190–195.

Salverda, M. L., Dellus, E., Gorter, F. A., Debets, A. J., van der Oost, J., Hoekstra, R. F., et al. (2011). Initial mutations direct alternative pathways of protein evolution. *PLoS Genetics*, *7*, e1001321.

Scanlon, T. C., Teneback, C. C., Gill, A., Bement, J. L., Weiner, J. A., Lamppa, J. W., et al. (2010). Enhanced antimicrobial activity of engineered human lysozyme. *ACS Chemical Biology*, *5*, 809–818.

Shim, J.-u., Olguin, L. F., Whyte, G., Scott, D., Babtie, A., Abell, C., et al. (2009). Simultaneous determination of gene expression and enzymatic activity in individual bacterial cells in microdroplet compartments. *Journal of the American Chemical Society*, *131*, 15251–15256.

Soskine, M., & Tawfik, D. S. (2010). Mutational effects and the evolution of new protein functions. *Nature Reviews Genetics*, *11*, 572–582.

Speck, J., Hecky, J., Tam, H. K., Arndt, K. M., Einsle, O., & Muller, K. M. (2012). Exploring the molecular linkage of protein stability traits for enzyme optimization by iterative truncation and evolution. *Biochemistry*, *51*, 4850–4867.

Stapleton, J. A., & Swartz, J. R. (2010). Development of an in vitro compartmentalization screen for high-throughput directed evolution of [FeFe] hydrogenases. *PloS One*, *5*, e15275.

Theriot, C. M., & Grunden, A. M. (2011). Hydrolysis of organophosphorus compounds by microbial enzymes. *Applied Microbiology and Biotechnology*, *89*, 35–43.

Tokuriki, N., Stricher, F., Serrano, L., & Tawfik, D. S. (2008). How protein stability and new functions trade off. *PLoS Computational Biology*, *4*, e1000002.

Tokuriki, N., & Tawfik, D. S. (2009). Chaperonin overexpression promotes genetic variation and enzyme evolution. *Nature*, *459*, 668–673.

Tokuriki, N., et al. (2012). Diminishing returns and tradeoffs constrain the laboratory optimization of an enzyme. *Nature Communications*, *3*, 1257.

Toth-Petroczy, A., & Tawfik, D. S. (2011). Slow protein evolutionary rates are dictated by surface-core association. *Proceedings of the National Academy of Sciences of the United States of America*, *108*, 11151–11156.

Tsai, P. C., Bigley, A., Li, Y., Ghanem, E., Cadieux, C. L., Kasten, S. A., et al. (2010). Stereoselective hydrolysis of organophosphate nerve agents by the bacterial phosphotriesterase. *Biochemistry*, *49*, 7978–7987.

Tsai, P. C., et al. (2012). Enzymes for the homeland defense: optimizing phosphotriesterase for the hydrolysis of organophosphate nerve agents. *Biochemistry*, *51*, 6463–6475. http://dx.doi.org/10.1021/bi300811t.

van Leeuwen, J. G., Wijma, H. J., Floor, R. J., van der Laan, J. M., & Janssen, D. B. (2012). Directed evolution strategies for enantiocomplementary haloalkane dehalogenases: From chemical waste to enantiopure building blocks. *Chembiochem*, *13*, 137–148.

Varadarajan, N., Cantor, J. R., Georgiou, G., & Iverson, B. L. (2009). Construction and flow cytometric screening of targeted enzyme libraries. *Nature Protocols*, *4*, 893–901.

Varadarajan, N., Pogson, M., Georgiou, G., & Verson, B. L. (2009). Proteases that can distinguish among different post-translational forms of tyrosine engineered using multicolor flow cytometry. *Journal of the American Chemical Society*, *131*, 18186–18190.

Wang, X., Minasov, G., & Shoichet, B. K. (2002). Evolution of an antibiotic resistance enzyme constrained by stability and activity trade-offs. *Journal of Molecular Biology*, *320*, 85–95.

Watts, K. T., Mijts, B. N., Lee, P. C., Manning, A. J., & Schmidt-Dannert, C. (2006). Discovery of a substrate selectivity switch in tyrosine ammonia-lyase, a member of the aromatic amino acid lyase family. *Chemistry & Biology*, *13*, 1317–1326.

Weinreich, D. M., Watson, R. A., & Chao, L. (2005). Perspective: Sign epistasis and genetic constraint on evolutionary trajectories. *Evolution*, *59*, 1165–1174.

Wu, S., Acevedo, J. P., & Reetz, M. T. (2010). Induced allostery in the directed evolution of an enantioselective Baeyer-Villiger monooxygenase. *Proceedings of the National Academy of Sciences of the United States of America*, *107*, 2775–2780.

Yang, G., Rich, J. R., Gilbert, M., Wakarchuk, W. W., Feng, Y., & Withers, S. G. (2010). Fluorescence activated cell sorting as a general ultra-high-throughput screening method for directed evolution of glycosyltransferases. *Journal of the American Chemical Society*, *132*, 10570–10577.

Zhang, Z. G., Yi, Z. L., Pei, X. Q., & Wu, Z. L. (2010). Improving the thermostability of Geobacillus stearothermophilus xylanase XT6 by directed evolution and site-directed mutagenesis. *Bioresource Technology*, *101*, 9272–9278.

CHAPTER THIRTEEN

Generation of High-Performance Binding Proteins for Peptide Motifs by Affinity Clamping

Shohei Koide[*,1], **Jin Huang**[†]

[*]Department of Biochemistry and Molecular Biology, The University of Chicago, Chicago, Illinois, USA
[†]Beijing Prosperous Biopharm Co., Ltd., Beijing, China
[1]Corresponding author: e-mail address: skoide@uchicago.edu

Contents

Abstract

We describe concepts and methodologies for generating "Affinity Clamps," a new class of recombinant binding proteins that achieve high affinity and high specificity toward short peptide motifs of biological importance, which is a major challenge in protein engineering. The Affinity Clamping concept exploits the potential of nonhomologous recombination of protein domains in generating large changes in protein function and the inherent binding affinity and specificity of the so-called modular interaction domains toward short peptide motifs. Affinity Clamping creates a clamshell architecture that clamps onto a target peptide. The design processes involve (i) choosing a starting modular interaction domain appropriate for the target and applying structure-guided modifications; (ii) attaching a second domain, termed "enhancer domain"; and (iii) optimizing the peptide-binding site located between the domains by directed evolution. The two connected domains work synergistically to achieve high levels of affinity

Methods in Enzymology, Volume 523
ISSN 0076-6879
http://dx.doi.org/10.1016/B978-0-12-394292-0.00013-8

and specificity that are unattainable with either domain alone. Because of the simple and modular architecture, Affinity Clamps are particularly well suited as building blocks for designing more complex functionalities. Affinity Clamping represents a major advance in protein design that is broadly applicable to the recognition of peptide motifs.

1. INTRODUCTION

Generating proteins that perform novel molecular-recognition functions has been a major goal of protein design. The central approach in the field has been to choose a starting "molecular scaffold" and to introduce modifications that create surfaces suitable for the new function (Koide, 2010). This concept is most evidently illustrated in the mechanism by which natural and synthetic antibodies are generated, where the amino acid sequences of a set of surface loops, or the complementarity determining regions, are diversified to generate a large ensemble from which functional molecules are selected (Sidhu & Fellouse, 2006; Winter, Griffiths, Hawkins, & Hoogenboom, 1994). Most structure-guided and computational design of binding sites also follows this concept (Kortemme & Baker, 2004).

Although the current protein design strategies are reasonably successful in generating binding interfaces that recognize highly structured protein molecules, or the "domainome" portion of the proteome (Colwill & Graslund, 2011), these strategies have not been particularly effective in generating binding interfaces for the "peptidome" portion, comprised of short, unstructured peptide segments (Cobaugh, Almagro, Pogson, Iverson, & Georgiou, 2008). Such short peptide motifs that are present in the context of larger proteins, in particular, those harboring a posttranslational modification, are important "hubs" in signal transduction and epigenetics. However, most affinity reagents currently in use for this class of targets are polyclonal antibodies that are inherently undefined, undercharacterized, and not renewable. The difficulty of designing binding interfaces for short peptides lies in the fundamental thermodynamics where a large entropic cost of immobilizing a flexible peptide must be offset by forming a small number of interactions.

Another challenge in protein design is to generate binding interfaces that recognize a predefined epitope within a targeted molecule. For example, when one generates an antibody to a protein target, the epitope of the antibody is typically poorly defined, and its precise definition requires laborious methods (Parmley & Smith, 1989; Rockberg, Lofblom, Hjelm, Uhlen, & Stahl, 2008). This situation stands in a stark contrast to the recognition of short nucleic acid motifs, where one can readily design an oligonucleotide that recognizes a predefined "epitope" within a much larger genome simply using sequence information. A major challenge in protein design has been to develop a platform technology that rapidly creates binding interfaces for a predefined peptide motif. Developing such a technology would enable systematic generation of affinity reagents on a proteomic scale with epitopes known *a priori*. Clearly, such a system would have a strong impact on biomedical research, biotechnology, and medicine.

2. THE AFFINITY CLAMPING CONCEPT

We have developed a new protein-engineering concept, termed "Affinity Clamping" to systematically generate capture reagents directed to predefined peptide motifs (Huang, Koide, Makabe, & Koide, 2008; Koide, 2009). This concept was inspired by a series of observations regarding structure–function relationships and the evolution of proteins. In the evolution of natural proteins, nonhomologous recombination of domains is considered the main driving force for a large leap in function (Bashton & Chothia, 2007). Comparative structural analyses suggest that more specific and complex protein functions are achieved by two or multidomain proteins with the active site frequently occurring at the interface between domains (Rossmann, Moras, & Olsen, 1974). Many proteins that act on peptides, for example, enzymes, fall in this category. There exist numerous modular interaction domains that bind to particular classes of peptide motifs, but their binding sites are small and shallow grooves located on the protein surface, which explains the low levels of affinity and specificity of the interactions between modular interaction domains and their respective target peptide motifs (Pawson & Nash, 2003).

The observations discussed above suggest that one could design a high-performance binding protein for peptide motifs by first connecting two nonhomologous domains and then optimizing the surfaces at the newly formed interface between them. Specifically, in Affinity Clamping, a natural peptide-binding domain (referred to as a "primary domain") is combined with another unrelated domain (an "enhancer domain"), followed by directed evolution of the surfaces of the "enhancer domain" that are expected to contact the peptide bound to the primary domain (Fig. 13.1) (Huang et al., 2008). The enhancer domain is essentially an antibody mimic, in that its role is to recognize the target peptide presented on the surface of

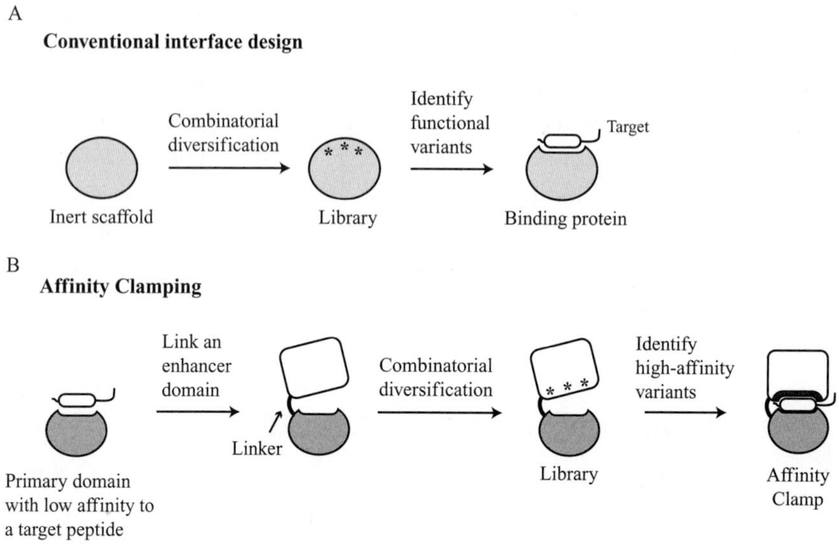

Figure 13.1 Schematics comparing conventional design of protein interaction interfaces (A) with Affinity Clamping (B). In conventional protein interface design, surfaces of an inert scaffold are diversified to produce an ensemble of variants (library) from which functional variants are identified. In Affinity Clamping, a framework for constructing new protein architecture is designed by attaching an inert enhancer domain to a primary domain that has weak affinity to a desired class of targets. Next, surfaces of the enhancer domain are improved by directed evolution, resulting in the formation of a clamshell architecture that provides a much larger interface for the peptide target than the interface between the primary domain and the target.

the primary domain. The resulting binding proteins, termed "Affinity Clamps," with their clamshell architecture would clamp on the target peptides.

We have successfully applied Affinity Clamping to a PDZ domain as the primary domain and dramatically enhanced its target recognition function (Huang, Makabe, Biancalana, Koide, & Koide, 2009; Huang et al., 2008). The binding affinity and specificity of the resulting Affinity Clamps were respectively >500-fold (to single nanomolar K_d values) and >2000-fold higher than those of the primary domain alone. The PDZ clamps were produced at high levels in *Escherichia coli* and were highly stable. They outperformed commercial monoclonal antibodies to the same target in Western blotting. These results validated the Affinity Clamping strategy in producing high-performance binding proteins.

Affinity Clamping addresses two complementary challenges in generating a high-affinity binding site for a short, flexible peptide. First, the Affinity

Clamp architecture creates peptide-interacting surfaces that are much larger than those of the primary domain alone, thereby enabling higher affinity and higher specificity. Second, the preexisting peptide-binding site of the primary domain provides weak but significant affinity and specificity for the targeted peptide motif, and thus Affinity Clamping converts the challenge of *de novo* generation of peptide-recognition surfaces into maturation of affinity and specificity.

3. DESIGN OF AFFINITY CLAMPS

The major breakthrough associated with Affinity Clamping is more conceptual than technical. Thus, we will focus this chapter primarily on descriptions of key concepts and precautions pertinent to setting up an Affinity Clamping project.

3.1. Choice of the primary domain

Affinity Clamping takes advantage of the fact that many small domains exist in nature that bind to a particular class of short peptide motif. Many complex signal transduction behaviors are mediated by regulatory proteins with highly modular architectures (Pawson & Nash, 2003). These proteins are built from linear combinations of individually folded domains (or "modules"), each of which performs a specific function. Close to 100 commonly occurring modular domains have been identified in eukaryotic proteins. Among these, the interaction domains are small proteins, usually less than 100 amino acid residues in length, and are independently folded and stable. Examples include PDZ, WW, and SH2 domains. These interaction domains usually bind to short peptides containing a chemical signature, for example, free C-terminus, phosphorylation, or other posttranslational modification. Importantly, the target peptide binds to a shallow cleft on the surface of these interaction domains (Pawson & Nash, 2003), and consequently, approximately half of the peptide surface is still exposed for making additional interactions with the enhancer domain.

Although it is straightforward to choose a particular domain family as a potential primary domain for Affinity Clamping (e.g., PDZ domain for the recognition of protein C-termini), it is not trivial to select a specific domain within a family. Because Affinity Clamps will ultimately be used in practical applications, they should be robust and easy to produce. Therefore, biophysical and chemical properties such as conformational stability, solubility, and the absence of reactive Cys residues are important aspects to consider, in addition to binding specificity. Such information is often unavailable in the literature and one may need to use guesswork from available, circumstantial data. For example, the availability of crystallographic and/or NMR structures is a good indication that the domain of interest has good biophysical properties. Several recent studies have systematically and comprehensively characterized functional and biophysical properties of domain family members (Machida et al., 2007; Tonikian et al., 2008). Clearly, such studies are highly valuable for choosing a primary domain.

3.2. Molecular display of the primary domain

Affinity Clamping requires the use of a directed evolution technology, such as phage display or yeast surface display (Koide, Koide, & Lipovsek, 2012), to produce large sequence diversity in the enhancer domain and to identify clones from such a repertoire. Therefore, a primary domain of choice must be compatible with the directed evolution technology of choice. Here, we will limit our discussion to phage display, although the concept is applicable to other directed evolution technologies.

The first test is to confirm that the primary domain of choice is robustly displayed on the surface of phages. Briefly, the gene for the domain is cloned in a phage display vector and an epitope tag, such as the V5 or FLAG tag, is added adjacent to the domain (Wojcik et al., 2010). The gene for the primary domain can be obtained from a natural source or can be synthesized. As the cost of gene synthesis continues to decline, it is often beneficial to use a synthetic gene that can be easily optimized for codon usage and for the inclusion and exclusion of convenient restriction enzyme sites. After confirming the DNA sequence of the phage display vector, phage particles are produced, and the level of surface display of the cloned domain is tested by performing phage ELISA against an anti-epitope tag antibody (Sidhu, Lowman, Cunningham, & Wells, 2000).

After surface display of the cloned primary domain is confirmed, phage ELISA is repeated using a peptide known to bind to the domain as the target. A peptide is synthesized with a biotin attached at a terminus (see Section 3.6 for target preparation) and the biotinylated peptide is immobilized to microtiter wells that have been coated with streptavidin and blocked with BSA. Because the interaction between the domain and peptide is generally weak, the phage ELISA signals may be much weaker than those for phage ELISA with an anti-epitope tag antibody. Still, it is important to confirm functional display of a domain of interest on the phage surface.

Although molecular display technologies for antibody fragments (single-chain Fv and Fab) have been refined over the years to be quite robust, it is important to recognize that one must fine-tune a molecular display system for individual proteins. Thus, extensive experimentation may be required to achieve a good level of surface display for the primary domain of choice. Important parameters include plasmid copy number, the promoter that drives the expression of the domain fused with a display partner (e.g., the phage p3 protein), signal sequence, growth temperature, aeration, induction method, and the length of expression. See Supplementary Materials of Wojcik et al. (2010) for an example of systematic optimization of a phage display system.

If functional display of the primary domain cannot be confirmed after adjusting experimental parameters, one needs to make a strategic decision. One may wish to try directed evolution of the chosen primary domain or test other family members with similar function. Also, it is useful to experimentally characterize the conformational stability of the chosen domain in the exact format used for molecular display, because the stability of protein domains is often sensitive to the choice of boundary positions (Hamill, Meekhof, & Clarke, 1998). The definition of a domain discerned based on homology can be too strict, and constructs designed based on such a definition may eliminate terminal residues critical for maintaining stability. Together, one must realize that establishing a molecular display system for a new protein can be a lengthy and laborious process, and it often requires intimate knowledge of the function, structure, and stability of the protein, before embarking an Affinity Clamping project.

3.3. Choice of the enhancer domain

The primary role of the enhancer domain is to present a binding surface of sufficient size and good complementarity for forming the interaction with a peptide target bound on the primary domain (Fig. 13.1). Because most Affinity Clamping projects would aim to generate high-affinity and high-specificity Affinity Clamps to the multitude of targets that the primary domain can recognize, the enhancer domain should be able to present diverse shapes and chemistry. This requirement is essentially identical to that for establishing an antibody library or an antibody-mimic library. Because Affinity Clamps contain a primary domain with preexisting affinity to a class of target peptides, the level of affinity enhancement required from an enhancer domain is much lower than that required to generate *de novo* target binding using the enhancer domain alone. This reduced requirement may give protein designers greater freedom in choosing an enhancer domain and/or in designing combinatorial libraries. Affinity Clamps to be developed should ideally have several attractive attributes so as to make them versatile and convenient tools. Such attributes include small size, high stability, the absence of disulfide bonds, ease of production, and compatibility with molecular display methods (Binz, Amstutz, & Pluckthun, 2005; Koide, 2010). Consideration of these attributes can help a protein designer select an enhancer domain suitable for her/his methodologies.

The so-called nonantibody scaffolds serve as particularly attractive candidates for enhancer domains. A number of such nonantibody scaffolds have been developed for the purpose of generating antibody-like, target-binding proteins that do not inherit undesirable properties associated with conventional antibodies (Binz et al., 2005; Koide, 2010). A diverse array of protein scaffolds, including fibronectin type III (FN3) domains (monobodies), designed ankyrin repeat proteins (DARPins), lipocalins (anticalins), and helix-bundle proteins (affibodies), have been established as viable platforms for generating high-performance binding proteins. In principle, any of these nonantibody scaffolds could be used as enhancer domains.

We have successfully used a human FN3 domain as the enhancer domain in constructing Affinity Clamps (Huang, Makabe, et al., 2009; Huang et al., 2008). FN3 is a small ($\sim$10 kDa) immunoglobulin-like β-sandwich protein with three loops located at one end of the molecule that can be extensively diversified to create a library of binding surfaces. Since the Koide group introduced it as a protein scaffold in 1998 (Koide, Bailey, Huang, & Koide, 1998), the FN3 domain has become the most widely used nonantibody scaffold and has been used to generate binders against a wide range of targets (Koide et al., 2012). FN3 has a highly stable core and it lacks disulfide bonds. FN3 domains are often found in multidomain proteins, and thus, FN3 is inherently capable of functioning in the presence of another domain fused to either of its termini. A large number of crystal structures of FN3-based binding proteins, termed monobodies, have revealed inherent plasticity of the three recognition loops (Gilbreth & Koide, 2012). At a more practical level, our group has invested significant effort in establishing effective molecular display systems for FN3 and in constructing high-performance combinatorial libraries that utilize the concept of highly biased amino acid compositions (Gilbreth, Esaki, Koide, Sidhu, & Koide, 2008; Koide, Gilbreth, Esaki, Tereshko, & Koide, 2007; Wojcik et al., 2010). Naturally, such know-how and the availability of reagents are invaluable in establishing an Affinity Clamp project.

One could use a protein domain that has not been validated as a molecular scaffold as an enhancer domain. However, one may need to expend considerable effort in establishing and customizing a molecular display format.

3.4. Linking the primary and enhancer domains

Once the primary and enhancer domains are chosen, their genes need to be connected to form a fusion protein. Most importantly, the two domains should be linked in such a way that the diversified surfaces of the enhancer domain can make extensive contacts with the target peptide presented on the primary domain (Fig. 13.1). This condition can be easily met if either terminus of the primary domain is located near its peptide-binding site. In such a case, a short linker comprising of a stretch of Gly and Ser residues can be added between the two domains to provide linkage and flexibility between them.

The study of the role of the linker in PDZ-based Affinity Clamps showed that the interdomain linker impacts the range of accessible interdomain geometries, and thus, it is an important parameter to consider in designing Affinity Clamps (Huang, Makabe, et al., 2009). If the linker is too short, the diversified surfaces of the enhancer domain may be able to make contacts only with a small portion of the target peptide. If the linker is too long, there may be too much conformational freedom, making it difficult to achieve high affinity. However, it is difficult to determine the single best length of the linker. Modeling utilizing molecular graphics is useful for estimating a reasonable range of linker lengths. One can introduce diversity in the linker length as a part of combinatorial library design.

Unfortunately, most of the modular domains found in nature have their termini located on the opposite side of the peptide-binding site (Pawson & Nash, 2003). This architecture is advantageous for ensuring the modularity of these domains because changes in their adjacent domains would minimally affect their target-binding function. However, it is clearly a major problem in constructing the clamp architecture because simple concatenation would place the enhancer domain far away from the peptide-binding site of the primary domain (Fig. 13.2). In Affinity Clamping, the goal is to design new architecture that substantially affects the binding function of the primary domain, a goal that is fundamentally in conflict with the

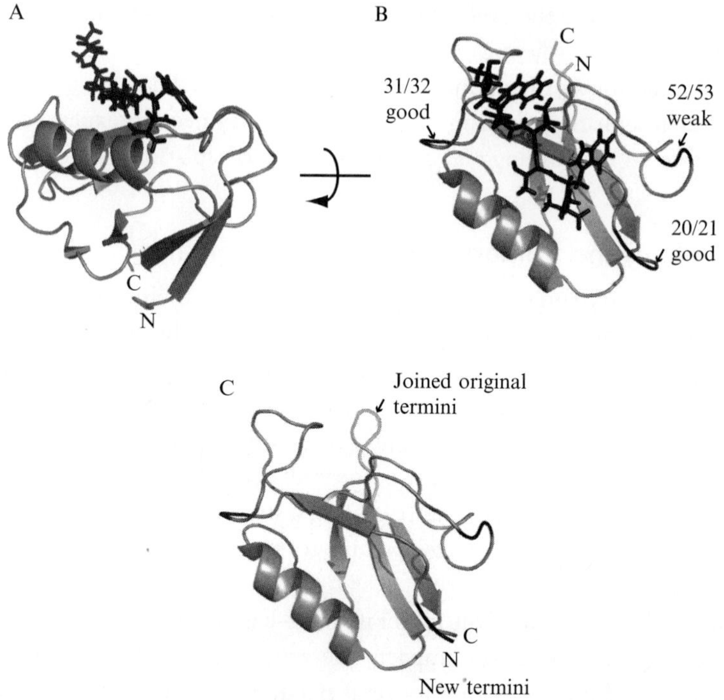

Figure 13.2 Solution of a common topological problem in constructing a clamshell framework with modular interaction domains by circular permutation. The erbin PDZ domain is used as an example (Huang et al., 2008). (A and B) Crystal structure of the erbin PDZ domain in two orthogonal views. Note that the termini are located on the opposite face of the molecule to the peptide-binding site. (B) Identification of new termini by insertion mutagenesis and functional characterization. The positions for Gly4 insertions are indicated with arrows and labeled with residue numbers. The effects of insertions at each location are also described. (C) The circular permutated erbin PDZ domain. The labels show the original termini that are now connected and the newly created N- and C-termini.

natural design of modular interaction domains. Therefore, although the functions of the modular interaction domains make them highly attractive primary domains, their architecture presents a design challenge.

Fortunately, another common attribute of the modular interaction domains provides a general solution to the topological challenge. The N- and C-termini of the modular interaction domains are usually juxtaposed (Pawson & Nash, 2003). Thus, one can relocate the termini by circular permutation, that is, by connecting the original termini and disjoining another location to create a new set of termini. For the PDZ domains, there is a subset of family members, the HtrA, that are naturally circularly permutated with respect to the rest of the PDZ family members (Runyon et al., 2007), indicating that circular permutation is evolutionarily accessible and so should be by protein design.

To perform circular permutation, one needs to identify a location at which the primary domain polypeptide may be disjoined so as to create the new termini. The location of the new termini needs to be carefully determined so that it minimally impacts the structure, function, and stability of the primary domain. Because disruption of a helix or strand most likely leads to a substantial loss of stability and also structural distortions, it makes sense to choose a position within a surface loop or turn for disjoining. Because a loop or a turn may also be important for stability, not all such positions are suitable. We have developed a simple but effective strategy for identifying such a location (Huang et al., 2008). We first analyze the three-dimensional structure of the chosen primary domain (or a close homolog) and the sequence variability among the domain family members. We choose several locations as candidates that are (i) close to the peptide-binding site, (ii) located within a loop/turn, and (iii) not highly conserved. We then insert a segment of four Gly residues to each of the candidate positions and examine the effects of the insertions on target binding using standard techniques such as ELISA and SPR. We select the location where the insertion mutation exhibits the minimal effect as the new termini. Figure 13.2 outlines these processes.

The next step is to design a sequence with which the original termini are linked. Even when a three-dimensional structure is available, it is often not obvious where exactly a domain "starts" and "ends." Although one could blindly link the termini with a flexible linker, for example, several Gly residues, a long, flexible linker can reduce the stability of the redesigned domain because a long linker may have large conformational entropy that is lost upon folding. If the linker is too short, or if one removes a residue near the termini that is important for stability, the redesigned protein may be distorted or severely destabilized. We examine the crystallographic B-factors of residues near the termini to assess the level of flexibility. NMR data are also valuable. Based on such examination, we remove residues that are likely to be highly mobile and connect the termini with a linker that has a high propensity to form a turn, such as Asn-Gly (Ramirez-Alvarado, Blanco, & Serrano, 1996). The function of the circularly permutated protein is then tested. In the case of erbin PDZ, this operation resulted in an $\sim$10-fold loss in binding affinity (Huang et al., 2008). Finally, the gene for the enhancer domain is attached to the gene for the redesigned primary domain. In most cases, the attachment of an inert enhancer domain causes little perturbation of the function of the primary domain.

3.5. Combinatorial library construction

Once molecular display of the fusion construct of the primary and enhancer domains is established, the system is ready for constructing a combinatorial library. Whereas amino acid diversity can be introduced into both primary and enhancer domains, our initial efforts have focused on diversification of the enhancer domain (Huang, Makabe, et al., 2009; Huang et al., 2008). Methods for constructing a combinatorial library depend on the format of molecular display and the enhancer domain. Those for the FN3 scaffold have recently been detailed in another volume of this series (Koide et al., 2012). Regardless, we recommend the use of highly biased amino acid compositions toward Tyr and varying the loop length, both of which are highly effective in generating high-affinity binding interfaces (Koide & Sidhu, 2009).

3.6. Choices and preparation of targets

Because Affinity Clamping is essentially affinity maturation of the primary domain, the protein designer already has a good idea of what peptide motifs to target by the time she/he has chosen the primary domain. In many cases, it is most convenient to produce a target peptide via chemical synthesis. Affinity Clamps can extend the recognition motif beyond that recognized by the primary domain, so it is important to prepare a peptide target that is extended from the core recognition motif of the primary domain in both N-terminal and C-terminal directions. For immobilization required for library sorting, we typically conjugate biotin to peptide targets, which can be readily achieved during chemical synthesis. Alternatively, simple peptide targets that do not contain posttranslational modification can be produced as a recombinant protein. For this purpose, we have found yeast small ubiquitin-like modifier (SUMO) fusion proteins with a His-tag to be particularly convenient because they are highly expressed, easily purified, and easily quantified (Huang et al., 2008; Malakhov et al., 2004). A SUMO fusion protein can be further biotinylated by chemical modification or by biosynthetic biotinylation using the Avi-tag and biotin ligase (www.avidity.com).

It is useful and important to prepare additional peptides that are highly homologous to the actual target peptide. They can be used for assessing the level of specificity of Affinity Clamps and also as competitors during library sorting. Such peptides can be designed by searching a proteome(s) of interest with the recognition pattern of the primary domain. Internet tools, such as Scansite (scansite.mit.edu) (Obenauer, Cantley, & Yaffe, 2003), are particularly useful for this purpose.

3.7. Library sorting

Phage display sorting is performed using a biotinylated peptide target and streptavidin-coated magnetic beads (Huang et al., 2008; Wojcik et al., 2010). Although direct immobilization of a target molecule to a polystyrene plate is commonly employed in antibody generation using phage display, it is not recommended for peptide targets for Affinity Clamps, because most of the peptide surface needs to be available for recognition by Affinity Clamps. One must carefully adjust library sorting conditions so as to be able to discriminate improved variants from the majority of clones that are not improved from the starting primary domain but, nevertheless, have detectable binding to the target. On the one hand, if the sorting conditions are too generous, one would simply be diluting the library without enriching the best clones. On the other hand, if the conditions are too stringent, even improved variants would be lost. This requirement is distinctly different from those for *de novo* generation of binding proteins in which a small number of functional variants are recovered in the presence of the vast majority of nonfunctional variants.

In order to find appropriate conditions, one needs to first perform a set of control experiments in which the target concentration is systematically varied and the recovery of phages displaying the primary domain and a functionally inert enhancer domain is determined by titering. The goal here is to find a target concentration where the phage recovery starts to increase above the background level (Fig. 13.3). The target concentration for actual library sorting can then be set at a value slightly lower than the identified threshold. Clearly, library sorting needs to be highly reproducible for this type of optimization to be effective. We have found that automated handling of magnetic beads using a robot (Thermo Kingfisher) makes library sorting highly reproducible (Fellouse et al., 2007).

Typically, a total of four rounds of library sorting are performed using decreasing target concentrations, for example, 100, 20, 4, and 1 nM, respectively, although conditions depend on each system, as described above. Successful selection is confirmed with two parameters, enrichment ratio and hit rate. The enrichment ratio is the ratio of the number of phages recovered

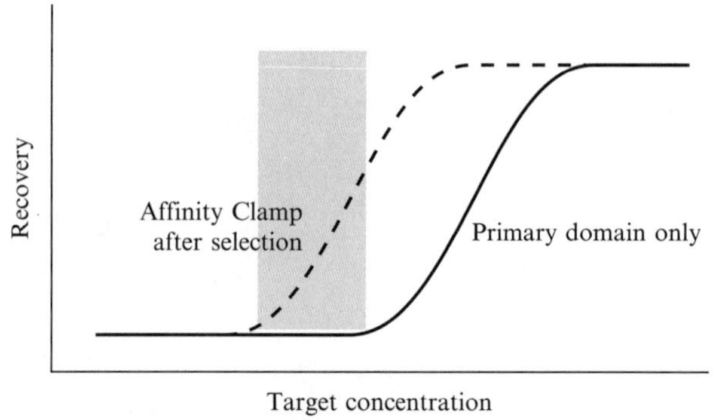

Figure 13.3 Schematic illustration of the importance of fine-tuning library sorting conditions for identifying Affinity Clamps. Recovery efficiency of phages displaying a low-affinity primary domain (solid line) and that of a high-affinity clone, corresponding to an evolved Affinity Clamp, is shown as a function of target concentration used in library sorting. The concentration range in which the Affinity Clamp can be effectively identified is shown as the shaded box.

from the selection with a target to the number of phages from a parallel selection without the target. Usually, an enrichment ratio of >10 indicates a successful selection. The hit rate is simply the number of the binding-positive clones in a set of the randomly picked clones as tested in phage ELISA, and it is the most relevant to successful generation of Affinity Clamps. It is important to perform these tests at a target concentration where the phage displaying the starting primary domain does not exhibit measurable binding and to include the phage displaying the primary domain only (or the primary domain with an inert enhancer domain) as a negative control. The amino acid sequences of the hits are deduced by DNA sequencing.

When extremely high affinity is desired, a cycle of affinity maturation can be performed (Huang et al., 2008). A secondary library is constructed from lead molecules obtained from the initial library sorting. A variety of methods are available for the design of such affinity maturation libraries, including extensive randomization of a single loop, DNA shuffling, and error-prone PCR (Koide et al., 2012). The resulting library is sorted with further reduced concentrations of the target.

4. PRODUCTION AND CHARACTERIZATION OF AFFINITY CLAMPS

The Affinity Clamp genes are transferred from phage clones to an expression vector, and individual clones are expressed in *E. coli* and purified as soluble proteins. Because Affinity Clamps are simple, single-polypeptide proteins, standard protocols for protein expression work well for them. Our default is to express Affinity Clamps as a fusion with a His_{10} tag. Protein samples are purified with Ni affinity chromatography followed by a size-exclusion chromatography, such as Superdex 75 (GE Life Sciences). It is critically important to ensure that Affinity Clamp proteins are monomeric, because the enhancer domain could induce oligomerization of an Affinity Clamp construct, which can lead to an apparent affinity increase due to mulitvalency.

To determine the binding affinity of selected Affinity Clamps using SPR, we express them as a fusion with a His_{10} tag for efficient immobilization on the Ni–NTA chip (BIAcore). Although short peptides can be used in SPR measurements, their small masses would result in low sensitivity. Thus, we utilized target peptides fused to the SUMO protein (Koide et al., 2012). Typically, an Affinity Clamp is immobilized on the chip at a level of 100–300 RU. High-affinity interactions are measured in a kinetic method, whereas for measuring weak binding affinity, such as for the starting primary domain, an equilibrium method may be required.

5. APPLICATIONS OF AFFINITY CLAMPS

Because of their high affinity, high specificity, and high stability, Affinity Clamps can be excellent affinity reagents in biological and biomedical research. Immediate and obvious utilities of Affinity Clamps are as antibody alternatives in applications such as immunoprecipitation and Western blotting. Because the target peptides for Affinity Clamps usually include a short peptide motif recognized by a modular interaction domain, they are likely to be accessible in the context of the full-length protein. Thus, unlike antibodies that are raised using a peptide derived from within a folded domain, Affinity Clamps are likely to recognize their targets in both native and denatured states, which were indeed the case for the PDZ clamps (Huang et al., 2008).

Affinity Clamps are single polypeptides and also modular, making them particularly suitable as building blocks for making fusion proteins. A matched pair of an Affinity Clamp and its target peptide can be used as a building block for forming tight and stable protein–protein interactions. We have developed such a pair specifically for this purpose by optimizing the peptide sequence for binding to one of the PDZ clamps. This pair has been demonstrated to be highly effective in protein immobilization and labeling in single-molecule measurements (Huang, Nagy, Koide, Rock, & Koide, 2009) and in SPR analysis of antibody–antigen interactions (Dyson et al., 2011). The target peptide was also incorporated within the LOV2 domain in such a way to confer light-dependent control of protein–protein interaction for synthetic biology experiments (Strickland et al., 2012).

By further expanding the idea of using Affinity Clamps as building blocks for higher functionality, we have developed ratiometric fluorescence sensors (Huang & Koide, 2010). Interdomain movements upon ligand binding are a common mechanism underlying allostery. The architecture of Affinity Clamps immediately suggests such ligand-induced conformational changes. By attaching a pair of FRET-optimized fluorescent proteins to a PDZ clamp, we were able to transduce conformational changes of the Affinity Clamp into changes in fluorescence emission.

6. CONCLUSION

The Affinity Clamp technology represents a major breakthrough in protein design, as it provides a rational pathway for producing high-performance affinity reagents for a predefined peptide motif. The presence of diverse families of modular interaction domains suggests that Affinity Clamps can be developed for a diverse panel of peptide motifs of high biological significance. Although the procedures for generating Affinity Clamps are still laborious and require high levels of skills in protein design, iterative refinement may lead to highly facile pipelines. The small and simple architecture of Affinity Clamps makes them ideal building blocks for designing functionalities beyond simple binding. We are confident that numerous Affinity Clamps will be developed for diverse purposes in the near future.

ACKNOWLEDGMENTS

We thank Drs. Ryan Gilbreth, Akiko Koide, Robert Wells, and Norihisa Yasui for critical reading of the chapter. This work was supported in part by the National Institutes of Health Grant R01-GM090324 to S. K. and by the University of Chicago Comprehensive Cancer Center.

REFERENCES

Bashton, M., & Chothia, C. (2007). The generation of new protein functions by the combination of domains. *Structure*, *15*, 85–99.

Binz, H. K., Amstutz, P., & Pluckthun, A. (2005). Engineering novel binding proteins from nonimmunoglobulin domains. *Nature Biotechnology*, *23*, 1257–1268.

Cobaugh, C. W., Almagro, J. C., Pogson, M., Iverson, B., & Georgiou, G. (2008). Synthetic antibody libraries focused towards peptide ligands. *Journal of Molecular Biology*, *378*, 622–633.

Colwill, K., & Graslund, S. (2011). A roadmap to generate renewable protein binders to the human proteome. *Nature Methods*, *8*, 551–558.

Dyson, M. R., Zheng, Y., Zhang, C., Colwill, K., Pershad, K., Kay, B. K., et al. (2011). Mapping protein interactions by combining antibody affinity maturation and mass spectrometry. *Analytical Biochemistry*, *417*, 25–35.

Fellouse, F. A., Esaki, K., Birtalan, S., Raptis, D., Cancasci, V. J., Koide, A., et al. (2007). High-throughput generation of synthetic antibodies from highly functional minimalist phage-displayed libraries. *Journal of Molecular Biology*, *373*, 924–940.

Gilbreth, R. N., Esaki, K., Koide, A., Sidhu, S. S., & Koide, S. (2008). A dominant conformational role for amino acid diversity in minimalist protein-protein interfaces. *Journal of Molecular Biology*, *381*, 407–418.

Gilbreth, R. N., & Koide, S. (2012). Structural insights for engineering binding proteins based on non-antibody scaffolds. *Current Opinion in Structural Biology*, *22*, 413–420.

Hamill, S. J., Meekhof, A. E., & Clarke, J. (1998). The effect of boundary selection on the stability and folding of the third fibronectin type III domain from human tenascin. *Biochemistry*, *37*, 8071–8079.

Huang, J., & Koide, S. (2010). Rational conversion of affinity reagents into label-free sensors for peptide motifs by designed allostery. *ACS Chemical Biology*, *5*, 273–277.

Huang, J., Koide, A., Makabe, K., & Koide, S. (2008). Design of protein function leaps by directed domain interface evolution. *Proceedings of the National Academy of Sciences of the United States of America*, *105*, 6578–6583.

Huang, J., Makabe, K., Biancalana, M., Koide, A., & Koide, S. (2009). Structural basis for exquisite specificity of affinity clamps, synthetic binding proteins generated through directed domain-interface evolution. *Journal of Molecular Biology*, *392*, 1221–1231.

Huang, J., Nagy, S. S., Koide, A., Rock, R. S., & Koide, S. (2009). A Peptide tag system for facile purification and single-molecule immobilization. *Biochemistry*, *48*, 11834–11836.

Koide, S. (2009). Generation of new protein functions by nonhomologous combinations and rearrangements of domains and modules. *Current Opinion in Biotechnology*, *20*, 398–404.

Koide, S. (2010). Design and engineering of synthetic binding proteins using nonantibody scaffolds. In S. J. Park & J. R. Cochran (Eds.), *Protein engineering and design* (pp. 109–130). Boca Raton, FL: CRC Press.

Koide, A., Bailey, C. W., Huang, X., & Koide, S. (1998). The fibronectin type III domain as a scaffold for novel binding proteins. *Journal of Molecular Biology*, *284*, 1141–1151.

Koide, A., Gilbreth, R. N., Esaki, K., Tereshko, V., & Koide, S. (2007). High-affinity single-domain binding proteins with a binary-code interface. *Proceedings of the National Academy of Sciences of the United States of America*, *104*, 6632–6637.

Koide, S., Koide, A., & Lipovsek, D. (2012). Target-binding proteins based on the 10th human fibronectin type III domain ((10)Fn3). *Methods in Enzymology*, *503*, 135–156.

Koide, S., & Sidhu, S. S. (2009). The importance of being tyrosine: Lessons in molecular recognition from minimalist synthetic binding proteins. *ACS Chemical Biology*, *4*, 325–334.

Kortemme, T., & Baker, D. (2004). Computational design of protein-protein interactions. *Current Opinion in Chemical Biology*, *8*, 91–97.

Machida, K., Thompson, C. M., Dierck, K., Jablonowski, K., Karkkainen, S., Liu, B., et al. (2007). High-throughput phosphotyrosine profiling using SH2 domains. *Molecular Cell*, *26*, 899–915.

Malakhov, M. P., Mattern, M. R., Malakhova, O. A., Drinker, M., Weeks, S. D., & Butt, T. R. (2004). SUMO fusions and SUMO-specific protease for efficient expression and purification of proteins. *Journal of Structural and Functional Genomics*, *5*, 75–86.

Obenauer, J. C., Cantley, L. C., & Yaffe, M. B. (2003). Scansite 2.0: Proteome-wide prediction of cell signaling interactions using short sequence motifs. *Nucleic Acids Research*, *31*, 3635–3641.

Parmley, S. F., & Smith, G. P. (1989). Filamentous fusion phage cloning vectors for the study of epitopes and design of vaccines. *Advances in Experimental Medicine and Biology, 251,* 215–218.

Pawson, T., & Nash, P. (2003). Assembly of cell regulatory systems through protein interaction domains. *Science, 300,* 445–452.

Ramirez-Alvarado, M., Blanco, F. J., & Serrano, L. (1996). De novo design and structural analysis of a model beta-hairpin peptide system. *Nature Structural Biology, 3,* 604–612.

Rockberg, J., Lofblom, J., Hjelm, B., Uhlen, M., & Stahl, S. (2008). Epitope mapping of antibodies using bacterial surface display. *Nature Methods, 5,* 1039–1045.

Rossmann, M. G., Moras, D., & Olsen, K. W. (1974). Chemical and biological evolution of nucleotide-binding protein. *Nature, 250,* 194–199.

Runyon, S. T., Zhang, Y., Appleton, B. A., Sazinsky, S. L., Wu, P., Pan, B., et al. (2007). Structural and functional analysis of the PDZ domains of human HtrA1 and HtrA3. *Protein Science, 16,* 2454–2471.

Sidhu, S. S., & Fellouse, F. A. (2006). Synthetic therapeutic antibodies. *Nature Chemical Biology, 2,* 682–688.

Sidhu, S. S., Lowman, H. B., Cunningham, B. C., & Wells, J. A. (2000). Phage display for selection of novel binding peptides. *Methods in Enzymology, 328,* 333–363.

Strickland, D., Lin, Y., Wagner, E., Hope, C. M., Zayner, J., Antoniou, C., et al. (2012). TULIPs: Tunable, light-controlled interacting protein tags for cell biology. *Nature Methods, 9,* 379–384.

Tonikian, R., Zhang, Y., Sazinsky, S. L., Currell, B., Yeh, J. H., Reva, B., et al. (2008). A specificity map for the PDZ domain family. *PLoS Biology, 6,* e239.

Winter, G., Griffiths, A. D., Hawkins, R. E., & Hoogenboom, H. R. (1994). Making antibodies by phage display technology. *Annual Review of Immunology, 12,* 433–455.

Wojcik, J., Hantschel, O., Grebien, F., Kaupe, I., Bennett, K. L., Barkinge, J., et al. (2010). A potent and highly specific FN3 monobody inhibitor of the Abl SH2 domain. *Nature Structural and Molecular Biology, 17,* 519–527.

CHAPTER FOURTEEN

Engineering Fibronectin-Based Binding Proteins by Yeast Surface Display

Tiffany F. Chen[*,†], **Seymour de Picciotto**[*,†], **Benjamin J. Hackel**[‡], **K. Dane Wittrup**[*,†,§,1]

[*]Department of Biological Engineering, Massachusetts Institute of Technology, Cambridge, Massachusetts, USA
[†]Koch Institute for Integrative Cancer Research, Massachusetts Institute of Technology, Cambridge, Massachusetts, USA
[‡]Department of Chemical Engineering and Materials Science, University of Minnesota, Minneapolis, Minnesota, USA
[§]Department of Chemical Engineering, Massachusetts Institute of Technology, Cambridge, Massachusetts, USA
[1]Corresponding author: e-mail address: wittrup@mit.edu

Contents

Abstract

Yeast surface display (YSD) presents proteins on the surface of yeast through interaction of the agglutinin subunits Aga1p and Aga2p. The human 10th type III fibronectin (Fn3) is a small, 10-kDa protein domain that maintains its native fold without disulfide bonds. A YSD library of Fn3s has been engineered with a loop amino acid composition similar to that of human antibody complementarity-determining region heavy chain loop 3

Methods in Enzymology, Volume 523
ISSN 0076-6879
http://dx.doi.org/10.1016/B978-0-12-394292-0.00014-X

(CDR-H3) and varying loop lengths, which has been shown to improve binding ability. There are many advantages of using these small, stable domains that maintain binding capabilities similar to that of antibodies. Here, we outline a YSD methodology to isolate Fn3 binders to a diverse set of target antigens.

1. INTRODUCTION

Yeast surface display (YSD) is a versatile platform for engineering proteins. Display of single-chain variable fragments (scFvs) for affinity maturation of already isolated scFvs or isolation of binders from naïve libraries are probably the most characterized applications of YSD (Boder, Raeeszadeh-Sarmazdeh, & Price, 2012; Boder & Wittrup, 1997; Chao et al., 2006; Colby et al., 2004). Several other proteins including T-cell receptors, cytokines, and green fluorescent protein among others have been engineered using YSD for improved characteristics of binding affinity, expression, stability, etc. (Gai & Wittrup, 2007). Yeast display libraries, which range from 10^7 to 10^9 transformants, are typically smaller than phage or ribosome display libraries, 10^{11} to 10^{13}, but have the advantage of a eukaryotic expression machinery. This allows for isolation of clones that would have been missed using other display systems, which suggests more proper display of eukaryotic proteins (Bowley, Labrijn, Zwick, & Burton, 2007).

In the YSD system, the protein to be engineered is fused to the C-terminal end of the Aga2p protein, which forms a covalent linkage to the Aga1p protein on the yeast surface through two disulfide bonds (Fig. 14.1). YSD requires two main components: the yeast, EBY100, and the plasmid, pCT-CON. EBY100 is deficient in the machinery to synthesize the amino acid tryptophan and contains the *Aga1* gene in the yeast genome. The pCT-CON plasmid encodes for (a) the gene *TRP1*, which is important for tryptophan synthesis; (b) the Aga2p protein fused to the protein of interest; and (c) ampicillin resistance, for plasmid production in *Escherichia coli* (Fig. 14.2). Both Aga1p- and Aga2p-mating proteins are controlled under a galactose-inducible promoter. When yeast are properly transformed with the pCT-CON plasmid, they can grow in selective media deficient in tryptophan, whereas untransformed EBY100 will not propagate. Switching the yeast from glucose-rich media to galactose-rich media will induce proper display of the protein of interest.

Engineering binders from scFvs and antibodies have been the standard practice, but over the past decade, the use of alternative scaffolds has been

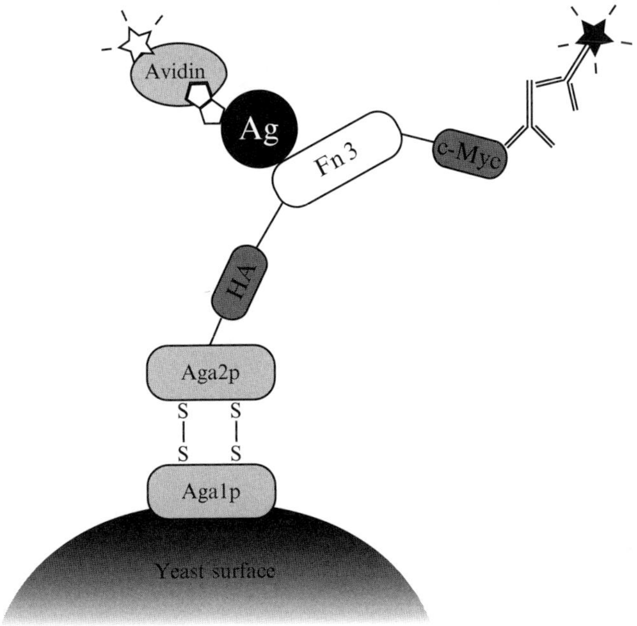

Figure 14.1 Schematic of the Fn3 scaffold displayed on the surface of yeast. The Fn3 is a C-terminal fusion to the Aga2p protein and flanked by two detection tags: hemagglutinin (HA) epitope tag (YPYDVPDYA) at the N-terminus and c-Myc epitope (EQKLISEEDL) at the C-terminus. The Aga2p protein forms two disulfide bonds with the Aga1p protein, which is anchored to the cell wall via β-glucan linkage. Biotinylated antigen is detected with a fluorophore-conjugated streptavidin. Full display of Fn3 is detected with a primary chicken anti-c-Myc antibody and a fluorophore-conjugated secondary antibody specific for chicken antibody. (For color version of this figure, the reader is referred to the online version of this chapter.)

an emerging field in protein engineering (Binz, Amstutz, & Pluckthun, 2005; Skerra, 2007). Antibodies (150 kDa) are quite complex protein structures comprised of several protein domains that require disulfide bonds and glycosylation for proper function. Alternative scaffolds such as anticalins, affibodies, darpins, and fibronectin domains are small proteins that have certain surface regions that can be highly diversified to bind to a variety of proteins (Koide & Koide, 2007; Löfblom et al., 2010; Skerra, 2008; Stumpp, Binz, & Amstutz, 2008). The advantage of alternative scaffolds lies in the fact that they can be quite small, single domains yet fairly stable while lacking disulfide bonds. The alternative scaffold most developed with the YSD platform is the 10th type III domain of the fibronectin (Fn3) protein; the primary domain involved in integrin binding (Hackel, Kapila, & Wittrup, 2008; Hackel & Wittrup, 2010; Lipovsek et al., 2007). This small,

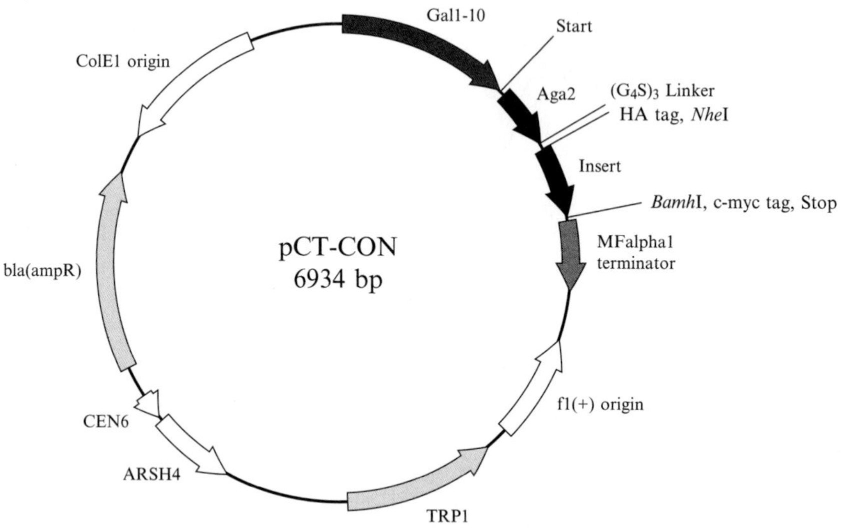

Figure 14.2 Vector map of pCT-CON. The *Fn3* gene would be in the place of the "insert" label on the plasmid. (For color version of this figure, the reader is referred to the online version of this chapter.)

approximately 10 kDa, cysteine-free yet stable domain is composed of two β-sheets comprising seven β-strands connected by three solvent-exposed loops on both sides, which resemble antibody CDRs (Fig. 14.3).

A library of Fn3s was previously generated that had varying loop lengths compared to that of wild-type Fn3, with an amino acid repertoire similar to that of antibody CDR-H3. This library known as G4 is the naïve library used for our selection processes. The library diversity comprises 2.5×10^8 transformants with 60% of them being full-length Fn3s, yielding approximately 1.5×10^8 clones (Hackel, Ackerman, Howland, & Wittrup, 2010). The current library size undersamples the immense sequence space created by the loop length and amino acid diversities. To compensate for this, diversity is constantly introduced into enriched sublibraries by several mechanisms: high mutagenesis through error-prone PCR focused on the three loop areas (BC, DE, FG loops), low mutagenesis by error-prone PCR for the entire *Fn3* gene to introduce framework mutations, and shuffling of the loops during homologous recombination (Hackel et al., 2008).

This library has yielded several binders to a variety of target proteins ranging from epidermal growth factor receptor (EGFR) (Hackel et al., 2010), immunoglobulins of various species (Hackel & Wittrup, 2010), carcinoembryonic antigen (CEA) (Pirie, Hackel, Rosenblum, & Wittrup, 2011), to human Fc gamma receptors (Hackel et al., 2010). Fn3 clones with high display levels correlate with increased stability as tested with circular

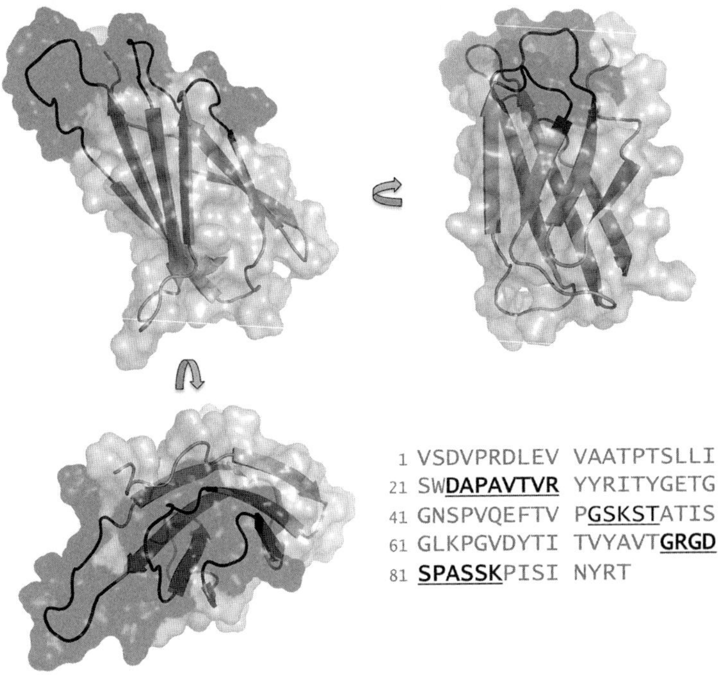

Figure 14.3 Fibronectin domain. The solution structure (PDB ID: 1TTG) of Fn3 is presented with 90° rotations. The wild-type sequence is indicated. The BC (red), DE (green), and FG (blue) loops are highlighted. (See Color Insert.)

dichroism thermal denaturation (Hackel & Wittrup, 2010; Hackel et al., 2010). Fusions of the EGFR binders have been shown to downregulate EGFR expression on various cancer cell lines *in vitro* (Hackel, Neil, White, & Wittrup, 2012) and inhibit tumor growth *in vivo* when fused as a triepitopic antibody (Spangler, Manzari, Rosalia, Chen, & Wittrup, 2012). Fusions of CEA binders to gelonin have served as potent immunotoxins *in vitro* (Pirie et al., 2011). Ongoing work in our laboratory demonstrates the versatility of this library in producing Fn3 binders to a large panel of EGFR ligands, perfringolysin O, DEC-205, among many other targets.

2. ENGINEERING AND SCREENING APPROACH OF Fn3s

The following outlines a detailed methodology of Fn3 binder selection using YSD (Fig. 14.4). The naïve G4 library is first screened using magnetic bead selection. The avidity of interaction between the yeast and

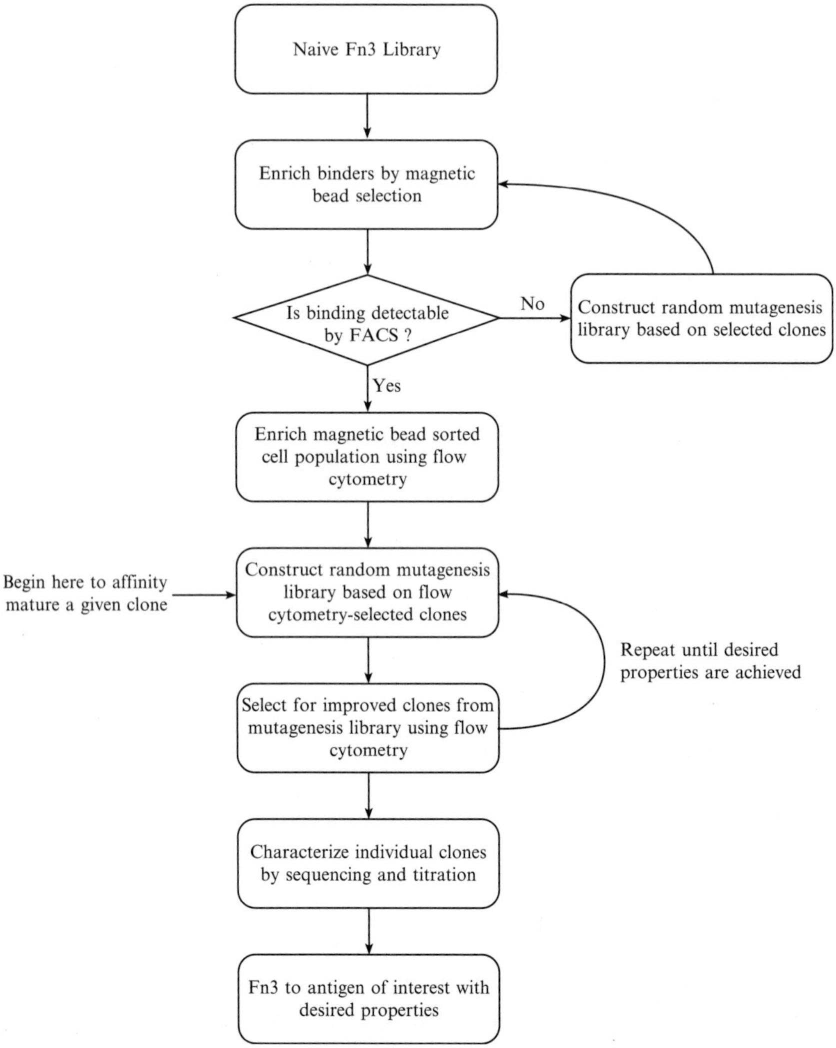

Figure 14.4 Flow chart for the methodology of engineering Fn3s from the naïve G4 library. Fn3 binders are first enriched using magnetic bead selection and then sorted using FACS. (For color version of this figure, the reader is referred to the online version of this chapter.)

multivalent antigen-coated beads allows for the enrichment of weak affinity binders (Ackerman et al., 2009). After two rounds of magnetic bead enrichment, the Fn3 sublibrary is subjected to mutagenesis to introduce diversity into the population. This new library is screened again using magnetic beads. Once binding of antigen is detectable on flow cytometry, sorting is switched to fluorescence-activated cell sorting (FACS). The selection process of the Fn3 binders remains the same with two rounds of enrichment followed by mutagenesis. This process of directed evolution has resulted in engineering Fn3 binders with picomolar affinity to antigens of interest (Hackel et al., 2008). The G4 library, EBY100, and associated pCT-CON vectors are available upon request.

2.1. Yeast media and plates

1. SD-CAA: 20 g/L D-glucose, 6.7 g/L yeast nitrogen base, 5 g/L casamino acids, 7.4 g/L citric acid monohydrate, 10.4 g/L sodium citrate, pH 4.5
2. SD-CAA plates: 20 g/L D-glucose, 6.7 g/L yeast nitrogen base, 5 g/L casamino acids, 5.4 g/L Na_2HPO_4, 8.6 g/L $NaH_2PO_4 \cdot H_2O$, 16 g/L agar, 182 g/L sorbitol
3. SG-CAA: 18 g/L galactose, 2 g/L D-glucose, 6.7 g/L yeast nitrogen base, 5 g/L casamino acids, 5.4 g/L Na_2HPO_4, 8.6 g/L $NaH_2PO_4 \cdot H_2O$, pH 6.0
4. YPD: 20 g/L dextrose, 20 g/L peptone, 10 g/L yeast extract
5. YPD plates: 20 g/L dextrose, 20 g/L peptone, 10 g/L yeast extract, 16 g/L agar

2.2. Naïve yeast library and culture conditions

1. To prepare the G4 library, thaw a frozen aliquot with 10× diversity (2.5×10^9 cells) into 1 L of SD-CAA minimal media. Grow at 30 °C with shaking at 250 rpm overnight to a typical OD_{600nm} of 6–8 ($OD_{600nm} = 1$ is $\sim 1 \times 10^7$ cells/ml).

Optional: If desired, cells can be passaged to a 1-L culture of fresh SD-CAA media to decrease the percentage of dead cells. Plating serial dilutions of the G4 library on SD-CAA plates can also test for viability. Frozen aliquots of the G4 library can be prepared by freezing 10× diversity in 15% glycerol in SD-CAA at −80 °C for long-term storage.

2. Inducing cells for display requires switching the media from glucose- to galactose-rich media. Pellet 30× diversity (7.5×10^9 cells) of the G4 library at 2500 × *g* for 5 min and resuspend the cells in 1 L of SG-CAA induction media for 12–24 h at 20 °C. Optimal induction occurs when cells are induced during exponential growth ($OD_{600nm} = 2$–7).

Optional: Cells can be induced at 30 °C, but cells will continue to grow, whereas induction at 20 °C should allow for less than one doubling from the initial OD_{600nm} during the incubation. After induction, cells can be used immediately or stored at 4 °C.

2.3. Library screening with magnetic beads

Initial library screening is performed using Biotin Binder Dynabeads (Invitrogen). The multivalency of the Dynabeads ($\sim 5 \times 10^6$ biotin-binding sites per bead) and the yeast ($\sim 10^5$ fibronectin per cell) allows for the isolation of weak affinity Fn3 binders through avidity interaction. The naïve library is first depleted of bare bead/streptavidin binders and then enriched for binders to the antigen of interest. Antigens of interest must be biotinylated before sorting.

2.3.1 Bead preparation

1. For the initial sort from the naïve library, two batches should be prepared to ensure adequate exposure of cells to the antigen of interest. Prepare 6.7–33 pmoles of biotinylated antigen for 10 μL of Dynabeads (4×10^5 beads/μL) in 100 μL of PBSA (1 × phosphate-buffered saline, 0.1% bovine serum albumin) in 2-mL microcentrifuge tubes. Antigen at 6.7 pmoles is sufficient for enrichment, but if the antigen is not limited, then 33 pmoles should be used.
2. Tubes should be incubated at 4 °C, rotating for at least 1 h.
3. Directly before the addition of cells, beads must be washed once with the addition of 1 mL of PBSA. Place the tube on a magnet for 2–5 min before removing the supernatant. After removal of supernatant, cells can be added to beads. This step is to remove any free antigen.

2.3.2 Initial cell sorting

1. Measure the cell density and pellet 15 × diversity of the induced G4 library (3.75×10^9 cells) for sorting.
2. Pellet cells at 3000 × *g* for 5 min and wash the cells with 1 mL of PBSA, then split into two aliquots in two 2-mL microcentrifuge tubes. Pellet cells at high speed (12,000 × *g*) for 1 min and resuspend cells in 1 mL PBSA.
3. Negative sort: Add 10 μL of bare beads in each tube and incubate cells and beads at 4 °C for at least 2 h rotating (Ackerman et al., 2009). This step serves as a negative selection by depleting yeast that display Fn3 binders to the bare beads/streptavidin. The initial library should not have many bare bead/streptavidin binders, so one depletion should be sufficient, but subsequent sorts should include at least two negative bare bead

depletions to prevent the isolation of bare bead/streptavidin binders. If your antigen of interest possesses a tag or fusion partner, we recommend including this tag or fusion partner in your negative selection process. Prepare the beads as recommended in the previous subsection using a biotinylated tag-only construct.

4. Positive sort: After 2 h of incubation at 4 °C, place the cells with beads on the magnet and transfer the unbound cells to the tubes with the washed, antigen-coated beads. Incubate at 4 °C for at least 2 h rotating.
5. Wash the negative sort with 1 mL PBSA on magnet, remove supernatant, and resuspend beads and cells in 5 mL of fresh SD-CAA. Store at 4 °C until all sorts are finished, only then grow up cultures at 30 °C.
6. Wash the positive sort with 1 mL PBSA on magnet, remove supernatant, and resuspend beads and cells in 5 mL of fresh SD-CAA. Perform serial dilutions of negative and positive sort cultures and plate on SD-CAA plates to determine and compare numbers of isolated yeast.

2.3.3 Cell growth and induction

1. Grow up cells at 30 °C with shaking at 250 rpm for at least 16 h.
2. Pellet culture and remove 4 mL of media. Resuspend cells and beads in remaining 1 mL of culture and transfer to 2-mL microcentrifuge tube. Place on magnet to remove the beads. Recover unbound cells and transfer back into original test tube.
3. Pellet at least 10 × diversity of cells at high speed for 1 min. Remove supernatant and resuspend cells in 5 mL of fresh SG-CAA media. Yeast double approximately every 4 h in SD-CAA minimal media; therefore, back-calculations of the hours of growth can determine volume of cells needed for 10 × diversity.
4. Incubate at 20 °C at 250 rpm for 8–24 h for the induction of protein expression.

2.3.4 Intermediate cell sorting

The newly induced population of yeast should be enriched in antigen binders and depleted of bare bead/streptavidin binders. To ensure that specific binders are isolated as opposed to nonspecific yeast, the stringency of negative sorts and washes is increased.

1. Measure the cell density and pellet 20 × diversity of the induced population for sorting.
2. Wash the cells with 1 mL of PBSA, pellet the cells at high speed (12,000 × *g*) for 1 min, and resuspend cells in 1 mL PBSA.

3. Negative sort 1: Add 10 μL of bare beads in the tube and incubate cells and beads at 4 °C for at least 2 h rotating.
4. Negative sort 2: After 2 h of incubation at 4 °C, place the cells with beads on the magnet and transfer the unbound cells to a new 2-mL microcentrifuge tube. Add 10 μL of bare beads and incubate at 4 °C for at least 2 h rotating.

Optional: Depending on the desired stringency and the generation of sort, negative sorts can be increased in number to prevent the isolation of bare bead/streptavidin binders. Negative sorts against nontarget molecules (such as epitope tags, fusion partners, or other undesired binding partners) can also be included.

5. Wash the negative sort 1 beads with 1 mL PBSA on the magnet and remove supernatant; repeat wash. Resuspend beads and cells in 5 mL of fresh SD-CAA. Store at 4 °C until all sorts are finished and only then grow up cultures at 30 °C.
6. Positive sort: After 2 h of incubation at 4 °C, place the cells with beads on the magnet and transfer the unbound cells to the tubes with the washed antigen-coated beads. Incubate at 4 °C for at least 2 h rotating. Repeat step 5 for negative sort 2.
7. Wash the positive sort with 1 mL PBSA on magnet, remove supernatant, and resuspend beads and cells in 5 mL of fresh SD-CAA. Perform serial dilutions of negative and positive sort cultures and plate on SD-CAA plates to determine and compare numbers of isolated yeast.

Optional: Depending on the desired stringency and the generation of sort, the number of washes can be increased to ensure the isolation of specific binders. The results from the serial dilutions on a plate are typically a good indication of whether the stringency should be increased. A good ratio of positive sort to negative sort cells is 10:1. Alternatively or additionally, incubations can be performed at room temperature.

8. Repeat Section 2.3.3 cell growth and induction protocol.

2.3.5 FACS of fully displaying yeast

After two rounds of magnetic-bead enrichment, sort for yeast displaying full-length Fn3 clones by isolating yeast that stain double positive for the N-terminal HA epitope tag and the C-terminal c-myc epitope tag. In truncations or frameshift mutants, the c-myc tag detection would be lost and, in the case of plasmid loss, the HA tag would not be detected. This sort ensures that all clones are full-length Fn3 clones, which are then used for mutagenesis.

1. Pellet at least 10 × library diversity or 1×10^7 cells and wash with 1 mL of PBSA.
2. Primary staining: Resuspend cells in 50 μL of PBSA and add 0.5 μL of mouse anti-HA as a 1:100 dilution (16B12, Covance) and 0.5 μL of chicken anti-c-myc (Invitrogen or Gallus Immunotech) as a 1:100 dilution. Incubate cells with the primary antibody at room temperature for at least 20 min. When labeling with primaries, two concerns must be taken into account: stoichiometric excess of the primary, and the concentration, which determines incubation periods.

Note: Yeasts typically display 1×10^5 Fn3s per cell.

3. Wash the cells with 1 mL of PBSA, pellet cells at high speed for 1 min, and remove supernatant. This step removes any unbound primary antibodies.
4. Secondary staining: Resuspend cells in 50 μL of PBSA and add 0.5 μL goat anti-mouse AlexaFluor® 488 (Invitrogen) and 0.5 μL goat anti-chicken AlexaFluor® 647 (Invitrogen). Incubate on ice for at least 15 min. Other fluorophore combinations can be used, but fluorophores that do not require compensation in flow cytometry are preferred (Shapiro, 2005).
5. Wash the cells with 1 mL of PBSA, pellet cells at high speed for 1 min, and remove supernatant. This step removes unbound secondary antibodies.
6. FACS: Resuspend cells in the desired sorting volume (0.5 mL) of PBSA and collect double-positive population. Collect cells in a small volume (2 mL) of SD-CAA media. Rinse the sides of the tube with 3 mL of SD-CAA media to collect any additional cells. Grow up cells at 30 °C with shaking at 250 rpm for at least 16 h. If the volume increases significantly after the sort, dilute cells in excess SD-CAA media, otherwise the yeast will flocculate.

Optional: In the case of increased volume of after sorting, cells can be rinsed off the sides of the tube and then gently pelleted at 2500 × *g* for 5 min. Remove as much media as possible and add 5 mL of fresh SD-CAA media.

2.4. Fn3 mutagenesis and electroporation

In order to reintroduce diversity into the enriched population of Fn3 binders, plasmids from yeast will be isolated and mutagenized. This will introduce clones with improved binding to the target of interest.

2.4.1 Zymoprep yeast

Extract the plasmid DNA from yeast using the Zymoprep™ Yeast Plasmid Miniprep II (Zymo Research). The following protocol is adapted from the original Zymo Research protocol.

1. Measure the cell density and pellet 1×10^8 cells in a microcentrifuge tube at $300 \times g$ for 1 min. Remove the supernatant.
2. Add 200 μL of Solution 1 to the pellet and resuspend the pellet by gentle pipetting. Add 3 μL of Zymolyase, mix, and incubate at 37 °C for 15–60 min. For cells in logarithmic growth ($OD_{600nm} = 2–7$), 15 min of incubation at 37 °C should be sufficient. For cells in the stationary phase, they need to be incubated for at least 30 min.
3. Add 200 μL of Solution 2 with gentle mixing. Then, add 400 μL of Solution 3 and mix gently. Pellet debris at high speed ($12{,}000 \times g$) for 8 min. Remove supernatant and transfer to a new microcentrifuge tube. Pellet remaining debris at high speed ($12{,}000 \times g$) for 3–5 min (depending on the amount of debris remaining).
4. Transfer the clear supernatant to a miniprep column (Epoch or Qiagen) and not the Zymo column. Centrifuge the column at high speed ($12{,}000 \times g$) for 1 min and discard the flow through.
5. Wash column using 550 μL of miniprep wash buffer (Qiagen buffer PE or Epoch WS). Spin the column at high speed for another 1 min and discard the flow through. Spin column again to remove any excess wash buffer.
6. Place column over new microcentrifuge tube and elute DNA with 40 μL of elution buffer (Qiagen or Epoch buffer EB). Spin at high speed for 1 min.

The expected yield is approximately five plasmids from each cell. With 1×10^8 cells, there should be approximately 1.25×10^7 plasmids per microliter.

2.4.2 Mutagenesis of Fn3 through error-prone PCR

Plasmids containing Fn3s will be mutagenized using error-prone PCR with two nucleotide analogues: 8-oxo-2′-deoxyguanosine-5′-triphosphate and 2′-deoxy-p-nucleoside-5′-triphosphate (8-oxo-dGTP and dPTP). Mutation rates can be adjusted by varying the concentration of nucleotide analogues and the number of PCR cycles. Mutations will be introduced using two methods: low levels of mutagenesis to the entire gene to introduce framework mutations and high levels of mutagenesis at the loop regions to improve binding.

1. Error-prone PCR preparation

Primers

Sample	Sequence	
Gene	5′-Primer	cgacgattgaaggtagatacccatacgacgttccagactacgctctgcag
	3′-Primer	atctcgagctattacaagtcctcttcagaaataagcttttgttcggatcc
BC loop	5′-Primer	gggacctggaagttgttgctgcgacccccaccagcctactgatcagctgg
	3′-Primer	tgaactcctggacagggctatttcctcctgtttctccgtaagtgatcctgtaata
DE loop	5′-Primer	caggatcacttacggagaaacaggaggaaatagccctgtccaggagttcactgtg
	3′-Primer	gcatacacagtgatggtataatcaactccaggtttaaggccgctgatggtagc
FG loop	5′-Primer	accatcagcggccttaaacctggagttgattataccatcactgtgtatgctgtc
	3′-Primer	gatccctgggatggtttgtcaatttctgttcggtaattaatggaaattgg

PCR in 50 μL volumes

Volume	Component	Final concentration
Gene reaction (one reaction total)		
5 μL	10 × *ThermoPol* buffer	1 × *ThermoPol* buffer
2.5 μL	5′-Primer (10 μ*M*)	0.5 μ*M* 5′-primer
2.5 μL	3′-Primer (10 μ*M*)	0.5 μ*M* 3′-primer
1 μL	dNTPs (10 m*M* each)	200 μ*M* dNTPs each
8 μL	Zymoprepped DNA	10^8 plasmids
5 μL	8-oxo-dGTP (20 μ*M*)	2 μ*M* 8-oxo-dGTP
5 μL	dPTP (20 μ*M*)	2 μL dPTP
20.5 μL	ddH_2O	
0.5 μL	*Taq* DNA polymerase	2.5 units
Loop reactions (three reactions total)		
5 μL	10 × *ThermoPol* buffer	1 × *ThermoPol* buffer
2.5 μL	5′ primer (10 μ*M*)	0.5 μ*M* 5′-primer
2.5 μL	3′ primer (10 μ*M*)	0.5 μ*M* 3′-primer
1 μL	dNTPs (10 m*M* each)	200 μ*M* dNTPs each
8 μL	Zymoprepped DNA	10^8 plasmids
5 μL	8-oxo-dGTP (200 μ*M*)	20 μ*M* 8-oxo-dGTP
5 μL	dPTP (200 μ*M*)	20 μL dPTP
20.5 μL	ddH_2O	
0.5 μL	*Taq* DNA polymerase	2.5 units

2. Thermal cycle setup

Step	Temperature (°C)	Time	Cycle
Initialization	94	3 min	1
Denaturation	94	45 s	2–16
Annealing	60	30 s	
Extension	72	90 s	
Final extension	72	10 min	17

Mutated DNA can be used immediately or stored at 4 °C.

3. Add 10 μL of 6 × DNA loading buffer to mutated DNA and run half of the reaction (30 μL) on a 1.5% agarose gel precast with GelGreen™ (Biotium, Inc.). Save the remaining half of mutated DNA at 4 °C for possible future use in the case that more DNA is needed. Run samples with ladder at 100 V for approximately 45–60 min. The gene PCR product should be visible, but loop PCR products may be hard to differentiate from primers. Excise gene band, and to ensure proper recovery of loops, excise entire loop bands including primers (Fig. 14.5). PCR products should be at 460 ± 18, 139 ± 6, 126 ± 6, 179 ± 6, base pairs for the gene, BC loop, DE loop, and FG loop bands.
4. Purify the PCR products individually (gene, BC loop, DE loop, and FG loop) using the Qiaquick Gel Extraction Kit (Qiagen) according to the manufacturer's instructions. Elute with 40 μL EB buffer.

2.4.3 Amplification of mutagenized Fn3

The amplification step of the mutagenized Fn3 is to generate sufficient quantities of DNA insert for yeast transformation. This step will amplify the loop regions and not the primers that were excised with the loop regions.

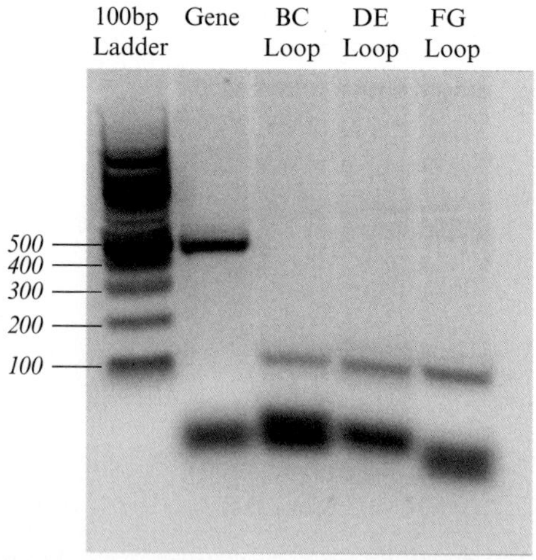

Figure 14.5 Gel of error-prone PCR products of the gene, BC loop, DE loop, and FG loop. The gene band should be at approximately 460 bp and the loops will run at approximately 120–190 bp. Primer bands run lower than the loop products, but if no differentiation can be detected, purify the entire primer and loop band.

1. Amplification PCR preparation

Primers

Sample	Sequence	
Gene	5′-Primer	cgacgattgaaggtagatacccatacg
	3′-Primer	atctcgagctattacaagtcctcttc
BC loop	5′-Primer	gggacctggaagttgttgctgcg
	3′-Primer	tgaactcctggacagggctatttcc
DE loop	5′-Primer	caggatcacttacggagaaacaggagg
	3′-Primer	gcatacacagtgatggtataatcaac
FG loop	5′-Primer	accatcagcggccttaaacctggag
	3′-Primer	gatccctgggatggtttgtcaatttc

Amplification primers are shorter versions of the error-prone PCR primers. Larger quantities of these primers are used; therefore, the shorter primers are more economically favorable.

PCR in 200 μL volumes

Volume	Component	Final concentration
20 μL	10× *ThermoPol* buffer	1× *ThermoPol* buffer
20 μL	5′-Primer (10 μ*M*)	1 μ*M* 5′-primer
20 μL	3′-Primer (10 μ*M*)	1 μ*M* 3′-primer
4 μL	dNTPs (10 m*M* each)	200 μ*M* dNTPs each
8 μL	Extracted mutated DNA	
126 μL	ddH_2O	
2 μL	*Taq* DNA polymerase	2.5 units

Split reactions into two 100 μL aliquots per sample for a total of eight tubes.

2. Thermal cycle setup

Step	Temperature (°C)	Time	Cycle
Initialization	94	3 min	1
Denaturation	94	45 s	2–31
Annealing	60	30 s	
Extension	72	90 s	
Final extension	72	10 min	32

Amplified DNA can be used immediately or stored at 4 °C.

3. Concentrate the amplified PCR products by ethanol precipitation. Combine the gene amplification products into a 1.5-mL microcentrifuge tube and combine all of the loop amplifications into a 2-mL microcentrifuge tube. *Optional*: 2 μL of pellet paint coprecipitant®

(Novagen, EMD Millipore) can be used to visualize the DNA pellet. Adjust the pH by adding 10% volume of 3 *M* sodium acetate to the PCR amplifications: 20 μL for the gene amplification and 60 μL for the loop amplifications. Add at least 2 × volume of 100% ethanol to each tube: 400 μL for gene amplification and 1200 μL for loops. Incubate at room temperature for 2 min and then spin at high speed (12,000 × *g*) for 5 min to pellet the DNA. Remove the supernatant and add 500 μL of 70% ethanol. Mix/vortex briefly and pellet again at high speed for 5 min. Remove the supernatant and add 500 μL of 100% ethanol. Mix/vortex briefly and pellet again at high speed for 5 min. Remove supernatant and air dry over night or for 10 min on a heat block at 48 °C. Dissolve dried pellet in 1 μL of ddH_2O in preparation for electroporation.

2.4.4 Preparation of vectors

Since there are two different amplification products: gene and loops, two different vectors that are variants of pCT-CON must be prepared for the electroporation. The pCT-Fn3-Gene vector has a sequence consisting of the *Pst*I, *Nde*I, and *Bam*HI sites separated by spacers replacing the DNA between the HA and c-myc tags. The pCT-Fn3-Loop vector contains the *Fn3* gene except the DNA between the BC loop and FG loop is replaced by *Nco*I, *Sma*I, and *Nde*I cut sites separated by spacers. The vectors are digested at three sites to ensure that the linearized vectors will only recircularize in the presence of insert. Vectors are available upon request.

1. To linearize the pCT-Fn3-Gene vector, digest with *Nde*I as specified by the manufacture's protocol at 37 °C overnight. To linearize the pCT-Fn3-Loop vector, digest with *Sma*I according to the manufacture's protocol at 25 °C overnight.
2. Adjust buffer volumes for both vectors. Digest pCT-Fn3-Gene with *Pst*I and *Bam*HI following the manufacture's protocol at 37 °C overnight. Digest pCT-Fn3-Loop with *Nco*I and *Nde*I at 37 °C overnight.
3. Concentrate digested vectors using ethanol precipitation as previously described (step 3 of Section 2.4.3). Dissolve vectors in ddH_2O at a concentration of 2 μg/μL. Store at 4 °C for short term or −20 °C for long term.

2.4.5 Electroporation protocol

Linearized vector and inserts will be introduced into the yeast through electroporation transformation. Yeast will naturally perform homologous recombination to recircularize linearized vector and inserts. The following electroporation protocol has been optimized for the Bio-Rad Gene Pulser Xcell (Bio-Rad) electroporation instrument. The electroporation protocol for the Bio-Rad Gene Pulser (Bio-Rad) can be performed as previously described (Chao et al., 2006).

1. An EBY100 starter culture in 5 mL of YPD can be prepared in the following ways: inoculate with a EBY100 colony from a freshly streaked YPD plate, a 1:50 dilution of EBY100 from a culture stored at 4 °C, or a streak from an EBY100 frozen stock. Grow culture overnight at 30 °C with shaking at 250 rpm.

Note: EBY100 doubles every 1.5–2 h in YPD.

2. Inoculate 50 mL of YPD at $OD_{600nm}=0.2$ and grow at 30 °C with shaking at 250 rpm until $OD_{600nm}=1.3–1.5$ (~4–6 h). The culture volume varies depending on the number of transformations to be performed; usually, 50 mL is sufficient for two electroporations (one for gene and another for loops).
3. Pellet cells in 50-mL conical tubes at 3000 × *g* for 3 min and remove supernatant.
4. Resuspend cells in half of the original YPD volume (25 mL) with 100 m*M* lithium acetate. Make fresh 1 *M* dithiothreitol, sterile filter, and add to cells to a final concentration of 10 m*M*. Incubate cells at 30 °C with shaking for 10 min.
5. Pellet cells at 3000 × *g* for 3min and remove supernatant. Place cells on ice; all of the following steps will be performed on ice and with chilled reagents.
6. Resuspend cells in chilled sterile ddH_2O using half of the original volume (25 mL). Pellet cells at 3000 × *g* for 3 min and remove supernatant.
7. Resuspend cells in 250 μL of chilled sterile ddH_2O. The total volume will be approximately 500 μL of cells for each 50 mL original culture volume.
8. In parallel with cell preparation, add 4 μg of vector (2 μL) to amplified inserts (digested gene vector to gene inserts and digested loop vector to loop inserts). Chill two 2 mm electroporation cuvettes on ice.
9. Mix 250 μL of cells with the gene preparation and the remaining 250 μL of cells with the loop preparation. Transfer mixtures to pre-chilled electroporation cuvettes and keep on ice until electroporation.
10. Use the square wave protocol on the Bio-Rad Gene Pulser Xcell. Perform a single pulse at 500 V with a 15-ms pulse duration. Typical "Droop" readings reported by the machine are within 5–6%.

11. Rescue cells with 1 mL of YPD and transfer cells to a 14-mL polypropylene round bottom tube (Falcon). Rinse the cuvette with an additional 1 mL of YPD and transfer cells to the same tube. Incubate cells at 30 °C without shaking for 1 h.
12. Plate dilutions of cells on SD-CAA plates to determine the number of transformants of the new library. Grow plates at 30 °C for 2–3 days. This protocol typically yields 10^7–10^8 transformants.
13. Pellet the remaining cells at 3000 × *g* for 3 min. Combine both gene and loop libraries in a single culture of 500 mL SD-CAA and grow at 30 °C with shaking at 250 rpm overnight.

Optional: The library can be passaged to decrease the number of untransformed cells (*Optional*: Step 1 of Section 2.2).

14. Induce cells at 10 × diversity with 500 mL of SG-CAA at 20 °C with shaking at 250 rpm for 8–24 h.

2.5. Library screening with FACS

After a few rounds of magnetic-bead enrichment, FACS can be used to quantitatively screen yeast-displayed Fn3 libraries. Libraries will be labeled for display through c-myc tag detection and binding with soluble antigen. Using this double labeling, antigen binding can be normalized by display levels.

2.5.1 Labeling protocol for FACS

1. Pellet 10 × diversity of induced yeast in a 1.5-mL microcentrifuge tube at max speed for 1 min and remove supernatant. Wash with 1 mL PBSA by pelleting again and remove the supernatant.
2. Resuspend cells in PBSA. Volumes vary depending on number of cells. Typically, a labeling volume of 50 μL is used for 1×10^6–1×10^7 cells and a volume of 0.5–1 mL is used for 1×10^8 cells.
3. Primary labeling: Add anti-c-myc antibody (chicken anti-c-myc (Invitrogen or Gallus Immunotech) or mouse anti-c-myc (9E10, Covance)) to at least 20 n*M* (3 mg/L) final concentration. Add desired amount of soluble biotinylated antigen. Concentrations will vary depending on the binding affinity of the entire population. Typical labeling concentrations when first switching from magnetic bead sorting to FACS range from 300 to 500 n*M*. The optimal concentration of labeling for subsequent sorts can be calculated as previously described (Boder & Wittrup, 1998).

Note: Always make sure primaries are in stoichiometric excess (10-fold) of Fn3. At lower labeling concentrations, volumes will need to be increased to maintain excess ligand.

4. Incubate at room temperature or 37 °C until equilibrium binding has been reached; approximately 30 min should suffice for labeling at 20 nM or higher.

Note: Time to half of equilibrium can be calculated by Eq. (14.1), where k_{on} is the rate of association (typical protein–protein k_{on} is $1 \times 10^5 M^{-1} s^{-1}$) and k_{off} is the rate of dissociation, which can be calculated from an approximation of the equilibrium constant, K_d, Eq. (14.2). $[L]_o$ is the initial ligand concentration operating under the assumption of no ligand depletion. Five half-lives is enough to reach 97% of equilibrium.

$$\tau_{1/2} = \frac{\ln 2}{k_{on}[L]_o + k_{off}} \qquad [14.1]$$

$$K_d = \frac{k_{off}}{k_{on}} \qquad [14.2]$$

5. Secondary labeling: Wash cells with 1 mL PBSA, pellet, and remove supernatant. Subsequent steps are performed on ice. Resuspend in desired volume of chilled PBSA and label with secondaries at 50 nM or higher. Typical secondaries used are anti-chicken or anti-mouse Alexa Fluor® 488 or Alexa Fluor® 647 conjugates (Invitrogen) for detection of the anti-c-myc antibody. Streptavidin Alexa Fluor® 488 or Alexa Fluor® 647 conjugates (Invitrogen) or anti-biotin PE conjugate (ebioscience) are used for biotinylated antigen detection. Incubate for approximately 15–30 min on ice. Once again, make sure secondaries are in stoichiometric excess of Fn3. Time to equilibrium can be calculated as mentioned previously.
6. Wash cells with 1 mL PBSA, pellet, and remove supernatant. Right before sorting, resuspend cells in PBSA at a concentration of 1×10^8 cells/mL and transfer to proper FACS tubes.

When sorting for high affinity binders with subnanomolar binding affinities, dissociation competition can be used by first labeling with a stoichiometric excess of biotinylated antigen, washing, and then competing with non-labeled unbiotinylated antigen.

2.5.2 Sorting for improved clones

1. In initial sorts, if no strong double-positive diagonal is detected, collect all cells that show detectable binding to the antigen of interest, as shown in Fig. 14.6A. This gate should cover the top 1–3% of displaying cells.
2. In subsequent sorts, in a diagonal sort window, collect the top 0.1–1% of cells that show strong display and binding as shown in Fig. 14.6B.

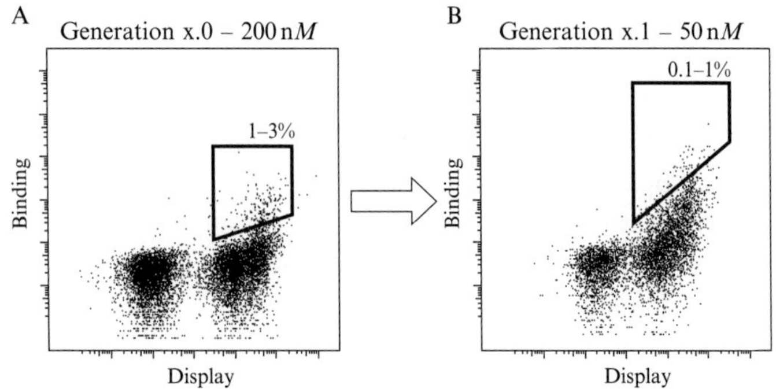

Figure 14.6 Representative FACS plots of sorting for improved clones. (A) In initial FACS sorts, collect the top 1–3% of displaying cells that show detectable binding, (B) in subsequent FACS sorts, collection gates should be more stringent (top 0.1–1% of population) to isolate binders with high affinity to the antigen of interest.

3. Collect cells in 2 mL of SD-CAA and follow the same procedure as outlined in step 6 of Section 2.3.5. Grow cells at 30 °C with shaking at 250 rpm overnight.
4. Induce cells with at least 10 × diversity starting at an initial OD_{600nm} of 1. Incubate cells at 20 °C with shaking at 250 rpm for 8–24 h. Induction time can be decreased to increase stringency in sorting by having less Fn3 displayed on the surface of yeast.

3. ANALYSIS OF INDIVIDUAL CLONES

After the population with binding of interest is isolated, individual clones can be identified and characterized with titrations.

3.1. Identification of individual clones

1. Zymoprep the population of interest as mentioned previously (Section 2.4.1). Transform 1 μL of zymoprep into XL1-Blue or DH5α supercompetent cells. Plate the entire reaction on LB + ampicillin plates and grow the plates at 37 °C overnight for at least 16 h.
2. Pick individual colonies into 5 mL of LB + ampicillin media at 37 °C with shaking at 250 rpm overnight.
3. Miniprep cultures to isolate plasmid DNA, elute in a volume of 50 μL. Typical yields with the plasmid pCT-CON are approximately 500 ng/μL. Send for sequencing.

Direction	Sequence
Forward	5′-gttccagactacgctctgcagg-3′
Reverse	5′-gattttgttacatctacactgttg-3′

3.2. Clonal yeast preparation

The following procedure is adapted from the Frozen-EZ Yeast Transformation II kit (Zymo Research) protocol.

1. Transform 1 μL (100–1000 ng) of miniprep DNA into 10 μL of prepared competent EBY100. Add 100 μL of Solution 3 and mix gently. Perform transformation according to the manufacturer's protocol.
2. Plate entire reaction on SD-CAA plates for selection. Grow plates at 30 °C for 2–3 days.

Note: If the cells are extremely competent, the amount of reaction plated can be decreased.

3. Pick a single colony and grow up in 5 mL of SD-CAA media at 30 °C with shaking at 250 rpm overnight.
4. Induce cells in 5 mL SG-CAA at starting an OD_{600nm} of 1. Incubate cells at 20 °C with shaking at 250 rpm for 8–24 h.

Note: Cultures can be stored at 4 °C until ready for analysis.

5. Titrations with the target of interest can be performed on the clonal yeast. The titration protocol is as previously described (Chao et al., 2006).

3.3. Expression of soluble Fn3

Soluble expression of Fn3s in *E. coli* has yielded approximately 50 mg/L of protein. Various expression vectors can be used ranging from the pET (Novagen, EMD Millipore), pMal (NEB), or pET SUMO (Invitrogen) vectors. These vectors can be altered to include a purification tag (i.e., His6 or FLAG) and the restriction sites *Nhe*I and *Bam*HI if not already included. We use a modified-pET vector, pEThk, which includes a C-terminal His6-KGSGK tag. The addition of two lysines provides primary amines that are available for chemical conjugation.

1. The Fn3 binder gene can be cut out using restriction enzyme digestion with *Nhe*I and *Bam*HI following the manufacturer's protocol. In parallel, the vector of interest should be digested with the same enzymes.
2. The digested product (~300 bp) should be run on a gel and extracted with the Qiaquick Gel Extraction Kit (Qiagen) according to the manufacturer's instructions.

3. The insert can then be ligated to the digested vector with Quick Ligase (NEB) or T4 DNA Ligase (NEB) according to the manufacturer's protocol. The ligation product should then be transformed into XL1-Blue or DH5α supercompetent cells. The entire transformation should be plated on LB+antibiotic plates (the antibiotic resistance varies depending on vector used).
4. Pick colonies from the plate into 5 mL cultures of LB+antibiotic and allow to grow overnight at 37 °C shaking at 250 rpm. Miniprep the cultures after approximately 16 h of growth and send the samples for sequencing along with the plasmid-specific sequencing primers.
5. Once the proper sequence is confirmed, the plasmid can be transformed into bacterial strains typically used for protein expression. Our preferred strain is Rosetta 2 (DE3) competent cells (Novagen, EMD Millipore). Transform 0.5 μL (50–250 ng) of the plasmid into 20 μL of cells using heat shock. Plate a fraction (~1/5) of the cells on an LB+antibiotic plate.
6. Pick a colony into a 5-mL overnight culture of LB+antibiotic for approximately 16 h at 37 °C shaking at 250 rpm.
7. Make a 1:1000 dilution of the overnight culture in fresh LB media with no antibiotics. Allow the culture to grow at 37 °C shaking at 250 rpm until log phase ($\sim OD_{600nm}=0.5$–1.0).
8. Induce with a 1:1000 dilution of 500 m*M* IPTG for a final concentration of 0.5 m*M* IPTG. Induce cells, shaking at 250 rpm at 37 °C for approximately 4 h, 30 °C for approximately 6 h, or 20 °C for overnight inductions.
9. Fn3 protein can be collected by pelleting cells and lysing them using sonication or several freeze–thaw cycles. After cell lysis, protein should be purified using the proper resin depending on the purification tag fused to the Fn3.

4. SUMMARY

YSD has been shown to be a powerful platform for protein engineering. In this chapter, we provided a detailed methodology for engineering binders from the 10th type III domain Fn3 protein scaffold. Engineering binders from this single-domain protein scaffold is different from typical scFv engineering. The inherent potential for convexity may alter preferred binding epitopes, while the potentially reduced diversified surface area relative to scFvs heightens the need for appropriate complementarity. Including a mutagenesis step after every two enrichments allows for the constant introduction of diversity and thoroughly utilizes directed evolution to affinity mature clones. These engineered Fn3 domains have a wide range of applications ranging from imaging (Hackel, Kimura, & Gambhir, 2012), *in vitro* detection and diagnosis (Gulyani et al., 2011), and *in vivo* therapeutics (Tolcher et al., 2011), and have many advantages because of their small size and single-domain architecture.

ACKNOWLEDGMENTS

We are grateful to all other current and former Wittrup lab members for informative discussions and daily help. We also thank the Koch Institute Flow Cytometry Core facility and Biopolymers Core facility for FACS experiments and sequencing, respectively.

Funding sources are from the Sanofi-aventis Biomedical Innovation award program, the NIH/NIGMS Biotechnology Training program, the Gordon & Adele Binder Fellowship, NIH Transformative R01 Program (R01EB010246-02), and Integrative Cancer Biology Program (ICBP) at MIT.

REFERENCES

Ackerman, M., Levary, D., Tobon, G., Hackel, B., Orcutt, K. D., & Wittrup, K. D. (2009). Highly avid magnetic bead capture: An efficient selection method for de novo protein engineering utilizing yeast surface display. *Biotechnology Progress*, *25*, 774–783.

Binz, H. K., Amstutz, P., & Pluckthun, A. (2005). Engineering novel binding proteins from nonimmunoglobulin domains. *Nature Biotechnology*, *23*, 1257–1268.

Boder, E. T., Raeeszadeh-Sarmazdeh, M., & Price, J. V. (2012). Engineering antibodies by yeast display. *Archives of Biochemistry and Biophysics*, *526*, 99–106.

Boder, E. T., & Wittrup, K. D. (1997). Yeast surface display for screening combinatorial polypeptide libraries. *Nature Biotechnology*, *15*, 553–557.

Boder, E. T., & Wittrup, K. D. (1998). Optimal screening of surface-displayed polypeptide libraries. *Biotechnology Progress*, *14*, 55–62.

Bowley, D. R., Labrijn, A. F., Zwick, M. B., & Burton, D. R. (2007). Antigen selection from an HIV-1 immune antibody library displayed on yeast yields many novel antibodies compared to selection from the same library displayed on phage. *Protein Engineering, Design and Selection*, *20*, 81–90.

Chao, G., Lau, W. L., Hackel, B. J., Sazinsky, S. L., Lippow, S. M., & Wittrup, K. D. (2006). Isolating and engineering human antibodies using yeast surface display. *Nature Protocols*, *1*, 755–768.

Colby, D. W., Kellogg, B. A., Graff, C. P., Yeung, Y. A., Swers, J. S., & Wittrup, K. D. (2004). Engineering antibody affinity by yeast surface display. *Methods in Enzymology*, *388*, 348–358.

Gai, S. A., & Wittrup, K. D. (2007). Yeast surface display for protein engineering and characterization. *Current Opinion in Structural Biology*, *17*, 467–473.

Gulyani, A., Vitriol, E., Allen, R., Wu, J., Gremyachinskiy, D., Lewis, S., et al. (2011). A biosensor generated via high-throughput screening quantifies cell edge Src dynamics. *Nature Chemical Biology*, *7*, 437–444.

Hackel, B. J., Ackerman, M. E., Howland, S. W., & Wittrup, K. D. (2010). Stability and CDR composition biases enrich binder functionality landscapes. *Journal of Molecular Biology*, *401*, 84–96.

Hackel, B. J., Kapila, A., & Wittrup, K. D. (2008). Picomolar affinity fibronectin domains engineered utilizing loop length diversity, recursive mutagenesis, and loop shuffling. *Journal of Molecular Biology*, *381*, 1238–1252.

Hackel, B. J., Kimura, R. H., & Gambhir, S. S. (2012). Use of (64)Cu-labeled fibronectin domain with EGFR-overexpressing tumor xenograft: Molecular imaging. *Radiology*, *263*, 179–188.

Hackel, B. J., Neil, J. R., White, F. M., & Wittrup, K. D. (2012). Epidermal growth factor receptor downregulation by small heterodimeric binding proteins. *Protein Engineering, Design & Selection*, *25*, 47–57.

Hackel, B. J., & Wittrup, K. D. (2010). The full amino acid repertoire is superior to serine/tyrosine for selection of high affinity immunoglobulin G binders from the fibronectin scaffold. *Protein Engineering, Design and Selection*, *23*, 211–219.

Koide, A., & Koide, S. (2007). Monobodies: Antibody mimics based on the scaffold of the fibronectin type III domain. *Methods in Molecular Biology*, *352*, 95–109.

Lipovsek, D., Lippow, S. M., Hackel, B. J., Gregson, M. W., Cheng, P., Kapila, A., et al. (2007). Evolution of an interloop disulfide bond in high-affinity antibody mimics based on fibronectin type III domain and selected by yeast surface display: Molecular convergence with single-domain camelid and shark antibodies. *Journal of Molecular Biology*, *368*, 1024–1041.

Löfblom, J., Feldwisch, J., Tolmachev, V., Carlsson, J., Ståhl, S., & Frejd, F. Y. (2010). Affibody molecules: Engineered proteins for therapeutic, diagnostic and biotechnological applications. *FEBS Letters*, *584*, 2670–2680.

Pirie, C. M., Hackel, B. J., Rosenblum, M. G., & Wittrup, K. D. (2011). Convergent potency of internalized gelonin immunotoxins across varied cell lines, antigens, and targeting moieties. *The Journal of Biological Chemistry*, *286*, 4165–4172.

Shapiro, H. M. (2005). *Frontmatter, in Practical Flow Cytometry*. Hoboken, NJ, USA: John Wiley & Sons, Inc. p. i–l.

Skerra, A. (2007). Alternative non-antibody scaffolds for molecular recognition. *Current Opinion in Biotechnology*, *18*, 295–304.

Skerra, A. (2008). Alternative binding proteins: Anticalins—Harnessing the structural plasticity of the lipocalin ligand pocket to engineer novel binding activities. *The FEBS Journal*, *275*, 2677–2683.

Spangler, J. B., Manzari, M. T., Rosalia, E. K., Chen, T. F., & Wittrup, K. D. (2012). Triepitopic antibody fusions inhibit cetuximab-resistant BRAF- and KRAS-mutant tumors via EGFR signal repression. *Journal of Molecular Biology*, *422*, 532–544.

Stumpp, M. T., Binz, H. K., & Amstutz, P. (2008). DARPins: A new generation of protein therapeutics. *Drug Discovery Today*, *13*, 695–701.

Tolcher, A. W., Sweeney, C. J., Papadopoulos, K., Patnaik, A., Chiorean, E. G., Mita, A. C., et al. (2011). Phase I and pharmacokinetic study of CT-322 (BMS-844203), a targeted adnectin inhibitor of VEGFR-2 based on a domain of human fibronectin. *Clinical Cancer Research*, *17*, 363–371.

CHAPTER FIFTEEN

Engineering and Analysis of Peptide-Recognition Domain Specificities by Phage Display and Deep Sequencing

Megan E. McLaughlin[*,†], **Sachdev S. Sidhu**[*,†,‡,1]

[*]Department of Molecular Genetics, University of Toronto, Toronto, Ontario, Canada
[†]Terrence Donnelly Centre for Cellular and Biomolecular Research, University of Toronto, Toronto, Ontario, Canada
[‡]Banting and Best Department of Medical Research, University of Toronto, Toronto, Ontario, Canada
[1]Corresponding author: e-mail address: sachdev.sidhu@utoronto.ca

Contents

Abstract

Protein interaction networks depend in part on the specific recognition of unstructured peptides by folded domains. Understanding how members of a domain family use a similar fold to recognize different peptide sequences selectively is a fundamental question. One way to advance our understanding of peptide recognition is to apply an existing model of peptide recognition for a particular domain toward engineering synthetic domain variants with desired properties. Successes, failures, and unintended outcomes can help refine the model and can illuminate more general principles of peptide recognition. Using the PDZ domain fold as an example, we describe methods for (1) structure-based combinatorial library design and directed evolution of domain variants

Methods in Enzymology, Volume 523
ISSN 0076-6879
http://dx.doi.org/10.1016/B978-0-12-394292-0.00015-1

and (2) specificity profiling of large repertoires of synthetic variants using multiplexed deep sequencing. Peptide-binding preferences for hundreds of variants can be decoded in parallel, enabling comparisons between different library designs and selection pressures. The tremendous depth of coverage of the binding peptide profiles also permits robust computational analysis. This approach to studying peptide recognition can be applied to other domains and to a variety of structural and functional models by tailoring the combinatorial library design and selection pressures accordingly.

1. INTRODUCTION

Protein interaction networks depend in part on the specific recognition of unstructured peptides by folded domains. The human genome encodes many peptide-recognition domain families, whose members may occur hundreds of times in different protein contexts. Understanding how such domains use a similar fold to recognize different peptide sequences selectively is fundamental to understanding how these networks function (Pawson, 2003).

One way to test a proposed model for peptide recognition is to apply it, by attempting to engineer synthetic domains with desired binding properties. First, structure–function data guide the design of a combinatorial library of domain variants, in which residues that are expected to contribute to specificity are randomized. Subsequently, positive and negative selection pressures can be applied to the library in order to recover synthetic variants that possess the desired properties.

Finally, synthetic variants are specificity profiled with a random peptide library to determine their binding preferences in a relatively unbiased way. Crucially, large repertoires of synthetic variants can be characterized using multiplexed deep sequencing to decode peptide profiles. This allows a comprehensive evaluation of the outcome of one or many experiments, enabling comparisons between different library designs or different selection pressures. Even when peptide profiles for hundreds of synthetic variants are multiplexed for a single deep sequencing run, the unprecedented depth of coverage enables robust computational analysis (Ernst et al., 2010). Successes, failures, and unintended outcomes can help to refine the initial model and can illuminate the general principles of peptide recognition.

The methods described in this chapter have been used to test a model of PDZ domain evolution (Ernst et al., 2009). PDZ domains typically bind specific C-terminal peptides of other proteins to assemble complexes (Songyang, 1997). More than 250 PDZ domains are encoded in the human genome, and they exhibit a broad range of specificities (Tonikian et al., 2008). To understand how a progenitor PDZ domain could give rise to such functional diversity, Ernst et al. (2009) designed a library in which the core binding site of one natural PDZ domain was randomized. Protease-resistant variants were selected from the library and several hundred of these well-folded domains were specificity profiled; nearly one quarter of them recognized peptides, and this repertoire of synthetic variants was as diverse and as specific as the natural set. This demonstrated directly what had been inferred from the rapid evolutionary expansion of natural PDZ domains: the PDZ fold supports diversity at key specificity-determining residues, enabling great functional complexity (Ernst et al., 2009).

The methods described in this chapter for high-throughput engineering and characterization of peptide-recognition domains can be applied to other families and to a variety of structural and functional hypotheses, by tailoring the combinatorial library design and selection pressures accordingly.

2. Directed Evolution of PDZ Variants

In previous work using the Erbin PDZ domain as progenitor (Fig. 15.1A) (Skelton, 2002), 10 binding-site positions that were predicted to determine specificity were diversified to generate a combinatorial library of PDZ variants (Ernst et al., 2009). The following sections provide protocols by which PDZ or other peptide-recognition domain variants can be displayed on phage particles as fusions to a coat protein; each displayed variant is encoded by the genome packaged inside (Fig.15.1B and C). PDZ variants with the desired properties are iteratively enriched using a selection process; PDZ variants that do not bind an immobilized substrate are washed away, while PDZ-phage that do bind are retained and amplified for the next round by passage through a bacterial host. After multiple rounds or enrichment, PDZ variants that possess the selected properties dominate the phage population, and these can be readily identified, sequenced, and characterized.

2.1. Preparation of PDZ variant library

Very large ($>10^{10}$) phage-displayed combinatorial libraries can be constructed using optimized procedures (Sidhu, Lowman, Cunningham, & Wells, 2000) that are based on the method of Kunkel, Roberts, and Zakour (1987) (Fig. 15.2). Mutagenic oligonucleotides with degenerate codons introduce tailored amino acid diversity at key positions of the PDZ domain. These oligonucleotides are annealed to uracil-containing

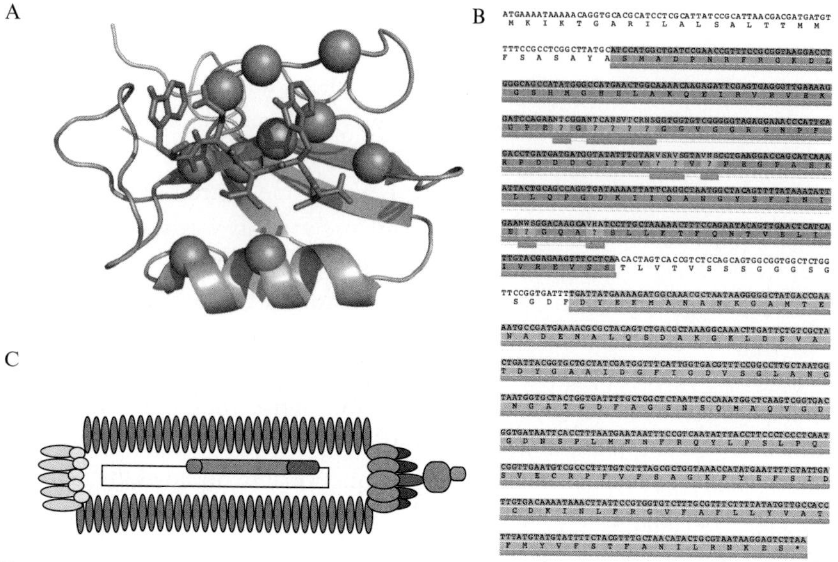

Figure 15.1 Design and display of a combinatorial library of PDZ variants on phage. (A) Structure of the Erbin PDZ domain (PDB:1N7T) (*gray cartoon*) bound to C-terminal peptide $WETWV_{COOH}$ (*magenta stick*). Ten domain residues predicted to determine binding specificity were diversified to generate a combinatorial library (*green spheres*). (B) Sequence of the recombinant phage coat protein p3 (*cyan*) displaying a PDZ variant (*gray with diversified residues in green*) with an N-terminal epitope tag (*orange*). Diversified codons are indicated (N = A/C/G/T, V = A/C/G, R = A/G, S = C/G, W = A/T, H = A/C/T). (C) Diagram of a phage particle, with enclosed phagemid genome encoding the displayed PDZ variant (*green PDZ variant, orange epitope tag, cyan coat protein p3*). Other major and minor coat proteins are shown as grey ellipses. (See Color Insert.)

single-strand DNA templates (dU-ssDNA) harvested from a *dut*$^-$/*ung*$^-$ *Escherichia coli* host. Separate oligonucleotides can be used to target different regions of the PDZ domain, provided that their annealing regions (15 bases on either side of the mutated region) do not overlap. The annealed mutagenic oligonucleotides prime the synthesis of a complementary DNA strand that is ligated to form a covalently closed circular, double-stranded DNA (CCC-dsDNA) heteroduplex. Electroporation of the CCC-dsDNA into a *dut*$^+$/*ung*$^+$ *E. coli* host, which preferentially inactivates the uracil-containing template strand, results in efficient mutagenesis (>80%). To ensure that only mutagenized PDZ variants are displayed, stop codons are first introduced into the parental template in the regions to be diversified. Mutation repairs the stop codon. Unless all regions are mutated, a functional

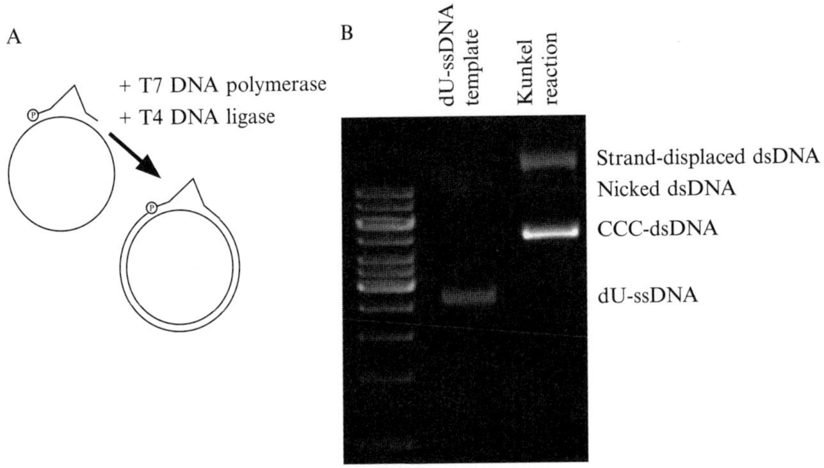

Figure 15.2 *In vitro* synthesis of heteroduplex CCC-dsDNA. (A) Mutagenic oligonucleotide annealing to dU-ssDNA template primes enzymatic synthesis of the complementary strand to yield heteroduplex CCC-dsDNA. (B) A successful Kunkel reaction converts all ssDNA to dsDNA, with the predominant species being CCC-dsDNA, which transforms *E. coli* efficiently. The two dsDNA species with slower mobilities result from undesirable strand displacement by T7 DNA polymerase or incomplete ligation due to insufficient ligase activity or incomplete oligonucleotide phosphorylation, and have much lower transformation efficiency.

fusion polypeptide is not expressed and the PDZ variant is cleared from the pool during selections.

The phagemid encoding the PDZ-coat protein has a dsDNA origin, so it can replicate inside the host as a plasmid. Coinfection with a helper phage (e.g., M13KO7) supplies genes encoding all the proteins necessary for ssDNA genome replication and phage particle assembly *in trans*, resulting in virion production.

2.1.1 Purification of dU-ssDNA template

This modified protocol for the QIAprep Spin M13 Kit (Qiagen, Valencia, CA) typically yields >20 μg of dU-ssDNA, which is sufficient for the construction of one library.

1. From a fresh LB/carb plate, pick a single colony of *E. coli* CJ236 (New England Biolabs, Beverly, MA) (or another *dut*$^{-}$/*ung*$^{-}$ strain) harboring the appropriate phagemid into 1 ml of 2YT medium (10 g bacto-yeast extract, 16 g bacto-tryptone, 5 g NaCl, water to 1.0 l, pH adjusted to 7.0 with NaOH, and autoclaved) supplemented with M13KO7 helper phage (New England Biolabs) (10^{10} pfu/ml) and appropriate antibiotics

to maintain the phagemid and the host F′ episome required for filamentous phage infection. For example, 2YT/carb/cmp medium contains carbenicillin (100 μg/ml) to select for phagemids that encode the recombinant coat protein and beta-lactamase gene, and chloramphenicol (10 μg/ml) to select for the F′ episome of *E. coli* CJ236.

2. Shake at 200 rpm and 37 °C for 2 h and add kanamycin (25 μg/ml) to select for clones that have been coinfected with M13KO7.
3. Shake at 200 rpm and 37 °C for 6 h and transfer the culture to 30 ml of 2YT/carb/kan medium supplemented with uridine (0.25 μg/ml).
4. Shake at 200 rpm and 37 °C for 16 h.
5. Centrifuge for 10 min at 17,600 × *g* and 4 °C to pellet bacteria. Transfer the supernatant to a new tube containing 1/4 volume of PEG/NaCl (7.5 ml) (20% PEG-8000 (w/v), 2.5 *M* NaCl, filter sterilized) and incubate for 5 min at room temperature.
6. Centrifuge for 10 min at 17,600 × *g* and 4 °C to pellet phage. Decant the supernatant; centrifuge for 1 min at 17,600 × *g* and 4 °C and aspirate the remaining supernatant.
7. Resuspend the phage pellet in 0.5 ml of phosphate-buffered saline (PBS) (137 m*M* NaCl, 3 m*M* KCl, 8 m*M* Na_2HPO_4, 1.5 m*M* KH_2PO_4, pH adjusted to 7.2 with HCl, and autoclaved) by gentle pipetting until the solution is homogeneous, and transfer to a 1.5-ml microcentrifuge tube.
8. Centrifuge for 5 min at 15,000 × *g* to pellet insoluble debris, and transfer the supernatant to a 1.5-ml microcentrifuge tube.
9. Add 7.0 μl of Buffer MP and mix. Incubate at room temperature for at least 2 min.
10. Apply the sample to a QIAprep spin column in a 2-ml microcentrifuge tube. Centrifuge for 30 s at 6000 × *g* in a microcentrifuge. Discard the flow-through.
11. Add 0.7 ml of Buffer MLB to the column. Centrifuge for 30 s at 6000 × *g* and discard the flow-through.
12. Add 0.7 ml of buffer MLB. Incubate at room temperature for at least 1 min.
13. Centrifuge for 30 s at 6000 × *g* and discard the flow-through.
14. Add 0.7 ml of Buffer PE. Centrifuge for 30 s at 6000 × *g* and discard the flow-through.
15. Repeat step 14.
16. Centrifuge the column for 30 s at 6000 × *g* to remove residual PE buffer.
17. Transfer the column to a fresh 1.5-ml microcentrifuge tube.

18. Add 100 μl of Buffer EB to the center of the column membrane. Incubate at room temperature for 10 min.
19. Centrifuge for 30 s at 6000 × *g*. Save the eluant, which contains the purified dU-ssDNA.
20. Analyze the DNA by electrophoresing 1.0 μl on a gel containing 0.8% agarose (w/v), 1×TAE (40 m*M* Tris–acetate, 1.0 m*M* EDTA; pH adjusted to 8.0 with NaOH), 1:5000 Sybr Safe DNA dye (Invitrogen, Carlsbad, CA). Faint bands with lower electrophoretic mobility are often visible. These are likely caused by secondary structure in the dU-ssDNA.
21. Determine the DNA concentration by measuring absorbance at 260 nm (A_{260} = 1.0 for 33 ng/ml of ssDNA). Typical DNA concentrations range from 200 to 500 ng/ml.

2.1.2 In vitro *synthesis of heteroduplex CCC-dsDNA*

The purified and desalted sample will contain 20 μg of highly pure, low-conductance CCC-dsDNA, which is sufficient for the construction of a library containing more than 10^{10} unique members.

2.1.2.1 Oligonucleotide phosphorylation with T4 polynucleotide kinase

1. In a 1.5-ml microcentrifuge tube, combine 0.6 μg of mutagenic oligonucleotide with 2.0 μl of 10 × TM buffer (0.1 *M* $MgCl_2$, 0.5 *M* Tris, pH 7.5), 2.0 μl of 10 m*M* ATP, and 1.0 μl of 100 m*M* DTT. Add water to a total volume of 20 μl. Mutagenic oligonucleotides targeting different regions should be phosphorylated separately.
2. Add 20 units of T4 polynucleotide kinase (New England Biolabs) to each tube. Incubate for 1 h at 37 °C and use immediately for annealing.

2.1.2.2 Annealing of the oligonucleotides to the template

1. To 20 μg of dU-ssDNA template, add 25 μl of 10×TM buffer, 20 μl of each phosphorylated oligonucleotide, and water to a final volume of 250 μl. These DNA quantities provide an oligonucleotide:template molar ratio of 3:1.
2. Incubate at 90 °C for 3 min, 50 °C for 3 min, and 20 °C for 5 min.

2.1.2.3 Enzymatic synthesis of CCC-dsDNA

1. To the annealed oligonucleotide/template mixture, add 10 μl of 10 m*M* ATP, 10 μl of 10 m*M* dNTP mix, 15 ml of 100 μ*M* DTT, 30 Weiss units of T4 DNA ligase (New England Biolabs), and 30 units of T7 DNA polymerase (New England Biolabs).

2. Incubate overnight at 20 °C.
3. Affinity purify and desalt the DNA using the Qiagen QIAquick Gel Extraction Kit (Qiagen), as follows:
4. Add 1.0 ml of Buffer QG and mix.
5. Apply the sample to two QIAquick spin columns placed in 2-ml microcentrifuge tubes. Centrifuge for 1 min at 11,000 × *g*. Discard the flow-through.
6. Add 750 μl Buffer PE to each column, and centrifuge for 1 min at 11,000 × *g*. Discard the flow-through.
7. Centrifuge the column for 1 min at 11,000 × *g* to remove excess Buffer PE.
8. Transfer the column to a fresh 1.5-ml microcentrifuge tube, and add 35 μl of ultrapure irrigation USP water (Braun Medical, Irvine, CA) to the center of the membrane. Incubate at room temperature for 2 min. Centrifuge for 1 min at 11,000 × *g* to elute the DNA. Combine the eluants and use immediately for *E. coli* electroporation, or freeze for later use.
9. Electrophorese 1.0 μl of the eluted reaction product alongside the ssDNA template. See Fig. 15.2B for expected results.

2.1.3 Conversion of CCC-dsDNA into a phage-displayed library

Phage-displayed library diversities are limited by methods for introducing DNA into *E. coli*, with the most efficient method being high-voltage electroporation. *E. coli* strain SS320 (Lucigen, Middleton, WI) is ideal for both high-efficiency electroporation and phage production (Sidhu et al., 2000). Preinfection of the cells with helper phage M13KO7 ensures that every transformed cell will produce phage particles.

2.1.3.1 Preparation of electrocompetent *E. coli* SS320

The following protocol yields ~12 ml of highly concentrated, electrocompetent *E. coli* SS320 (~3×10^{11} cfu/ml) infected by M13KO7 helper phage.

1. Inoculate 25 ml 2YT/tet (10 μg/ml) medium with a single colony of *E. coli* SS320 from a fresh LB/tet plate. Incubate at 37 °C with shaking at 200 rpm to mid-log phase ($OD_{600} = 0.8$).
2. Make 10-fold serial dilutions of M13KO7 by diluting 20 μl stock into 180 μl of PBS.
3. Mix 500 μl aliquots of mid-log phase *E. coli* SS320 with 200 μl of each M13KO7 dilution and 4 ml of 2YT top agar (16 g tryptone, 10 g yeast

extract, 5 g NaCl, 7.5 g granulated agar, water to 1 l, pH adjusted to 7.0 with NaOH, and autoclaved).

4. Pour the mixtures onto prewarmed LB/tet plates and grow overnight at 37 °C.
5. Pick an average-sized single plaque and place in 1 ml of 2YT/kan/tet medium. Grow for 8 h at 37 °C with shaking at 200 rpm.
6. Transfer the culture to 250 ml of 2YT/kan medium in a 2-l baffled flask. Grow overnight at 37 °C with shaking at 200 rpm.
7. Inoculate six 2-l baffled flasks containing 900 ml of superbroth/tet/kan medium (12 g tryptone, 24 g yeast extract, 5 ml glycerol, autoclaved, and supplemented with 100 ml of autoclaved 0.17 *M* KH_2PO_4, 0.72 *M* K_2HPO_4) with 5 ml of the overnight culture. Incubate at 37 °C with shaking at 200 rpm to mid-log phase ($OD_{600}=0.8$).
   The following steps should be done in a cold room, on ice, with pre-chilled solutions and equipment.
8. Chill three of the flasks on ice for 5 min with occasional swirling.
9. Centrifuge for 10 min at 5000 × *g* and 4 °C. During this spin, chill the second set of flasks on ice with occasional swirling.
10. Decant the supernatant and add culture from the remaining chilled flasks.
11. Repeat the centrifugation and decant the supernatant.
12. Fill the tubes with 1.0 m*M* HEPES, pH 7.4, and add sterile magnetic stir bars to facilitate pellet resuspension. Swirl to dislodge the pellet from the tube wall and stir at a moderate rate to resuspend the pellet completely.
13. Centrifuge for 10 min at 5000 × *g* and 4 °C. Decant the supernatant, being careful to retain the stir bar. To avoid disturbing the pellet, maintain the position of the centrifuge tube when removing from the rotor.
14. Repeat steps 12 and 13.
15. Resuspend each pellet in 150 ml of 10% ultrapure glycerol (Invitrogen). Use stir bars and do not combine the pellets. Centrifuge for 15 min at 5000 × *g* and 4 °C.
16. Decant the supernatant and remove the stir bar. Remove remaining traces of supernatant with a pipet.
17. Add 3.0 ml of 10% ultrapure glycerol to one tube and resuspend the pellet by pipetting. Transfer the suspension to the next tube and repeat until all of the pellets are resuspended. Transfer 350 μl aliquots into 1.5-ml microcentrifuge tubes. Flash freeze in liquid nitrogen and store at −80 °C.

2.1.3.2 *E. coli* electroporation and phage propagation

1. Chill the purified CCC-dsDNA (20 μg in a maximum volume of 100 μl) and a 0.2-cm-gap electroporation cuvette on ice.
2. Thaw a 350 ml aliquot of electrocompetent *E. coli* SS320 on ice. Add the cells to the DNA and mix by pipetting several times, but avoid introducing bubbles.
3. Transfer the mixture to the cuvette, tap the cuvette to release any trapped air bubbles and electroporate. For electroporation, follow the manufacturer's instructions, preferably using a BTX ECM-600 electroporation system with the following settings: 2.5 kV field strength, 125 ohms resistance, and 50 mF capacitance.
4. Immediately rescue the electroporated cells by adding 1 ml SOC medium (5 g bacto-yeast extract, 20 g bacto-tryptone, 0.5 g NaCl, 0.2 g KCl, water to 1 l, pH adjusted to 7.0 with NaOH, and autoclaved; supplemented with 5.0 ml of autoclaved 2.0 *M* $MgCl_2$ and 20 ml of filter sterilized 1.0 *M* glucose) and transferring to 10 ml SOC medium in a 250-ml baffled flask. Rinse the cuvette twice with 1 ml SOC medium. Add SOC medium to a final volume of 25 ml.
5. Incubate for 30 min at 37 °C with shaking at 200 rpm.
6. To determine the practical library diversity, plate serial dilutions on LB/carb plates (5 μl of 10^{-1} to 10^{-8} dilutions).
7. Transfer the culture to a 2-l baffled flask containing 500 ml 2YT medium, supplemented with antibiotics for phagemid and M13KO7 helper phage selection (e.g., 2YT/carb/kan medium).
8. Incubate for 16 h at 37 °C with shaking at 200 rpm.
9. Centrifuge the culture for 10 min at 16,000 × *g* and 4 °C.
10. Transfer the supernatant to a fresh tube and add 1/4 volume of PEG/NaCl solution to precipitate the phage. Incubate for 20 min on ice.
11. Centrifuge for 10 min at 16,000 × *g* and 4 °C. Decant the supernatant. Spin briefly and remove the remaining supernatant with a pipet.
12. Resuspend the phage pellet in 80 ml of PBS.
13. Centrifuge for 5 min at 27,000 × *g* and 4 °C to pellet insoluble matter.
14. Transfer the supernatant to a fresh tube and add 1/4 volume of PEG/NaCl solution to precipitate the phage. Incubate for 20 min on ice.
15. Centrifuge for 10 min at 16,000 × *g* and 4 °C. Decant the supernatant. Spin briefly and remove the remaining supernatant with a pipet.
16. Resuspend the phage pellet in 10 ml of PBS. Add protease inhibitors PMSF and EDTA, each to a final concentration of 1 m*M*. Add sterile ultrapure glycerol to a final concentration of 50% (v/v) and invert to

mix. Store library at −20 °C for up to 1 month, or aliquot and flash freeze for longer term storage at −80 °C.

17. Titer the library stock by infecting 90 μl volumes of log phase *E. coli* XL1Blue (Agilent Technologies, Santa Clara, CA) grown in 2YT/tet (OD_{600} = 0.4–0.8) with 10 μl of 10-fold serial dilutions of the library in PBS (10^{-1} to 10^{-16}) for 30 min at 37 °C. Serial dilution of the phage prior to infection is necessary for an accurate titer, as the titer is anticipated to be greater than the concentration of cells (~10^8 cells/ml at OD_{600} = 0.8). Plate 10 μl of each infection to LB/carb plates and grow overnight at 37 °C.

2.2. Selection of PDZ variants

Structural selection pressure for well-folded (i.e., protease resistant) PDZ variants can be imposed using an immobilized anti-tag antibody; the phage particle is not retained by its epitope tag if the intervening PDZ variant is proteolytically cleaved. The *E. coli* strains used for viral production express outer membrane proteases, so this structural selection is intrinsic to all selections, even if additional selection pressures are applied. Functional selection pressure for peptide binding can be imposed by immobilizing a GST-peptide with a particular sequence.

2.2.1 Round 1

1. Coat Maxisorp immunoplate wells with 100 μl of selection target (10 μg/ml of anti-tag antibody or 100 μg/ml GST-peptide, diluted in sterile PBS) for 2 h at room temperature or overnight at 4 °C. The number of wells required for each target depends on the diversity of the library. The phage concentration should not exceed 10^{13} cfu/ml, and the total number of phages should exceed the library diversity by 1000-fold. Thus, for a library with 10^{10} diversity, 10 wells will be required.
2. Remove the coating solution and block for 1 h at room temperature with 200 μl of sterile PBS + 0.5% BSA (PB).
3. During blocking, resuspend PDZ-phage library as follows:
4. Dilute an appropriate volume of the PDZ-phage library 10-fold with sterile PBS and add 1/4 volume of PEG/NaCl solution to precipitate the phage. Incubate for 20 min on ice.
5. Centrifuge for 10 min at 16,000 × *g* and 4 °C. Decant the supernatant. Spin briefly and remove the remaining supernatant with a pipet.

6. Resuspend the phage pellet by gentle pipetting in an appropriate volume of sterile PBS + 0.5% BSA + 0.05% Tween-20 (PBT). Store resuspended PDZ phage on ice until use.
7. Remove the blocking solution and wash three times with sterile PBS + 0.5% Tween-20 (PT).
8. Add 100 μl of PDZ-phage library to all blocked wells. Incubate at 4 °C for 2 h with gentle shaking.
9. Remove the phage solution and wash eight times with 4 °C PT.
10. To elute bound phage, add 100 μl of 100 m*M* HCl to each well, incubate for 5 min at room temperature with gentle shaking and then transfer the HCl solution to a 1.5-ml microfuge tube. Pool eluted phage from multiple wells coated with the same target.
11. To neutralize the eluted phage, add 15 μl of 1.0 *M* Tris–HCl, pH 11.0 for every 100 μl of eluted phage.
12. Add all of the neutralized phage solution to 10 volumes of log phase *E. coli* strain XL1Blue (OD_{600} = 0.4–0.8) in 2YT/tet medium. Incubate for 30 min at 37 °C with shaking at 200 rpm.
13. Remove 10 μl of infected cells, prepare 10-fold serial dilutions in 2YT (10^{-1} to 10^{-8}), and plate 5 μl of each dilution to LB/carb plates. Incubate plates at 37 °C overnight. The concentration of eluted phage is anticipated to be less than the concentration of cells, so an accurate titer can be obtained by direct dilution of the infected cells.
14. Add M13KO7 helper phage to the selection culture at a final concentration of 10^{10} cfu/ml. Incubate for 45 min at 37 °C with shaking at 200 rpm.
15. Transfer the selection culture to 25 volumes of 2YT/carb/kan medium and incubate overnight (minimum 16 h) at 37 °C with shaking at 200 rpm.
16. Titer the resuspended input library, to ensure that the desired library coverage has been achieved, as described for library production (Section 2.1).

2.2.2 Rounds 2 to 5

Subsequent rounds follow the same procedure as Round 1, with the exceptions noted below.

A. Coat Maxisorp immunoplate wells with 100 μl of selection target overnight at 4 °C, using the same concentration as for the first round. The number of wells can be reduced, as long as a minimum of 1000-fold coverage of the first round output diversity is maintained. In practice, we typically use three wells per target for subsequent selection rounds.

B. Coat an equivalent number of wells with an appropriate negative control protein for the selection target; for example, GST if the selection target is GST-tagged. Treat these wells identically to the selection target wells and titer the PDZ-phage eluted from both sets of wells (as described in Section 2.2.1, steps 12 and 13). Selection progress can be tracked by calculating the enrichment, which is the ratio of PDZ-phage eluted from the selection target compared to the negative control protein.

C. Precipitate phage from the culture supernatant as described for harvesting the library (Section 2.1.3.2, steps 9–11) but resuspend the phage pellet in PBT using 1/10th of the total overnight culture volume. A ten-fold concentration of the phage supernatant should result in a PDZ-phage titer of approximately 10^{12} to 10^{13} cfu/ml, if the overnight culture has grown to saturation.

D. Infect cells using only half the neutralized phage volume and store the rest (as well as the remaining input PDZ-phage) at 4 °C for up to a week, or indefinitely at −80 °C after addition of sterile ultrapure glycerol to 50% (v/v).

E. Amplify only the selection target cultures overnight, as described for Round 1. The cells infected with phage eluted from the negative control proteins can be discarded.

2.3. Binding validation of PDZ variants by phage ELISA

PDZ-phage clones from enriched populations are tested for specific binding to the selection target, compared to at least one negative control, in an enzyme-linked immunosorbent assay (ELISA). PDZ-phage bound to proteins immobilized in a microtiter plate are detected with an enzyme-conjugated antibody that recognizes M13 filamentous phage and cleaves a colorimetric substrate. Ratios greater than two indicate specific binding by a PDZ-phage clone, provided the absolute signal for negative control binding is close to the background level.

1. Infect log phage *E. coli* XL1Blue and plate for single colonies from the selection rounds with the highest enrichment ratios.
2. Inoculate 450 μl aliquots of 2YT/carb/M13KO7 medium in 96-well microtubes with single colonies harboring phagemids and grow overnight (minimum 16 h) at 37 °C with shaking at 200 rpm.
3. For each clone, coat one well of a 384-well Maxisorp immunoplate with 30 μl of selection target at the same concentration used to coat the selection wells. Coat one adjacent well with 30 μl of an appropriate negative control. For GST-peptide selections, the best negative control is GST

alone, coated at the same concentration as the GST-peptides. For anti-tag antibody selections, wells coated with PB can be used.

4. Centrifuge cultures for 10 min at 3400 × *g* and 4 °C to pellet bacteria.
5. Transfer 360 μl of phage supernatant to a clean 96-well plate and add 40 μl of 10 × PBS to adjust the pH.
6. Block wells with 60 μl of PB for 1 h at room temperature with gentle shaking.
7. Wash wells three times with PT.
8. Transfer 30 μl of phage supernatant to selection and control wells. Incubate at 4 °C for 1 h.
9. Wash wells six times with cold PT.
10. Add 30 μl of horseradish peroxidase/anti-M13 antibody conjugate (GE Healthcare, Piscataway, NJ) (diluted 1:8000 in cold PBT). Incubate at 4 °C for 30 min.
11. Wash wells six times with 4 °C PT and twice with 4 °C PBS.
12. Add 25 μl of freshly prepared TMB substrate (Kirkegaard & Perry Laboratories, Gaithersburg, MD) to every well. Allow color to develop for 5–10 min with gentle shaking at room temperature.
13. Stop the reaction by adding 25 μl of 1 *M* H_3PO_4 and read spectrophotometrically at 450 nm in a microtiter plate reader.

2.4. Sequencing of positive PDZ-phage clones

Clonal PDZ-phage supernatant (i.e., from the phage ELISA) can be used as template in a PCR reaction for subsequent Sanger sequencing, using primers with 5′ extensions that are complementary to standard sequencing primers.

1. Add 2 μl of phage supernatant to 23 μl of the following PCR mix: 1 × PCR buffer, 200 μ*M* dNTPs, 400 μ*M* forward primer , 400 μ*M* reverse primer, 0.1 unit of Taq polymerase.
2. Amplify the DNA fragment with the following program:
   1: 95 °C 5 min, 2: 95 °C 30 s, 3: 55 °C 30 s, 4: 72 °C 30 s, 5: GOTO2, 24×, 6: 72 °C 5 min, 7: 4 °C forever.
3. Confirm expected size of PCR product by agarose gel electrophoresis.
4. Clean up PCR product by mixing 2 μl of PCR reaction with 7.6 μl distilled water, 0.2 μl Exonuclease I (2 units) (United States Biochemical, Cleveland, OH), and 0.2 μl shrimp alkaline phosphatase (0.2 units) (United States Biochemical). Incubate at 37 °C for 30 min followed by 80 °C for 15 min to inactivate the enzymes.
5. The sample can be used directly as template in standard Big-Dye terminator sequencing reactions.

3. PEPTIDE PROFILING OF PDZ VARIANTS

To evaluate the peptide-binding preferences of engineered PDZ variants, they are produced as recombinant proteins and allowed to select peptide-phage from a random peptide library (Fig.15.3). Over iterative rounds of selection, peptides that bind specifically become enriched, until they dominate the peptide-phage population. The binding peptide sequences are decoded using deep sequencing, which yields millions of reads and can accommodate multiplexing of hundreds of enriched peptide-phage pools (Fig. 15.4).

3.1. High-throughput purification of PDZ variants

PDZ variants are subcloned from the display phagemid into an IPTG-inducible expression vector that adds an N-terminal 6×His tag (for purification) and a GST tag (for direct immobilization in Maxisorp immunoplates). Our preferred cloning method uses PCR instead of restriction enzymes to generate an insert encoding the PDZ variant, because it is not compromised by spurious restriction enzyme sites in the diversified regions of the selected

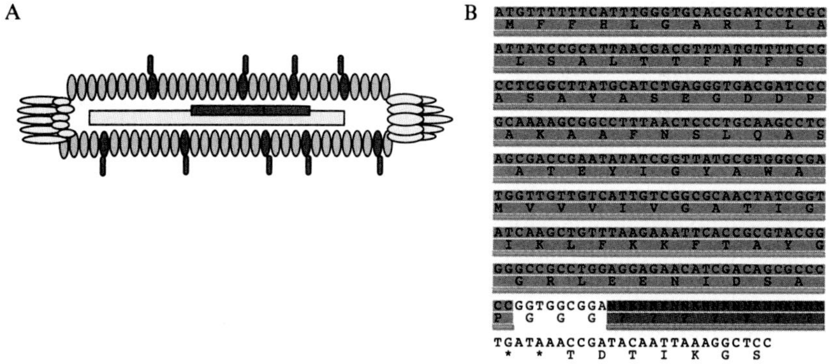

Figure 15.3 Design and display of a random peptide library on phage. (A) Diagram of a phage particle displaying C-terminal peptides (*magenta*) fused to an engineered major coat protein p8 (*dark gray*) encoded by the enclosed phagemid genome. Most of the phage coat is composed of wild-type p8 (*light gray*). Minor coat proteins are shown as white ellipses. (B) Sequence of the engineered major coat protein p8 (*gray*) and seven peptide residues randomized with degenerated NNK codons (*magenta*) (N=A/C/G/T, K=G/T). The untranslated region 3′ of the stop codons is the reverse primer annealing region for barcoding PCR reactions. (For interpretation of the references to color in this figure legend, the reader is referred to the online version of this chapter.)

A

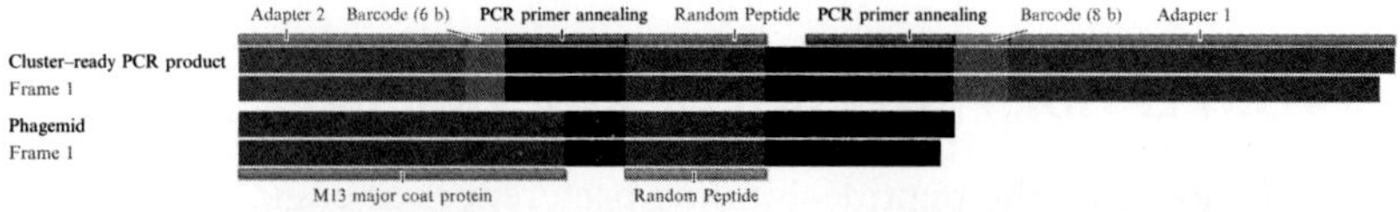

B

ID	Adapter 2	Barcode (6 b)	PCR primer annealing
F#	CAAGCAGAAGACGGCATACGAGATCGGTCTCGGCATTCCTGCTGAACCGCTCTTCCGATCT	NNNNNN	AGCGCCCCCGGTGGCGGA

ID	Adapter 1	Barcode (8 b)	PCR primer annealing
R#	AATGATACGGCGACCACCGAGATCTACACTCTTTCCCTACACGACGCTCTTCCGATCT	NNNNNNNN	AGGAGCCTTTAATTGTATCGGT

ID	Barcode (6 b)
F01	CGTGAT
F02	ACATCG
F03	GCCTAA
F04	TGGTCA
F05	CACTGT
F06	ATTGGC
F07	GATCTG
F08	TCAAGT
F09	CTGATC
F10	AAGCTA
F11	GTAGCC
F12	TACAAG
F13	CAGTGG
F14	AGCTCT
F15	GTCTGC
F16	TGACTG
F17	CCTATC
F18	AGCAAT
F19	GATGTA
F20	TTGACG
F21	CGACTT
F22	AGACAA
F23	GTACTC
F24	TGTGTC

ID	Barcode (8 b)
R01	GAAGAGAT
R02	CCAGCTTG
R03	TCAGCAGC
R04	CAGGGAAT
R05	TGCTCTCC
R06	ATCTTGCT
R07	GAGAGATC
R08	CCCTGAAG
R09	CTTGATTG
R10	CCGATGCG
R11	GGCCTGGG
R12	GGATCTTC
R13	TCTTGAGG
R14	AGGAGGCT
R15	TGAATCTC
R16	CTCAGCCC
R17	CTTTCCTG
R18	AGCCATGC
R19	CAGAAATT
R20	TCAGAATG
R21	CAGATGCT
R22	GGCTCCTG
R23	TCCATTCT
R24	TCTTGCCT

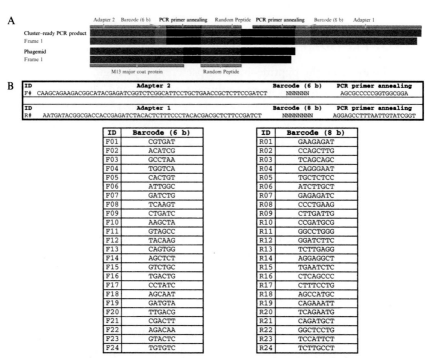

Figure 15.4 Barcoding strategy for multiplexed Illumina sequencing of peptide profiles. (A) Alignment of cluster-ready PCR product to the portion of the phagemid encoding the displayed peptide. (B) PCR primer sequences used to amplify the region encoding the peptide; add two unique barcodes and add adapter sequences necessary for single- or paired-end Illumina sequencing. Each set of 24 barcodes has equal representation of all bases at all positions and were designed such that more than two sequencing errors would be necessary to interconvert their sequences. (For color version of this figure, the reader is referred to the online version of this chapter.)

PDZ variants (Walker, Taylor, Rowe, & Summers, 2008). Sequence-verified expression clones are arrayed in a 96-well plate and stored as glycerol stocks in *E. coli* XL1Blue.

1. Inoculate overnight cultures from the master glycerol stock plate in deep 96-well plates containing 1.4 ml 2YT/carb and grow overnight at 37 °C with shaking at 200 rpm.
2. Inoculate two duplicate expression plates containing 1.4 ml/well rich autoinducing medium (1% N-Z amine AS, 0.5% yeast extract, 25 m*M* Na_2HPO_4, 25 m*M* KH_2PO_4, 50 m*M* NH_4Cl, 5 m*M* Na_2SO_4, 2 m*M* $MgSO_4$, 54 m*M* glycerol, 2.8 m*M* glucose, 5.6 m*M* alpha-lactose) (Studier, 2005) with 10 µl of overnight culture per well and grow for 24–48 h at 37 °C with shaking at 200 rpm.

3. Centrifuge cultures for 10 min at 3400 × *g* and 4 °C to pellet bacteria. Discard supernatant by inverting the deep-well plate.
4. Transfer cultures from second plate to first deep-well plate and repeat step 3 to combine both cultures into one pellet.
5. Freeze bacterial pellets at −20 °C overnight or up to 1 week.
6. Aliquot 250 μl of resuspension buffer (50 m*M* HEPES, pH 7.5, 500 m*M* NaCl, 5 m*M* imidazole, 5% glycerol, 1 m*M* PMSF, 1 m*M* benzamidine) to frozen pellets and shake plate at room temperature until the pellets have thawed.
7. If necessary, pipet up and down until the pellets are completely resuspended.
8. Aliquot 750 μl of lysis buffer (50 m*M* HEPES, pH 7.5, 500 m*M* NaCl, 5 m*M* imidazole, 5% glycerol, 1 m*M* PMSF, 1 m*M* benzamidine, 1% Triton X-100, 1 mg/ml lysozyme, 10 units/ml benzonase) to resuspended pellets and shake plate at room temperature for 30 min to allow for complete lysis and DNA digestion.
9. Clear lysate by centrifugation for 30 min at 3400 × *g* and 4 °C.
10. During spin, equilibrate 7 ml of packed NiNTA resin (GE Healthcare) by washing three times with at least 10× volume of resuspension buffer, pulsing to 1500 × *g* to pellet resin between each wash. After the last wash, resuspend resin in 21 ml of lysis buffer and aliquot 200 μl to each well of a 96-well filter plate (2-ml volume, 0.45-μm filter) (Seahorse Bioscience, North Billerica, MA) whose bottom is sealed with two layers of parafilm.
11. Transfer 800 μl of cleared lysate to the resin in the filter plate and seal the top with foil plate seal.
12. Incubate the lysate with resin for 1 h at room temperature with end-over-end mixing to keep the resin in suspension.
13. Remove the unbound lysate by unsealing the filter plate and centrifuging for 5 min at 1000 × *g* and 4 °C over an empty deep-well plate.
14. Reseal the bottom of the filter plate with parafilm, aliquot 1 ml of wash buffer (50 m*M* HEPES, pH 7.5, 500 m*M* NaCl, 30 m*M* imidazole, 5% glycerol) per well, seal the top of the filter plate with adhesive foil and shake plate vigorously by hand to resuspend the resin.
15. Centrifuge for 5 min at 1500 × *g* and 4 °C over an empty deep-well plate to remove wash buffer.
16. Repeat steps 14 and 15 twice more, for a total of three washes.
17. Seal bottom of the filter plate with parafilm, aliquot 150 μl of elution buffer (50 m*M* HEPES, pH 7.5, 500 m*M* NaCl, 250 m*M*

imidazole, 5% glycerol) per well, and incubate at room temperature for 10 min.

18. During elution buffer incubation, aliquot protease inhibitors to non-binding 96-well plate that will be used to collect the eluate (1.5 μl of 100 × protease inhibitor cocktail and 1.5 μl of 500 m*M* EDTA, pH 8.0).
19. Remove bottom parafilm seal from filter plate and place immediately over a nonbinding 96-well plate.
20. Centrifuge for 5 min at 1500 × *g* and 4 °C to collect eluate.
21. Measure eluted protein concentrations using denaturing Bradford assay. Aliquot 10 μl of 1 *M* NaOH to a 96-well nonbinding plate. Add 5 μl of protein standard or eluted protein and to mix. Dilute Bradford reagent 1:5 in water and add 200 μl to denatured protein samples. Read plate spectrophotometrically at 595 nm.
22. Transfer 15 μl of eluted protein to a 96-well PCR plate containing 5 μl of 4× SDS sampler buffer, seal with foil, and heat at 95 °C in thermocycler for 5 min to denature. Confirm protein size, integrity, and overall purity by SDS-PAGE.
23. Store the eluted proteins at 4 °C, typically for up to 1 month.

3.2. Preparation of random peptide-phage library

Peptides can be displayed as fusions to the major bacteriophage coat protein, p8. There are thousands of copies of p8 per phage particle, which allows for multivalent display and selection of lower affinity peptides through avidity (Fig. 15.3C). The natural p8 C-terminus is buried, so an inverted version of p8 was engineered to display an exposed C-terminus (Fuh et al., 2000; Held & Sidhu, 2004), which is typically necessary for PDZ domain binding.

The random peptide library is constructed using the same method described earlier, but using a phagemid-encoding engineered p8 (Tonikian, Zhang, Boone, & Sidhu, 2007) The mutagenic oligonucleotide randomizes seven C-terminal residues using the NNK codon, which encodes all 20 genetically encoded amino acids and the amber stop codon. The theoretical diversity of this library (10^9) is smaller than typical electroporation diversities using the methods detailed here (10^{10}), so it can in principle be oversampled in the initial round of a selection.

3.3. High-throughput peptide profiling selections

This high-throughput method uses one well per GST-PDZ variant, which still allows a 100-fold coverage of the peptide library in the first round of selection (Ernst et al., 2009).

3.3.1 Round 1

1. Coat Maxisorp immunoplate wells with GST-PDZ variants by adding 20 μl of eluted protein to 80 μl of sterile PBS. Concentrations typically range from 0.5 to 5 μg/ml. Incubate for 2 h at room temperature or overnight at 4 °C.
2. Remove the coating solution and block for 1 h at room temperature with 200 μl of sterile PB.
3. During blocking, resuspend peptide-phage library as follows:
4. Dilute an appropriate volume of peptide-phage library (10^{12} cfu per target) 10-fold with sterile PBS and add 1/4 volume of PEG/NaCl solution to precipitate the phage. Incubate for 20 min on ice.
5. Centrifuge for 10 min at 16,000 × *g* and 4 °C. Decant the supernatant. Spin briefly and remove the remaining supernatant with a pipet.
6. Resuspend the phage pellet by gentle pipetting in an appropriate volume of PBT to yield 100 μl per target at 10^{13} cfu/ml. Store resuspended peptide-phage on ice until use.
7. Remove the blocking solution and wash the plate three times with PT.
8. Add 100 μl of peptide-phage library to all blocked wells. Incubate at 4 °C for 2 h with gentle shaking.
9. Remove the phage solution and wash eight times with 4 °C PT.
10. Elute bound phage by adding 100 μl of *E. coli* XL1Blue (OD_{600} = 0.4–0.8) in 2YT/tet medium. Seal the plate and incubate for 30 min at 37 °C with shaking at 200 rpm.
11. Add M13KO7 helper phage to each well at a final concentration of 10^{10} cfu/ml. Seal the plate and incubate for 45 min at 37 °C with shaking at 200 rpm.
12. Transfer the cultures to a 96-well plate containing 1.4 ml 2YT/carb/kan medium. Seal the plate with a breathable filter and incubate for 20 h at 37 °C with shaking at 200 rpm.
13. Titer the input library to confirm that the intended level of coverage has been achieved.

3.3.2 Rounds 2 to 5

Subsequent rounds follow the same procedure as Round 1, with the exceptions noted below.

A. Coat Maxisorp immunoplate wells with 10 μl of eluted GST-PDZ variant protein, diluted in 90 μl of sterile PBS, for 2 h at room temperature or overnight at 4 °C.

B. Prepare the amplified peptide-phage as follows:

a. Centrifuge overnight cultures for 10 min at 3400 × *g* to pellet bacteria.
b. Aliquot 100 µl of sterile 10 × PBS stock to every well of a sterile 96-well plate.
c. Transfer 900 µl of phage supernatant to 10 × PBS to adjust pH.
d. Seal plate with foil and incubate at 65 °C for 30 min to heat sterilize, then chill to 4 °C before transferring 100 µl of phage supernatant to blocked wells.

3.3.3 Peptide-phage pool ELISAs

Enrichment of specific peptide-phage can be monitored using an ELISA to compare binding of the peptide-phage pool to immobilized GST-PDZ variant or GST alone, essentially as described in Section 2.2. The progress of the selections, based on ELISA enrichment ratios, can indicate how many rounds are necessary and which rounds should be sequenced to yield peptide-binding profiles. Typically, enrichment ratios greater than two are observed by round three.

3.4. Preparation of Illumina sequencing libraries

Although we have used other deep sequencing technologies to decode peptide profiles (Ernst et al., 2010), the Illumina platform has proven most fruitful, due to its simple sample preparation and low error rate. The sequencing depth (>30 million reads per lane) permits hundreds of samples to be multiplexed in a single run, which is cost-effective. Sequencing library preparation entails a PCR reaction, using an enriched peptide-phage pool as template, to add the necessary adapter sequences, along with two barcodes to uniquely identify each peptide-phage pool. The PCR reaction conditions are designed to minimize the introduction of bias, by using the maximum amount of enzyme and an excess of phage genomes as template (Quail et al., 2008). A set of 24 forward and 24 reverse barcodes can label up to 576 pools uniquely (Fig. 15.4).

The barcoded PCR products are quantitated and pooled, and then purified by agarose gel extraction (Smith et al., 2010; Quail et al., 2008). The resulting product is a cluster-ready sequencing library, suitable for sequencing with a single-end 125-base read as shown here, or using shorter paired-end reads, which are standard Illumina protocols available from many service providers.

This sequencing strategy reads the noncoding strand; consequently, the reads must be reverse complemented before analysis and translation. Reads with low-confidence base calls are removed, and then the high quality reads are demultiplexed based on the barcodes. Duplicate reads are removed from each pool so that each unique DNA sequence is represented once. Then the peptide sequences are translated and summarized as logos (Schneider & Stephens, 1990) or as position-weight matrices for further analysis.

1. Array unique combinations of barcoded forward and reverse primers (5 μl of 5 μ*M* stock of each primer) and corresponding peptide-phage pools (20 μl) in a 96-well PCR plate.
2. Prepare Phusion PCR master mix and add to each well, such that the total volume is 50 μl and the final concentrations are 1 × High-Fidelity buffer, 200 μ*M* dNTPs, 0.5 μ*M* forward primer, 0.5 μ*M* reverse primer, with 1U Phusion enzyme (New England Biolabs) per 50 μl PCR reaction.
3. Amplify the DNA fragment with the following thermocycler program: 1:98 °C 180 s, 2:98 °C 10 s, 3:68 °C 10 s, 4:72 °C 10 s, 5: GOTO2—24×, 6:72 °C, 300 s, 7:4 °C forever.
4. Confirm PCR product size by agarose gel electrophoresis (expected size is 173 bp).
5. Quantitate the concentration of each PCR product using a dsDNA specific method, for example, a fluorescent dye such as PicoGreen (Invitrogen). Dilute PicoGreen reagent 1:40 in TE, add 1 μl or PCR product (or a dilution in TE) to an optically clear 96-well PCR plate, centrifuge briefly, and read in a qPCR machine using SYBR excitation and emission wavelengths. Lambda phage DNA (Invitrogen), serially diluted in TE, can be used to generate a standard curve for quantitation and to establish the linear range.
6. Pool equal quantities of PCR products. For highly multiplexed samples to yield sufficient reads for all pools, it is imperative to pool equal quantities accurately. The resulting pool should contain several μg of dsDNA.
7. Concentrate pooled PCR products using column-based kit (Qiagen) according to instructions, and elute in TE.
8. Purify the PCR product by excision from an agarose gel. It is particularly important to remove lower molecular weight contaminants, as they preferentially hybridize to the flow cell and reduce the number of full-length reads. Purify the extracted PCR product using a kit (Qiagen). Use an extended incubation in denaturing buffer QG (rather than heating) to completely dissolve the gel slice; heating in denaturing buffer can introduce GC bias (Quail et al., 2008).
9. Quantitate the purified PCR product using a dsDNA specific method, for example, PicoGreen, and dilute to the required concentration for submission to the sequencing service provider.

10. Sequencing data is generally exported in fastq format. Free Web-based tools such as Galaxy can be used to evaluate the quality of the data and to analyze it (galaxy.psu.edu) (Blankenberg et al., 2010). Alternatively, software designed to identify multiple specificities in peptide-binding profiles will perform quality filtering, demultiplexing, translation, and logo generation (Kim et al., 2012).

4. SUMMARY

The methods described here for directed evolution and high-throughput characterization of synthetic PDZ variants enable engineering approaches that can provide insight into the molecular basis and evolution of selective peptide recognition in this and other domain families.

ACKNOWLEDGMENT

This work was supported by funding from the Canadian Institutes of Health Research awarded to S. S. S.

REFERENCES

Blankenberg, D., Gordon, A., Von Kuster, G., Coraor, N., Taylor, J., & Nekrutenko, A. (2010). Manipulation of FASTQ data with Galaxy. *Bioinformatics*, *26*, 1783–1785.

Ernst, A., Gfeller, D., Kan, Z., Seshagiri, S., Kim, P. M., Bader, G. D., et al. (2010). Coevolution of PDZ domain-ligand interactions analyzed by high-throughput phage display and deep sequencing. *Molecular BioSystems*, *6*, 1782–1790.

Ernst, A., Sazinsky, S. L., Hui, S., Currell, B., Dharsee, M., Seshagiri, S., et al. (2009). Rapid evolution of functional complexity in a domain family. *Science Signaling*, *2*, ra50.

Fuh, G., Pisabarro, M. T., Li, Y., Quan, C., Lasky, L. A., & Sidhu, S. S. (2000). Analysis of PDZ domain-ligand interactions using carboxyl-terminal phage display. *The Journal of Biological Chemistry*, *275*, 21486–21491.

Held, H. A., & Sidhu, S. S. (2004). Comprehensive mutational analysis of the M13 major coat protein: Improved scaffolds for C-terminal phage display. *Journal of Molecular Biology*, *340*, 587–597.

Kim, T., Tyndel, M. S., Huang, H., Sidhu, S. S., Bader, G. D., Gfeller, D., et al. (2012). MUSI: An integrated system for identifying multiple specificity from very large peptide or nucleic acid data sets. *Nucleic Acids Research*, *40*, e47.

Kunkel, T. A., Roberts, J. D., & Zakour, R. A. (1987). Rapid and efficient site-specific mutagenesis without phenotypic selection. *Methods in Enzymology*, *154*, 367–382.

Pawson, T. (2003). Assembly of cell regulatory systems through protein interaction domains. *Science*, *300*, 445–452.

Quail, M. A., Kozarewa, I., Smith, F., Scally, A., Stephens, P. J., Durbin, R., et al. (2008). A large genome center's improvements to the Illumina sequencing system. *Nature Methods*, *5*, 1005–1010.

Schneider, T. D., & Stephens, R. M. (1990). Sequence logos: A new way to display consensus sequences. *Nucleic Acids Research*, *18*, 6097–6100.

Sidhu, S. S., Lowman, H. B., Cunningham, B. C., & Wells, J. A. (2000). Phage display for selection of novel binding peptides. *Methods in Enzymology*, *328*, 333–363.

Skelton, N. J. (2002). Origins of PDZ Domain Ligand Specificity. Structure determination and mutagenesis of the erbin PDZ domain. *The Journal of Biological Chemistry, 278,* 7645–7654.

Smith, A. M., Heisler, L. E., St Onge, R. P., Farias-Hesson, E., Wallace, I. M., Bodeau, J., et al. (2010). Highly-multiplexed barcode sequencing: An efficient method for parallel analysis of pooled samples. *Nucleic Acids Research, 38,* e142.

Songyang, Z. (1997). Recognition of unique carboxyl-terminal motifs by distinct PDZ domains. *Science, 275,* 73–77.

Studier, F. W. (2005). Protein production by auto-induction in high density shaking cultures. *Protein Expression and Purification, 41,* 207–234.

Tonikian, R., Zhang, Y., Boone, C., & Sidhu, S. S. (2007). Identifying specificity profiles for peptide recognition modules from phage-displayed peptide libraries. *Nature Protocols, 2,* 1368–1386.

Tonikian, R., Zhang, Y., Sazinsky, S. L., Currell, B., Yeh, J.-H., Reva, B., et al. (2008). A specificity map for the PDZ domain family. *PLoS Biology, 6,* e239.

Walker, A., Taylor, J., Rowe, D., & Summers, D. (2008). A method for generating sticky-end PCR products which facilitates unidirectional cloning and the one-step assembly of complex DNA constructs. *Plasmid, 59,* 155–162.

CHAPTER SIXTEEN

Efficient Sampling of SCHEMA Chimera Families to Identify Useful Sequence Elements

Pete Heinzelman*, Philip A. Romero†, Frances H. Arnold†,1

*Department of Chemical, Biological & Materials Engineering, University of Oklahoma, Norman, Oklahoma, USA

†Division of Chemistry and Chemical Engineering, California Institute of Technology, Pasadena, California, USA

[1]Corresponding author: e-mail address: frances@cheme.caltech.edu

Contents

Abstract

SCHEMA structure-guided recombination is an effective method for producing families of protein chimeras having high sequence diversity, functional diversity, and thermostabilities greater than any of the parent proteins from which the chimeras are made. A key feature of SCHEMA chimera families is their amenability to a "sample, model, and predict" operation that allows one to characterize members of a small chimera sample set and use those data to construct models that accurately predict the properties of every member of the family. In this chapter, we describe applications of this "sample, model, and predict" approach and outline methods for designing chimera sample sets that enable efficient construction of models to identify useful sequence elements. With these models we can also predict the sequences and properties of the most desirable chimeras.

Methods in Enzymology, Volume 523
ISSN 0076-6879
http://dx.doi.org/10.1016/B978-0-12-394292-0.00016-3

1. INTRODUCTION

SCHEMA structure-guided recombination is an effective method for producing large families of enzyme chimeras having high sequence and functional diversity. The chimeras are made by recombining a set of homologous parent proteins at crossover locations specifically chosen to minimize structural disruption. We have shown that members of these chimera families can have thermostabilities and maximum catalytic temperatures (T_{opt}) higher than those of any of the parent enzymes while retaining high catalytic activity (Heinzelman et al., 2009; Li et al., 2007; Smith et al., 2012). Additionally, for chimera families in which the residues that impact catalytic activity and substrate specificity are not highly conserved across the parent enzymes, it has proven possible to generate chimeras that are simultaneously thermostable and have substrate specificity profiles that are distinct from those of the parents (Li et al., 2007).

The ability to identify the sequences of the most desirable chimeras in a given family by using predictive modeling approaches contributes greatly to SCHEMA recombination's utility as a protein engineering tool. Such modeling allows one to design and construct a small sample set of chimera sequences (perhaps a few dozen), characterize their properties, and then use those data to predict the sequences of the chimera family members that have the most desirable property profiles. In this era of rapid and inexpensive gene synthesis, the construction of highly informative chimera sample sets has become accessible to virtually every laboratory. In this chapter, we describe some successful applications of this "sample, model, and predict" approach, whose main steps are illustrated and described in Fig. 16.1. We also outline methods for designing SCHEMA chimera family sample sets which with relatively moderate time and labor inputs can translate to the accurate prediction of dozens of useful new chimera sequences.

2. SCHEMA CHIMERA FAMILY DESIGN OVERVIEW

SCHEMA chimera families are constructed by recombining contiguous stretches of amino acids, or "blocks," taken from (structurally related) protein homologs, or "parents." In SCHEMA recombination, the crossover locations are chosen to maximize the number of chimeras that will be folded and functional. The design of SCHEMA chimera families uses the recombination as a shortest path problem (RASPP) algorithm to identify blocks

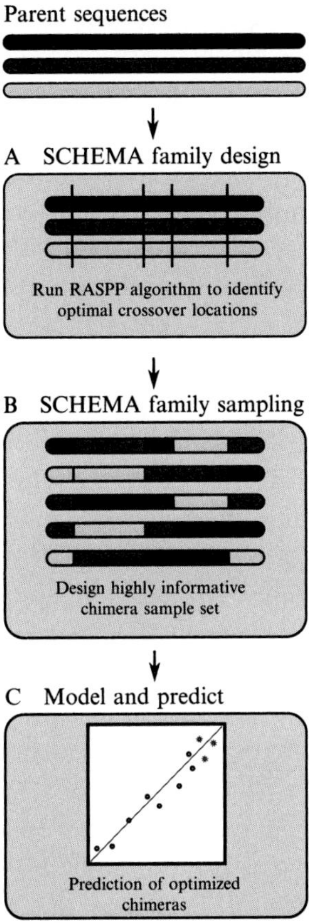

Figure 16.1 Schematic of "sample, model, and predict" approach for a SCHEMA chimera family. Once parent genes are chosen, the SCHEMA software is used to design a chimera family containing P^B unique sequences, where B is the number of blocks and P is the number of parent enzymes (Step A). A two-step active learning algorithm is then used to design a highly informative sample set of chimeras. Genes corresponding to these chimeras are synthesized, the chimeras are expressed in a recombinant host, and the properties of interest are measured experimentally (Step B). Linear regression is used to construct a model that can predict the sequences of the chimeras with the most desirable properties. Genes corresponding to these improved chimeras can then be synthesized, or constructed by standard cloning procedures, and the forecasted property improvements validated experimentally. This prediction and validation process is depicted in the bottom panel (Step C), where the abscissa denotes the predicted value for the chimera property of interest, for example, thermostability, while the experimentally measured property values appear on the ordinate. The lighter, starred points in this plot illustrate chimeras that are predicted and experimentally validated to be most desirable with respect to the chimera property that was modeled, for example, thermostability.

that minimize the number of amino acid side chain interactions, or contacts, that are disrupted when the blocks are swapped to generate new sequences (Endelman, Silberg, Wang, & Arnold, 2004). A given chimera is characterized by the number of parental contacts that have been broken, E. RASPP seeks to minimize the library average broken contacts $<E>$, which increases the number of enzyme chimera family members that are folded and functional (Meyer, Hochrein, & Arnold, 2006). The greater the sequence identity among the parents, the fewer the contacts that can be broken when they are recombined. Of course, the sequence diversity in the chimera family increases with the sequence diversity of the parents. Therefore, chimera sequence diversity and the fraction of chimeras in a library that are functional tend to trade off. The mutation level of a given chimera is measured by the number of amino acids that are different from the closest parent sequence, m. The library average mutation level $<m>$ is a measure of sequence diversity.

The SCHEMA algorithm source code, which is publicly available (http://www.che.caltech.edu/groups/fha/Software.htm), requires as inputs an alignment of the homologous parent sequences, a minimum desired block size, the number of blocks into which the parents are to be divided, and a set of crystal structure or structure model coordinates for one of the parent enzymes. A contact map, or listing of all the amino acid pairs with side chain heavy atoms lying within 4.5 Å of one another, is generated by the source code. Although the majority of SCHEMA families to date have been constructed by dividing three parents into eight blocks, there are no fundamental constraints on either the number of blocks or the number of parents; for example, active fungal cellobiohydrolase class I (CBH I) chimeras with improved thermostabilities have been obtained by recombining eight blocks from each of five parents (Heinzelman et al., 2010). However, the required sampling for model building and prediction increases with the number of blocks and parents (see Section 3.1).

Most SCHEMA enzyme families have been constructed from such parents having ~60–75% primary sequence identity; as many as half of the chimeras of such parents have been found to be functional (Otey et al., 2006). SCHEMA recombination can also be applied to parents with much lower sequence identities, although the fraction of sequences that encode functional enzymes is expected to decrease. More than 20% of the members of a beta-lactamase chimera family constructed from such parents sharing just 34–42% identity were found to be catalytically active (Meyer et al., 2006).

3. PREDICTION OF THERMOSTABLE CHIMERAS BY LINEAR REGRESSION MODELING

It has been demonstrated that the blocks comprising a chimera make linearly additive contributions to the chimera's thermostability (Heinzelman et al., 2009a; Li et al., 2007; Smith et al., 2012) as well as its temperature optimum for catalytic activity, T_{opt} (Smith et al., 2012). Thus, linear regression can be used to construct quantitative models that can accurately predict the thermostabilities of all of the members of the chimera family, where the chimeras with the greatest thermostability or the highest T_{opt} values are typically of particular interest. Constructing such a model requires thermostability measurements, for example, in the form of T_{50} values (the temperature at which 50% of enzyme activity is lost after incubation for a specified time interval), for an appropriate sample set of chimeras. As few as 35 chimera T_{50} measurements enabled construction of an accurate predictive linear regression model for a 6561-member cytochrome P450 chimera family (Li et al., 2007).

As shown in Eq. (16.1), linear thermostability models allow the T_{50} value for a given chimera to be expressed as the sum of the T_{50} for a reference parent sequence (T_{P1}) and contributions of each of its blocks ($B=1$–8 for eight-block recombination), which can be positive, negative, or zero, that are derived from the other parents ($P=2$, 3 for three-parent recombination).

$$T_{50} = T_{\text{P1}} + \sum_{B}\sum_{P} A_{BP} Q_{BP} \qquad [16.1]$$

A_{BP} is a dummy variable coded such that if a chimera contains block 1 from parent 2, $A_{12}=1$ and $A_{13}=0$. The regression coefficients Q_{BP} represent the thermostability contributions of the blocks A_{BP} relative to corresponding blocks from the reference parent, P1. Since the thermostability contributions of the blocks from P1 are accounted for in the T_{P1} term, no $A_{BP}Q_{BP}$ addition or subtraction is made to a chimera's predicted T_{50} value for block positions at which P1 appears. The predictive accuracy of these linear models can be determined using cross-validation.

Linear regression modeling was first applied in the context of cytochrome P450 chimeras to predict chimera thermostability (Li et al., 2007). T_{50} measurements were made for a sample set of 184 catalytically active chimeras chosen from a large pool of clones that featured all or most of the possible $3^8=6561$ unique sequences that can be constructed by recombining eight blocks from three parents. Linear regression provided a model with good correlation between observed and predicted T_{50} values (10-fold cross-validated $R^2=0.73$). This model also provided high correlation ($R^2=0.90$) between observed and predicted T_{50} values for a set of 20 new P450 chimeras not included among the first 184 that were tested. Most importantly, the model was 100% accurate in predicting the sequences of chimeras with T_{50} values greater than those of any of the three parent enzymes, at least among the 11 tested. The predicted most stable chimera was in fact more stable than any other chimera tested, with a T_{50} value 9.5 °C higher than that of the most thermostable parent (Fig. 16.2).

In addition to giving rise to new thermostable P450 chimeras, SCHEMA recombination yielded new P450 enzymes with high sequence and functional diversity. The 184 sampled chimeras differed from each other by an average of 46 amino acid mutations and contained up to 99 mutations relative to the closest parent enzyme. This sequence diversity led to useful new activities. In particular, the thermostable P450 chimeras were able to hydroxylate drug compounds on which none of the parent enzymes were active (Landwehr et al., 2007; Sawayama et al., 2009).

Many fewer than 184 measurements were required to build useful predictive models of the P450 thermostability. In fact, linear regression models based on just 35 of the data points from the set of 184 measurements accurately predicted the T_{50} values of the P450 chimeras in the 20 chimera test set referenced above (Li et al., 2007). A considerable amount of cloning effort is required to construct a complete or near complete chimera family such as the one from which the randomly chosen P450 sample set chimeras were selected. Furthermore, at least half of these chimeras were expected to be unfolded and not functional. Thus, we have been considering alternative approaches to generating sample sets. Taking advantage of modern total gene synthesis capabilities we can rapidly and cost effectively obtain a small number of sample set chimera genes with sequences that have been designed to provide a high level of information content for use in linear regression modeling.

3.1. Chimera sample set design approaches

The construction and characterization of designed SCHEMA chimera family sample sets were first applied in the context of fungal cellobiohydrolase class II (CBH II) enzymes, processive glycosyl hydrolases that play a key role in industrial biomass-to-fuel conversion processes (Heinzelman et al., 2009).

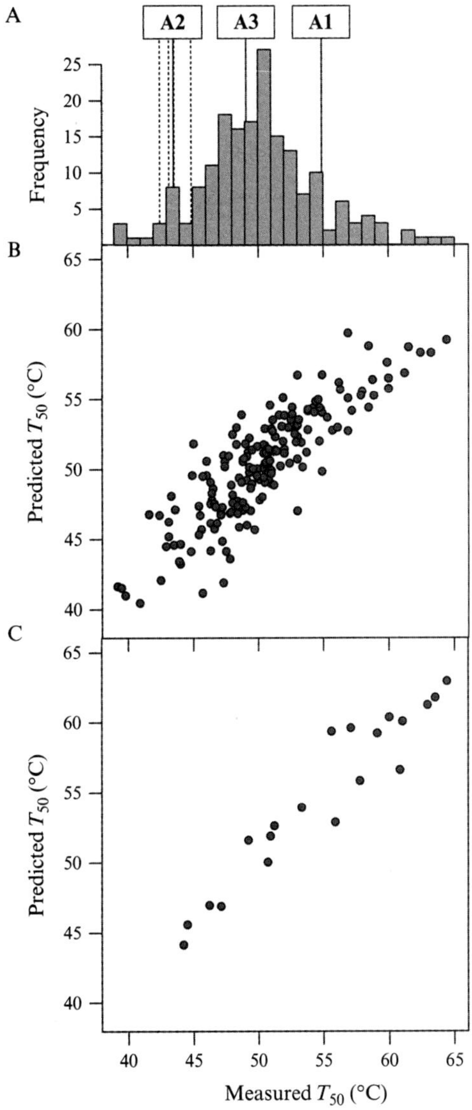

Figure 16.2 Sample, model, and predict allows accurate identification of chimeras that are more stable than the parent enzymes, illustrated here for a family of cytochrome P450 chimeras (Li et al., 2007). (A) Panel shows the distribution of T_{50} values for 184 chimeric cytochrome P450s, with T_{50}s for the three parent enzymes A1, A2, and A3 indicated (solid lines). Four experimental replicate measurements for parent A2 are shown to illustrate the high level of T_{50} measurement reproducibility that was achieved (dotted lines, standard deviation of 1.0 °C). (B) P450 chimera blocks make additive contributions to thermostability, allowing the use of linear regression to construct an accurate predictive model. Panel shows the good correlation between measured T_{50} values and T_{50} values predicted by an ordinary linear regression model ($R^2=0.73$). (C) Linear regression model from (B) accurately predicts T_{50} values for 20 new thermostable P450 chimeras ($R^2=0.90$), including the chimera predicted to be the most thermostable family member (top right-most point).

Gene synthesis company DNA 2.0 (Menlo Park, CA) provided a sample set of 48 synthetic CBH II chimera genes, which were codon-optimized for the expression host organism, *Saccharomyces cerevisiae*. To maximize the utility of the sample set chimera T_{50} dataset for constructing a linear regression model to predict chimera thermostability, the sample set was designed to provide equal representation to each of the three parents at each of the eight block positions. Specifically, 16 of the 48 chimeras featured block 1 (B1) from parent 1 (P1), 16 featured block 1 from parent 2 (P2), and 16 featured block 1 from parent 3 (P3). This equal representation was achieved by using each respective parent as a background in 16 of the 48 sample set chimeras and then substituting blocks from either one or both of the other two parents at three of the eight block positions.

Although this sample set design provided equal representation to each parent at each block position, it was not guided by any structural criteria or design algorithm. This approach was found to have a considerable drawback in that 25 of the 48 chimeras were not secreted by the *S. cerevisiae* expression host. This result is consistent with the expectation that about half of the 6561-member chimera family would be nonfunctional. Block 4 from parent 2 (B4P2) was particularly detrimental to CBH II chimera secretion; only 1 of the 16 sample set chimeras containing this block was secreted and active, the remaining 15 represented wasted gene synthesis effort. (We note here that all 23 of the CBH II chimeras that were secreted were also catalytically active.)

Measured thermostabilities of the 23 secreted chimeras showed that the sample set already contained CBH II chimeras that were more thermostable than any of the parents. This dataset, however, did not allow us to produce a quantitative linear model for predicting the thermostabilities of every member of the CBH II chimera family due to insufficient representation of some blocks in the sample set. Subsequent synthesis and characterization of an additional 41 CBH II chimeras, 31 of which were secreted (Heinzelman et al., 2009; Heinzelman et al., 2009a), led to the construction of an accurate (10-fold cross-validated $R^2=0.88$) linear regression model for predicting chimera T_{50} values. Although this was a highly desirable outcome, it appeared likely that the time, labor, and resources, specifically the number of synthesized genes required to build this model, could have been markedly reduced by choosing chimera sample set members that also had a high probability of being secreted. (We observed that more stable chimeras also tended to be more highly secreted, consistent with what has been reported for heterologous expression of other proteins in *S. cerevisiae*; Shusta, Kieke, Parke, Kranz, & Wittrup, 1999.)

The aim of maximizing the fraction of synthesized chimera genes corresponding to catalytically active enzymes was pursued in the context of a SCHEMA arginase chimera family built by recombining eight blocks from this enzyme's two human isoforms, both of which are candidate cancer therapeutics (Romero et al., 2012). To minimize the number of synthesized chimera genes needed to construct an accurate model for predicting arginase stability in human serum, a two-step active learning algorithm was employed.

The first step was to find an "informative" set of chimeras for use in constructing a logistic regression model that could predict the probability that a given arginase chimera would be catalytically active. A key objective in designing this "informative" sample set was to achieve adequate representation of each block from all of the parent enzymes so that we could accurately assess each block's impact on chimera expression.

The second step of the active learning algorithm then used this predictive model to identify a set of chimeras that were both highly informative, that is, provided adequate representation of each block from all of the parents and were likely to be functional. Characterizing the chimeras in this second sample set provided data that were used to construct a model for predicting the chimeras that were most desirable with respect to the enzyme property of interest, which in the case of the arginases was stability in human serum at physiological temperatures. In the following sections, we detail the procedures for applying the two-step active learning algorithm to construct accurate regression models for predicting the properties of every member of a SCHEMA enzyme chimera family.

3.1.1 Members of chimera families are represented with a binary vector $\mathbf{x}$, which codes for the parent identity at every block position. The covariance matrix of any collection of chimeras is given by the dot product between all pairs of sequences

$$\Sigma_{ij} = \mathbf{x}_i \cdot \mathbf{x}_j \qquad [16.2]$$

3.1.2 The first step of the active learning algorithm involves finding a set of sequences S that maximizes the mutual information between that candidate set and the other members of the chimera family ($L \backslash S$)

$$I(S; L\backslash S) = H(L\backslash S) - H(L\backslash S|S) \qquad [16.3]$$

where $H(L\backslash S)$ is the entropy of the other members of the family, and $H(L\backslash S \mid S)$ is the entropy of the same sequences after the candidate set S has been observed. The mutual information for a given candidate sample set quantifies how much evaluating the properties of the members of the candidate sample set reduces the uncertainty (Shannon entropy) in predictions for the other members of the family.

Due to the fact that the Shannon entropy for a logistic regression model cannot be calculated in closed form, the logistic response is approximated with a Gaussian likelihood. This approximation allows the properties of collections of sequences and their relationships to be represented with a multivariate Gaussian distribution. Calculating the mutual information requires finding the covariance matrix of the sequences in $L\backslash S$

$$\Sigma_{L\backslash S} = X_{L\backslash S} X_{L\backslash S}{}^{\mathrm{T}} \qquad [16.4]$$

where $X_{L\backslash S}$ is composed of all $\mathbf{x}$ in $L\backslash S$ ($X = [\mathbf{x}_1, \mathbf{x}_2, \ldots, \mathbf{x}_n]^{\mathrm{T}}$) and T is the matrix transpose. The mutual information calculation also requires finding the covariance matrix of the sequences in $L\backslash S$ conditioned on S

$$\Sigma_{L\backslash S|S} = X_{L\backslash S} X_{L\backslash S}{}^{\mathrm{T}} - X_{L\backslash S} X_S{}^{\mathrm{T}} \left(X_S X_S{}^{\mathrm{T}}\right)^{-1} X_S X_{S\backslash L}{}^{\mathrm{T}} \qquad [16.5]$$

With these covariance matrices, the Shannon entropy is given by

$$H = \frac{1}{2} \ln\left[(2\pi e)^k |\Sigma|\right] \qquad [16.6]$$

where k is the dimensionality of the covariance matrix. The mutual information is the difference between the entropies before and after the sequences in S are observed (Eq. 16.3).

3.1.3 Having defined the mutual information objective function of Eq. (16.3), we now carry out an operation to find a set of sequences that maximizes it. Gaussian mutual information is a submodular set function (Krause & Guestrin, 2007) and therefore can be efficiently maximized using a greedy approximation algorithm that sequentially selects the most informative sequence (Nemhauser, Wolsey, & Fisher, 1978). To perform the greedy optimization, the mutual information of every sequence in the chimera family is evaluated and the sequence with the highest mutual information is used in the next iteration. Next, the mutual information of this sequence and all other sequences in the family is evaluated, and the set of two sequences with the highest mutual information is chosen. This greedy sequence selection process is repeated until the sample set has the desired number of sequences. For large chimera families, the speed of the greedy algorithm can be significantly increased by using "lazy" evaluations (Minoux, 1978). We typically use this algorithm to select a chimera sample set containing a number of sequences that is equal to the number of parameters in the logistic regression model. The MATLAB Toolbox for Submodular Function Optimization has useful functions for calculating Gaussian mutual information and maximizing submodular functions using accelerated greedy algorithms (Krause, 2010). We note here that greedy maximization of the objective function in Eq. (16.3) will not always find the global optimum but is guaranteed to find a solution near the optimum (Nemhauser et al., 1978). We also note that for every solution there are multiple, equivalent sample sets with the same level of mutual information.

3.1.4 After one of the maximally informative sample sets of chimera sequences has been identified, the corresponding genes are synthesized and expressed in a recombinant host. The chimera genes' functional status is then determined. This set of sequence/functional status data is then used to train a Bayesian logistic regression model that can predict the probability of forming a catalytically active enzyme (or otherwise functional protein) for all of the chimeras in the family. Chapter 4 of Bishop (2006) provides a detailed account of how to perform Bayesian logistic regression using the Laplace approximation, and specific methods we have used in applying Bayesian regression to the analysis of chimera families are further discussed in Section 3.4.

For the arginase SCHEMA chimera family, the first step of the active learning algorithm identified a set of eight (out of $2^8=256$ total chimeras in the family) arginase chimera sequences possessing maximum information content. As noted above, performing iterations of the greedy algorithm would have yielded additional sample sets with identical information content but differences in sequences for at least some of the chimeras contained therein. After identifying one of the maximally informative chimera sample sets, in particular, the sample set that was returned after the first time that the greedy algorithm was executed, codon-optimized genes encoding these eight arginase chimeras were synthesized and expressed in a recombinant *Escherichia coli* host. Three of the eight sample set chimeras were functional arginases. These data were sufficient to train a Bayesian logistic regression model for predicting the probability of function for each of the 256 chimeras in the family.

3.1.5 The second step of the active learning algorithm consists of identifying a set of highly informative and functional chimeras. The predictions from the logistic regression model can be used to calculate the expected value of the mutual information between the chosen set of sequences and the other members of the chimera family

$$E[I(S;L\backslash S)] = \sum_{A\in\mathcal{P}(S)} \left[I(A;L\backslash A)\prod_{\mathbf{c}\in A} p_{\mathbf{c}} \prod_{\mathbf{c}\in(S\backslash A)} 1-p_{\mathbf{c}} \right], \qquad [16.7]$$

where the sum is over all subsets A in the power set of S and $p_{\mathbf{c}}$ represents the predicted probability of being functional for chimera **c** from the catalytic activity logistic regression model.

3.1.6 Chimera sample sets with maximized expected mutual information can be identified by applying the expression appearing in Eq. (16.7). Maximizing this expected mutual information criterion identifies sets of chimera sequences that are both highly informative and likely to be functional. As the expected value of the mutual information is submodular, it can be efficiently maximized using a greedy algorithm, as described in Step 3.1.3 of Section 3.1. In performing this maximization, covariance matrices are conditioned (Eq. 16.5) on all functional sequences that were observed in the first step of the active learning algorithm so as to encourage the exploration of new regions of sequence space.

In the case of the arginase SCHEMA chimera family, a single execution of the greedy algorithm was performed to identify one of the possible sets of four additional chimeras that maximized the expected value of the mutual information. Genes corresponding to these four arginase chimeras were then synthesized and expressed in *E. coli*. All four of these new arginase chimeras were found to be highly active, validating the functional status logistic regression model's utility for predicting the sequences of catalytically active arginase chimeras.

3.2. Measurement of sample set chimera stability data

A number of techniques are available for obtaining chimera sample set stability data for regression model construction. Stability can be given in terms of free energies of (un)folding, as determined by standard methods, but this is not useful for the many proteins that do not unfold reversibly. A convenient thermostability data collection technique for enzymes is a T_{50} (temperature at which 50% activity is lost after a specified incubation time) measurement, which is useful for the many enzymes that undergo irreversible thermal denaturation.

3.2.1 In measuring chimera T_{50}s, one begins by specifying a thermal denaturation incubation time interval, which should be at least 10 min to allow for temperature equilibration. The sample set chimeras are incubated for the specified time interval across a range of temperatures, in aqueous buffered solution that does not contain substrate. Each thermal denaturation interval is halted by cooling the chimera samples in an ice water bath.

3.2.2 After completing the thermal denaturations, an appropriate substrate is added to all of the heat-treated samples as well as to an untreated reference sample, and the chimeras are assayed for activity. An appropriate enzymatic activity assay is performed to determine each heat-treated sample's residual activity, which is defined as the measured catalytic activity for a heat-treated sample divided by measured catalytic activity of the unheated reference sample.

3.2.3 The T_{50} value, defined as the temperature at which 50% of enzyme activity is lost after the specified incubation time, can then be determined by using nonlinear regression (we typically use the Microsoft Excel solver feature) to fit the residual activity-temperature data pairs. In order to ensure accurate T_{50} measurements, the thermal denaturation temperature range should be such that all of the chimeras evaluated have residual activities ≤25% at the maximum temperature tested. Furthermore, we have observed that using a circulating water bath to make T_{50} measurements results in greater reproducibility (to within 1 °C) than is typically obtained by using a thermal cycler for the incubation step. Finally, one must be certain that enzyme denaturation is irreversible, where some disulfide-containing enzymes might require the addition of a reducing agent to prevent refolding after heat treatment (Heinzelman et al., 2010).

Although T_{50} measurement is convenient for obtaining sizable thermostability datasets for regression model construction in a short period of time, there may be cases in which it does not adequately break out differences in chimera stability or does not reflect the desired property well. Chimera stability can also be measured in terms of retention of catalytic activity over long time intervals. Such an approach was used to measure arginase chimera stability, where stability was quantified by area-under-curve (AUC) determination for activity as a function of time at moderate temperature (Romero et al., 2012). Thermal denaturation half-life ($t_{1/2}$) measurement (Heinzelman et al., 2009) and T_{opt} determination are additional ways (Smith et al., 2012) to describe chimera stability and obtain data for regression model construction.

3.3. Simple linear regression modeling of thermostability data

As noted above, the observation that the blocks which comprise a given enzyme chimera make additive contributions (Eq. 16.1) to that chimera's stability, e.g., T_{50} and/or T_{opt} value, enables the use of linear regression models to predict the sequences of the "best" chimera family members, at least with respect to that criterion. In this section, we describe the approaches that have been used to construct linear regression models for predicting chimera properties. This description features an emphasis on modeling chimera datasets using Bayesian linear regression, an approach that has proven particularly useful for modeling datasets much smaller than those obtained for the P450, CBH II, and family 48 glycosyl hydrolase chimera families, all of which contained at least 51 data points.

For the P450 and CBH II chimera families, simple least squares regression was used to obtain a linear model for predicting chimera T_{50} values (Heinzelman et al., 2009a; Li et al., 2007). The coefficient of determination (R^2 value) between observed and predicted T_{50} values for these respective models was confirmed using 10-fold cross-validation (Dietterich, 1998).

This approach was later improved upon in the context of modeling the thermostabilities of the members of the family 48 glycosyl hydrolase chimera family. In particular, the predictive accuracy of the model was improved by training the model on T_{50} data and E, the number of broken contacts in a given chimera (Smith et al., 2012). Including E as a parameter improved the correlation between observed and predicted T_{50}, increasing the model's R^2 value from 0.82 to 0.88.

3.4. Bayesian linear regression modeling of chimera sample set thermostability data

As noted above, the T_{50} and/or T_{opt} datasets for the P450, CBH II, and family 48 glycosyl hydrolase chimera families all contained 51 or more data points that could be used for linear regression analysis. In contrast, just seven data points (three expressed chimeras from the first sample set and all four from the second) were available for modeling arginase chimera stability. The relatively small size of this dataset motivated the construction of a Bayesian linear regression model, which was expected to improve the ability to accurately fit the limited number of arginase stability data points relative to the previously used linear least squares models (Bishop, 2006).

Ordinary least squares regression finds the set of model parameter values that minimizes the squared difference between experimentally measured and predicted thermostability values, for example, T_{50}s. This parameter set is known as the maximum likelihood estimate, since it is the parameter set that is most probable given the observed data. The maximum likelihood estimate approach is convenient to use as the model parameters can be found by using simple, well-known formulas. While this modeling approach is effective when relatively large datasets, that is, 30 or more data points, are available, it tends to "overfit" the data, or may even be unsolvable, when only small numbers of data points are available for model construction (Bernardo & Smith, 1994).

For cases in which the dataset to be modeled is of limited size, Bayesian linear regression can be used to include additional information. This additional information is specified by a prior probability distribution over the regression parameters.

3.4.1 In the case of modeling chimera stability, before any experimental measurements are made, the effect of block substitutions is assumed to be small and all blocks are taken as being equally likely to be stabilizing. This prior information is encoded with a zero mean, isotropic Gaussian prior. With this prior, the Bayesian parameter estimates β can be found in closed form and are given by Eq. (16.8)

$$\beta = \left(X^{\mathrm{T}} X + \frac{\sigma_{\mathrm{b}}^2}{\sigma_{\mathrm{n}}^2} I \right)^{-1} X^{\mathrm{T}} y \qquad [16.8]$$

where X is the matrix that codes a chimera's block identities ($X = [\mathbf{x}_1, \mathbf{x}_2, \ldots, \mathbf{x}_n]^{\mathrm{T}}$), I is the identity matrix, y is the vector of corresponding stability measurements, σ_{b}^2 is the prior block variance, and σ_{n}^2 is the variance of the measurement noise. We have found MATLAB to be useful for performing linear algebra calculations.

3.4.2 The values of the two variance hyperparameters, σ_{b}^2 and σ_{n}^2, that minimize the squared error are estimated by performing cross-validation on the chimera sequence/stability measurement dataset. To do this, a range of hyperparameters is scanned and for each combination the model is evaluated for its cross-validated mean squared error (MSE) of prediction. The hyperparameter combination that has the lowest MSE is then used for prediction.

3.4.3 If there are insufficient data to perform cross-validation, the Eq. (16.8) hyperparameters can be estimated directly from the data using the empirical Bayes method (Bernardo & Smith, 1994). In this case, the likelihood function is integrated over all possible regression coefficients (β in Eq. 16.8) to yield a marginalized likelihood function. This marginal likelihood is then maximized with respect to σ_{b}^2 and σ_{n}^2 using gradient descent, Newton's method, or other iterative approaches (Bishop, 2006). Finally, hyperparameters found by cross-validation or empirical Bayes are substituted into Eq. (16.8) to provide Bayesian parameter estimates that are used to construct the predictive chimera stability model.

In the case of modeling the arginase chimera sample set stability data, the Bayesian linear regression approach yielded a model that provided an excellent fit between observed and predicted arginase chimera log AUC values, returning a R^2 value of 0.96 despite being trained on only nine data points. This strong correlation between measured and predicted stability suggests that Bayesian linear regression will be valuable in reducing the numbers of synthetic chimera sample set genes and experimental measurements needed to construct accurate models for predicting the stabilities and possibly other properties of members of SCHEMA protein chimera families.

4. SUMMARY

The application of linear regression analysis to stability data obtained by characterizing small, designed sample sets of SCHEMA chimeras enables the efficient construction of predictive models that accurately identify the sequences of chimera family members whose stabilities are greater than those of the parent enzymes. This "sample, model, and predict" approach allows the sequences of hundreds of enzymes with improved properties and high sequence diversity to be identified and offers an avenue for improving enzymes that are not amenable to engineering using high-throughput screening methods.

A recently developed, two-step active learning algorithm that accounts for both the probability of a chimera being catalytically active and the information content of the chimera sample set markedly reduces the number of sample set genes needed for constructing robust predictive chimera property models. The ability of this approach to decrease the size of the chimera sample set needed to build accurate models will make it possible to model the properties of the members of very large SCHEMA chimera families without undue requirements for gene synthesis and data sampling. Furthermore, the Bayesian linear regression method used in the context of this learning algorithm has enabled the development of a robust predictive model based on data collected for an extremely small chimera sample set that would be extremely difficult to accurately characterize using linear regression. This improved ability to construct accurate regression models from very small datasets could be extrapolated to efficiently identifying the sequences of chimeras with improvements in properties other than (thermo)stability, such as specific catalytic activity or product inhibition, thus further increasing the utility of SCHEMA recombination in allowing the generation of large numbers of improved enzymes that feature high levels of sequence diversity.

ACKNOWLEDGMENTS

The authors acknowledge funding from the Institute of General Medical Sciences of the National Institutes of Health (ARRA grant 2R01-GM068664-05A1) for work on cytochrome P450s and the U.S. Army Research Office Institute for Collaborative Biotechnologies (grant W911NF-09-D-0001) for technology development and cellulase engineering. The contents of this chapter are solely the responsibility of the authors and do not necessarily represent the official views of the sponsors.

REFERENCES

Bernardo, J. M., & Smith, A. M. (1994). *Bayesian theory*. (1st ed.). New York: Wiley & Sons.

Bishop, C. M. (2006). *Pattern recognition and machine learning*. (1st ed.). New York: Springer.

Dietterich, T. G. (1998). Approximate statistical tests for comparing supervised classification learning algorithms. *Neural Computation, 10*, 1895–1923.

Endelman, J. B., Silberg, J. J., Wang, Z.-G., & Arnold, F. H. (2004). Site-directed protein recombination as a shortest-path problem. *Protein Engineering, Design & Selection, 17*, 589–594.

Heinzelman, P., Komor, R., Kannan, A., Romero, P. A., Yu, X., Mohler, S., et al. (2010). Efficient screening of fungal cellobiohydrolase class I enzymes for thermostabilizing sequence blocks by SCHEMA structure-guided recombination. *Protein Engineering, Design & Selection, 23*, 871–880.

Heinzelman, P., Snow, C. D., Wu, I., Nguyen, C., Villalobos, A., Govindarajan, S., et al. (2009). A family of thermostable fungal cellulases created by structure-guided recombination. *Proceedings of the National Academy of Sciences of the United States of America, 106*, 5610–5615.

Heinzelman, P., Snow, C. D., Smith, M. A., Yu, X., Kannan, A., Boulware, K., et al. (2009a). SCHEMA recombination of a fungal cellulase uncovers a single mutation that contributes markedly to stability. *Journal of Biological Chemistry, 284*, 26229–26233.

Krause, A. (2010). SFO: A toolbox for submodular function optimization. *Journal of Machine Learning Research, 11*, 1141–1144.

Krause, A., & Guestrin, C. (2007). Near-optimal observation selection using submodular functions. *Proceedings of 22nd conference on artificial intelligence (AAAI)*. Nectar Track 22, (pp. 1650–1654).

Li, Y., Drummond, D. A., Sawayama, A. M., Snow, C. D., Bloom, J. D., & Arnold, F. H. (2007). A diverse family of thermostable cytochrome P450s created by recombination of stabilizing fragments. *Nature Biotechnology, 25*, 1051–1056.

Landwehr, M., Carbone, M., Otey, C. R., Li, Y., & Arnold, F. H. (2007). Diversification of catalytic function in a synthetic family of chimeric cytochrome p450s. *Chemisty and Biology, 14*, 269–278.

Meyer, M. M., Hochrein, L., & Arnold, F. H. (2006). Structure-guided SCHEMA recombination of distantly related beta-lactamases. *Protein Engineering, Design & Selection, 19*, 563–570.

Minoux, M. (1978). Accelerated greedy algorithms for maximizing submodular set functions. *Proceedings of the 8th IFIP conference on optimization techniques* (pp. 234–243), New York: Springer.

Nemhauser, G. L., Wolsey, L. A., & Fisher, M. L. (1978). An analysis of approximations for maximizing submodular set functions—I. *Mathematical Programming, 14*, 265–294.

Otey, C. R., Landwehr, M., Endelman, J. B., Hiraga, K., Bloom, J. D., & Arnold, F. H. (2006). Structure-guided recombination creates an artificial family of cytochromes P450. *PLoS Biology, 4*, e112.

Romero, P., Stone, E., Lamb, C., Chantranupong, L., Krause, A., Miklos, A., et al. (2012). SCHEMA designed variants of human arginase I & II reveal sequence elements important to stability and catalysis. *ACS Synthetic Biology, 1*, 221–228.

Sawayama, A. M., Chen, M. M., Kulanthaivel, P., Kuo, M. S., Hemmerle, H., & Arnold, F. H. (2009). A panel of cytochrome P450 BM3 variants to produce drug metabolites and diversify lead compounds. *Chemistry: A European Journal, 15*, 11723–11729.

Shusta, E. V., Kieke, M. C., Parke, E., Kranz, D. M., & Wittrup, K. D. (1999). Yeast polypeptide fusion surface display levels predict thermal stability and soluble secretion efficiency. *Journal of Molecular Biology, 292*, 949–956.

Smith, M. A., Rentmeister, A., Snow, C. D., Wu, T., Farrow, M. F., Mingardon, F., et al. (2012). A diverse set of family 48 bacterial glycoside hydrolase cellulases created by structure-guided recombination. *FEBS Journal, 279*, 4453–4465.

CHAPTER SEVENTEEN

Protein Switch Engineering by Domain Insertion

Manu Kanwar, R. Clay Wright, Amol Date, Jennifer Tullman, Marc Ostermeier[1]

Department of Chemical and Biomolecular Engineering, Johns Hopkins University, Baltimore, Maryland, USA

[1]Corresponding author: e-mail address: oster@jhu.edu

Contents

Abstract

The switch-like regulation of protein activity by molecular signals is abundant in native proteins. The ability to engineer proteins with novel regulation has applications in biosensors, selective protein therapeutics, and basic research. One approach to building proteins with novel switch properties is creating combinatorial libraries of gene fusions between genes encoding proteins that have the prerequisite input and output

Methods in Enzymology, Volume 523
ISSN 0076-6879
http://dx.doi.org/10.1016/B978-0-12-394292-0.00017-5

functions of the desired switch. These libraries are then subjected to selections and/or screens to identify those rare gene fusions that encode functional switches. Combinatorial libraries in which an insert gene is inserted randomly into an acceptor gene have been useful for creating switches, particularly when combined with circular permutation of the insert gene. Methods for creating random domain insertion libraries are described. Three methods for creating a diverse set of insertion sites in the acceptor gene are presented and compared: DNase I digestion, S1 nuclease digestion, and multiplex inverse PCR. A PCR-based method for creating a library of circular permutations of the insert gene is also presented.

1. INTRODUCTION

The ability of proteins to be regulated by molecular signals is a hallmark of biological systems. Such proteins function as switches in which recognition of an input signal (e.g., small molecule or protein) regulates an output function (e.g., enzyme activity or DNA affinity). The ability to create novel protein switches would have applications in sensing and therapeutics. One approach to building new switches is to reengineer existing natural switches. A more challenging approach, but perhaps more powerful in the long term, is to build new switches from existing protein domains with the prerequisite input and output functions of the desired switches. Domain insertion, the insertion of one protein domain into another, has been used to couple the functions of two proteins to create protein switches (Ostermeier, 2005). The design challenge lies in the fusion of the two protein domains such that communication is established between the two functions.

One approach is using evolution as the design algorithm. The first step in this algorithm is creating protein diversity, from which selections or screens can be used to identify the rare variants with switching properties. One approach to creating this diversity is creating combinatorial libraries in which an insert gene is randomly inserted into the acceptor gene. The insert gene can encode either the output function (Guntas, Mansell, Kim, & Ostermeier, 2005; Guntas, Mitchell, & Ostermeier, 2004; Guntas & Ostermeier, 2004; Tullman, Guntas, Dumont, & Ostermeier, 2011) or the input function (Edwards, Busse, Allemann, & Jones, 2008; Wright, Wright, Eshleman, & Ostermeier, 2011) of the switch.

For example, insertion of β-lactamase (BLA) into sugar-binding proteins can create switch proteins in which BLA activity is modulated by the respective sugar (Guntas et al., 2005, 2004; Guntas & Ostermeier, 2004; Tullman et al., 2011). We have extensively created and characterized switches comprising fusions of BLA and maltose-binding protein (MBP). These studies have shown that switches can be created that provide >100-fold change in activity upon effector binding (Guntas et al., 2005) and that switching is reversible (Guntas et al., 2004). The switches are modular, allowing mutations that increase effector affinity (Kim & Ostermeier, 2006) or provide the switch with the ability to be activated by new effectors (Guntas et al., 2005). The MBP–BLA switch genes function in *E. coli* cells and confer a new phenotype: maltose-dependent resistance to β-lactam antibiotics (Guntas et al., 2005, 2004). Many of the MBP–BLA switches function as heterotropic allosteric enzymes, and NMR and crystallographic studies of one switch are consistent with the expectation that the individual domain structures of RG13 are substantially conserved from MBP and BLA (Ke et al., 2012; Wright, Majumdar, Tolman, & Ostermeier, 2010). More recently, we have identified MBP–BLA switch genes that do not encode allosteric proteins but rather encode a protein whose cellular accumulation increases in the presence of the effector, thereby conferring an effector-dependent switching phenotype to cells (Heins, Choi, Sohka, & Ostermeier, 2011; Sohka et al., 2009). Finally, our MBP–BLA switch work has shown how domain fusion can result in emergent properties. One of our MBP–BLA switches is negatively, allosterically regulated by Zn^{2+}, which is unexpected since neither MBP nor BLA has significant affinity for Zn^{2+} (Ke et al., 2012; Liang, Kim, Boock, Mansell, & Ostermeier, 2007).

Recently, we have also used domain insertion to create prodrug-activating enzymes that are turned on in the presence of a protein cancer marker and render cancer cells selectively sensitive to the prodrug in a cancer marker-dependent fashion (Wright et al., 2011). From random domain insertion libraries comprising the CH1 domain of *p*300 (which binds the cancer marker HIF-1α) inserted into yeast cytosine deaminase, which can activate the prodrug 5-fluorocytosine, we selected two switch genes that conferred on 5-fluorocytosine an increased sensitivity to *E. coli* cells and human colon and breast cancer cell lines in a HIF-1α-dependent manner. Evidence suggested that the better performing switch functioned primarily through a mechanism in which HIF-1α caused the switch protein to accumulate to higher levels in the cell. This strategy offers a platform for the development of inherently selective protein therapeutics for cancer and other diseases, although significant delivery challenges will need to be overcome for success to be achieved.

This chapter covers the methodology for creating random domain insertion libraries using nonspecific endonucleases or PCR-based methods

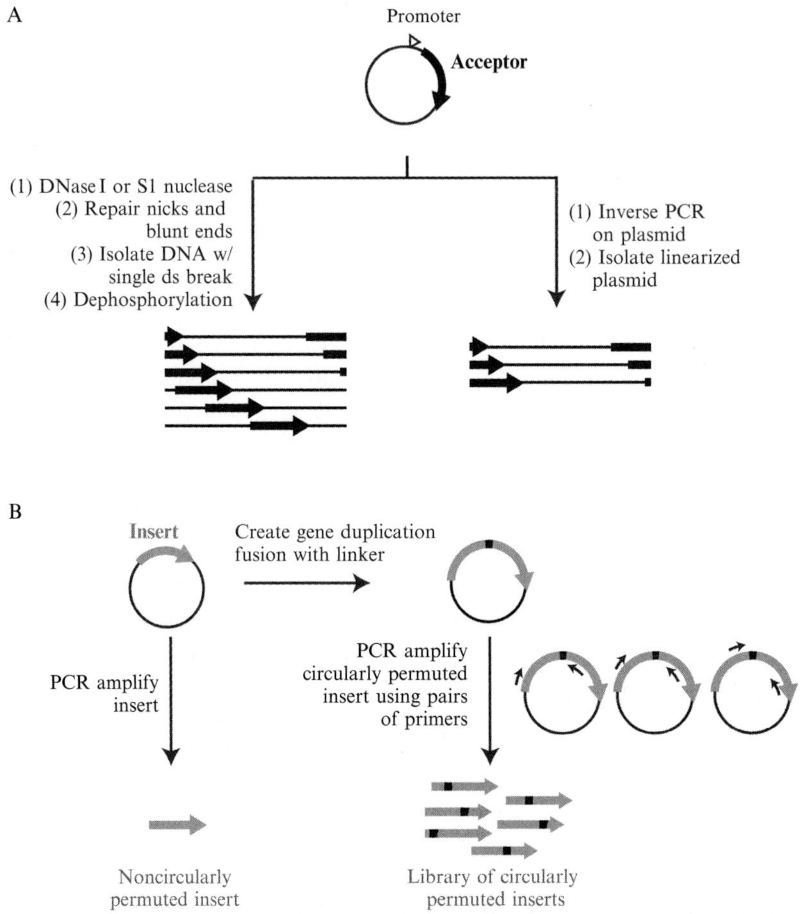

Figure 17.1 Overview of methods for random domain insertion with optional circular permutation of insert. (A) Methods for making a linear plasmid DNA containing the acceptor gene in which the site of linearization is created randomly throughout the plasmid by DNase I or S1 nuclease or targeted to occur in the acceptor gene by multiplex inverse PCR. (B) Methods for preparing the insert DNA either without circular permutation or with circular permutation via a PCR on a gene duplication fusion.

(Fig. 17.1) but does not cover random insertions using transposons (Edwards et al., 2008). Although the methods are presented within the context of switch construction, they are applicable to any study in which random insertion of one DNA segment into another is useful (e.g., structure–function studies, studies of promoter or chromosomal structure, and studies of oligomeric or repeat proteins).

2. CREATION OF RANDOM DOUBLE-STRANDED BREAKS IN PLASMIDS CONTAINING THE ACCEPTOR DNA

The acceptor gene is cloned into an expression vector of choice. This plasmid is then targeted for the creation of a random double-stranded break by endonuclease digestion with DNase I (Section 2.1) or S1 nuclease (Section 2.2), or for the amplification of the desired linear double-stranded plasmid DNA with "breaks" at select locations by inverse PCR (Ochman, Gerber, & Hartl, 1988) in a multiplex fashion. The three methods (DNase I digestion, S1 nuclease digestion, and multiplex inverse PCR) have different strengths and weaknesses (Table 17.1). Because the DNase I and S1 nuclease methods create a lot of "junk" library members (e.g., insertions outside the acceptor gene, insertions out of frame or backward), these methods should be used only with a powerful selection or screen that can handle large libraries. Targeting the insertion using multiplex inverse PCR makes for a highly focused library enriched in the types of fusions that might be switches. The drawbacks are the cost of the sets of primers and the trade-off between increased diversity and the large number of PCR reactions to perform. Methods that create short deletions or tandem duplications at the insertion site can be beneficial, as this added diversity will adjust the distance and geometric orientation between the two fused domains (i.e., tandem duplications can serve as pseudo-natural "linkers" between the two domains).

2.1. Creation of random breaks using DNase I

1. Prepare approximately 100 μg of plasmid DNA containing the gene encoding for the acceptor protein using the Qiagen DNA Miniprep Kit (Qiagen, Valencia, CA) according to the manufacturer's instructions. We have found that the Miniprep kits have better yield with less genomic DNA contamination than the Midiprep or Maxiprep kits.
2. Prepare 50 μL of working stock solution of DNase I (1 U/μL) (New England Biolabs, Ipswich, MA) in 25 m*M* Tris–HCl pH 7.5 containing 50% glycerol and store it at −20 °C.
3. Prepare 1.2 mL of diluent solution in 50 m*M* Tris–HCl pH 7.5 containing 1 m*M* $MnCl_2$ and 6 μL of 100× BSA (New England Biolabs, Ipswich, MA). The diluent solution must be freshly prepared and kept at room temperature.
4. Determine the DNase I concentration that will result in the highest yield of single cut double-stranded DNA for a 5 μg plasmid digestion for

Table 17.1 Comparison of methods for creating random double-strand breaks in acceptor DNA in order to create domain insertion libraries

	DNase I	S1 nuclease	Multiplex inverse PCR
Pros	• More random than S1 nuclease • Some tandem duplications produced	• Easier library creation than DNase I • Unlike DNase I, not prone to large deletions	• Creates focused library • The distribution of direct insertions, tandem duplications, and deletions is user-defined • Easier library creation • Allows for simultaneous creation of linker libraries at insertion site • Shorter series of steps
Cons	• Library construction is challenging • Prone to large deletions • Large fraction of library inserted outside gene or out of frame	• May not be as random as DNase I • Prone to insertions at sequences that produce stem-loop structures, such as inverted repeats • May not produce tandem duplications at insertion site • Large fraction of library inserted outside gene or out of frame	• Requires large sets of primers and many PCR reactions • Library is less diverse • Creation of large tandem duplications problematic
References	Guntas et al. (2005, 2004), Guntas and Ostermeier (2004), and Wright et al. (2011)	Tullman et al. (2011) and Wright et al. (2011)	This work

8 min at room temperature in a total reaction volume of 100 μL. In brief, 5 μg of plasmid DNA in 95 μL of diluent solution is incubated with varying concentrations of DNase I (5 μL) for 8 min at room temperature. The reaction is stopped by the addition of 1.2 μL of 1.0 *M* EDTA and the mixture incubated at 75 °C for 10 min to heat inactivate the DNase I. The reaction product is visualized on a 0.8% agarose, lithium borate (LB) (Faster Better Media, Hunt Valley, MD) gel to determine the optimal DNase I dilution. The optimal DNase I concentration is the one that produces a sharp band for linearized plasmid (no smearing). Achievement of this outcome usually requires that 30–50% of the plasmid remain supercoiled while 30% of the supercoiled plasmid DNA is converted to the desired linear DNA product.

5. Once the optimal DNase I dilution is determined, prepare 1140 μL of plasmid DNA solution (60 μg plasmid DNA in 1140 μL of diluent solution) and dispense in 12 microtubes each containing 95 μL (i.e., 5 μg DNA) of plasmid DNA solution. Incubate for 10 min at room temperature.
6. Add 5 μL of the optimal DNase I dilution (determined earlier) to the first microtube containing 95 μL plasmid DNA solution, and flick to mix. Add 5 μL of the optimal DNase I dilution to the next tube after 30 s and repeat the step until DNase I has been added to each of the 12 microtubes. Add 1.2 μL of 1.0 *M* EDTA to each tube after it has incubated for 8 min. Incubate all 12 tubes at 75 °C for 10 min to heat inactivate the DNase I.
7. Combine the contents of all 12 tubes and purify the DNA using 25 μg DNA clean and concentrator (Zymo Research Corp., Irvine, CA). Use at least three spin columns. Elute the DNA in 100 μL of 0.1 × elution buffer prewarmed to 50 °C per column and combine the eluate from each spin column. The prepared DNA can be stored at −20 °C.

2.2. Creation of random double-stranded breaks using S1 nuclease

Unlike DNase I digestion, S1 nuclease digestion of supercoiled DNA essentially halts after the creation of linear dsDNA with blunt ends; hence, libraries are easier to construct. S1 nuclease is typically known as a single strand-specific-nuclease. However, S1 nuclease has been shown to digest supercoiled plasmids especially in regions containing inverted repeats or forming cruciform structures (Lilley, 1980; Panayotatos & Wells, 1981). Although inverted repeats are common in the origins of replication of plasmids, plasmids digested at these sites are generally excluded from the combinatorial library upon transformation into *E. coli* due to the inability of the plasmid to be replicated when the origin is disrupted by an inserted sequence. We established that S1 nuclease also creates dsDNA breaks elsewhere in the plasmid as well, and that these breaks appear fairly random, although we have not extensively studied the distribution of cut sites (Tullman et al., 2011). If, however, inverted repeat sequences occur in the gene sequence of the acceptor protein domain, one should choose another method to create the random double-stranded breaks. We have used a plasmid containing the p15a origin of replication to successfully create ribose-activated BLAs using S1 nuclease digestions to create sites for insertion into the acceptor gene encoding ribose-binding proteins (Tullman et al., 2011).

1. Prepare 50 μg of plasmid containing the acceptor gene sequence via standard procedures (e.g., Qiagen Qiaquick Miniprep Kit, Qiagen, Valencia, CA). It is important that the DNA be of high quality (RNA and genomic DNA-free). We have found that the Miniprep kits have better yield with less genomic DNA contamination than the Midiprep or Maxiprep kits.
2. Aliquot 2 μg of plasmid into each of 12 tubes and incubate in 50 m*M* sodium acetate, 280 m*M* NaCl, and 4.5 m*M* $ZnSO_4$ at pH 4.5 (at 25 °C) with 10 units of S1 nuclease (Promega, Madison, WI) in a final volume of 25 μL for 20 min at 37 °C.
3. Stop the reaction by pooling the 12 tubes and adding the DNA-binding buffer from the DNA clean and concentrator 25 μg kit (Zymo Research Corp.). Continue the purification of the DNA using the binding column and wash buffer as per the manufacturer's instructions and elute in 50 μL of nuclease-free DI water prewarmed to 50 °C. The prepared DNA can be stored at −20 °C.

2.3. Creation of random "breaks" using multiplex inverse PCR

For multiplex inverse PCR (Fig. 17.2A), divergent abutting primers that amplify the entire plasmid need to be designed for every desired insertion site (e.g., at every codon in the acceptor gene or a subset of these codons). These primers are used to "open up" the plasmid at desired locations for insertion. Mixing and matching of nearby primers can be used to open up the plasmid such that deletions or tandem duplications of portions of the acceptor gene occur (Fig. 17.2B). For example, a pair of divergent primers that exclude the codons between them can be used to create deletions. Using primers that have three overlapping bases (one codon overlap) will create a one amino

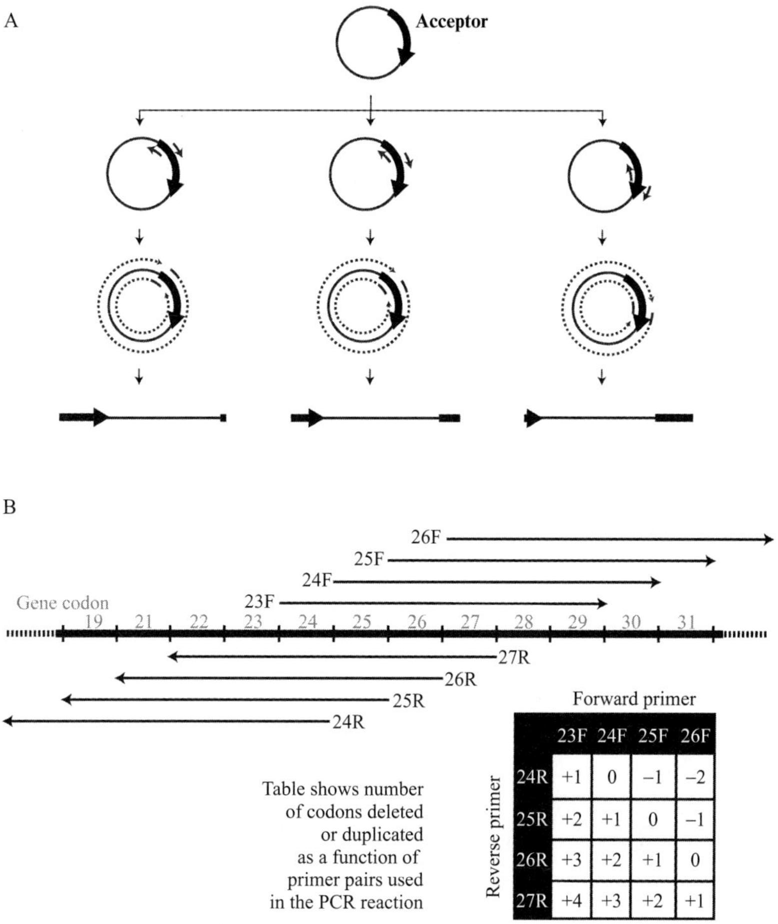

	23F	24F	25F	26F
24R	+1	0	−1	−2
25R	+2	+1	0	−1
26R	+3	+2	+1	0
27R	+4	+3	+2	+1

Figure 17.2 Schematic depiction of multiplex inverse PCR. (A) Basic mechanism. PCR is performed with abutting sets of primers designed to amplify in opposite directions around the plasmid resulting in a set a linear plasmid dsDNA molecules whose site of linearization is defined by the primers. (B) Details of primer design and method for creation of deletions and duplications at the site of linearization. As an example, amplification using a set of four forward and four reverse primers is shown in the region around codons 19–31 of a gene. Different combinations of forward and reverse primers results in codon deletions or tandem duplications as indicated in the table.

acid repeat on either side of the insert gene in the final library, primers with six base overlap (two codon overlap) will create two amino acid repeat and so on. However, the number of amino acid tandem duplications that can be obtained is limited. The yield of the correct product for PCR reactions using primers corresponding to more than two codons of overlap can be significantly reduced due to the primer dimerization.

This method not only defines the location of insertion of target DNA, but also ensures a library with all in-frame fusions (although the insert gene will still be inserted backward 50% of the time). Additionally, the insertion of DNA is targeted only in the gene of interest, and therefore, unproductive insertions in the backbone of the plasmid that do not produce combinatorial fusions of the target genes can be eliminated. The highly focused library that is produced is ideal for creating protein switches for which selection or screening methods are limited and cumbersome. However, there are some limitations and disadvantages of the PCR-based method that need to be considered while selecting the best method for construction of a DNA library. Two oligos are required for every insertion location, and a separate PCR reaction needs to be performed for every desired insertion site. The design of all the primers and execution of all the reactions is time consuming and may require optimization.

2.3.1 Multiplex inverse PCR

PCR reactions can be executed in a 96-well PCR plate using a 96-well plate thermocycler. For the PCR reactions to be performed effectively in parallel, it is crucial that all the primers have similar melting temperatures.

1. Prepare a master mix that consists of polymerase buffer, dNTPs, polymerase, DNA template, and water for the required reactions. For 100 PCR reactions each of 50 μL total volume, the following master mix can be prepared:
   10 × Polymerase buffer—1000 μL
   10 m*M* dNTP—150 μL
   DNA template—1000 ng (10 ng each reaction)
   Phusion HF polymerase—100 units (New England Biolabs, Ipswitch, MA)
   Add water to a final volume of 5 mL.
2. Dispense 50 μL of the master mix into each well of the 96-well PCR plate. Add 1 μL of 10 μ*M* forward and reverse primers to each well (final concentration 0.2 μ*M*).
3. Perform the thermocycling reaction. A typical cycling protocol for a 4 kb plasmid with primers designed to have approximately a 60 °C melting temperature is as follows:
   Initial incubation at 98 °C for 3 min
   30 cycles of
      98 °C for 60 s
      60 °C for 60 s
      72 °C for 30 s/kb

After 30 cycles, incubate at 72 °C for 5 min and hold at 4 °C.

Note that longer extension times are required than in typical PCR reactions since the whole plasmid is amplified. Extension times should be at least 30 s per kb of PCR product.

4. Electrophorese the PCR reactions on an agarose gel to confirm successful amplification of the desired product. We typically use 0.8% agarose, LB (Faster Better Media) gel and load 5 μL of DNA or less per lane. Perform electrophoresis at 275 V for about 10 min (adjust depending on the size of the plasmid).
5. For some reactions, optimization of the amount of template DNA, the extension time, the annealing temperature and/or the number of cycles may be necessary to obtain amplification of the desired product. In general, try a change in annealing temperature first, followed by a change in the extension time, and finally the amount of template DNA. The addition of DMSO and/or betaine to a final concentration of 1–5% and 1 *M*, respectively, may help decrease nonspecific products.

2.3.2 Purification of PCR products

1. Combine all successful PCR reactions.
2. Prepare LB or TAE deep-well 0.8% agarose gel and load up to 10 μg of DNA per lane in the deep wells of the gel.
3. Electrophorese at 95 V for about 50 min to ensure good separation between linearized PCR product and supercoiled plasmid. Isolate the desired DNA from the gel using Qiaquick gel extraction kit (Qiagen) according to the manufacturer's instructions.
4. Determine the concentration of the recovered DNA using UV/Vis spectrophotometry.

3. REPAIR, PURIFICATION AND DEPHOSPHORYLATION OF ACCEPTOR DNA

This step is necessary for DNase I- and S1 nuclease-digested DNA only. For acceptor DNA prepared by multiplex inverse PCR, skip to Section 4.

3.1. Evaluate quality and determine quantity of linear DNA

1. Determine the total DNA concentration of the DNase I or S1 nuclease digested DNA using a UV/Vis spectrophotometer.
2. Analyze 200–500 ng of the recovered DNA via agarose gel electrophoresis on a 0.8% agarose, LB gel. The use of LB rather than TAE or TBE

facilitates the separation of the linear and nicked circle DNA. On an LB gel, linear runs the fastest, followed by nicked circles, and then supercoiled.

3. Quantify the fraction of linear DNA using gel-imaging software and determine the concentration of linear plasmid DNA. This is a particularly important step for linear DNA prepared by DNase I digestion. With DNase I digested DNA, we have had good success with digestions in which ~30% of the plasmid has been converted to linear DNA. If the sample is over digested, the resulting combinatorial library will be dominated by members in which large segments of the acceptor plasmid are deleted. In the case in which plasmid is over digested, repeat the supercoiled plasmid DNA I digestion using less DNase I.

3.2. Repair the DNA (blunt the ends and seal any nicks that might be present in the DNA)

1. Repair the digested DNA (to create blunt ends) using T4 DNA ligase and T4 DNA polymerase (New England Biolabs, Ipswich, MA). The amount of ligase and polymerase employed is calculated based on the amount of linear DNA determined earlier. Incubate multiple tubes each containing not more than 75 μL of total reaction mixture containing DNA, T4 DNA ligase (160 cohesive end units/μg linear DNA), T4 DNA polymerase (1 unit/μg linear DNA), 200 μ*M* dNTPs, 1 × T4 ligase buffer, and 1 × BSA in water at 12 °C for 20 min.
2. Add EDTA to 10 m*M* and incubate at 75 °C for 15 min to stop the reaction.
3. Allow the solution to cool to room temperature. Purify and isolate the repaired DNA using DNA clean and concentrator 25 μg kit (Zymo Research Corp.) as described in Section 2.2 in step 3.

3.3. Isolate the repaired, linear DNA

1. Make a 0.8% agarose, LB (Faster Better Media) gel. The use of LB rather than TAE or TBE buffer facilitates the separation of the linear and nicked circle DNA.
2. Load 1.5 μg of DNA or less per lane. If the DNA has been frozen, heat to 65 °C for approximately 2 min before loading the gel to ensure better separation on the gel.
3. Run the gel at 275 V for about 30 min (adjust depending on the size of the plasmid), until the DNA is separated into three distinct bands.

The slowest migrating band is supercoiled DNA, the fastest is linear DNA, and the band in the middle is nicked circles.

4. Extract the band containing the linear DNA and purify using the Qiagen gel extraction kit (Qiagen) per the manufacturer's instructions.

3.4. Dephosphorylate the acceptor DNA

1. For every microgram of DNA, add 20 units of Antarctic phosphatase (New England Biolabs, Ipswitch, MA) in 1 × Antarctic phosphatase buffer. Other phosphatases such as calf intestinal or shrimp alkaline can be used; however, Antarctic phosphatase is the only one of these that can be easily heat inactivated, and the buffer does not interfere with the subsequent ligation reaction. This property is advantageous as it avoids any DNA loss during a purification step at this stage.
2. Incubate for 1 h at 37 °C.
3. Heat to 65 °C for 10 min to inactivate the phosphatase. The dephosphorylated vector does not require purification and can be directly used in the ligation step.

4. PREPARATION OF INSERT DNA

We have successfully created protein switches with circularly permuted inserts (Guntas et al., 2005, 2004; Tullman et al., 2011) and noncircularly permuted inserts (Guntas & Ostermeier, 2004; Wright et al., 2011). The protocol for a noncircularly permuted insert is the easiest (Section 4.1), as it is just a simple PCR reaction. However, our switches with the largest differences between their on and off states contain circularly permuted insert proteins (Guntas et al., 2005). Although we have previously used protocols for random circular permutation of the insert gene with DNase I and S1 nuclease (Guntas et al., 2005, 2004; Wright et al., 2011), the protocols are very challenging. We have recently switched to a parallel PCR method to create the circularly permuted library of insert genes (Fig. 17.1B) and present this recommended protocol here (Section 4.2). This method has the additional advantage that circular permutations in the library can be researcher-defined. For example, the loci for circular permutation can be limited to regions that are solvent accessible, flexible, and not functionally relevant (Lo et al., 2012)—sites that are most likely to tolerate circular permutation. Successful domain insertion and circular permutation generally requires proximity between the N- and C-termini of the insert domain or longer linkers to span the added distance. Although there is no requirement for the protein to have known structure in order to construct the library, the choice of appropriate linker length is greatly informed by structural information. We have not attempted switch construction from proteins of unknown structure.

4.1. Creation of noncircularly permuted insert DNA

1. Design primers to anneal to the 5′ and 3′ ends of the insert DNA with approximately 15–24 base pairs and 55 °C melting temperatures. To the extent that the termini of the insert protein are not proximal, longer linkers will be required. To add linker amino acids between the insert and acceptor proteins, add the corresponding nucleotides to the 5′ ends of the primers.
2. Phosphorylated primers can be ordered from most manufacturers or unphosphorylated primers can be phosphorylated using T4 kinase (New England Biolabs, Ipswich, MA) following the manufacturer's directions.
3. Perform PCR using Phusion HF (New England Biolabs, Ipswich, MA) or another high fidelity polymerase according to the manufacturer's directions using a DNA containing the desired insert gene as the template. A sample protocol is provided.

    10 μL 5 × HF Buffer (or GC buffer if high GC content)
    2.5 μL each primer (0.5 μ*M* final concentration)
    1 μL 10 m*M* dNTPs
    0.5 μL Phusion HF polymerase
    1 μL template DNA (10 ng)[1]
    32.5 μL nuclease-free water.

Temperature cycling:

1. 98 °C for 30 s
2. 30 cycles of 98 °C for 5 s/55 °C for 15 s (or other appropriate annealing temperature)/72 °C for 15 s (15–30 s per kb)
3. 72 °C for 10 min
4. 4 °C hold.

4. Electrophorese the PCR product on an agarose gel and extract and purify the band that appears at the correct size using Qiaquick gel extraction kit (Qiagen) according to the manufacturer's instructions.
5. Determine the concentration of the recovered DNA using a UV/Vis spectrophotometer.

[1] It is beneficial to use template DNA with a different resistance marker than that of your linearized acceptor plasmid to avoid carrying the template plasmid forward into the library.

4.2. Creation of circularly permutation libraries as insert DNA

Circular permutation of the insert protein can drastically increase the diversity of domain insertion libraries thereby increasing the chance of isolating a hybrid enzyme with the desired properties. Below we describe in general the creation of circular permutation libraries by parallel PCR using primers corresponding to the beginning and end of the desired circular permutant.

4.2.1 Design criteria for circular permutation

The length and composition of the linker spanning the distance between the original N- and C-termini must be rationally designed. However, one is not necessarily limited to a single linker choice as several linker variants may be included in the library design since the linkers are encoded by the PCR primers used to create the template construct.

1. Linker length—Determine the distance between the alpha carbons of the N- and C- terminal residues in the insert protein using Pymol (Schrödinger, 2010) or similar molecular visualization software. Depending on the secondary structure of the linker residues, one amino acid will span a distance of 1.5–3 Å.
2. Linker composition—In most cases, flexible linkers rich in glycine, serine, and threonine are used to span the original termini in a manner that does not constrain the termini location or distort the domain's structure. Linkers of more diverse composition or higher rigidity can also be used.
3. Choice of the set of new termini—In the case of small domains, every residue may be queried; however, larger domains may require some limitation to minimize both library size and primer costs. Residues that are solvent accessible, flexible, loosely packed, and between secondary structure elements are most favorable for successful circular permutation (Lo et al., 2012). However, a variety of studies have shown that sites for successful circular permutation can lie within secondary structure elements (Meister, Kanwar, & Ostermeier, 2009).

4.2.2 Creation of template constructs and primer pairs

1. Create template construct containing an end-to-end fusion of the desired insert domain spanned by the designed linker(s) using standard cloning techniques (Fig. 17.1B).
2. Mix each template construct (i.e., one could use a set of templates with different circular permutation linkers between the N- and C-termini) in equimolar quantities to be used as a template for each PCR reaction (Section 4.2.3).

3. For each codon or select codons within the insert gene, design a forward primer starting at the first base of the codon and a reverse primer starting at the last base of the previous codon and extending in the 3′ direction until the individual melting temperatures are between 60 and 65 °C, and other standard primer design criteria are optimized. We have written a MATLAB script to design such primer pairs, which is freely available.
4. For each primer pair, make a primer mix containing 10 μ*M* each (forward and reverse) primer. Duplications or deletions at the new termini can be made by shifting the reverse primer ahead or back one codon, respectively, analogous to how the multiplex inverse PCR can create similar diversity.

4.2.3 Parallel PCR reactions

1. Make the following template mixture and aliquot 9 μL to each well:
   Water—4 μL/reaction
   Betaine (5.5 *M* stock)—4 μL/reaction
   DMSO—0.6 μL/reaction
   Template (10 ng/μL stock)—0.4 μL/reaction.
2. To each well, add 1 μL of the 10 μ*M* primer pair mix.
3. Immediately before putting the PCR plate into a thermocycler preheated to 98 °C, add 10 μL of Phusion HF Master Mix (New England Biolabs, Ipswich, MA) to each well.
4. After an initial 30 s 98 °C incubation, perform the following cycle 30 ×
   98 °C—10 s
   63 °C—20 s
   72 °C—15 s/kb.
5. Incubate at 72 °C for 5 min.
6. The success of each reaction can be verified by electrophoresing 5 μL in an agarose gel.

4.2.4 Purification and preparation of insert for cloning

1. Combine 12 μL from each reaction in a polypropylene tube and mix.
2. Purify this library mixture by phenol/chloroform extraction followed by ethanol precipitation (Sambrook & Russell, 2001).
3. Electrophorese the DNA on an agarose gel and isolate the desired DNA from the gel using Qiaquick gel extraction kit (Qiagen) according to the manufacturer's instructions.
4. Phosphorylate the PCR product library mixture using NEB's Quick Blunting Kit (E1201), per the manufacturer's instructions. It is important

that T4 polymerase be included in this reaction to chew back 3′ ends leaving free 5′ ends, on which the kinase acts much more readily than on blunt ends. Using T4 polynucleotide kinase alone, even in the presence of crowding factors, does not yield enough 5′ phosphorylated DNA for efficient library creation. The products of this reaction, after heat treatment at 70 °C for 10 min, may be directly added to the ligation reaction (Section 5.1).

5. LIGATION, TRANSFORMATION, RECOVERY, AND STORAGE OF THE LIBRARY

5.1. Ligation of the insert and acceptor DNA

The following reaction may be scaled up after initial transformation tests to yield a library of the desired number of transformants. The main parameter to optimize to obtain better ligations is the insert:vector ratio.

1. Mix the following:
   Acceptor DNA—50 ng
   Insert DNA—threefold molar excess over acceptor DNA
   PEG-8000 (30% w/v)—5 μL
   T4 ligase buffer—2 μL
   Water—up to 20 μL
   T4 DNA Ligase (2,000,000 cohesive end units/mL)—0.5 μL.
2. Place ligation mixture in a thermocycler and cycle between 10 and 30 °C every 30 s for 30 min (Lund, Duch, & Pedersen, 1996).
3. Dilute ligation 10-fold to prevent PEG coprecipitation and purify over a clean and concentrator column (Zymo Research Corp.) and elute in 20 μL of water preheated to 50 °C. Optionally, concentrate the DNA by vacuum centrifugation to a final volume of 2 μL.

5.2. Transformation

1. Add the 2 μL of ligation product to 40 μL of high efficiency ($>1\times10^9$ transformants/μg) DH10B cells (Life Technologies, Grand Island, NY) or any other highly electrocompetent strain.
2. Incubate on ice for 15 s and electroporate in a 0.2-cm cuvette (e.g., in a GenePulser II electroporator (Biorad Laboratories, Hercules, CA) set to 25 μF capacitance, 200 L/500H Ω resistance, and 2.5 kV).
3. Add 1 mL SOC (2% w/v bacto-tryptone, 0.5% w/v yeast extract, 8.56 m*M* NaCl, 2.5 m*M* KCl, 10 m*M* $MgCl_2$, 10 m*M* $MgSO_4$, 20 m*M* glucose) that has been prewarmed to 37 °C and mix.

4. Dispense into a 15 mL centrifuge tube and incubate with shaking at 37 °C for 1 h. Multiple electroporations can be combined at this step.
5. Make four serial dilutions of 10 μL of the cells into 90 μL of SOC and plate 50 μL of each dilution on 10 cm LB plates containing the appropriate selective antibiotic. Place the remaining cells on large LB plates (e.g., 245 × 245 mm bioassay dish; Nalgene-Nunc) containing the appropriate selective antibiotic. If multiple electroporations are performed, one can either plate aliquots of 1 mL on multiple plates or concentrate the cells by centrifugation as above followed by resuspension in 1 mL SOC and plating on a single plate. Although single isolates are not required at this point, overcrowding on a plate can lead to bias within one's library. The colonies that grow on the 10 cm plates are used to determine the number of transformants (total transformants = (number of colonies on the 10 cm plate)/(volume of transformed cells plated on 10 cm plate) × (volume of cells plated on large plates)).
6. Incubate overnight at 37 °C.

5.3. Recovery and storage of library

1. Recover cells from 245 × 245 mm bioassay dish by adding 2 × 15 mL storage media (18 mL LB, 9 mL 50% glycerol, 3 mL 20% w/v glucose) to the top of the plate, scrape cells from media using a cell spreader, and then pipette cells into a 50 mL polypropylene centrifuge tube.
2. Spin cells in a centrifuge at 2000 × *g* at 4 °C for 20 min.
3. Decant supernatant and add 2 mL of storage media.
4. Resuspend the pelleted cells by gentle shaking.
5. Store in aliquots in 1.5 mL cryovials at −80 °C.

6. CHARACTERIZATION OF THE LIBRARY

The colonies appearing on the small plate of the library can be used to characterize the library. There are many ways to access the library quality, from analyzing plasmid size, to restriction digestions, to PCR screens of individual colonies. The appropriate and most useful method(s) depend on the nature of the library created. The following restriction digest assay on purified plasmid from 10 colonies is straightforward and provides information on both the fraction of the library that received an insert and the diversity of insertion site. Choose a restriction enzyme (or pair of enzymes) that digests once within the insert DNA and once within the acceptor DNA. Digest the prepared DNA from each of the 10 colonies following manufacturer's instructions. Analyze the size of the digestion products via agarose gel electrophoresis.

ACKNOWLEDGMENTS

This work was supported by the National Institute of General Medicine at the National Institutes of Health (R01 GM066972), the National Science Foundation (CBET-0828724), and the Defense Threat Reduction Agency (HDTRA1-09-1-0016).

REFERENCES

Edwards, W. R., Busse, K., Allemann, R. K., & Jones, D. D. (2008). Linking the functions of unrelated proteins using a novel directed evolution domain insertion method. *Nucleic Acids Research*, *36*, e78.

Guntas, G., Mansell, T. J., Kim, J. R., & Ostermeier, M. (2005). Directed evolution of protein switches and their application to the creation of ligand-binding proteins. *Proceedings of the National Academy of Sciences of the United States of America*, *102*, 11224–11229.

Guntas, G., Mitchell, S. F., & Ostermeier, M. (2004). A molecular switch created by in vitro recombination of nonhomologous genes. *Chemistry & Biology*, *11*, 1483–1487.

Guntas, G., & Ostermeier, M. (2004). Creation of an allosteric enzyme by domain insertion. *Journal of Molecular Biology*, *336*, 263–273.

Heins, R. A., Choi, J. H., Sohka, T., & Ostermeier, M. (2011). In vitro recombination of non-homologous genes can result in gene fusions that confer a switching phenotype to cells. *PLoS One*, *6*(11), e27302.

Ke, W., Laurent, A. H., Armstrong, M. D., Chen, Y., Smith, W. E., Liang, J., et al. (2012). Structure of an engineered β-lactamase maltose binding protein fusion protein: Insights into heterotropic allosteric regulation. *PLoS One*, *7*, e39168.

Kim, J. R., & Ostermeier, M. (2006). Modulation of effector affinity by hinge region mutations also modulates switching activity in an engineered allosteric TEM1 β-lactamase. *Archives of Biochemistry and Biophysics*, *446*, 44–51.

Liang, J., Kim, J. R., Boock, J. T., Mansell, T. J., & Ostermeier, M. (2007). Ligand binding and allostery can emerge simultaneously. *Protein Science*, *16*, 929–937.

Lilley, D. M. (1980). The inverted repeat as a recognizable structural feature in supercoiled DNA molecules. *Proceedings of the National Academy of Sciences of the United States of America*, *77*, 6468–6472.

Lo, W. C., Dai, T., Liu, Y. Y., Wang, L. F., Hwang, J. K., & Lyu, P. C. (2012). Deciphering the preference and predicting the viability of circular permutations in proteins. *PLoS One*, *7*, e31791.

Lund, A. H., Duch, M., & Pedersen, F. S. (1996). Increased cloning efficiency by temperature-cycle ligation. *Nucleic Acids Research*, *24*, 800–801.

Meister, G. E., Kanwar, M., & Ostermeier, M. (2009). Circular permutation of proteins. In S. Lutz & U. Bornscheuer (Eds.), *Protein engineering handbook*. Weinheim: Wiley-VCH Verlag GmbH & Co. KGaA.

Ochman, H., Gerber, A. S., & Hartl, D. L. (1988). Genetic applications of an inverse polymerase chain reaction. *Genetics*, *120*, 621–623.

Ostermeier, M. (2005). Engineering allosteric protein switches by domain insertion. *Protein Engineering, Design & Selection*, *18*, 359–364.

Panayotatos, N., & Wells, R. D. (1981). Cruciform structures in supercoiled DNA. *Nature*, *289*, 466–470.

Sambrook, J., & Russell, D. (2001). *Molecular cloning: A laboratory manual* (3rd ed.). Cold Spring Harbor, NY: Cold Spring Harbor Laboratory Press.

Schrödinger, L.L.C., 2010. The PyMOL Molecular Graphics System, Version 1.3r1.

Sohka, T., Heins, R. A., Phelan, R. M., Greisler, J. M., Townsend, C. A., & Ostermeier, M. (2009). An externally-tunable bacterial band-pass filter. *Proceedings of the National Academy of Sciences of the United States of America*, *106*, 10135–10140.

Tullman, J., Guntas, G., Dumont, M., & Ostermeier, M. (2011). Protein switches identified from diverse insertion libraries created using S1 nuclease digestion of supercoiled-form plasmid DNA. *Biotechnology and Bioengineering*, *108*, 2535–2543.

Wright, C. M., Majumdar, A., Tolman, J. R., & Ostermeier, M. (2010). NMR characterization of an engineered domain fusion between maltose binding protein and TEM1 β-lactamase provides insight into its structure and allosteric mechanism. *Proteins*, *78*, 1423–1430.

Wright, C. M., Wright, R. C., Eshleman, J. R., & Ostermeier, M. (2011). A protein therapeutic modality founded on molecular regulation. *Proceedings of the National Academy of Sciences of the United States of America*, *108*, 16206–16211.

CHAPTER EIGHTEEN

Design of Chimeric Proteins by Combination of Subdomain-Sized Fragments

José Arcadio Farías Rico, Birte Höcker[1]
Max Planck Institute for Developmental Biology, Tübingen, Germany
[1]Corresponding author: e-mail address: birte.hoecker@tuebingen.mpg.de

Contents

Abstract

Hybrid proteins or chimeras are generated by recombination of protein fragments. In the course of evolution, this mechanism has led to major diversification of protein folds and their functionalities. Similarly, protein engineers have taken advantage of this attractive strategy to build new proteins. Methods that use homologous recombination have been developed to (semi) randomly create chimeras from which the best can be selected. We wanted to recombine very divergent or even unrelated fragments, which is not possible with these methods. Consequently, based on the observation that nature evolves new proteins also through illegitimate recombination, we developed a strategy to design chimeras using protein fragments from different folds. For this approach, we employ detailed structure comparisons, and based on structural similarities, we choose the fragments used for recombination. Model building and minimization can be used to assess the design, and further optimization can be performed using established computational design methodologies. Here, we outline a general approach to rational

Methods in Enzymology, Volume 523
ISSN 0076-6879
http://dx.doi.org/10.1016/B978-0-12-394292-0.00018-7

protein chimera design based on our experience, and provide considerations for the selection of the fragments, the evaluation, and possible redesign of the constructs.

1. INTRODUCTION

A chimera, according to Greek mythology, is a monstrous creature that is composed of parts of different animals. The term is used to describe a mythical creature that is believed to be an impossible and foolish fantasy. Interestingly, protein engineers have picked up this term to describe hybrid proteins created from parts of two or more different proteins. But hybrid proteins are not as unusual as one might think. In fact dramatic gene reorganizations are observed in nature and believed to play a crucial role in the natural evolution of the protein universe.

Horizontal gene transfer and gen(om)e duplication produced the material that nature tinkered with (Lynch & Conery, 2000; Ochman, Lawrence, & Groisman, 2000). Newly acquired or copied proteins, which are free from selection pressure right after the replication event, accumulate mutations and diversify through random drift until they become fixed through natural selection or are lost over time. Point mutations result in small changes that diversify protein function. In contrast, more dramatic changes that can lead to significantly different protein structures are caused by gene rearrangements such as fusion and recombination (Eisenbeis & Höcker, 2010). Two types of recombination have been described: *homologous recombination* takes place between closely related sequences, while *illegitimate recombination* describes recombination between nonhomologous or very divergent sequences. All these mechanisms can lead to different chimeric proteins with new characteristics and that is why we imitate them to design new proteins, as described in this chapter.

1.1. Natural and laboratory protein evolution

In protein evolution, the most common operational unit is the protein domain. A protein domain refers to a segment of the polypeptide chain that is able to fold autonomously and that can be found in diverse protein architectures (Fig. 18.1A). Recombination of protein domains has led to the development of large multidomain proteins, whose domains evolved to interact and accomplish one or more functions together.

Similarly, domains themselves are built up from smaller subunits. Many protein domains share structurally similar fragments that are found in various

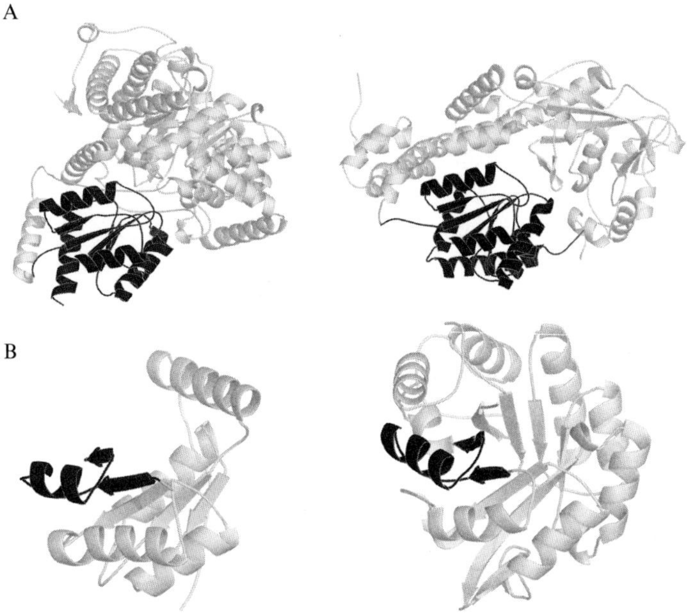

Figure 18.1 Evolutionary units. (A) Combination of protein domains led to the emergence of diverse multidomain proteins: The methionine synthase (MetH) from *Escherichia coli* (left; PDB-ID 1K7Y) and the Methylmalonyl-CoA mutase (MmCoA) from *Propionibacterium freudenreichii* (right; PDB-ID 7REQ) share a B12-binding domain (black). (B) The domains themselves are also composed of smaller subunits. Diverse protein folds share structurally similar fragments that are found in various structural contexts: HisF from *Thermotoga maritima* (left; PDB-ID 1THF) and the B12-binding domain of MmCoA from *P. freudenreichii* (right; PDB-ID 7REQ) are built from βαβ elements (one element highlighted in black).

structural contexts. This observation led to the proposition that domains evolved through the assembly of smaller gene fragments that encode intrinsically stable subunits (Grishin, 2001; Söding & Lupas, 2003). Such subunits can be structurally stable supersecondary structural elements such as a βαβ element (Fig. 18.1B) and can also be associated with a certain function, for example, a DNA binding motif. Combinatorial shuffling of these subunits would enable a faster diversification of protein architecture, as has been tested by Riechmann and Winter. They combined a domain fragment from the cold-shock protein CspA that is incapable of independent folding with peptides encoded by random pieces of *Escherichia coli* DNA in order to find new protein domains (Riechmann & Winter, 2000, 2006). These experiments support the role of nonhomologous recombination as the source of stable folded protein domains.

Inspired by Nature's strategies, protein engineers construct chimeric proteins to create new functions. New multidomain proteins have been generated for multiple purposes. Probably the most popular protein used for fusion constructs is the green fluorescent protein (GFP), which has been fused to a vast variety of protein domains. The capability of GFP to provide a readout makes these fusions useful for tracking a protein and testing the properties (e.g., ligand binding or proper folding) present in the target protein connected to the GFP reporter. A larger challenge, however, is the recombination of subdomain-sized fragments to create protein domains with novel functions. For this, biologists developed *in vitro* recombination strategies that, when coupled to a selection or screening technique, imitate Nature's ability to create new function through evolution. Methods for protein chimeragenesis have been developed with the introduction of DNA shuffling (Stemmer, 1994), where closely related proteins are recombined in order to obtain a library of improved chimeras. More powerful methods that allow a deeper sequence space exploration were established after the introduction of this pioneering work. One example is iterative truncation for the creation of hybrid enzymes (ITCHY): first, the progressive truncation of coding sequences generates amino or carboxy-terminal fragment libraries of two genes, which is followed by ligation of the products to create a single-crossover chimeric library (Ostermeier, Shim, & Benkovic, 1999). SCRATCHY is a subsequent improvement of this recombination strategy; in combination with DNA shuffling, the number of crossovers present in the chimeric proteins becomes enriched (Lutz, Ostermeier, Moore, Maranas, & Benkovic, 2001). Further development of recombination techniques led to a homologous recombination approach restricted by structural information with the SCHEMA algorithm. This method was developed in order to avoid the disruption of native interactions in the parental proteins (Voigt, Martinez, Wang, Mayo, & Arnold, 2002). The use of SCHEMA yielded novel chimeric cytochrome P450 proteins with improved biophysical features (Otey et al., 2006).

1.2. Probing evolutionary concepts with protein design

Rational design of proteins from subdomain-sized fragments should be feasible if gene duplication and fusion are a source of folded domains. We put this to a test when we tried to mimic enzyme evolution scenarios proposed for the emergence of the ubiquitous TIM- or $(\beta\alpha)_8$-barrel protein fold. The fold was suggested to have evolved from a precursor half its size, based on the striking twofold symmetry observed in the sequences and crystal structures of two related $(\beta\alpha)_8$-barrel proteins from the histidine biosynthesis pathway, HisA and HisF (Fani, Lio, Chiarelli, & Bazzicalupo, 1994; Lang, Thoma, Henn-Sax, Sterner, & Wilmanns, 2000). Further evidence for this hypothesis was provided by the fact that separately produced halves of HisF behaved as independent folding units that could form a functional heterodimer (Höcker, Beismann-Driemeyer, Hettwer, Lustig, & Sterner, 2001). According to the proposed first step, two identical copies of the same half-barrel (the C-terminal half, named HisF-C) were fused *in tandem*. Only minimal optimization guided by rational design, namely, the reconstruction of a salt-bridge network at the half-barrels' interface, led to the production of a stable, monomeric structure that unfolded cooperatively (Höcker, Claren, & Sterner, 2004). The construct was further optimized through directed evolution selecting for improved solubility upon expression (Seitz, Bocola, Claren, & Sterner, 2007) until it was stable enough to allow structure determination by crystallography of this highly symmetrical artificial protein (Höcker, Lochner, Seitz, Claren, & Sterner, 2009).

In another approach, the genes for the N- and C-terminal $(\beta\alpha)_4$-half-barrels of HisF and HisA were fused crosswise to yield the chimeric proteins HisFA and HisAF. One of these $(\beta\alpha)_8$-barrels (HisAF) was expressed as a very stable, compact monomer that unfolds with high cooperativity (Höcker, Claren, & Sterner, 2004). Using directed evolution, a few mutations were identified that established efficient TrpF activity, which is mechanistically similar to HisA activity, for the chimeric protein HisAF (Claren, Malisi, Höcker, & Sterner, 2009). Together, these experiments illustrate how chimeric proteins can be engineered in order to better understand and explore the diversification of enzymatic activities in the $(\beta\alpha)_8$-barrel fold.

The effect of duplication and fusion as a mechanism of fold evolution has been evaluated also in other folds. For instance, the DNA-methyltransferase superfamily is a good example of how multistep gene rearrangements, via circular permutation, can lead to new protein topologies. Starting from two copies of the HaeIII methyltransferase gene cloned *in tandem*, Peisajovich et al. digested the construct from its 5′ and 3′ ends and selected for truncated but functional intermediate proteins that resembled natural methyltransferases (Peisajovich, Rockah, & Tawfik, 2006). Analysis of the circularly permuted variants indicated that although different topologies can evolve through multistep rearrangements, there are intrinsic protein modules that restrict the possibilities of achievable arrangements. On the other hand, this means that with the right modules, new proteins and new topologies are designable.

1.3. Designing a protein chimera by combining fragments from different folds

Thus far, we had designed chimeric proteins in order to explore, understand, and validate basic aspects of protein evolution. The primary aim of our experiments was to test the importance of combinatorial assembly in the evolution of protein domains. However, we quickly recognized the potential application of this approach in generating novel protein functions using chimeric protein scaffolds. It would be advantageous to combine protein fragments that individually carry specific functional features, and that once merged in a single protein domain would have an extended functional repertoire.

As a first test, we decided to design a protein by combining fragments from different folds. By exchanging a large part of the protein HisF with structurally similar fragments from the response regulator protein CheY, we generated the chimera CheYHisF. CheY belongs to the flavodoxin-like fold and we have hypothesized that its superfamily might be related to the $(\beta\alpha)_8$-barrel fold because of a striking structural similarity to a half-barrel (Höcker, Schmidt, & Sterner, 2002). No matter whether these fragments are very divergent but share a common origin or whether they are the product of convergent evolution, they can be usefully combined to form a protein domain with novel features as the analysis of CheYHisF showed (Bharat, Eisenbeis, Zeth, & Höcker, 2008). The crystal structure of the chimera revealed interesting aspects of the chimera. On the one hand, it showed that the overall structure of the fragments is extremely similar to the structure in the parent proteins. On the other hand, it became apparent that the interfaces were not optimal: the C-terminus of the protein inserted into the core of the β-barrel forming an unexpected ninth β-strand, which alleviated the strain at the interface. In a subsequent study, we analyzed the interface of the fragments using the computational design program ROSETTA and identified five mutations that adjusted the interface so that the intended 8-stranded structure could be formed (Eisenbeis et al., 2012). Both the first and second-generation chimera carried functional features of at least one of the parents, namely, the ability to bind phosphate or phosphorylated ligands. Based on this functional feature, we easily improved the binding affinity of the 8-stranded CheYHisF chimera for a product analog of a related reaction to the level of natural enzymes by the introduction of two rational mutations. We further built two more chimeras using fragments from HisF and the flavodoxin-like proteins NarL and methylmalonyl CoA mutase. One of these, namely NarlHisF, turned out to be highly stable (Shanmugaratnam, Eisenbeis, & Höcker, 2012).

Based on this experience of rationally designing protein fold chimeras, we have developed our own general chimeragenesis strategy. Here, we outline considerations for the selection of the parental structures, and for the evaluation and possible redesign of chimeric constructs, using CheYHisF as an example.

2. SELECTING THE STARTING STRUCTURES FOR CHIMERA DESIGN

A chimera is, by definition, built up from fragments of different proteins. Accordingly, the choice of the starting structures that will be used in the design is critical. The overall success, including stability, folding, and proper function are dependent on the parent structures. Thus, they should be selected carefully.

2.1. Selection of the parental structures

The starting structures for the construction of a novel chimeric protein should be selected based on the specific goal of the experiment. We designed CheYHisF in order to test whether evolution could have generated contemporary protein diversity using subdomain-sized fragments from different folds. Given that all proteins are built from the same secondary structure elements, local similarities can be mainly found among members of the same structural class (all α, all β, α/β, or α+β). Bioinformatic tools, such as PDBeFold (Krissinel & Henrick, 2004), DALI (Holm & Rosenström, 2010), or GANGSTA (Guerler & Knapp, 2008) can be used to screen the structural repositories for structures with similarities to any desired query. In PDBeFold, for instance, it is possible to select the minimum aligned length of the query and the hit; another handy feature of the server is the possibility to sort the results based on sequence identity or *P*-value, among other scores. Using the program DALI, we originally detected the high structural similarity between the $(\beta\alpha)_8$-barrel HisF and the flavodoxin-like CheY (Höcker, Schmidt, & Sterner, 2002). The resulting chimera CheYHisF turned out to be a well-folded and monomeric protein domain (albeit with an additional secondary structural element), which could be reengineered in successive rounds of rational design to become more compact and functional. This result indicates that our approach can be used to engineer new proteins. For example, we envision that in a similar manner, two complementary activities could be engineered into a single domain or a binding site could be transplanted onto a different scaffold protein.

CheY and HisF from *Thermotoga maritima* were used as parental proteins to build the chimera CheYHisF. The following considerations were taken into account when these structures were chosen as starting points:

1. Despite the fact that both protein structures belong to different folds, we detected a strong structural similarity that covered one β-strand and three consecutive βα elements (Fig. 18.2).

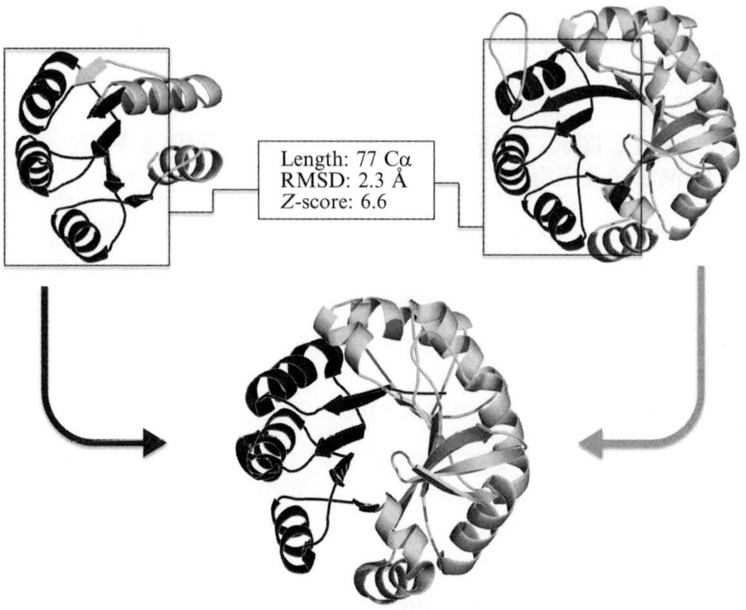

Figure 18.2 Chimera design based on structural superposition. A structural superposition between CheY (top left; PDB-ID 1TMY) and HisF (top right; PDB-ID 1THF) was used to guide the design of the chimera CheYHisF. The superposed areas from both proteins are highlighted (in black). The superposition covered 77 Cα atoms with a *Z*-score of 6.6 and a RMSD of 2.3 Å. In the final chimera (bottom center; PDB-ID 2LLE), the black region from CheY replaced the black region from HisF.

2. CheY and HisF are single domain proteins, which is probably an advantage. When using a fragment from a multidomain protein, one needs to consider interdomain interactions that might contribute to the stability of the single domains. To predict and reengineer such interactions in the novel chimeric environment is difficult. Therefore we favor the use of single domain proteins as starting structures, or we try to use domains with as little interface to other domains as possible.
3. High protein stability is another desirable feature of the parental proteins. The fragments need to be both robust and plastic enough to accommodate the new environment. The proteins we used to build CheYHisF originated from the thermophile *Thermotoga maritima*. We reasoned that the enhanced stability of the parents would contribute to the stability of the new protein.
4. In order to rationally design the chimera, high-quality atomic structures are needed. In an optimal case, one would like to have a set of structures from the same parental protein. Valuable information for the design can

be gained from such data, for example, which parts of the protein are flexible or involved in certain functions.

2.2. Structural comparisons of the parents

There are a number of online servers for protein structural comparison and alignment. One easy-to-use server is PDBeFold (http://www.ebi.ac.uk/msd-srv/ssm) because of its extended functionalities (Krissinel & Henrick, 2004). It is possible to structurally compare the protein pairs in different ways: matching individual chains, matching connectivity, or matching specific ranges of the protein pairs, etc. Here, we describe the procedure to superpose the starting structures and define the elements to be used in the combinatorial assembly:

1. Launch the PDBeFold tool.
2. Specify the protein domains to be compared, using the PDB identifiers: 1TMY (CheY) and 1THF (HisF).
3. Check the boxes to match individual chains, match connectivity, and enforce sequence identity, best matches only, and unique matches. Also, select the highest precision in the matching; this setting may force the program to return a short but precise superposition. It might be necessary to adjust the parameters for every experiment.
4. An inspection of the match list reveals several scores per match returned: *Q*-score (quality based on alignment length and root mean square deviation (RMSD)), *P*-score (probability for an equal hit by chance), and Z-score (statistical significance in terms of Gaussian statistics), as well as length of alignment, RMSD, and percent of sequence identity of the aligned region. A detailed explanation for these scores can be found in the documentation of the software.
5. The structural alignment between 1TMY (CheY) and 1THF (HisF) shows that four of the five βα elements in CheY have a topologically equivalent match in HisF. 77 Cα atoms were superposed with an RMSD of 2.3 Å: β1 from CheY matches β1 from HisF, the next βα element from CheY did not have a match in HisF, and the alignment extends over 3 βα modules deviating only in the position of the α-helix of the last βα module. This alignment highlights how nature evolved two different structural solutions to shield a hydrophobic core: in the flavodoxin-like fold, the α-helices 1 and 5 are packed against the curved β-sheet, whereas in the $(\beta\alpha)_8$-barrel fold, two $(\beta\alpha)_4$ elements are fused to shield a central hydrophobic area.

6. Guided by structural superposition, we decided to engineer CheYHisF in the following way: the fragments β1 and α2–β5 from the flavodoxin-like fold were fused to α4–α8 from the $(\beta\alpha)_8$-barrel (Fig. 18.2). In this case, we chose to swap the largest fragment because we wanted to explore how nature could generate fold diversity by combinatorial assembly. If the objective is, for example, to introduce a functional feature into an alternative scaffold, only the functional motif (perhaps an α-helix involved in protein–protein interaction) could be swapped. So far, we have mainly chosen junctions outside of secondary structural elements. If a junction is placed inside a secondary structural element, one needs to take care that the geometry and orientation of the residues in that element are the same, for example, the register of an α-helix or the pattern of a β-strand. When we select the regions to combine, we always try to place the fragments in the same environment. If a fragment has a surface shielded from solvent, we try to conserve the orientation of that surface in the novel chimeric context.

3. EVALUATION AND OPTIMIZATION OF THE CHIMERA

Once a combination of structural fragments is selected, the chimera design has to be evaluated. Before testing the construct experimentally, it can be evaluated *in silico*. Especially, if a large number of combinations are possible, computational methods allow filtering for the most promising constructs.

3.1. *In silico* evaluation of the chimeric protein

In order to visualize the possible new interactions that the fragments might establish, a homology model of the chimeric protein can be generated. We used the program MODELLER (Eswar, Eramian, Webb, Shen, & Sali, 2008) to generate a first model. In this specific experiment, we used information from only the two parent structures CheY and HisF. As an input, we provided a sequence alignment of the chimera with the two parental proteins as well as the PDB-files 1TMY and 1THF. The model will strongly depend on the quality of the alignment. Therefore, it is crucial to pay special attention to the alignment and, if necessary, adjust it by hand. Also, it is useful to generate multiple models that vary, for example, in the cut sites of the parental fragments by a few amino acids. Providing chimeric sequences with varying cut points in the above-mentioned alignment will generate the different models.

Here is a protocol for building a homology model of a chimera using the programs HHpred and MODELLER as provided by the Bioinformatics Toolkit of the Max Planck Institute for Developmental Biology (Biegert, Mayer, Remmert, Söding, & Lupas, 2006):

1. Launch the HHpred web-server from the MPI Toolkit (http://toolkit.tuebingen.mpg.de/hhpred); this interface will allow you to easily use the web-version of the MODELLER algorithm. The homology detection software HHpred is widely used to detect structural templates to build homology models of protein sequences without known structure (Söding, Biegert, & Lupas, 2005).
2. Copy and paste your sequence in the input field. We recommend using default parameters and to search the Protein Data Bank.
3. Give your job a name or store the automatically assigned id number. This allows you to access your job again later.
4. Once the job ran through, you can screen the visual output of your results. As expected, the parental proteins used for building the chimeric sequence will be at the top of the search.
5. Click the button "create model." In the visual output there are many hits. You will have to check the boxes of the hits you want to use as templates. We chose only the parents CheY (1tmy) and HisF (1thf). The use of many templates would add sequence diversity to the homology building procedure. However, for the model of a chimera, this is not necessary; it might, in fact, add an unwanted bias due to an uneven distribution of templates.
6. Once the templates for both parental proteins are selected, click on the button "Create model from manual template selection." This will take you to the web-interface of MODELLER at the MPI Toolkit and will also paste the multiple sequence alignment created by HHpred.
7. Review the alignment carefully and adjust it if necessary. This is a crucial step: the quality of the model strongly depends on the quality of the alignment!
8. Submit your job.
9. In the output, you can open different tabs in order to save the PDB-file of the model, view your structure with Chemis3D (a Java applet structure viewer), and verify the quality of your models with the scoring algorithms VERIFY3D, SOLVX, and ANOLEA.
10. The output of the evaluation programs can be helpful to detect weak points in the chimera. The scores of all programs should converge. If there is an area in the model that has a bad scoring in the evaluation

outputs of all three programs, the chimera needs to be improved in this area. It is worth generating multiple models, for example, by varying the cut sites of the parental fragments, and then comparing the performance of these models in the evaluation programs.

It is further recommended to perform a detailed inspection of the fragment interface area in the models in order to evaluate whether side-chain or backbone atoms overlap (resulting in a steric clash) in the new chimeric context. Postprocessing of the homology model using the webserver MolProbity (http://molprobity.biochem.duke.edu/) can be used to get a first impression. Here, hydrogens will be added and a detailed analysis of clashes and irregularities of the model will be carried out (Chen et al., 2010). Another useful tool to display the hydrogen-added homology model is the Crystallographic Object-Oriented Toolkit (COOT; Emsley, Lohkamp, Scott, & Cowtan, 2010). It provides an intuitive visualization of the clashes, possible hydrogen bonds, and salt-bridges (displayed as pink, green, and blue clouds, respectively). Generally, a protein domain that is potentially able to reach a proper folded state will feature a continuous blue and green cloud between the interacting lateral chains, representing a favorable network of electrostatic interactions. However, it is important to keep in mind that one is dealing with a model; especially, the positions of the lateral chains tend to be less accurate than the backbone modeling. There is not a defined metric of when a model is good enough. The visualization with COOT can be useful to spot big problems, such as nonphysical overlaps, but it is only an accessory approach. In practice, during the analysis of another chimeric design, we did observe a very bad overlap between the aromatic ring of a tyrosine and the backbone of an α-helix prompting us to mutate that particular tyrosine residue.

For a more detailed and energy-based analysis, the program ROSETTA (Rohl, Strauss, Misura, & Baker, 2004) can be used to perform a minimization of the homology model and detect problematic regions that may need further rounds of optimization by design. The algorithm of the ROSETTA relax mode reduces the energy of a structure by slightly moving backbone and side-chain torsion angles (Simons, Kooperberg, Huang, & Baker, 1997). The homology models can be minimized using an iterative energy minimization protocol. It is necessary to minimize the parental structures as well, in order to obtain basal energies relative to the energies of the chimeric proteins. Comparison with the calculated energies for residues in the parental context allows us to estimate whether these residues will interact favorably in the new chimeric environment. We used eight alternating rounds of backbone perturbation with side-chain repacking and a gradient-based energy minimization. The iterative minimization of each structure should be repeated, for example, 20 times, and the calculated energies of each residue should be averaged. Residues with unusually high energies might indicate problematic areas that can be visualized in the model. If clusters of high-energy residues are observed at the fragments' interface, these areas will need to be optimized (see Section 3.3). The different models with their varying cut sites should be compared, to identify the best chimera constructs. When comparing the overall ROSETTA energies of the designs, it must be kept in mind that they need to be adjusted to the number of amino acids in the polypeptide chain.

3.2. Experimental characterization of the chimera

Once a chimeric construct is chosen, it needs to be tested experimentally. For this purpose, it is necessary to clone, express, and purify the chimeric design, followed by biophysical characterization to determine whether the designed chimera is able to reach a folded state.

The chimeric gene can be constructed by PCR using the genes from the parental proteins as templates. Alternatively, the gene can be ordered from a company specialized in gene synthesis. Generally, we utilize *E.coli* for heterologous expression. Whether the protein is found in the soluble or insoluble fraction after protein expression can already be seen as an indirect indication of the overall robustness of the design. However, we have observed that many aggregated proteins can be refolded to a stable native state. This was the case with CheYHisF, which was found mainly in the insoluble fraction. Therefore, a refolding procedure was performed to obtain enough soluble protein to carry out biophysical as well as structural characterization. Alternatively, different expression systems can be tested or solubility tags fused to the chimeric protein in order to increase the amount of soluble protein. If none of these strategies allow the production of soluble protein, the chimeric construct needs to be redesigned.

Solubly expressed protein that contains, for example, a histidine-tag can be easily purified using an affinity chromatography column, whereas the refolding protocol also works as a purification step if insolubly expressed protein needs to be extracted from inclusion bodies. In any of these scenarios, it is necessary to determine whether the protein adopts specific oligomerization states or ill-defined aggregates, and whether further purification is required. Size exclusion chromatography is a valuable tool that can be used for such purposes: analytical gel filtration experiments were conducted with CheYHisF, HisF, and CheY in order to compare the oligomerization state and apparent sizes of the parental proteins with the chimeric design. A homogenous protein solution is crucial for further structural characterization.

To evaluate the formation of secondary structural content, we use far-UV circular dichroism (CD) spectroscopy. In the case of CheYHisF, we observed a slightly lower α-helical content than in both parental proteins. To test whether the novel protein also assumed a well-defined tertiary structure, we compared the tryptophan fluorescence of the parental proteins and the chimeric design. This method can be used to test whether the indole chromophore of tryptophan is shielded from solvent, and thus, is in a comparable environment. Further biophysical tests involve measuring protein stability, for example, using reversible unfolding by guanidinum chloride or temperature-induced unfolding. We observed that the CheYHisF chimera unfolded cooperatively, which indicated that the protein forms a compact, well-folded structure.

More detailed information is gained by solving an atomic-resolution structure of the designed chimera. How do the fragments derived from the different folds fit together? Do the fragments retain their original three-dimensional conformations, or do they undergo a conformational change in order to reach a better fit in the new structural context? Where are the shortcomings of the chimeric design? These questions can only be answered if an atomic model of the chimera is available. We used X-ray crystallography to determine the structure of the CheYHisF chimera. After screening multiple crystallization conditions, diffracting protein crystals of the chimera CheYHisF were obtained. The structure was determined by molecular replacement using corresponding parts from the high-resolution structures of the parental proteins as search models.

3.3. Analysis of the structure and further redesign

The X-ray structure of our first CheYHisF design revealed the presence of a ninth β-strand in the core of the βα-barrel. The strand invades between β1 and β2, and consists of residues from the C-terminus of HisF and some residues from the histidine-tag introduced for purification. The biophysical evidence as deduced from CD, fluorescence, and analytical gel filtration suggested that the structures of HisF and CheYHisF were very similar. But only the atomic model could reveal the presence as well as the interactions of the extra strand in the barrel (Bharat et al., 2008).

To test the importance of the ninth β-strand, a shortened variant, in which the extra β-strand forming residues were removed, was constructed. This variant had the tendency to aggregate, indicating the importance of the ninth strand for the integrity of the protein. Another interesting feature observed in the atomic model was the overall high B-factor distribution, which may indicate fragile crystal packing caused by high flexibility within the protein. Moreover, the highest B-factors were observed around the ninth β-strand as well as closely situated helices α1 and α2. Despite this unexpected insertion, the parental fragments for the most part retain their structure in the new context. Even the phosphate binding site contributed by the HisF fragment is still intact as can be seen by a sulfate bound in the crystal structure. However, the CheY and HisF fragments do not fit as closely together as anticipated, leaving space for the insertion of the additional β-strand. In summary, these observations underscore the importance of an experimentally determined structure for the analysis of a novel designed domain.

In a second approach, we optimized the area where the fragments interact by introducing five mutations identified by computational design with the program ROSETTA (Eisenbeis et al., 2012). For the redesign, we again used the iterative approach described above, which included eight alternating rounds of backbone perturbation with side-chain redesign and gradient-based minimization. Furthermore, a mutation had to overcome a certain energy threshold to be considered for further analysis. This was achieved by using the -favor_native_residue flag, which is an option in ROSETTA version 2.3 and which we varied from 0.75 to 2.75 ROSETTA energy units per amino acid thus increasing the threshold. With this approach, we wanted to ensure that only mutations that significantly improved the energy were proposed. We allowed mutations all over the protein instead of restricting design to the suboptimal interface because amino acids in the second and third shell might relieve tension at the interface.

The atomic structure of this optimized chimera determined by NMR spectroscopy revealed a good agreement with the design (Eisenbeis et al., 2012). This result emphasizes that a detailed analysis of the new interface between the recombined fragments as described in Section 3.1 is advantageous during the design process. Problematic areas might already be detected and adjustments introduced before the first round of experimental tests. Unanticipated problems, however, such as the presence of the ninth strand in CheYHisF, can be targeted in a redesign based on the solved structure of a first design. This could also include possible optimization of functional features.

4. SUMMARY AND FINAL CONSIDERATIONS

It has been proposed that protein domains evolved by combination of subdomain size fragments. Structural alignments between divergent pairs of proteins suggest that nature has assembled these fragments to create the structural diversity we observe today. Based on this observation, we developed a general strategy comprised of rational observations and computational design for the construction of protein chimeras. Our methodology can be used to build stable and functional protein domains. We envisage the construction of protein chimeras that combine different functionalities inherited from their parent proteins in a single domain. Production of new enzymes by recombination of naturally occurring functions will clearly constitute a valuable resource for applied research.

REFERENCES

Bharat, T. A., Eisenbeis, S., Zeth, K., & Höcker, B. (2008). A βα-barrel built by the combination of fragments from different folds. *Proceedings of the National Academy of Sciences of the United States of America*, *105*, 9942–9947.

Biegert, A., Mayer, C., Remmert, M., Söding, J., & Lupas, A. N. (2006). The MPI Bioinformatics Toolkit for protein sequence analysis. *Nucleic Acids Research*, *34*, W335–W339.

Chen, V. B., Arendall, W. B., 3rd., Headd, J. J., Keedy, D. A., Immormino, R. M., Kapral, G. J., et al. (2010). MolProbity: All-atom structure validation for macromolecular crystallography. *Acta Crystallographica. Section D, Biological Crystallography*, *66*, 12–21.

Claren, J., Malisi, C., Höcker, B., & Sterner, R. (2009). Establishing wild-type levels of catalytic activity on natural and artificial (beta alpha)8-barrel protein scaffolds. *Proceedings of the National Academy of Sciences of the United States of America*, *106*, 3704–3709.

Eisenbeis, S., & Höcker, B. (2010). Evolutionary mechanism as a template for protein engineering. *Journal of Peptide Science*, *16*, 538–544.

Eisenbeis, S., Proffitt, W., Coles, M., Truffault, V., Shanmugaratnam, S., Meiler, J., et al. (2012). Potential of fragment recombination for rational design of proteins. *Journal of the American Chemical Society*, *134*, 4019–4022.

Emsley, P., Lohkamp, B., Scott, W. G., & Cowtan, K. (2010). Features and development of Coot. *Acta Crystallographica Section D – Biological Crystallography*, *66*, 486–501.

Eswar, N., Eramian, D., Webb, B., Shen, M. Y., & Sali, A. (2008). Protein structure modeling with MODELLER. *Methods in Molecular Biology*, *426*, 145–159.

Fani, R., Lio, P., Chiarelli, I., & Bazzicalupo, M. (1994). The evolution of the histidine biosynthetic genes in prokaryotes: A common ancestor for the hisA and hisF genes. *Journal of Molecular Evolution*, *38*, 489–495.

Grishin, N. V. (2001). Fold change in evolution of protein structures. *Journal of Structural Biology*, *134*, 167–185.

Guerler, A., & Knapp, E. W. (2008). Novel protein folds and their nonsequential structural analogs. *Protein Science*, *17*, 1374–1382.

Höcker, B., Beismann-Driemeyer, S., Hettwer, S., Lustig, A., & Sterner, R. (2001). Dissection of a (betaalpha)8-barrel enzyme into two folded halves. *Nature Structural Biology*, *8*, 32–36.

Höcker, B., Claren, J., & Sterner, R. (2004). Mimicking enzyme evolution by generating new (βα)8-barrels from (βα)4-half-barrels. *Proceedings of the National Academy of Sciences of the United States of America, 101*, 16448–16453.

Höcker, B., Lochner, A., Seitz, T., Claren, J., & Sterner, R. (2009). High-resolution crystal structure of an artificial (βα)8-barrel protein designed from identical half-barrels. *Biochemistry, 48*, 1145–1147.

Höcker, B., Schmidt, S., & Sterner, R. (2002). A common evolutionary origin of two elementary enzyme folds. *FEBS Letters, 510*, 133–135.

Holm, L., & Rosenström, P. (2010). Dali server: Conservation mapping in 3D. *Nucleic Acids Research, 38*, W545–W549.

Krissinel, E., & Henrick, K. (2004). Secondary-structure matching (SSM), a new tool for fast protein structure alignment in three dimensions. *Acta Crystallographica. Section D, Biological Crystallography, 60*, 2256–2268.

Lang, D., Thoma, R., Henn-Sax, M., Sterner, R., & Wilmanns, M. (2000). Structural evidence for evolution of the beta/alpha barrel scaffold by gene duplication and fusion. *Science, 289*, 1546–1550.

Lutz, S., Ostermeier, M., Moore, G. L., Maranas, C. D., & Benkovic, S. J. (2001). Creating multiple-crossover DNA libraries independent of sequence identity. *Proceedings of the National Academy of Sciences of the United States of America, 98*, 11248–11253.

Lynch, M., & Conery, J. S. (2000). The evolutionary fate and consequences of duplicate genes. *Science, 290*, 1151–1155.

Ochman, H., Lawrence, J. G., & Groisman, E. A. (2000). Lateral gene transfer and the nature of bacterial innovation. *Nature, 405*, 299–304.

Ostermeier, M., Shim, J. H., & Benkovic, S. J. (1999). A combinatorial approach to hybrid enzymes independent of DNA homology. *Nature Biotechnology, 17*, 1205–1209.

Otey, C. R., Landwehr, M., Endelman, J. B., Hiraga, K., Bloom, J. D., & Arnold, F. H. (2006). Structure-guided recombination creates an artificial family of cytochromes P450. *PLoS Biology, 4*, e112.

Peisajovich, S. G., Rockah, L., & Tawfik, D. S. (2006). Evolution of new protein topologies through multistep gene rearrangements. *Nature Genetics, 38*, 168–174.

Riechmann, L., & Winter, G. (2000). Novel folded protein domains generated by combinatorial shuffling of polypeptide segments. *Proceedings of the National Academy of Sciences of the United States of America, 97*, 10068–10073.

Riechmann, L., & Winter, G. (2006). Early protein evolution: Building domains from ligand-binding polypeptide segments. *Journal of Molecular Biology, 363*, 460–468.

Rohl, C. A., Strauss, C. E., Misura, K. M., & Baker, D. (2004). Protein structure prediction using Rosetta. *Methods in Enzymology, 383*, 66–93.

Seitz, T., Bocola, M., Claren, J., & Sterner, R. (2007). Stabilisation of a (betaalpha)8-barrel protein designed from identical half barrels. *Journal of Molecular Biology, 372*, 114–129.

Shanmugaratnam, S., Eisenbeis, S., & Höcker, B. (2012). A highly stable protein chimera built from fragments of different folds. *Protein Engineering Design and Selection, 25*, 699–703.

Simons, K. T., Kooperberg, C., Huang, E., & Baker, D. (1997). Assembly of protein tertiary structures from fragments with similar local sequences using simulated annealing and Bayesian scoring functions. *Journal of Molecular Biology, 268*, 209–225.

Söding, J., Biegert, A., & Lupas, A. N. (2005). The HHpred interactive server for protein homology detection and structure prediction. *Nucleic Acids Research, 33*, W244–W248.

Söding, J., & Lupas, A. N. (2003). More than the sum of their parts: On the evolution of proteins from peptides. *Bioessays, 25*, 837–846.

Stemmer, W. P. (1994). DNA shuffling by random fragmentation and reassembly: in vitro recombination for molecular evolution. *Proceedings of the National Academy of Sciences of the United States of America, 91*, 10747–10751.

Voigt, C. A., Martinez, C., Wang, Z. G., Mayo, S. L., & Arnold, F. H. (2002). Protein building blocks preserved by recombination. *Nature Structural Biology, 9*, 553–558.

CHAPTER NINETEEN

α-Helix Mimicry with α/β-Peptides

Lisa M. Johnson, Samuel H. Gellman[1]

Department of Chemistry, University of Wisconsin, Madison, Wisconsin, USA

[1]Corresponding author: e-mail address: gellman@chem.wisc.edu

Contents

Abstract

We describe a general strategy for creating peptidic oligomers that have unnatural backbones but nevertheless adopt a conformation very similar to the α-helix. These oligomers contain both α- and β-amino acid residues (α/β-peptides). If the β content reaches 25–30% of the residue total, and the β residues are evenly distributed along the backbone, then substantial resistance to proteolytic degradation is often observed. These α/β-peptides can mimic the informational properties of α-helices involved in protein–protein recognition events, as documented in numerous crystal structures. Thus, these unnatural oligomers can be a source of antagonists of undesirable protein–protein interactions that are mediated by natural α-helices, or agonists of receptors for which the natural polypeptide ligands are α-helical. Successes include mimicry of BH3 domains found in proapoptotic proteins, which leads to ligands for antiapoptotic Bcl-2 family proteins, and mimicry of the gp41 CHR domain, which leads to inhibition of HIV infection in cell-based assays.

Methods in Enzymology, Volume 523
ISSN 0076-6879
http://dx.doi.org/10.1016/B978-0-12-394292-0.00019-9

1. INTRODUCTION

Proteins evolve to display a specific set of properties that are advantageous to the organism in which they are produced. Scientists often seek molecules that mimic only a subset among the properties of a particular protein. Such mimics can be used as research tools, diagnostic agents, or medicines; some applications require the introduction of properties that are not manifested by the original protein.

Starting from a prototype protein, researchers have traditionally had access to only a few types of modification. (1) *Mutation*: the side chain can be altered at one position, or multiple side chains can be altered. (2) *Truncation*: portions of the polypeptide that are not necessary for the properties of interest can be removed. (3) *Augmentation*: polypeptide segments can be grafted onto the prototype protein to confer or enhance characteristics that are necessary for the intended application. (4) *Decoration*: nonpeptide moieties (e.g., a carbohydrate, a fluorophore, a synthetic polymer) can be attached to side chains or termini. These different modification strategies can be implemented in tandem.

Here, we focus on a different type of modification strategy, in which the polypeptide backbone is altered. This approach can be implemented in a way that does not affect the identity or sequence of side chains relative to the prototype. Alternatively, one can take advantage of unique property-modification opportunities that become available when α-amino acid residues are replaced with other subunits; however, this version of the strategy usually involves loss of native side chains. The backbone modification strategy can provide protein mimics that manifest unusual property profiles. Maintaining the recognition properties of a prototype polypeptide while diminishing or abolishing the susceptibility to degradation by proteases has been of particular interest (Seebach et al., 1998). We show how this goal has been achieved with oligomers that contain both α- and β-amino acid residues (α/β-peptides).

It should be noted that the use of subunits other than those derived from α-amino acids necessitates chemical synthesis of polypeptides at present. One can hope, however, that modification of the biosynthetic machinery responsible for protein synthesis (e.g., the ribosome, tRNA synthetases) will ultimately allow incorporation of building blocks beyond α-amino acids via RNA-templated synthesis. Hybrid approaches, such as using expressed protein ligation to connect a synthetic peptide that contains β-amino acid residues to a larger biosynthetic poly-α-peptide, offer considerable latitude for current efforts (Arnold et al., 2002; David et al., 2008).

2. HELICAL SECONDARY STRUCTURES FROM β-PEPTIDES AND α/β-PEPTIDES

Fundamental studies in several laboratories over the past two decades have revealed that oligomers of β-amino acids (β-peptides) can adopt a variety of helical secondary structures (Cheng, Gellman, & DeGrado, 2001; Gellman, 1998; Martinek & Fulop, 2003; Seebach & Matthews, 1997). β-Amino acids have two carbon atoms between the carboxyl and amino groups, and both carbons can bear side chains. Altering β-amino acid substitution patterns enables considerable variation in the nature and the extent of the secondary structure propensity. In this regard, β-amino acid residues are more versatile than α-amino acid residues, that is, there appears to be a larger number of easily accessed regular secondary structures among β-peptides than among α-peptides (Cheng et al., 2001; Gellman, 1998; Seebach & Matthews, 1997). β-Peptides constructed entirely from β^3-homoamino acid residues (Fig. 19.1E) generally seem to favor a helical secondary structure that contains a C=O(*i*)–N–H(*i* – *2*) H-bonding pattern among backbone amide groups. As these H-bonds occur in 14-atom rings, this conformation is sometimes referred to as the "14-helix" (Fig. 19.2) (Gellman, 1998; Seebach & Matthews, 1997; Seebach et al., 1996). The intrinsic 14-helical propensity of β^3 residues seems comparable to the α-helical propensity of most α-amino acid residues (Cheng & DeGrado, 2001), but the 14-helical propensity can be dramatically increased by using the cyclic β residue derived from *trans*-2-aminocyclohexanecarboxylic acid (ACHC) (Fig. 19.1A) (Appella, Barchi, Durell, & Gellman, 1999; Appella, Christianson, Karle, Powell, & Gellman, 1999; Lee et al., 2007; Vaz, Pomerantz, Geyer, Gellman, & Brunsveld, 2008). The nature of β-peptide helical propensity can be fundamentally altered by using a different ring constraint, as found in *trans*-2-aminocyclopentanecarboxylic acid (ACPC) (Appella et al., 1997; Appella, Christianson, Klein, et al., 1999; Lee, Syud, Wang, & Gellman, 2001), which promotes a secondary structure that contains C==O(*i*)–H—N(*i* + 3) H-bonds ("12-helix"; Fig. 19.1B). The strong conformation-directing effects of ring-containing β-amino acid residues have no parallel among α-amino acid residues because in the latter case one cannot use a cyclic constraint to influence the torsional preferences of a backbone bond without removing an H-bonding site, as illustrated by proline.

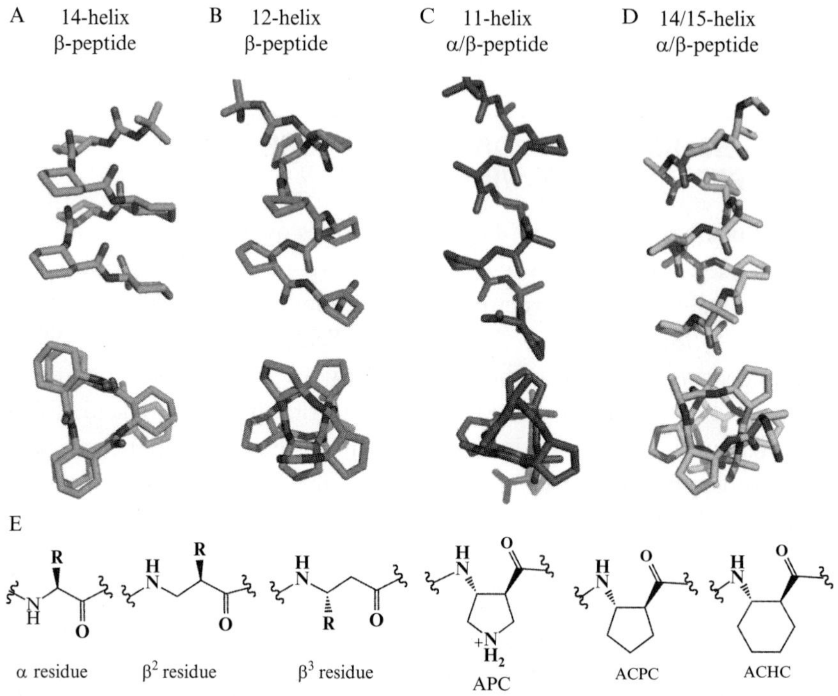

Figure 19.1 Crystal structures: (A) the β-peptide 14-helix (ACHC hexamer; CSD ID: REFPUN01; Appella, Christianson, Karle, et al., 1999), (B) the β-peptide 12-helix (ACPC hexamer; CSD ID: WELMAB; Appella, Christianson, Klein, et al., 1999), (C) the α/β-peptide 11-helix (1:1 α/β octamer; alternation of ACPC and Ala or Aib residues; CSD ID: OGATUM; Choi, Guzei, Spencer, & Gellman, 2008), (D) the α/β-peptide 14/15-helix (1:1 α/β decamer; alternation of ACPC and Ala or Aib residues; CSD ID: OGAVEY; Choi et al., 2008), and (E) α-Amino acid residue and several types of β-amino acid residues. (For the color version of this figure, the reader is referred to the online version of this chapter.)

Oligomers that contain both α- and β-amino acid residues display distinctive helical secondary structures (Choi, et al., 2008; Choi, Guzei, Spencer, & Gellman, 2009; De Pol, Zorn, Klein, Zerbe, & Reiser, 2004; Hayen, Schmitt, Ngassa, Thomasson, & Gellman, 2004). The heterogeneous backbone of α/β-peptides introduces a new parameter of structural variation: in addition to selecting the side-chain identities and the nature and extent of conformational preorganization (for the β residues), one can choose from among diverse α/β backbone patterns. For oligomers with 1:1 α:β alternation, the nature of the helix can be changed by altering the cyclic constraint embedded in the β residues and other structural parameters. Two helices that are favored by ACPC residues in β sites are shown in

Fig. 19.1C and D. Varying the α/β pattern, beyond 1:1, leads to additional helical secondary structures (Berlicki et al., 2012; Choi et al., 2009).

3. BIOLOGICAL FUNCTION FROM HELICAL β-PEPTIDES

As the rules governing helix formation have been elucidated for β-peptides and α/β-peptides, efforts to parlay this folding behavior into biological function have been pursued in a number of laboratories. Initial studies focused on functions that depend on general structural features of natural α-helical prototypes rather than on a highly specific three-dimensional arrangement of side chains projected from an α-helical scaffold. The first example involved inhibition of cholesterol absorption by intestinal cells. This activity was known to be manifested by α-peptides that form a globally amphiphilic α-helix (Boffelli et al., 1997) (i.e., a helix that projects lipophilic side chains along one side and hydrophilic side chains along the other), but the activity does not depend on a specific amino-acid sequence. Seebach et al. showed that β-peptides that form a short, globally amphiphilic 14-helix (Fig. 19.2) display weak inhibition of cholesteryl ester uptake in a cell-based assay (Werder, Hauser, Abele, & Seebach, 1999).

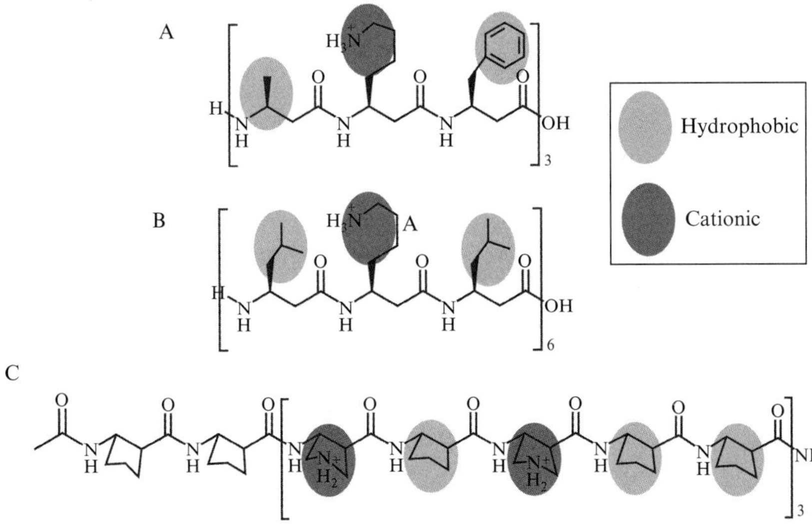

Figure 19.2 Three examples of β-peptides designed to form globally amphiphilic helices: (A) 14-helical (3 residues per turn) cholesterol uptake inhibitor (Werder et al., 1999), (B) 14-helical antimicrobial peptide (Hamuro et al., 1999), and (C) 12-helical (2.5 residues per turn) antimicrobial peptide (Porter et al., 2000). (For the color version of this figure, the reader is referred to the online version of this chapter.)

Globally, amphiphilic helical β-peptides were used by Hamuro, Schneider, and DeGrado, (1999) and Porter, Wang, Lee, Weisblum, and Gellman (2000) to achieve functional mimicry of α-helical host-defense peptides, which constitute part of the innate immune response to bacterial infection (Hancock & Sahl, 2006). The membrane-disruption activity of these peptides does not depend on a specific amino acid sequence or on absolute configuration (Wade et al., 1990). β-Peptides were shown to inhibit bacterial growth if the sequence pattern of lipophilic and cationic residues allowed formation of a globally amphiphilic 12- or 14-helix (Godballe, Nilsson, Petersen, & Jenssen, 2011). Comparable behavior was subsequently demonstrated for α/β-peptides, although in some cases, selective antibacterial activity did not seem to depend upon the ability to form an amphiphilic helix (Schmitt, Weisblum, & Gellman, 2004, 2007). Ultimately, it was found that relatively small molecules (Doerksen et al., 2004) displaying globally amphiphilic conformations could manifest comparable antibacterial activity, as could sequence-random poly-β-peptides (Mowery et al., 2007) and other polymers (Kenawy, Worley, & Broughton, 2007).

Subsequent β-peptide work focused on more specific functions that nevertheless may not require a particular spatial arrangement of side chains, as illustrated by the inhibition of γ-secretase (Imbimbo & Giardina, 2011). This membrane-embedded protease processes amyloid precursor protein to generate Aβ40 and Aβ42, the peptides that form plaques associated with Alzheimer's disease. Hydrophobic α-helix-forming α-peptides can inhibit γ-secretase, and the enantiomers of these peptides (containing D- rather than L-α-amino acid residues) are very active (Das et al., 2003). Imamura et al. (2009) then found that 12-helical β-peptides, oligomers of ACPC, are potent γ-secretase inhibitors.

In another example that does not seem to require a specific three-dimensional arrangement of side chains, Beck-Sickinger et al. have examined analogues of interleukin-8 (IL-8) in which the C-terminal α-helix is replaced by a β-peptide segment intended to form a globally amphiphilic 14-helix four to five turns in length (David et al., 2008). These chimeric polypeptides contained the human IL-8 sequence through residue 60, which forms a β-sheet. The C-terminal α-helix packs against one face of the β-sheet, thereby stabilizing the entire tertiary structure. Signaling through the CXCR1 and CXCR2 receptors relies mainly on contacts with the N-terminus of IL-8, but IL-8 must be properly folded in order for signal transduction to occur. Beck-Sickinger et al. found that signaling activity via CXCR1 was largely retained when the native C-terminal segment was replaced with a completely different α-peptide sequence known to form a globally amphiphilic α-helix. Reduced activity was observed for an IL-8 chimera containing a C-terminal β-peptide segment designed to adopt a globally amphiphilic 14-helical conformation. This level of success in α-helix mimicry may reflect the system's ability to tolerate considerable sequence variation in the C-terminal segment, which suggests that a signaling-competent IL-8 fold does not require a precise arrangement of specific side chains within the helical portion.

Several groups have explored the use of β-peptides to mimic α-helices that are crucial for specific protein–protein recognition events. Success has been reported for the N-terminal domain of the tumor suppressor p53, which forms a short, distorted α-helix that binds to a cleft on DM2. This interaction centers on three side chains (Phe, Trp, and Leu) aligned along one side of the p53 helix (i, $i+4$, $i+7$ sequence relationship) (Kussie et al., 1996). β-Peptides that project these three side chains along one side of a 14-helix (i, $i+3$, $i+6$ sequence relationship) can bind to the DM2 cleft as well (Kritzer, Lear, Hodsdon, & Schepartz, 2004), even though the β-peptide 14-helix and the α-helix differ significantly as scaffolds for side chain arrangement. The recognition cleft of DM2 appears to be somewhat tolerant in terms of the geometry of the three side chains on the binding partner, because even macrocyclic α-peptides that contain these side chains but adopt a β-sheet conformation can bind to DM2 (Fasan et al., 2004). In addition, the DM2 cleft can be effectively targeted by small molecules (Zhao et al., 2002), including the nutlins (Vassilev et al., 2004) and oligomers engineered to display protein-like side chains in an α-helix-mimetic fashion (Desai, Pfeiffer, & Boger, 2003; Hara, Durell, Myers, & Appella, 2006; Plante et al., 2009; Yin et al., 2005).

It has recently been reported that helical β-peptides can be designed to interact with α-helical partners embedded in a membrane, specifically a transmembrane helix of the integrin $\alpha_{IIb}\beta^3$ (Shandler et al., 2011). Many aspects of this example are intriguing, including the observation that some all-β^3 sequences can apparently adopt 12-helical conformations rather than 14-helical conformations (Korendovych, Shandler, Montalvo, & DeGrado, 2011; Raguse, Porter, Weisblum, & Gellman, 2002). The helix–helix interaction mode in this system features a relatively large crossing angle, which means that the helices make contact over only a relatively small region (in contrast to a coiled-coil dimer, in which the helices are nearly parallel).

As the α-helix to be mimicked grows longer, and the number of side chains arrayed along one side of that helix and recognized by a binding partner grows larger, the negative impact of deviations from the α-helical spacing among side chains is likely to increase. This observation may explain why nearly all attempts so far to mimic informational α-helices with β-peptides, peptoids, or aryl oligomers have been limited to examples involving just three side chains (typically displaying an i, $i+4$, $i+7$ sequence relationship in the natural prototype), as summarized above. We devoted considerable effort to trying to mimic the arrangement of *four* aligned hydrophobic side chains that is characteristic of BH3 domain α-helices (i, $i+4$, $i+7$, $i+11$ sequence relationship) (Chen et al., 2005; Lessene, Czabotar, & Colman, 2008; Petros, Olejniczak, & Fesik, 2004), but we found that neither the β-peptide 14-helix nor the β-peptide 12-helix was an effective scaffold for this purpose (Sadowsky et al., 2007, 2005). Also unsuccessful was the 11-helical scaffold formed by α/β-peptides with a 1:1 α:β backbone pattern. The 14/15-helix formed by 1:1 α/β-peptides was somewhat more effective but still imperfect for BH3 domain α-helix mimicry. Part of the challenge in this system is that a properly folded BH3 domain displays a side-chain carboxylate on the opposite side of the α-helix relative to the linear array of four hydrophobic side chains (Asp at position $i+9$). Thus, in this case, and many others, successful mimicry of an α-helical "message" requires more than recreating an isolated side-chain stripe. Unnatural scaffolds will be of greatest use for α-helix mimicry if they allow recapitulation of the complete three-dimensional arrangement of groups that make energetically important contacts with the partner protein.

4. α-HELIX MIMICRY WITH α/β-PEPTIDES

4.1. Sequence-based design

Our failure to identify effective BH3 domain mimics via structure-based design, starting from established β-peptide or 1:1 α/β-peptide helices, led us to explore a new approach to α-helix mimicry that has proven to be very fruitful. This approach begins with "sequence-based design," in which a subset of the α residues in a prototype sequence is replaced with homologous β^3 residues. α→β Replacements are made throughout the sequence according to simple patterns that result in 25–33% β residue incorporation. At this stage, β residue choice is "automatic" because the original side chain is retained, that is, leucine is replaced with β^3-homoleucine (β^3-hLeu), serine is replaced with β^3-hSer, and so on. In a second step, which is necessary in only some cases, the helix-forming propensity of the α/β-peptide is enhanced by β^3→cyclic β replacements. Fundamental α/β-peptide structural studies showed that incorporation of the C_α—C_β bond into a five-membered ring, with *trans* disposition of the amino and carboxyl groups, promotes a local conformation consistent with α-helix-like secondary structure (Choi et al., 2008; Horne, Price, & Gellman, 2008; Price, Horne, & Gellman, 2010). Therefore, residues derived from the β-amino acids ACPC and APC (Fig. 19.1E) are useful for residue-based preorganization of α-helix-mimetic α/β-peptides.

Initial evaluation of the sequence-based design approach involved self-recognizing α-helices based on the dimerization domain of yeast transcriptional regulator GCN4. GCN4-pLI is a designed variant that forms a parallel helix-bundle tetramer (Harbury, Zhang, Kim, & Alber, 1993). Figure 19.3 compares the crystal structure of GCN4-pLI with those of analogues containing $\alpha \rightarrow \beta^3$ replacements in three regular patterns, ααβαααβ, ααβ, and αααβ (Horne, Price, et al., 2008). Each of the α/β-peptides retains the side-chain sequence of the α-peptide prototype because for each $\alpha \rightarrow \beta$ replacement, the β^3 residue is homologous to the original α residue. All three α/β-peptides adopt conformations very similar to the α-helix. Because the ααβαααβ pattern is tailored to the heptad residue repeat characteristic of the α-helix, in this case, the β^3 residues are aligned along one side of the helix. By design, this "β-stripe" is diametrically opposed to the hydrophobic side-chain stripe that provides the basis for self-assembly; thus, the β^3 residues reside exclusively on the exterior of the four-helix

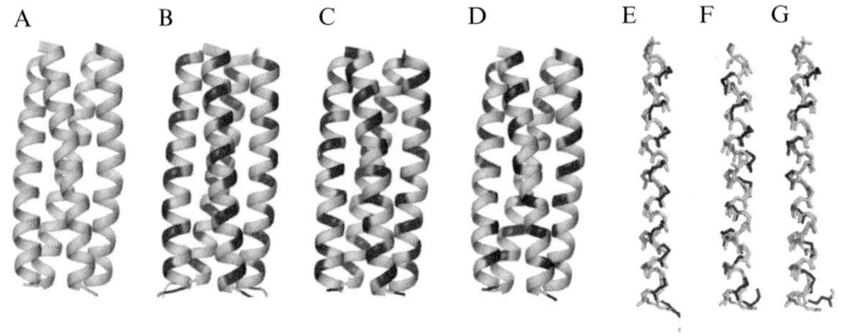

Figure 19.3 Helix bundles formed by α-peptide GCN4-pLI (A) (PDB ID: 1GCL; Harbury et al., 1993) and three α/β-peptide homologues with varying backbone patterns: (B) ααβαααβ (PDB ID: 2OXK), (C) ααβ (PDB ID: 3C3G), and (D) αααβ (PDB ID: 3C3F). Each image is based on a crystal structure. α Residues are shown in yellow, and β^3 residues are shown in blue. Backbone overlays between the α peptide GCN4-pLI and (E) ααβαααβ, (F) ααβ, and (G) αααβ homologues (Horne, Price, et al., 2008). (See Color Insert.)

bundle for the ααβαααβ version. In contrast, the ααβ or αααβ patterns cause the β residues to spiral around the helix periphery. Two of the β^3 side chains in each case form part of the tetramer core (Horne, Price, et al., 2008).

The α/β-peptide helix-bundle crystal structures reveal that the ααβ, αααβ, and ααβαααβ backbones all adopt conformations that adhere closely to the α-helical prototype over eight helical turns, despite the presence of approximately one extra backbone carbon atom per turn in the α/β-peptides. Accommodation of these extra atoms appears to be smoothly distributed along the entire backbone (Horne, Price, et al., 2008).

The excellent structural mimicry of α-helical GCN4-pLI displayed by α/β-peptide homologues containing $\alpha \rightarrow \beta^3$ replacements in various periodic patterns was accompanied by destabilization of the tetrameric quaternary structure. We hypothesize that the lower stability of the α/β-peptide helix bundles relative to the α-peptide helix bundle results from conformational entropy. Each $\alpha \rightarrow \beta^3$ replacement introduces an "extra" flexible bond into the peptidic backbone, and there are 8–11 such replacements among the α/β-peptide homologues of GCN4-pLI. Thus, these α/β-peptides must suffer a greater loss of conformational entropy upon helical folding than does the α-peptide (Horne, Price, et al., 2008).

4.2. BH3 domain mimicry

Successful structural mimicry of self-recognizing α-helices by GCN4-inspired α/β-peptides that contain periodic, side chain-preserving $\alpha \rightarrow \beta^3$ replacements led us to explore comparable approaches for mimicry of α-helical "messages" that are read out by complementary proteins. BH3 domain α-helices represented an interesting starting point for such studies since we had previously encountered difficulties in mimicking BH3 domains with β-peptide or 1:1 α/β-peptide helices (Sadowsky et al., 2007, 2005). Initial sequence-based efforts focused on the Puma BH3 domain, which binds tightly to many antiapoptotic members of the Bcl-2 family (Chen et al., 2005). There are seven ways to incorporate the ααβαααβ pattern into a given amino acid sequence, which give rise to seven different positions of the β residue stripe on the helix (Fig. 19.4B) (Lee et al., 2011). We evaluated all seven of these $\alpha \rightarrow \beta^3$ replacement registers in the context of a Puma-derived 26-mer; binding to Bcl-x_L and to Mcl-1 was assessed for each α/β-peptide with a competition fluorescence polarization (FP) assay (Horne, Boersma, et al., 2008). Although these seven α/β-peptides are isomers or nearly isomers of one another (some have one more CH_2 unit than others, because some have eight β^3 residues while others have seven), and

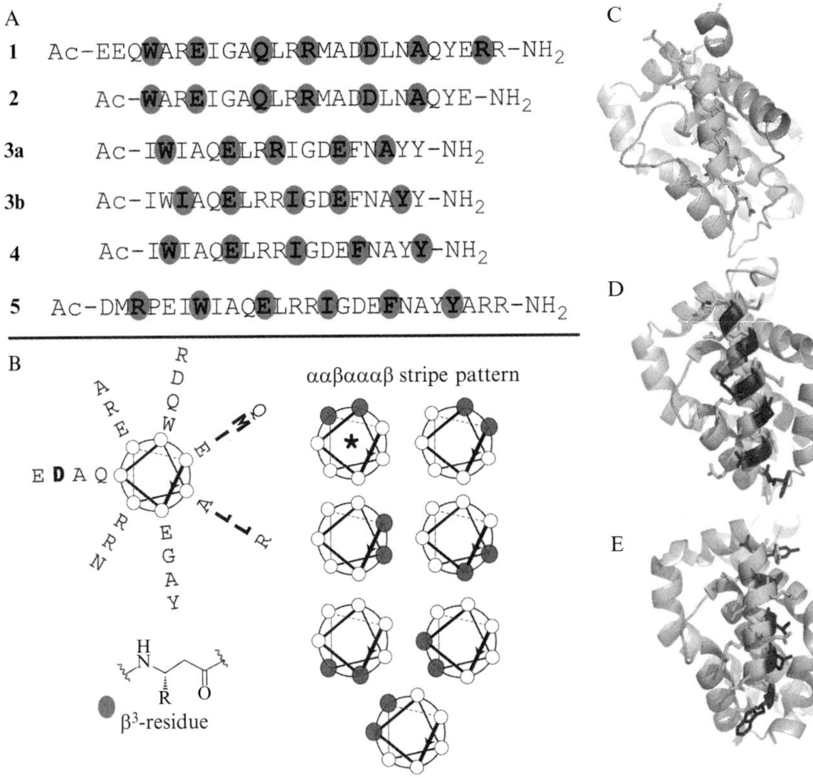

Figure 19.4 (A) α/β-Peptides with side chain sequences derived from the Puma BH3 domain (**1** and **2**) or the Bim BH3 domain (**3a-b**, **4**, and **5**). The standard single-letter code is used; letters covered with a blue dot indicate β^3-homologues of the α residue designated by the letter. Compounds **1**, **2**, and **3a-b** feature the ααβαααβ backbone repeat, while **4** and **5** have the αααβ backbone repeat. (B) Helical wheel diagram of the Puma BH3 domain with the four key hydrophobic residues and the key Asp in bold. The smaller helix-wheel diagrams show how the stripe of β^3 residues shifts incrementally around the helix periphery among the seven different versions of the ααβαααβ backbone repeat (Horne, Boersma, Windsor, & Gellman, 2008); the asterisk (*) indicates the α/β registry found in Puma-derived α/β-peptides **1** and **2**. (C–E) Images illustrating the complexes between the BH3 domain-derived α- or α/β-peptides (α-residues are yellow, β residues are blue) and Bcl-x_L (light purple): (C) Bak BH3 domain (NMR structure; PDB ID: 1BLX; Sattler et al., 1997); (D) α/β-peptide **2** (crystal structure; PDB ID: 2YJ1; Lee et al., 2011); (E) α/β-peptide **4** (crystal structure; PDB ID: 4A1W; Boersma et al., 2012). (See Color Insert.)

they all contain the same sequence of side chains, their affinities for each protein spanned several orders of magnitude (as measured by K_i). One ααβαααβ registry (**1**) showed the tightest binding to both Bcl-x_L and Mcl-1; if we use the standard *abcdefg* parlance of the coiled-coil field to define positions in each sequence heptad, and designate the positions of the four key hydrophobic residues as *a* and *d*, then the β residue positions in the tightest-binding α/β-peptide are *c* and *g*. For Bcl-x_L, α/β-peptide **1** bound too tightly to be quantified, as was true of the Puma BH3 α-peptide ($K_i < 1$ n*M* in each case). For Mcl-1, however, α/β-peptide **1** bound at least 15-fold less tightly than the Puma α-peptide ($K_i = 150$ n*M* vs. $K_i < 10$ n*M*).

In light of the conformational entropy hypothesis offered above, it is intriguing that an α/β-peptide such as **1** could match a homologous α-peptide in binding to a partner proteins (Bcl-x_L) when the former has seven additional flexible backbone bonds relative to the latter. A crystal structure of Bcl-x_L complexed with α/β-peptide **2**, a slightly truncated analogue of **1**, showed that the stripe of β^3 residues is largely oriented toward solvent, as expected, but that some of the β^3 residues make direct contact with the protein (Fig. 19.4D) (Lee et al., 2011). Thus, it is possible that the α/β-peptide could serendipitously form improved contacts with the complementary Bcl-x_L cleft relative to the original BH3 domain. As the best α/β-peptide is substantially inferior to the BH3 domain α-peptide in binding to Mcl-1, this hypothetical serendipity would be protein-specific.

Our interest in mimicking BH3 domain α-helices is twofold: we seek to recapitulate the features required for tight binding to the BH3-recognition clefts of antiapoptotic proteins in the Bcl-2 family, but we want to *discourage* recognition by proteases. One well-appreciated problem associated with the use of peptides as drugs is the very high susceptibility of α-amino acid-based oligomers to proteolytic degradation *in vivo* (Lee, 1988). Replacing a single α residue with a β residue has long been known to inhibit proteolytic cleavage at nearby amide bonds, presumably because the β residue disrupts recognition of the peptide backbone by the protease active site (Steer, Lew, Perlmutter, Smith, & Aguilar, 2002). We wondered whether an α/β-peptide such as **1** has a sufficient β residue density to provide significant resistance across its entire length to the action of an aggressive protease.

We used proteinase K, a relatively nonspecific enzyme, to assess proteolytic susceptibility of α/β-peptide **1** (Wu & Kim, 1997). Although **1** contains mostly α residues (19 of 26), the half-life for cleavage of this α/β-peptide by proteinase K is >4000-fold longer than the half-life for cleavage of the corresponding α-peptide (Puma BH3 domain). This striking level of protection is not universal among oligomers with the ααβαααβ backbone pattern; for another member of this series, the half-life in the presence of proteinase K is only ~240-fold longer than that of the α-peptide. Thus, the precise positioning of α→β replacements is important with regard to proteolytic susceptibility. However, these results and others discussed below suggest that incorporation of 25–30% β residues, with even distribution along the sequence, often delivers significant improvements in half-life in the presence of proteinase K.

In order to determine whether sequence-based design is broadly useful for generating BH3 domain mimics that resist proteolysis, we turned to the Bim BH3 domain, which binds to all antiapoptotic Bcl-2 family members (Chen et al., 2005). Starting from a Bim BH3 18-mer α-peptide, we examined all seven α/β-peptide homologues with the ααβαααβ backbone pattern as well as all three with the ααβ pattern and all four with the αααβ pattern for binding to Bcl-x_L and Mcl-1 via competition FP assays. This set of 14 α/β-peptides displayed a large range of K_i values for each protein. Among the ααβαααβ set, the one with the highest affinity for Mcl-1 (**3a**) had the same α→β registry that led to the highest Mcl-1 affinity among Puma-based α/β-peptides (i.e., β residues at heptad positions *c* and *g*). However, in contrast to the Puma series, a different α→β registry (**3b**; heptad positions *d* and *g*) provided the highest Bcl-x_L affinity among the Bim-derived oligomers with the ααβαααβ pattern. α/β-Peptide **3a** bound to both proteins examined, that is, to Bcl-x_L as well as Mcl-1, but **3b** was highly selective for Bcl-x_L (Boersma et al., 2012).

The strongest binding to both Bcl-x_L and Mcl-1 among the 14 Bim-derived α/β-peptides was observed for **4**, which has the αααβ backbone pattern. α/β-Peptide **4** bound almost as tightly to Bcl-x_L as did the corresponding Bim BH3 α-peptide, but the affinity of **4** for Mcl-1 was considerably reduced relative to the α-peptide. Two of the four key hydrophobic side chains in the Bim BH3 sequence are contributed by β^3 residues in **4**, and a crystal structure of **4** bound to Bcl-x_L confirmed that these β^3 side chains make intimate contacts with the protein (Fig. 19.4E). Most studies of the Bim BH3 domain have involved 26-residue α-peptides rather than the 18-mer we used as a basis for the α/β-peptide Bim BH3 homologues. We therefore prepared **5**, the 26-mer that maintains the α/β backbone pattern of **4**. Comparison of **5** with the Bim BH3 26-mer α-peptide via competition surface plasmon resonance assays revealed that these two oligomers have nearly identical affinities for Bcl-x_L, Bcl-w, and Mcl-1, and that the α/β-peptide binds significantly more tightly than the α-peptide to Bcl-2. The half-life of α/β-peptide **5** in the presence of proteinase K is >180-fold longer than the half-life of the Bim BH3 α-peptide (Boersma et al., 2012).

Overall, our findings in the Bim system show that it is not always necessary to avoid α→β replacement at sites that contribute key side chains to a protein-recognition interface, as β residues contribute key side chains to the interface between Bcl-x_L and α/β-peptide **4**. In addition, these findings raise the possibility that α→β replacements need not be limited to the simple backbone patterns we have explored so far. These patterns are helpful, however, in allowing a researcher to examine diverse α+β arrangements in small oligomer libraries (e.g., the set of 14 from which **3a**, **3b**, and **4** were obtained) while ensuring a sufficient proportion and distribution of β residues to inhibit protease action.

4.3. Mimicry of the gp41 CHR domain: Inhibitors of HIV infection

We turned to mimicry of a long α-helix that is essential for infection of cells by HIV in order to assess the versatility of our α/β-peptide design strategy. The protein gp41 occurs on the surface of the HIV particle; after the virus has recognized a target cell, gp41 orchestrates the fusion of the viral envelope with the cell membrane (Eckert & Kim, 2001). This fusion process is believed to require formation of six-helix bundle from three gp41 molecules, each of which contributes one C-terminal heptad repeat (CHR) and one N-terminal heptad repeat (NHR) segment (Chan, Fass, Berger, & Kim, 1997). The CHR segment forms an α-helix containing >10 turns. Long α-peptides derived from either the CHR or NHR domain of gp41 inhibit HIV infection in cell-based assays (Sodroski, 1999), and one such peptide, the 36-mer enfuvirtide, is an FDA-approved drug for AIDS patients. Enfuvirtide has a half-life of just a few hours in the bloodstream, and the daily dose is two 90 mg injections (Matthews et al., 2004).

Our efforts began with the 38-residue α-peptide T-2635 (Fig. 19.5), developed by researchers at Trimeris (Dwyer et al., 2007). T-2635 contains many mutations relative to the native CHR sequence that are intended to enhance α-helical propensity, including replacement of native residues with Ala and introduction of acid/base residue pairs intended to form salt bridges in the α-helical conformation (i, $i+4$ spacing). These mutations were introduced in positions that do not form interhelical contacts upon six-helix bundle formation. Figure 19.5B shows the crystal structure of the six-helix assembly that forms when T-2635 is mixed with the NHR-derived peptide N36 (Horne et al., 2009).

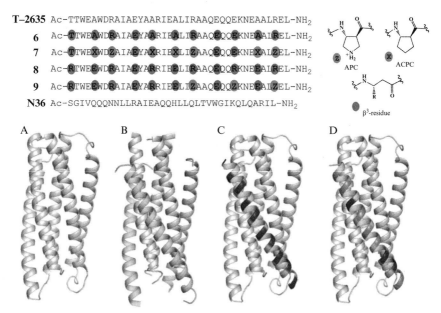

T–2635	Ac-TTWEAWDRAIAEYAARIEALIRAAQEQQEKNEAALREL-NH_2
6	Ac-TTWEAWDRAIAEYAARIEALIRAAQEQQEKNEAALREL-NH_2
7	Ac-TTWEXWDZAIAEYAXRIEXLIZAAQEQQEKNEXALZEL-NH_2
8	Ac-RTWEEWDRAIAEYARRIEELIRAAQEQQRKNEEALREL-NH_2
9	Ac-RTWEEWDRAIAEYARRIEELIZAAQEQQZKNEEALZEL-NH_2
N36	Ac-SGIVQQQNNLLRAIEAQQHLLQLTVWGIKQLQARIL-NH_2

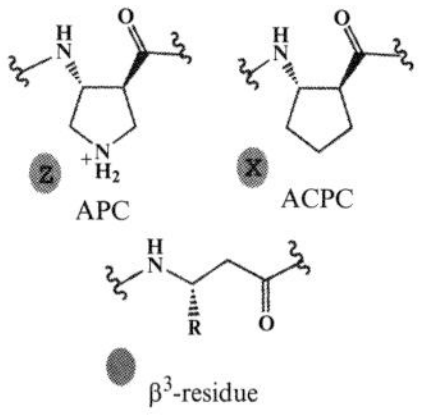

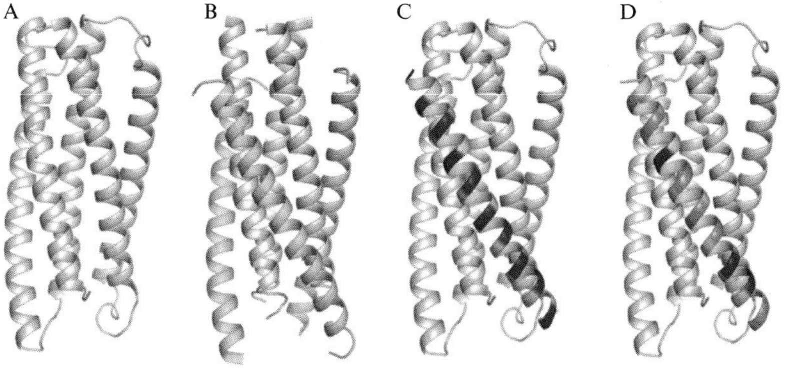

Figure 19.5 α- and α/β-Peptides related to gp41. The color scheme key for the β residues in the sequences is shown at the right: blue for β^3 residues and peach for cyclic β residues. Crystal structures: (A) gp41-5 shown in gray ribbon cartoon (PDB ID: 3O3X; Johnson et al., 2011); (B) 6-helix bundle formed by α-peptides T-2635 (yellow) and N36 (gray) (PDB ID: 3F4Y; Horne et al., 2009); (C) α/β-peptide **6** (yellow for α residues, blue for β residues) bound to gp41-5 (gray) (PDB ID: 3O42; Johnson et al., 2012); (D) α/β-peptide **7** (yellow for α residues, blue and peach for β residues) bound to gp41-5 (gray) (PDB ID: 3O43; Johnson et al., 2011). (See Color Insert.)

We used designed protein gp41-5 to compare the recognition properties of α-peptide T-2635 with α/β-peptide homologues. This protein contains five of the six α-helices of the six-helix bundle, with short, flexible linkers between each pair of helix-forming segments (Frey et al., 2006). The crystal structure of gp41-5 (Fig. 19.5A) shows that this protein adopts a five-helix bundle tertiary structure that displays a long groove into which a CHR helix can bind (Johnson, Horne, & Gellman, 2011). gp41-5 provides the basis for a competition FP assay that can be used for initial assessment of α/β-peptides intended to serve as CHR mimics.

The CHR helix is significantly longer than BH3 domain helices; therefore, CHR domain mimicry is more challenging than BH3 domain mimicry, because any small mismatch between a prototype α-helix and the similar helix formed by an α/β-peptide will be magnified as length increases. This difference in helix length presumably underlies an important functional difference between these two systems: small molecules can serve as potent inhibitors of BH3 domain recognition (Desai et al., 2003; Hara et al., 2006; Plante et al., 2009; Vassilev et al., 2004; Yin et al., 2005; Zhao et al., 2002); however, extensive efforts directed toward blocking gp41-mediated HIV infection with small molecules (including short α-peptides or β-peptides) have met with limited success so far (Frey et al., 2006; Johnson et al., 2011).

The length of the gp41 CHR domain discouraged us from exploring all of the diverse patterns of α → β replacement that had been examined in the context of BH3 domain mimicry. We focused on the ααβαααβ pattern in a registry that would orient the β residue stripe away from the surfaces of partner helices in the six-helix bundle, resulting in α/β-peptide **6**. This T-2635 homologue displays much lower affinity for gp41-5 than does T-2635 itself ($K_i = 3800$ n*M* vs. $K_i < 2$ n*M*, in the competition FP assay) (Horne et al., 2009). This profound drop in affinity is tentatively ascribed to the conformational entropy associated with the 11 additional flexible bonds in the backbone of **6** relative to T-2635. This hypothesis is supported by the crystal structure of **6** bound to gp41-5 (Fig. 19.5C), which does not reveal any obvious mismatch at the interface between the α/β-peptide and the complementary cleft on the protein. The conformational entropy hypothesis led us to examine analogues of **6** in which flexible β^3 residues were replaced with appropriate ring-constrained residues. "Appropriate" in this context means that the constraint is provided by incorporation of the C_α—C_β bond into a five-membered ring (as discussed earlier), and that the chemical properties of the ring reflect those of the side chain that is replaced. Thus, **7** was generated by replacing the four β^3-hAla residues in **6** with ACPC (hydrophobicity maintained) and the three β^3-hArg residues in **6** with APC (positive charge maintained). Preorganized α/β-peptide **7** showed considerable improvement in binding to gp41-5 ($K_i = 9$ n*M*) relative to flexible α/β-peptide **6**, although **7** did not match the affinity of T-2635. A crystal structure of **7** bound to gp41-5 shows that the rigidified β residues are well accommodated within a long α-helix-like conformation (Fig. 19.5D). Rigidified α/β-peptide **7** was quite resistant to degradation by proteinase K: the half-life for **7** was >280-fold longer than for T-2635 (in contrast, the half-life for flexible α/β-peptide **6** was only 20-fold longer than for T-2635). α/β-Peptide **7** is nearly as potent as T-2635 at inhibiting HIV infection in cell-based assays (Horne et al., 2009).

Recent efforts based on T-2635 have identified a new strategy, based on engineered ion pairs, for improving functional α-helix mimicry that is complementary to sequence-based design and the use of preorganized β residues. α/β-Peptide **8** has the same α → β replacement sites as **6** or **7**, but the sequence of **8** has been altered to maximize the number of possible intrahelical side chain ion pairs (i, $i+3$ and/or i, $i+4$). Despite the fact that **8** contains only flexible β^3 residues, this α/β-peptide matches preorganized α/β-peptide **7** in terms of affinity for gp41-5. In addition, **7** and **8** are equipotent in terms of inhibiting HIV infection in cell-based assays (Johnson et al., 2012).

The half-life of α/β-peptide **8** in the presence of proteinase K is quite similar to that of the corresponding α-peptide, which was surprising in light of the resistance to proteolysis of other α/β-peptides with comparable β residue proportion and distribution. Mass spectrometric analysis indicated that there are four major cleavage sites in the α-peptide, three of which are suppressed in the α/β-peptide, but that the remaining cleavage site in α/β-peptide **8** (after Ala-23) is very susceptible. We were pleased to find that replacement of three β^3-hArg residues near the cleavage site in **8** with ring-preorganized APC residues, to generate **9**, delivered an α/β-peptide with extraordinary resistance to proteinase K action (half-life ~2.5 days). Moreover, **9** is the tightest-binding α/β-peptide ligand for gp41-5 found to date ($K_i < 0.2$ nM) (Johnson et al., 2012). Overall, the results obtained with **8** and **9** show that conformation-specific ion pairing and β residue preorganization can be employed in a coordinated manner to produce very effective α-helix mimics.

5. TOWARD A GENERAL APPROACH FOR α-HELIX MIMICRY WITH PROTEASE-RESISTANT α/β-PEPTIDES

The results summarized above suggest a straightforward strategy for developing α/β-peptides that mimic the recognition properties of an α-helical prototype. This strategy begins with the preparation of a small library of α/β-peptides in which subsets of α-amino acid residues are replaced with homologous β^3-amino acid residues (i.e., each α residue is replaced with the β^3 residue bearing the same side chain). These initial oligomers can have the ααβ, αααβ, or ααβαααβ backbone patterns discussed earlier; however, it is possible that other β residue arrangements, not necessarily regular, will produce superior properties. One key goal for achieving biological activity is to have a density and specific placement of β residues that inhibit the action of proteases encountered by the α/β-peptide *in vivo*. Peptides containing a combination of α- and β^3-amino acid residues can be prepared via conventional stepwise solid-phase methodology, in either manual or automated mode. We have found that microwave irradiation enhances reaction rates of difficult couplings, especially with cyclic β residues. β^3-Amino acid building blocks that contain nearly all of the proteinogenic side chains are commercially available in protected forms amenable for Fmoc-based solid-phase synthesis.

If the α/β^3 set does not produce oligomers with sufficient efficacy, then one can modify the best member(s) of this set by replacing flexible β^3 residues with ring-preorganized residues. Ideally, these replacements would be made in a way that preserves the chemical character of the original side chain. The Fmoc-protected derivative of ACPC is commercially available (the (*S*,*S*) configuration is necessary to mimic the right-handed α-helix formed by L-α-amino acid residues), and this building block can be used to replace hydrophobic β^3 residues. Unfortunately, protected APC derivatives cannot currently be purchased. Recent results suggest that engineering of intrahelical ion pairs into solvent-oriented positions can be useful as well (Johnson et al., 2012).

We continue to refine strategies for the mimicry of signal-bearing α-helices with unnatural peptidic oligomers. Topics of ongoing interest include the following: (1) Development of ring-rigidified β-amino acids that provide acidic or neutral-polar side chains (for replacing residues such as β^3-hGlu or β^3-hGln). (2) Exploration of β^2-amino acid residues, particularly at positions that tolerate $\alpha \rightarrow \beta^3$ replacement and contribute a side chain to the interface between the prototype α-helix and the partner protein. β^2-Amino acids have the side chain adjacent to the carbonyl; therefore, replacing a β^3 residue with the isomeric β^2 residue causes a subtle shift in the spatial position of the side chain. Unfortunately, few β^2-amino acid building blocks are commercially available at present. (3) Incorporation of γ-amino acid residues (Guo et al., 2009). This prospect is attractive because an αβααγα hexad contains the same number of backbone atoms as an ααααααα heptad, and we recently found that use of properly preorganized β and γ residues leads to formation of a helix very similar to the α-helix (Sawada & Gellman, 2011). (4) Extension of α/β- or α/β/γ-peptide design strategies to mimic poly-α-peptide conformations other than the α-helix so that this approach can be used to develop antagonists for a wider range of protein–protein interactions (Haase et al., 2012). Although improvements in protein surface mimicry should emerge from these ongoing efforts, current strategies for α-helix mimicry with α/β-peptides seem sufficiently mature for broad application.

ACKNOWLEDGMENTS

This work was supported by NIH grants GM056414 and GM061238. L. M. J. was supported in part by an NIH Chemistry-Biology Interface Training Grant (T32 GM008293).

REFERENCES

Appella, D. H., Barchi, J. J., Durell, S. R., & Gellman, S. H. (1999). Formation of short, stable helices in aqueous solution from β-amino acid hexamers. *Journal of the American Chemical Society, 121*, 2309–2310.

Appella, D. H., Christianson, L. A., Karle, I. L., Powell, D. R., & Gellman, S. H. (1999). Synthesis and characterization of trans-2-aminocyclohexanecarboxylic acid oligomers: An unnatural helical secondary structure and implications for β-peptide tertiary structure. *Journal of the American Chemical Society, 121*, 6206–6212.

Appella, D. H., Christianson, L. A., Klein, D. A., Powell, D. R., Huang, X. L., Barchi, J. J., et al. (1997). Residue-based control of helix shape in β-peptide oligomers. *Nature, 387*, 381–384.

Appella, D. H., Christianson, L. A., Klein, D. A., Richards, M. R., Powell, D. R., & Gellman, S. H. (1999). Synthesis and structural characterization of helix-forming β-peptides: Trans-2-aminocyclopentanecarboxylic acid oligomers. *Journal of the American Chemical Society, 121*, 7574–7581.

Arnold, U., Hinderaker, M. P., Nilsson, B. L., Huck, B. R., Gellman, S. H., & Raines, R. T. (2002). Protein prosthesis: A semisynthetic enzyme with a β-peptide reverse turn. *Journal of the American Chemical Society, 124*, 8522–8523.

Berlicki, L., Pilsl, L., Weber, E., Mandity, I. M., Cabrele, C., Martinek, T. A., et al. (2012). Unique α, β- and α, α, β, β-peptide foldamers based on cis-β-aminocyclopentanecarboxylic acid. *Angewandte Chemie, International Edition, 51*, 2208–2212.

Boersma, M. D., Haase, H. S., Peterson-Kaufman, K. J., Lee, E. F., Clarke, O. B., Colman, P. M., et al. (2012). Evaluation of diverse α/β-backbone patterns for functional α-helix mimicry: Analogues of the Bim BH3 domain. *Journal of the American Chemical Society, 134*, 315–323.

Boffelli, D., Compassi, S., Werder, M., Weber, F. E., Phillips, M. C., Schulthess, G., et al. (1997). The uptake of cholesterol at the small-intestinal brush border membrane is inhibited by apolipoproteins. *FEBS Letters, 411*, 7–11.

Chan, D. C., Fass, D., Berger, J. M., & Kim, P. S. (1997). Core structure of gp41 from the HIV envelope glycoprotein. *Cell, 89*, 263–273.

Chen, L., Willis, S. N., Wei, A., Smith, B. J., Fletcher, J. I., Hinds, M. G., et al. (2005). Differential targeting of prosurvival Bcl-2 proteins by their BH3-only ligands allows complementary apoptotic function. *Molecular Cell, 17*, 393–403.

Cheng, R. P., & DeGrado, W. F. (2001). De novo design of a monomeric helical β-peptide stabilized by electrostatic interactions. *Journal of the American Chemical Society, 123*, 5162–5163.

Cheng, R. P., Gellman, S. H., & DeGrado, W. F. (2001). β-peptides: From structure to function. *Chemical Reviews, 101*, 3219–3232.

Choi, S. H., Guzei, I. A., Spencer, L. C., & Gellman, S. H. (2008). Crystallographic characterization of helical secondary structures in α/β-peptides with 1:1 residue alternation. *Journal of the American Chemical Society, 130*, 6544–6550.

Choi, S. H., Guzei, I. A., Spencer, L. C., & Gellman, S. H. (2009). Crystallographic characterization of helical secondary structures in 2:1 and 1:2 α/β-peptides. *Journal of the American Chemical Society, 131*, 2917–2924.

Das, C., Berezovska, O., Diehl, T. S., Genet, C., Buldyrev, I., Tsai, J. Y., et al. (2003). Designed helical peptides inhibit an intramembrane protease. *Journal of the American Chemical Society, 125*, 11794–11795.

David, R., Gunther, R., Baurmann, L., Luhmann, T., Seebach, D., Hofmann, H. J., et al. (2008). Artificial chemokines: Combining chemistry and molecular biology for the elucidation of interleukin-8 functionality. *Journal of the American Chemical Society, 130*, 15311–15317.

De Pol, S., Zorn, C., Klein, C. D., Zerbe, O., & Reiser, O. (2004). Surprisingly stable helical conformations in α/β-peptides by incorporation of cis-β-aminocyclopropane carboxylic acids. *Angewandte Chemie, International Edition, 43*, 511–514.

Desai, P., Pfeiffer, S. S., & Boger, D. L. (2003). Synthesis of the chlorofusin cyclic peptide: Assignment of the asparagine stereochemistry. *Organic Letters, 5*, 5047–5050.

Doerksen, R. J., Chen, B., Liu, D. H., Tew, G. N., DeGrado, W. F., & Klein, M. L. (2004). Controlling the conformation of arylamides: Computational studies of intramolecular hydrogen bonds between amides and ethers or thioethers. *Chemistry, 10*, 5008–5016.

Dwyer, J. J., Wilson, K. L., Davison, D. K., Freel, S. A., Seedorff, J. E., Wring, S. A., et al. (2007). Design of helical, oligomeric HIV-1 fusion inhibitor peptides with potent activity against enfuvirtide-resistant virus. *Proceedings of the National Academy of Sciences of the United States of America, 104*, 12772–12777.

Eckert, D. M., & Kim, P. S. (2001). Mechanisms of viral membrane fusion and its inhibition. *Annual Review of Biochemistry, 70*, 777–810.

Fasan, R., Dias, R. L. A., Moehle, K., Zerbe, O., Vrijbloed, J. W., Obrecht, D., et al. (2004). Using a β-hairpin to mimic an alpha-helix: Cyclic peptidomimetic inhibitors of the p53-HDM2 protein-protein interaction. *Angewandte Chemie, International Edition, 43*, 2109–2112.

Frey, G., Rits-Volloch, S., Zhang, X. Q., Schooley, R. T., Chen, B., & Harrison, S. C. (2006). Small molecules that bind the inner core of gp41 and inhibit HIV envelope-mediated fusion. *Proceedings of the National Academy of Sciences of the United States of America, 103*, 13938–13943.

Gellman, S. H. (1998). Foldamers: A manifesto. *Accounts of Chemical Research, 31*, 173–180.

Godballe, T., Nilsson, L. L., Petersen, P. D., & Jenssen, H. (2011). Antimicrobial β-peptides and α-peptoids. *Chemical Biology & Drug Design, 77*, 107–116.

Guo, L., Chi, Y., Almeida, A. M., Guzei, I. A., Parker, B. K., & Gellman, S. H. (2009). Stereospecific synthesis of conformationally constrained γ-amino acids: New foldamer building blocks that support helical secondary structure. *Journal of the American Chemical Society, 131*, 16018–16020.

Haase, H. S., Peterson-Kaufman, K. J., Lan Levengood, S. K., Checco, J. W., Murphy, W. L., & Gellman, S. H. (2012). Extending foldamer design beyond α-helix mimicry: α/β-Peptide inhibitors of vascular endothelial growth factor signaling. *Journal of the American Chemical Society, 134*, 7652–7655.

Hamuro, Y., Schneider, J. P., & DeGrado, W. F. (1999). De novo design of antibacterial β-peptides. *Journal of the American Chemical Society, 121*, 12200–12201.

Hancock, R. E. W., & Sahl, H. G. (2006). Antimicrobial and host-defense peptides as new anti-infective therapeutic strategies. *Nature Biotechnology, 24*, 1551–1557.

Hara, T., Durell, S. R., Myers, M. C., & Appella, D. H. (2006). Probing the structural requirements of peptoids that inhibit HDM2-p53 interactions. *Journal of the American Chemical Society, 128*, 1995–2004.

Harbury, P. B., Zhang, T., Kim, P. S., & Alber, T. (1993). A switch between two-, three-, and four-stranded coiled coils in GCN4 leucine zipper mutants. *Science, 262*, 1401–1407.

Hayen, A., Schmitt, M. A., Ngassa, F. N., Thomasson, K. A., & Gellman, S. H. (2004). Two helical conformations from a single foldamer backbone: "Split personality" in short α/β-peptides. *Angewandte Chemie, International Edition, 43*, 505–510.

Horne, W. S., Boersma, M. D., Windsor, M. A., & Gellman, S. H. (2008). Sequence-based design of α/β-peptide foldamers that mimic BH3 domains. *Angewandte Chemie, International Edition, 47*, 2853–2856.

Horne, W. S., Johnson, L. M., Ketas, T. J., Klasse, P. J., Lu, M., Moore, J. P., et al. (2009). Structural and biological mimicry of protein surface recognition by α/β-peptide foldamers. *Proceedings of the National Academy of Sciences of the United States of America, 106*, 14751–14756.

Horne, W. S., Price, J. L., & Gellman, S. H. (2008). Interplay among side chain sequence, backbone composition, and residue rigidification in polypeptide folding and assembly. *Proceedings of the National Academy of Sciences of the United States of America, 105*, 9151–9156.

Imamura, Y., Watanabe, N., Umezawa, N., Iwatsubo, T., Kato, N., Tomita, T., et al. (2009). Inhibition of γ-secretase activity by helical β-peptide foldamers. *Journal of the American Chemical Society, 131*, 7353–7359.

Imbimbo, B. P., & Giardina, G. A. M. (2011). γ-Secretase inhibitors and modulators for the treatment of Alzheimer's disease: Disappointments and hopes. *Current Topics in Medicinal Chemistry, 11*, 1555–1570.

Johnson, L. M., Horne, W. S., & Gellman, S. H. (2011). Broad distribution of energetically important contacts across an extended protein interface. *Journal of the American Chemical Society, 133*, 10038–10041.

Johnson, L. M., Mortenson, D. E., Yun, H. G., Horne, W. S., Ketas, T. J., Lu, M., et al. (2012). Enhancement of α-helix mimicry by an α/β-peptide foldamer via incorporation of a dense ionic side-chain array. *Journal of the American Chemical Society, 134*, 7317–7320.

Kenawy, E. R., Worley, S. D., & Broughton, R. (2007). The chemistry and applications of antimicrobial polymers: A state-of-the-art review. *Biomacromolecules, 8*, 1359–1384.

Korendovych, I. V., Shandler, S. J., Montalvo, G. L., & DeGrado, W. F. (2011). Environment- and sequence-dependence of helical type in membrane-spanning peptides composed of β^3-amino acids. *Organic Letters, 13*, 3474–3477.

Kritzer, J. A., Lear, J. D., Hodsdon, M. E., & Schepartz, A. (2004). Helical β-peptide inhibitors of the p53-hDM2 interaction. *Journal of the American Chemical Society, 126*, 9468–9469.

Kussie, P. H., Gorina, S., Marechal, V., Elenbaas, B., Moreau, J., Levine, A. J., et al. (1996). Structure of the MDM2 oncoprotein bound to the p53 tumor suppressor transactivation domain. *Science, 274*, 948–953.

Lee, V. H. (1988). Enzymatic barriers to peptide and protein absorption. *Critical Reviews in Therapeutic Drug Carrier Systems, 5*, 69–97.

Lee, M. R., Raguse, T. L., Schinnerl, M., Pomerantz, W. C., Wang, X., Wipf, P., et al. (2007). Origins of the high 14-helix propensity of cyclohexyl-rigidified residues in β-peptides. *Organic Letters, 9*, 1801–1804.

Lee, E. F., Smith, B. J., Horne, W. S., Mayer, K. N., Evangelista, M., Colman, P. M., et al. (2011). Structural basis of Bcl-xL recognition by a BH3-mimetic α/β-peptide generated by sequence-based design. *Chembiochem, 12*, 2025–2032.

Lee, H. S., Syud, F. A., Wang, X. F., & Gellman, S. H. (2001). Diversity in short beta-peptide 12-helices: High-resolution structural analysis in aqueous solution of a hexamer containing sulfonylated pyrrolidine residues. *Journal of the American Chemical Society, 123*, 7721–7722.

Lessene, G., Czabotar, P. E., & Colman, P. M. (2008). BCL-2 family antagonists for cancer therapy. *Nature Reviews. Drug Discovery, 7*, 989–1000.

Martinek, T. A., & Fulop, F. (2003). Side-chain control of β-peptide secondary structures—Design principles. *European Journal of Biochemistry, 270*, 3657–3666.

Matthews, T., Salgo, M., Greenberg, M., Chung, J., DeMasi, R., & Bolognesi, D. (2004). Enfuvirtide: The first therapy to inhibit the entry of HIV-1 into host CD4 lymphocytes. *Nature Reviews. Drug Discovery, 3*, 215–225.

Mowery, B. P., Lee, S. E., Kissounko, D. A., Epand, R. F., Epand, R. M., Weisblum, B., et al. (2007). Mimicry of antimicrobial host-defense peptides by random copolymers. *Journal of the American Chemical Society, 129*, 15474–15476.

Petros, A. M., Olejniczak, E. T., & Fesik, S. W. (2004). Structural biology of the Bcl-2 family of proteins. *Biochimica et Biophysica Acta*, *1644*, 83–94.

Plante, J. P., Burnley, T., Malkova, B., Webb, M. E., Warriner, S. L., Edwards, T. A., et al. (2009). Oligobenzamide proteomimetic inhibitors of the p53-hDM2 protein-protein interaction. *Chemical Communications*, 5091–5093.

Porter, E. A., Wang, X. F., Lee, H. S., Weisblum, B., & Gellman, S. H. (2000). Antibiotics—Non-haemolytic β-amino-acid oligomers. *Nature*, *404*, 565.

Price, J. L., Horne, W. S., & Gellman, S. H. (2010). Structural consequences of β-amino acid preorganization in a self-assembling α/β-peptide: Fundamental studies of foldameric helix bundles. *Journal of the American Chemical Society*, *132*, 12378–12387.

Raguse, T. L., Porter, E. A., Weisblum, B., & Gellman, S. H. (2002). Structure-activity studies of 14-helical antimicrobial β-peptides: Probing the relationship between conformational stability and antimicrobial potency. *Journal of the American Chemical Society*, *124*, 12774–12785.

Sadowsky, J. D., Fairlie, W. D., Hadley, E. B., Lee, H. S., Umezawa, N., Nikolovska-Coleska, Z., et al. (2007). (α/β+α)-Peptide antagonists of BH3 domain/Bcl-xL recognition: Toward general strategies for foldamer-based inhibition of protein-protein interactions. *Journal of the American Chemical Society*, *129*, 139–154.

Sadowsky, J. D., Schmitt, M. A., Lee, H. S., Umezawa, N., Wang, S., Tomita, Y., et al. (2005). Chimeric (α/β+α)-peptide ligands for the BH3-recognition cleft of Bcl-xL: Critical role of the molecular scaffold in protein surface recognition. *Journal of the American Chemical Society*, *127*, 11966–11968.

Sattler, M., Liang, H., Nettesheim, D., Meadows, R. P., Harlan, J. E., Eberstadt, M., et al. (1997). Structure of Bcl-x_L-Bak peptide complex: Recognition between regulators of apoptosis. *Science*, *275*, 983–986.

Sawada, T., & Gellman, S. H. (2011). Structural mimicry of the α-helix in aqueous solution with an isoatomic α/β/γ-peptide backbone. *Journal of the American Chemical Society*, *133*, 7336–7339.

Schmitt, M. A., Weisblum, B., & Gellman, S. H. (2004). Unexpected relationships between structure and function in α, β-peptides: Antimicrobial foldamers with heterogeneous backbones. *Journal of the American Chemical Society*, *126*, 6848–6849.

Schmitt, M. A., Weisblum, B., & Gellman, S. H. (2007). Interplay among folding, sequence, and lipophilicity in the antibacterial and hemolytic activities of α/β-peptides. *Journal of the American Chemical Society*, *129*, 417–428.

Seebach, D., Abele, S., Schreiber, J. V., Martinoni, B., Nussbaum, A. K., Schild, H., et al. (1998). Biological and pharmacokinetic studies with β-peptides. *Chimia*, *52*, 734–739.

Seebach, D., & Matthews, J. L. (1997). β-peptides: A surprise at every turn. *Chemical Communications*, 2015–2022.

Seebach, D., Overhand, M., Kuhnle, F. N. M., Martinoni, B., Oberer, L., Hommel, U., et al. (1996). β-peptides: Synthesis by Arndt-Eistert homologation with concomitant peptide coupling. Structure determination by NMR and CD spectroscopy and by X-ray crystallography. Helical secondary structure of a β-hexapeptide in solution and its stability towards pepsin. *Helvetica Chimica Acta*, *79*, 913–941.

Shandler, S. J., Korendovych, I. V., Moore, D. T., Smith-Dupont, K. B., Streu, C. N., Litvinov, R. I., et al. (2011). Computational design of a beta-peptide that targets transmembrane helices. *Journal of the American Chemical Society*, *133*, 12378–12381.

Sodroski, J. G. (1999). HIV-1 entry inhibitors in the side pocket. *Cell*, *99*, 243–246.

Steer, D. L., Lew, R. A., Perlmutter, P., Smith, A. I., & Aguilar, M. I. (2002). β-amino acids: Versatile peptidomimetics. *Current Medicinal Chemistry*, *9*, 811–822.

Vassilev, L. T., Vu, B. T., Graves, B., Carvajal, D., Podlaski, F., Filipovic, Z., et al. (2004). In vivo activation of the p53 pathway by small-molecule antagonists of MDM2. *Science*, *303*, 844–848.

Vaz, E., Pomerantz, W. C., Geyer, M., Gellman, S. H., & Brunsveld, L. (2008). Comparison of design strategies for promotion of β-peptide 14-helix stability in water. *Chembiochem, 9*, 2254–2259.

Wade, D., Boman, A., Wahlin, B., Drain, C. M., Andreu, D., Boman, H. G., et al. (1990). All-D amino acid-containing channel-forming antibiotic peptides. *Proceedings of the National Academy of Sciences of the United States of America, 87*, 4761–4765.

Werder, M., Hauser, H., Abele, S., & Seebach, D. (1999). β-peptides as inhibitors of small-intestinal cholesterol and fat absorption. *Helvetica Chimica Acta, 82*, 1774–1783.

Wu, L. C., & Kim, P. S. (1997). Hydrophobic sequence minimization of the α-lactalbumin molten globule. *Proceedings of the National Academy of Sciences of the United States of America, 94*, 14314–14319.

Yin, H., Lee, G. I., Park, H. S., Payne, G. A., Rodriguez, J. M., Sebti, S. M., et al. (2005). Terphenyl-based helical mimetics that disrupt the p53/HDM2 interaction. *Angewandte Chemie, International Edition, 44*, 2704–2707.

Zhao, J. H., Wang, M. J., Chen, J., Luo, A. P., Wang, X. Q., Wu, M., et al. (2002). The initial evaluation of non-peptidic small-molecule HDM2 inhibitors based on p53-HDM2 complex structure. *Cancer Letters, 183*, 69–77.

AUTHOR INDEX

Note: Page numbers followed by "*f*" indicate figures, "*t*" indicate tables, and "*np*" indicate footnotes.

B

C

D

H

I

L

N

O

Q

R

S

U

V

W

X

Y

Z

SUBJECT INDEX

Note: Page numbers followed by "*f*" indicate figures, and "*t*" indicate tables.

D

E

I

K

L

M

N

Q

R

S

T

U

V

W

Y

Z

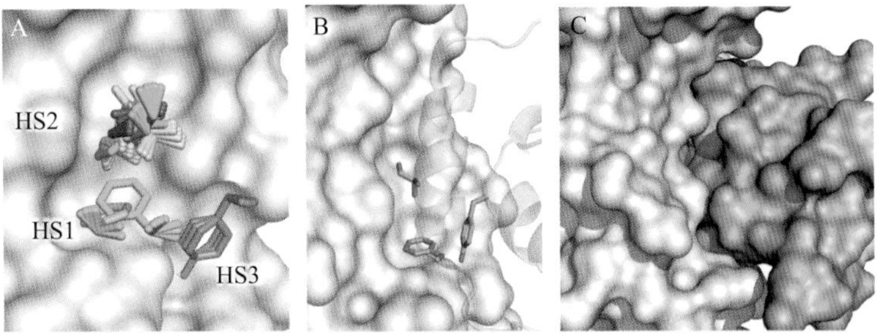

Timothy A. Whitehead *et al.*, Figure 1.1 The computational design procedure realizes three features of natural protein–protein interactions: cores of high-affinity interactions with the target surface (A), favorable interactions among core residues (B), and high shape complementarity (C). (A) Three hotspot residue libraries (HS1, HS2, and HS3) were computed to form the idealized core of the interaction with the influenza hemagglutinin (HA) surface (gray). HS1 comprises two major configurations for a Phe aromatic ring and is supported by HS2, which contains the hydrophobic residues Phe, Leu, Ile, Met, and Val (green, purple, navy blue, cyan, and light brown, respectively), and HS3 comprises Tyr conformations. In the specific case of design of HA binders, the geometric constraint on HS2 is laxer than on the other hotspot positions and many different residues can be accommodated there. Residues from each hotspot-residue library interact favorably both with the target HA surface and the other hotspot-residue libraries, recapitulating two features common to many natural complexes: a core of highly optimized interactions with the target and internally stabilized contacts between key sidechains. (B) Cocrystal structure of HB80 and the HA surface. The hotspot residues are shown in dark green, realizing one of the energetically favorable combinations seen in panel a (comprising a Phe, Ile, and Tyr, for HS1, HS2, and HS3, respectively). (C) Common to many natural protein interactions, the surfaces of the designed and target proteins fit together snugly in a high shape complementary configuration. All molecular graphics were generated using PyMol (DeLano, 2002).

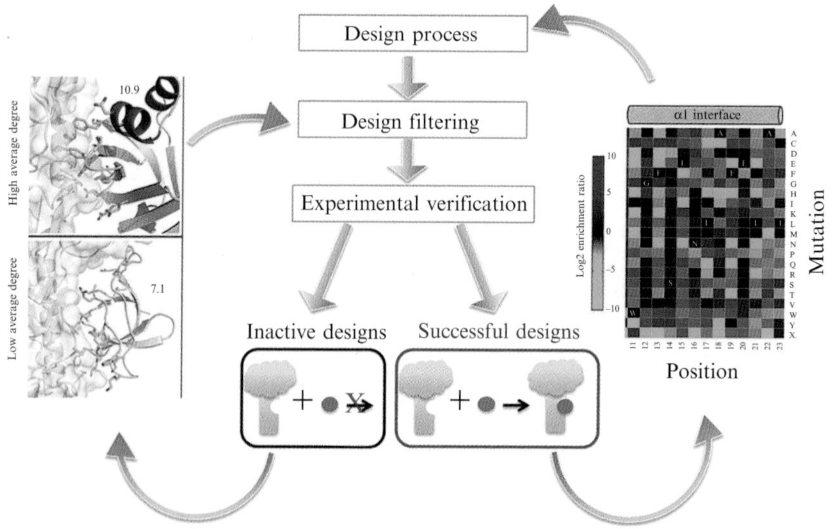

Timothy A. Whitehead *et al.*, Figure 1.3 A schematic for how computation and experiment have been integrated to probe the physical basis for molecular recognition and to generate novel proteins of biomedical potential. Experimental screening of designs using yeast cell-surface display leads to their classification as active or inactive. Inactive designs can be compared to native protein interfaces (arrow pointing left): metrics that discriminate inactive designs from natural proteins can be used to formulate automated computational filters to prune unpromising designs in future design efforts and highlight areas for future methodological improvements in design. By contrast, successful designs (arrow pointing right) can be experimentally probed for detailed structure–function relationships using newly developed next generation sequencing technologies. Here, a plot showing enrichment ratio (a proxy for affinity) as a function of point mutation is shown for the interface stretch of one of the designs. This wealth of information can be used to identify limiting approximations in the energetic potential underlying the design calculations. By identifying mutations that clearly improve binding, this dataset can also be used to program high affinity and specificity binders from these initial designs that could subsequently be used as therapeutics. *The left-hand side of panel was reproduced with permission from Fleishman, Whitehead, Strauch, et al. (2011), and right-hand side with permission from Whitehead et al. (2012).*

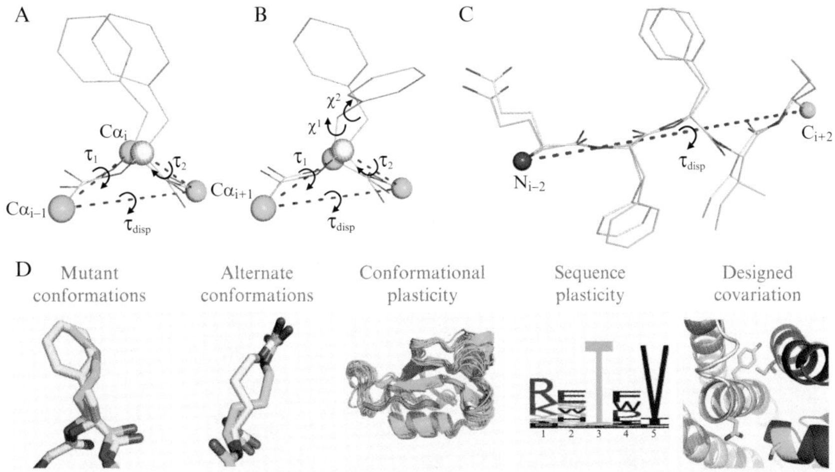

Noah Ollikainen *et al.*, Figure 4.1 The backrub move and its applications in Rosetta. (A) The Richardson group originally described the "Backrub" move as a rotation around the $C\alpha_{i-1}$ and $C\alpha_{i+1}$ axis by τ_{disp}, along with simultaneous peptide plane rotations (τ_1 and τ_2), without disturbing other surrounding atom coordinates. (B) By changing the position of the $C\alpha_i$–$C\beta_i$ bond vector, this move can couple side-chain rotameric changes with small local backbone adjustments. (C) In Rosetta, the generalized backrub move is a single rotation that can also include longer intervals and other backbone atom types as pivots for the rotations. (D) Implementing backrubs as a Monte Carlo move in Rosetta enables a variety of flexible backbone prediction and design applications that are described in this chapter: predicting mutant conformations (Fig. 4.2), modeling alternative conformations (Figs. 4.3 and 4.4), coupling conformational and sequence plasticity (Fig. 4.5), and designing amino acid covariation at protein interfaces (Fig. 4.6).

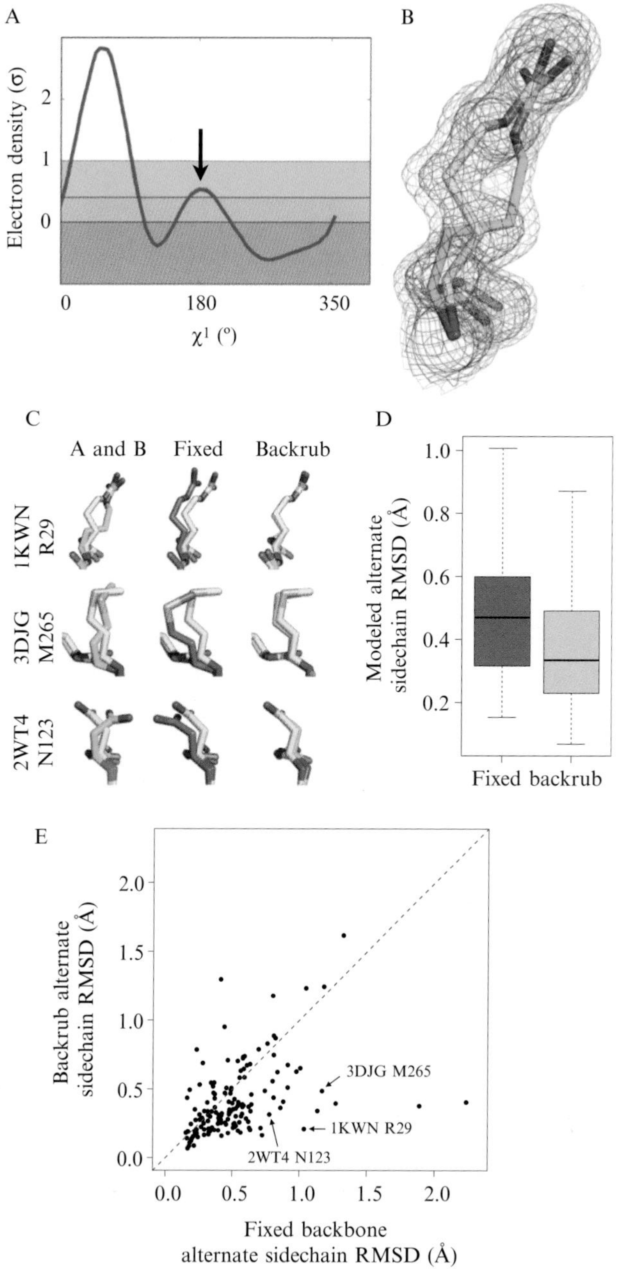

Noah Ollikainen *et al.*, Figure 4.3 *Continued*

Noah Ollikainen *et al.*, Figure 4.3 Backrub sampling improves the prediction of alternative side-chain conformations observed in protein crystal structures. (A) Electron density sampling by Ringer around the $\chi 1$ of R29 from PDB 1KWN reveals high electron density for the primary conformation 60° and a secondary peak (indicated by the black arrow), above the 0.3σ threshold that enriches for alternative conformations (shaded green area), near the 180° rotameric bin. (B) 2mFo-DFc electron density surrounding R29 from PDB 1KWN contoured at 1σ (blue mesh) and 0.3σ (cyan mesh). The original PDB model is shown in yellow, with an alternative conformation identified by Ringer and modeled with qFit at 25% occupancy shown in green. (C) Example predictions with Rosetta, with the indicated PDB codes and residues. Sampling of side-chain conformations (yellow) starting from alternative conformations (green, right) is improved by flexible backbone backrub moves (cyan, right) compared to fixed backbone side chain only sampling (magenta, center). (D) The overall quantification of the results, showing that backrub sampling increases identification of discrete side-chain local minima modeled as alternative conformations by qFit compared to fixed backbone models over a set of 152 side chains with solvent accessibility less than 30%. The median RMSD decreases from 0.47 to 0.33. Box plots are shown as in Fig. 4.2C. (E) A scatter plot representation of the data in (D) shows that backrub leads to large improvements compared to fixed backbone for many alternative conformation predictions.

A

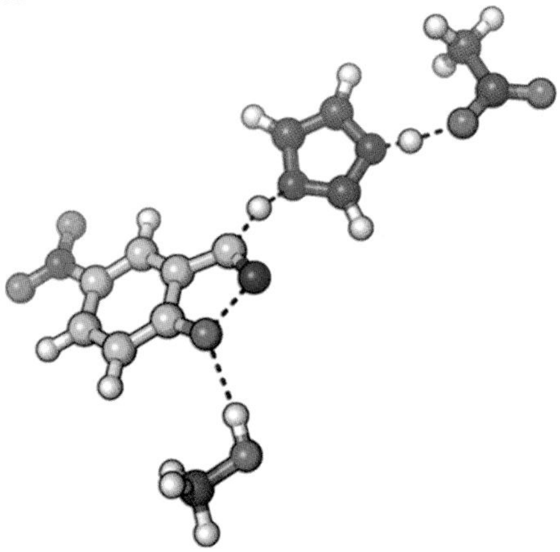

B

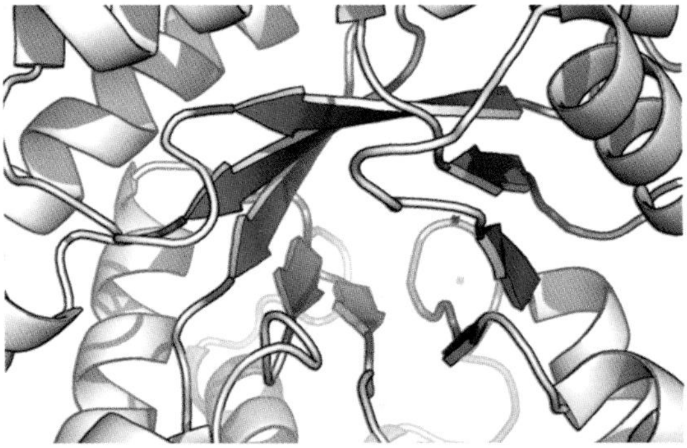

Gert Kiss *et al.*, Figure 7.1 *Continued*

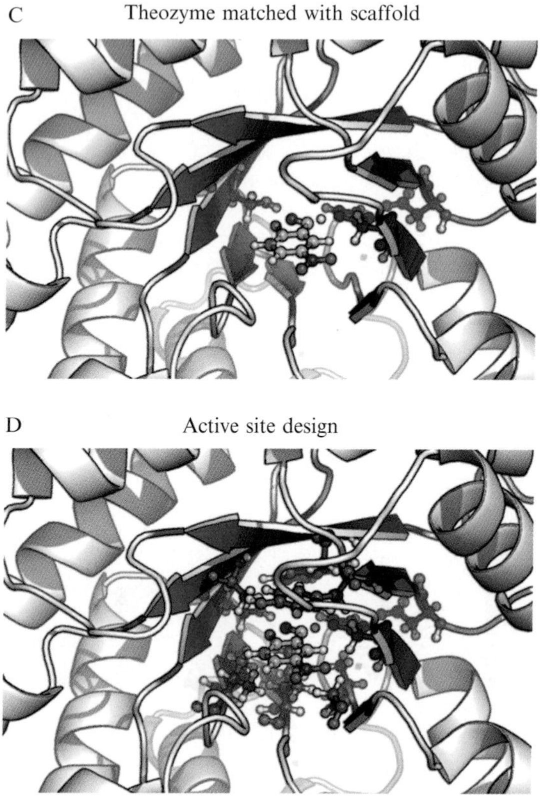

Gert Kiss *et al.*, Figure 7.1 Key steps in the inside-out enzyme design protocol for the Kemp elimination reaction. *Reprint from Kiss et al. (2011).*

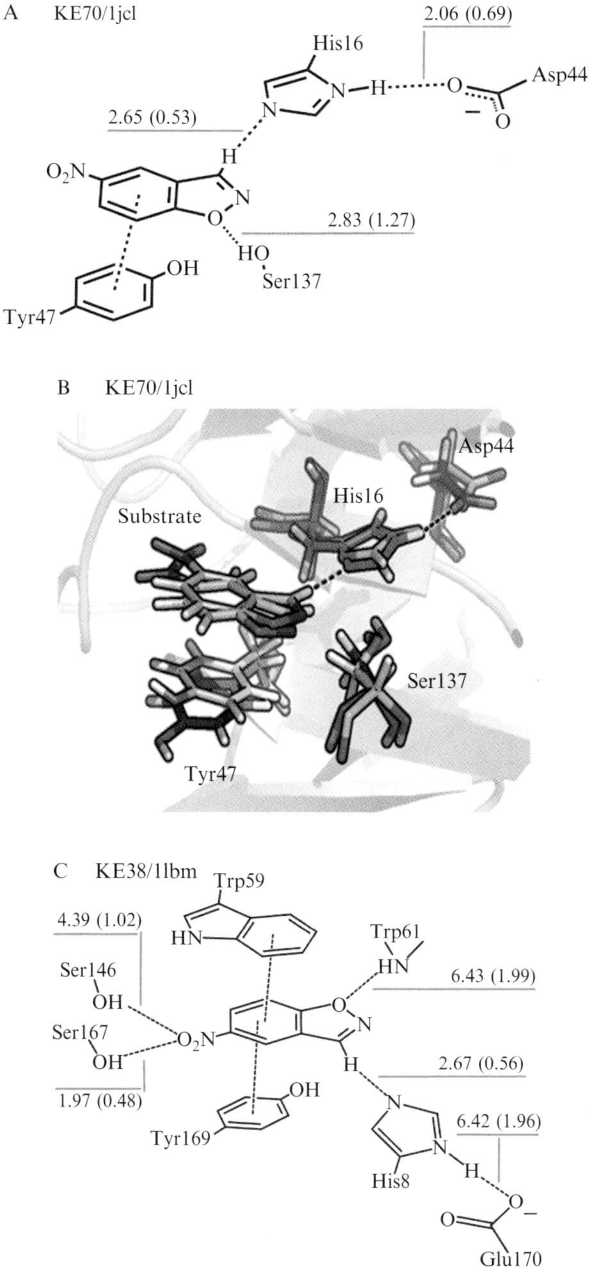

Gert Kiss *et al.*, Figure 7.4 *Continued*

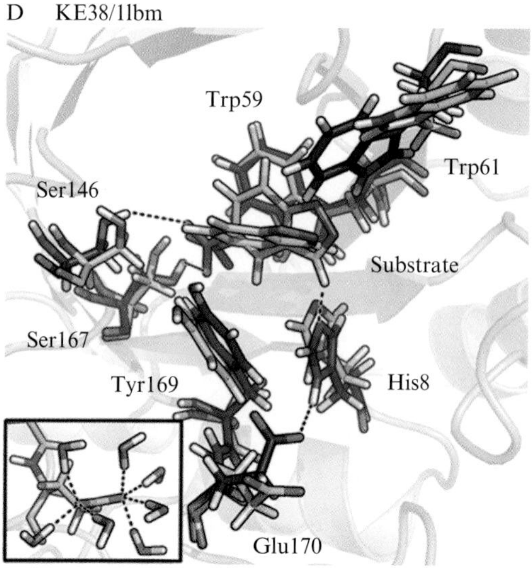

Gert Kiss *et al.*, Figure 7.4 Design vs. MD. Schematic representation of the catalytic unit (A, C) and representative MD geometry (light blue) over Rosetta design geometry (black with orange substrate) (B, D). Bond labels in (A) and (C) are maxima of distance distributions with full widths at half-maximum in parentheses. All values are in Å. The backbone RMSD of the catalytic unit of KE70 and KE38 is 0.57 and 0.76 Å; the side chain RMSD is 0.95 and 2.24 Å, respectively. The inset in (D) shows Glu170 in direct contact with seven water molecules on average. *Reprint from Kiss et al. (2010).*

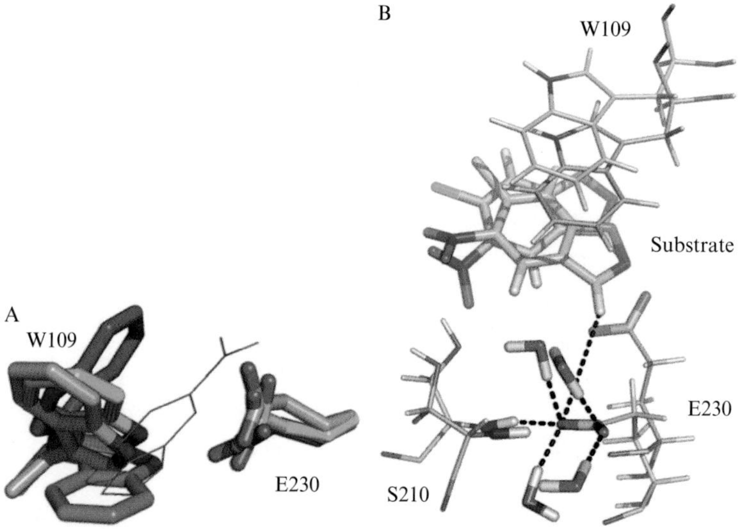

Gert Kiss *et al.*, Figure 7.7 Trp109 and Glu230 rotamers in the KE59 structures. (A) Trp109 and Glu230 rotamers in the KE59 model (magenta) and in the apo structures of R1-7/10H (pink), R5-11/5F (green), and R8-2/7A (violet). The 5-nitrobenzisoxazole substrate (magenta lines) is overlayed from the KE59 model. (B) The predominant conformation of Glu230 observed in MD simulations (blue) vs. the conformation observed in the crystal structures (green), shown here for R13-3/11H variant. *Reprint from Khersonsky et al. (2012).*

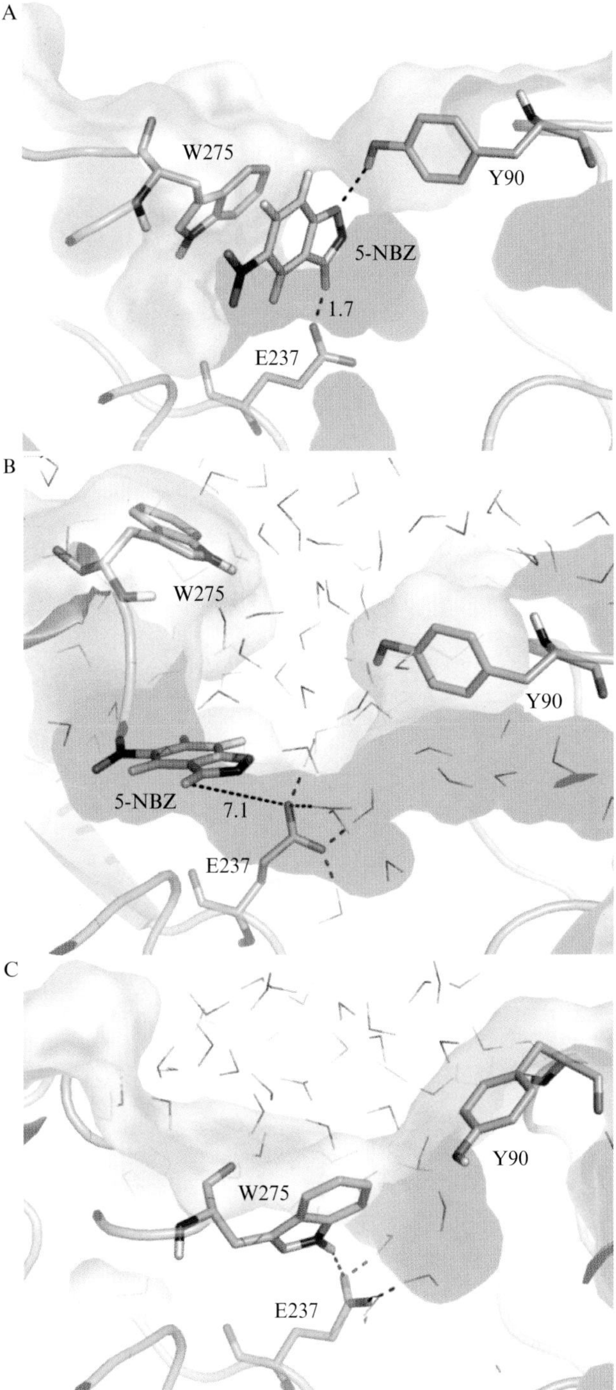

Gert Kiss *et al.*, Figure 7.10 The active site of HG-1 (A) prior to MD, (B) after MD on the 5-NBZ:HG-1 complex, and (C) after MD in the absence of 5-NBZ. All distance values are in Å.

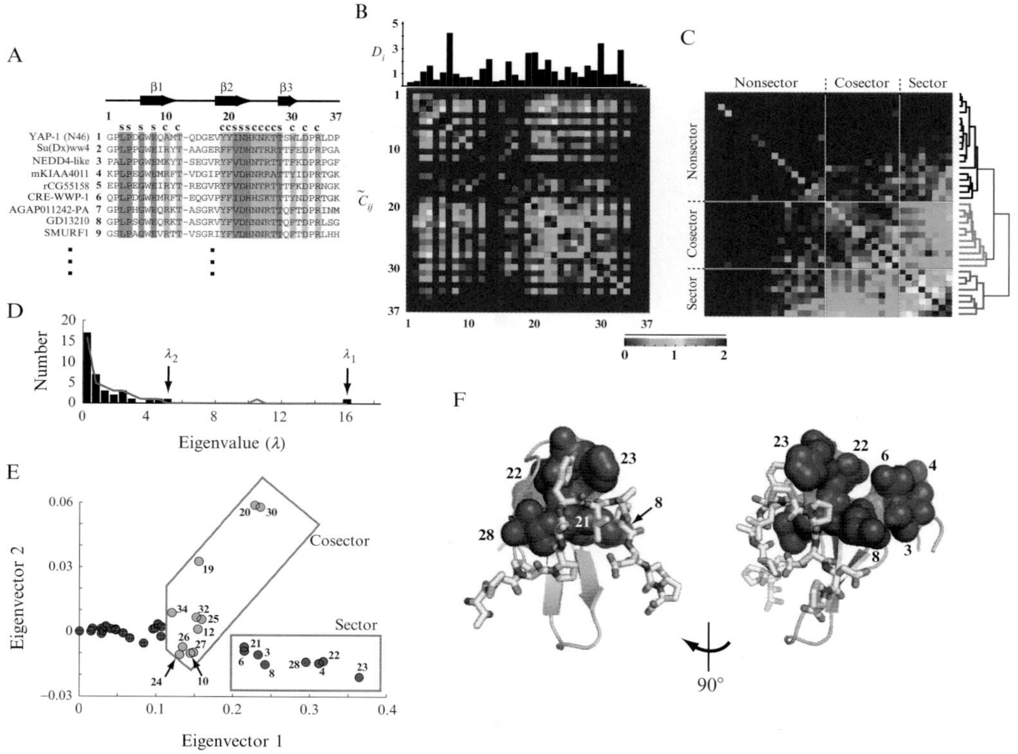

Kimberly A. Reynolds *et al.*, Figure 10.1 Statistical coupling analysis (SCA). (A) A portion of the alignment for the WW domain family. Sector (s), cosector (c), or nonsector (unmarked) positions (defined below) show no obvious arrangement in primary or secondary structure. (B) The site-independent conservation (D_i, bar graph) and the SCA matrix of coevolution between all pairs of amino acids $\left(\tilde{C}_{ij}\right)$. Values in the matrix are as indicated by the color bar. (C) Clustering in the matrix reveals three main groups of residues: sector, cosector, and nonsector. (D) Comparison of the eigenspectrum of the SCA matrix generated from the natural alignment (bars) to eigenspectra for randomized versions of the alignment (line) indicates that just the top two eigenvalues, λ_1 and λ_2, are distinguished from noise. (E) The corresponding eigenvectors reveal the positions that contribute the most to the top eigenvalues and define the sector positions (red) and cosector positions (blue). (F) The sector shown as red space-filling spheres on a representative WW domain structure (PDB ID: 2LAW, gray cartoon) in complex with a peptide ligand (stick bonds).

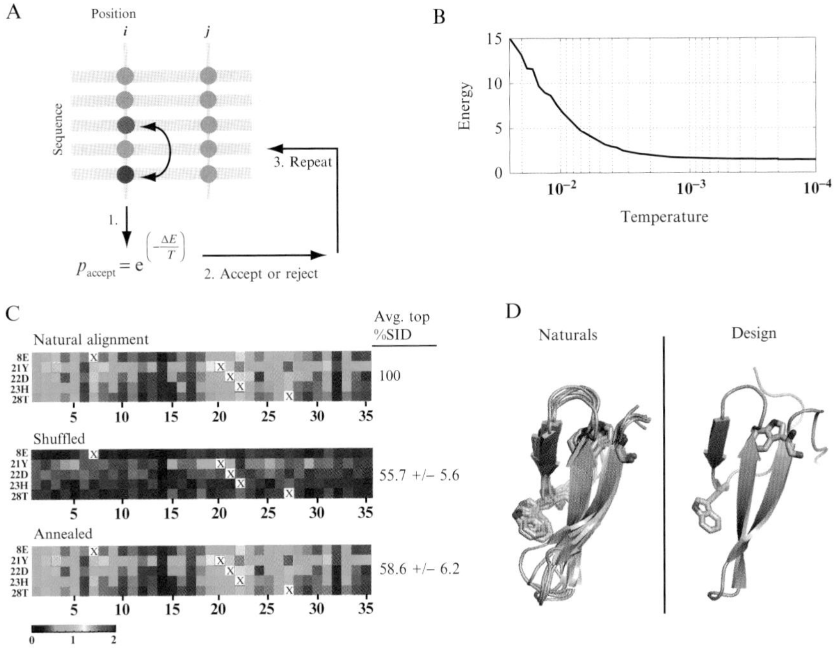

Kimberly A. Reynolds *et al.*, Figure 10.3 Evolution-based design of the WW domain. (A) The Monte Carlo design process. Each iteration involves two steps: (1) a swap of amino acids between two randomly chosen sequences at one randomly chosen position and (2) a decision to either accept or reject the swap based on the difference in the objective function (ΔE, see text) and a computational "temperature" (T). If $\Delta E > 0$, the swap is accepted with a probability determined by the Boltzmann distribution. (B) A MCSA trajectory for the WW domain family. The process starts with a high temperature and exponentially cools the MSA, converging toward a minimum for the objective function. (C) In Socolich et al. (2005), the objective function involved correlations for five sector positions. This portion of the SCA matrix (with self-correlations blanked) is shown for the natural WW alignment, the vertically shuffled (high temperature) alignment, and the annealed alignment. At right is indicated the average sequence identity ($\pm$ SD) to the closest sequence in the natural WW domain MSA. (D) Comparison of experimental structures for a collection of natural WW domains and one of the synthetic WW domains. The eponymous tryptophan residues are indicated in stick bonds.

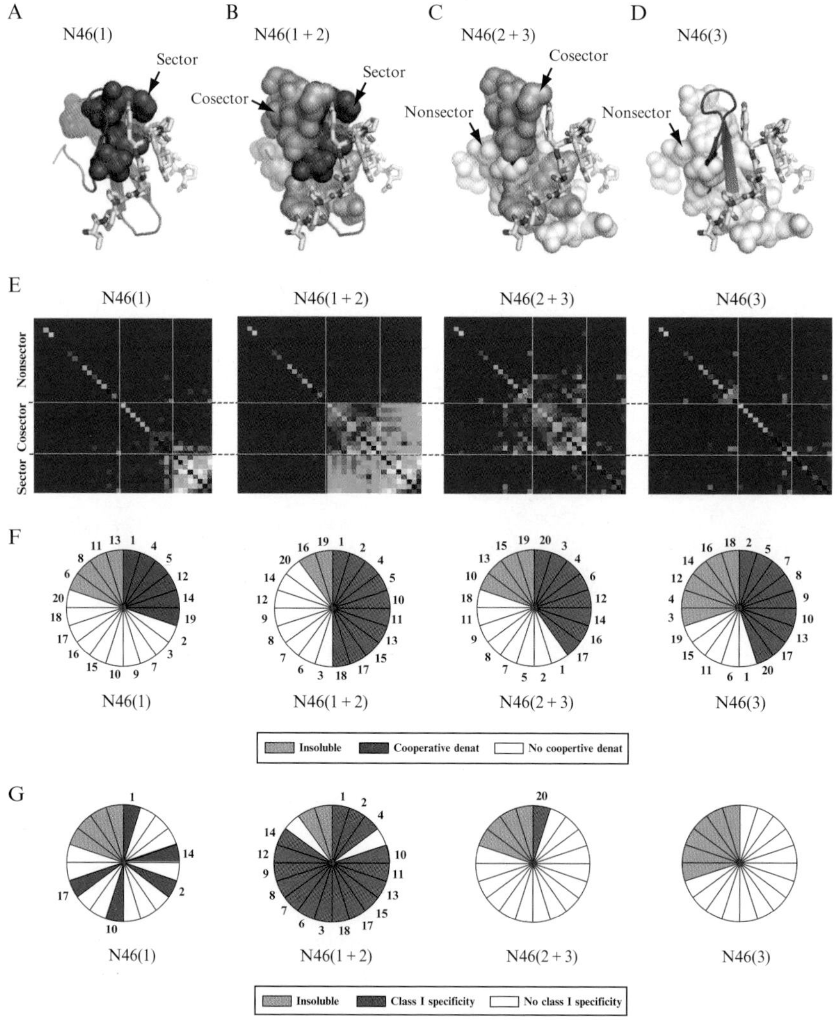

Kimberly A. Reynolds *et al.*, Figure 10.4 SCA-based design of shuffled WW domains. (A–D) In each panel, the group of residues for which couplings were preserved (non-shuffled positions) are shown in space-filling spheres on the structure of the "N46" WW domain in complex with a Class I peptide ligand (PDB 2LAW). Sector, cosector, and nonsector positions are labeled. (E) SCA matrices for each set of shuffled N46 sequences, visually illustrating the correlations that are scrambled. (F–G) Ordered pie charts showing the outcome of thermal denaturation (F) and Class I peptide binding specificity (G) experiments for each of the four groups of synthetic WW domains. The order of slices is the same in the two panels to facilitate comparison.

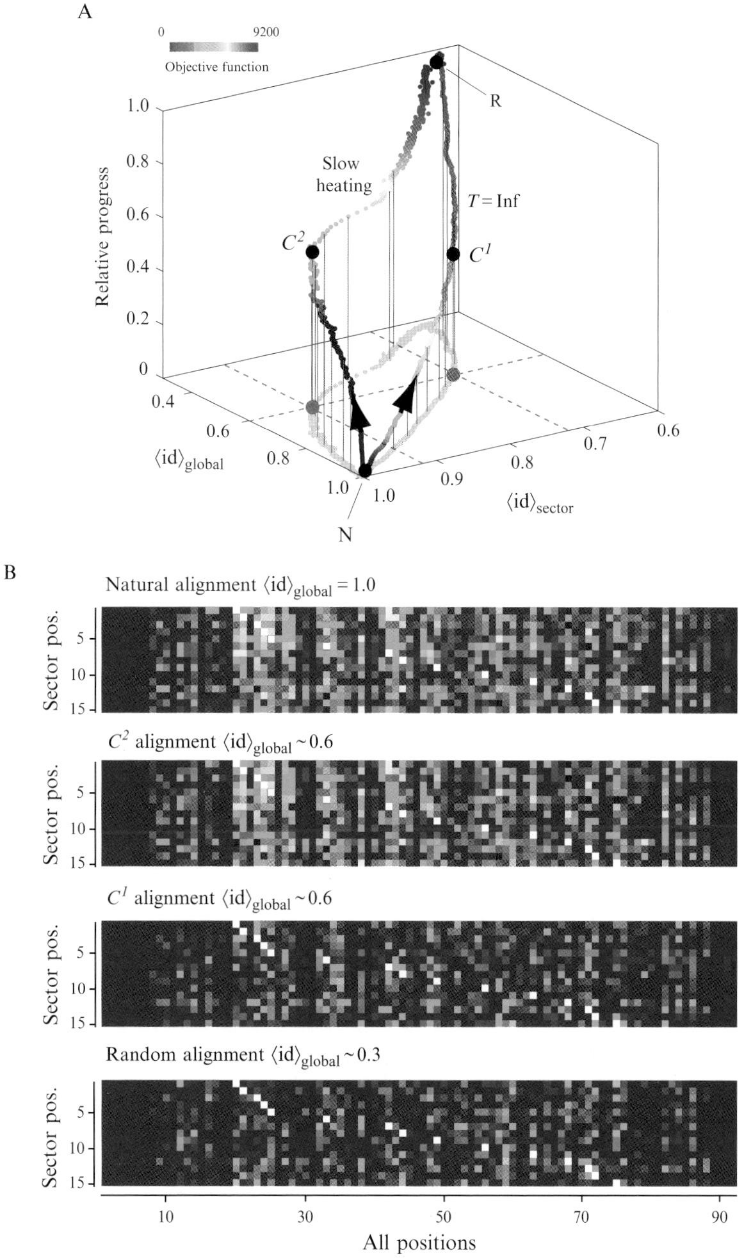

Kimberly A. Reynolds *et al.*, Figure 10.5 *Continued*

Kimberly A. Reynolds *et al.*, Figure 10.5 MCSH trajectories for the PDZ domain family. (A) A plot mapping the progress of two different heating trajectories against the average "top-hit" sequence identities of designed sequences calculated for the full-length sequence ($\langle id \rangle_{global}$) or for just sector positions ($\langle id \rangle_{sector}$). In the $T = \text{Inf}$ trajectory, each position is allowed to mutate within its conservation pattern without regard to correlations, and global and sector identity drop together. In the slow heating trajectory, the temperature is gradually increased to "melt out" couplings between positions in an order that depends on the strength and collective nature of the correlations. The value of the objective function along the trajectories is indicated by the color bar, and four points are marked for reference with panel B: N, the natural MSA; R, the fully randomized, vertically shuffled MSA; and C^1 and C^2, two intermediary points that share the same global sequence divergence but differ significantly in sector divergence. (B) Subset of the SCA matrix $\tilde{C}_{ij}$ for 15 sector positions in the PDZ family to illustrate the property of the heating trajectories. Despite identical global sequence divergence, C^1 sequences show a pattern of correlations that nearly approaches the fully randomized case while C^2 sequences show correlations that are nearly the same as for the natural MSA. Experimental analysis of C^2 and C^1 sequences or more generally sequences drawn from both trajectories represent a systematic investigation of how properties of natural proteins are stored in the pattern of correlations.

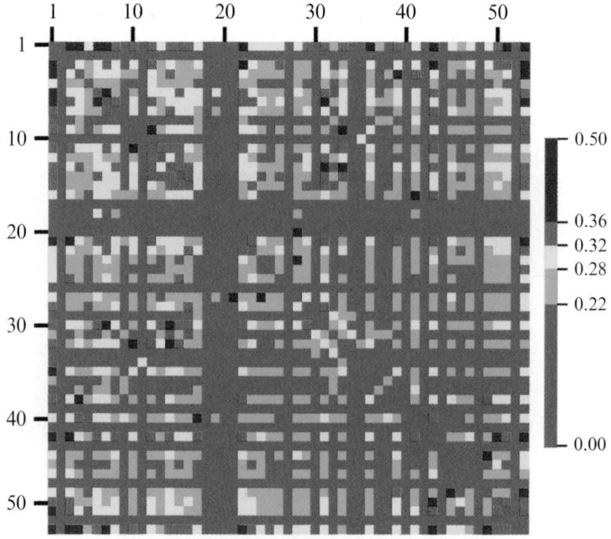

Venuka Durani and Thomas J. Magliery, Figure 11.6 A heat map representing the MI data of BPTI: The position numbers as per the MSA are specified on the left and top of the heat map. The color scale is shown on the right side.

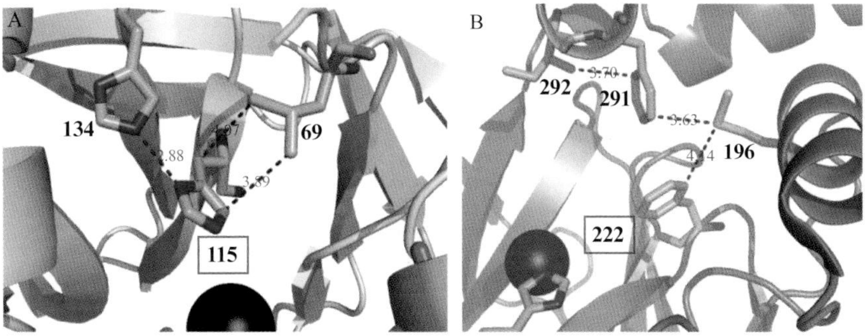

Moshe Goldsmith and Dan S. Tawfik, Figure 12.4 Neighbor joining and revisiting key positions in library design. Example taken from the directed evolution of serum paraoxonase-1 (Goldsmith et al., 2012; Gupta et al., 2011). Two key active-site positions that are close to PON1's catalytic calcium (gray sphere) and are in direct contact with substrates were explored: His115 (panel A) and Phe222 (panel B). The selection of advantageous mutations in these two first-shell positions led us to explore adjacent first- and second-shell residues (e.g., 134, 69, 292, 291, and 196) in subsequent rounds. Positions 115, 69, and 222 were then reexplored for their optimal compositions.

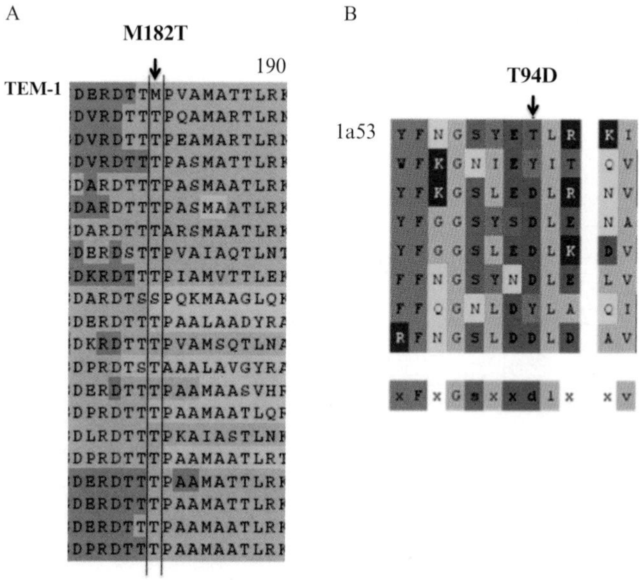

Moshe Goldsmith and Dan S. Tawfik, Figure 12.6 The identification of ancestral/consensus mutations. (A) A segment from the alignment of the TEM-1 β-lactamase family is displayed (for complete alignment and characterization see Bershtein, Goldin, & Tawfik, 2008). Highlighted is position 182 that differs in the wild-type gene (Met, top sequence) from nearly all other family members. The mutation M182T restores the wild-type sequence both to its family consensus and to its predicted family ancestor sequence at this position. The incorporation of this mutation, and of other consensus mutations, into wild-type TEM-1 increases its stability and its evolvability (Bershtein et al., 2008). (B) A segment from the alignment of IGPS from *S. Solfataricus* used for the computational design of enzyme KE59 (Khersonsky et al., 2012) and of its related family members. Highlighted is position 94 (Thr, top sequence) in which the template and the designed KE59 carry a Thr, whereas the consensus is Asp. The mutation Thr94Asp was spiked into random mutation libraries of KE59 and was identified in every active variant isolated from these libraries, thus indicating its essential compensatory role.

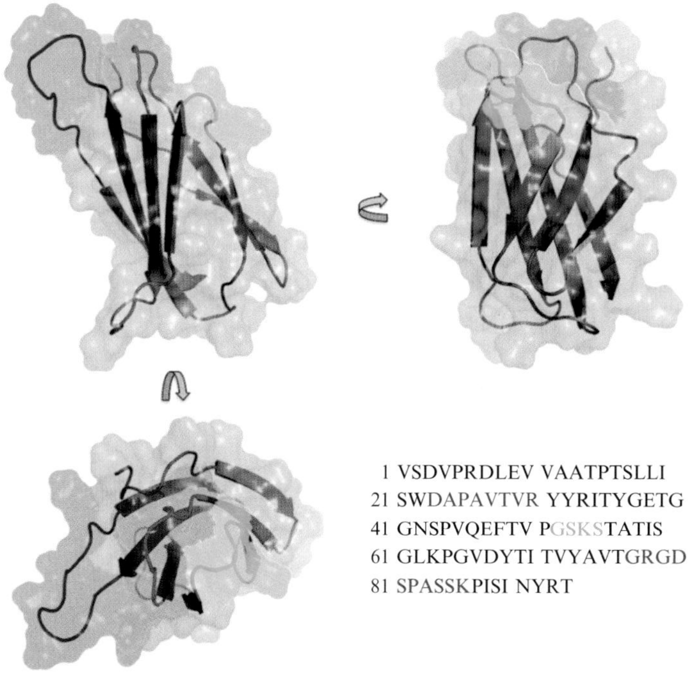

Tiffany F. Chen *et al.*, Figure 14.3 Fibronectin domain. The solution structure (PDB ID: 1TTG) of Fn3 is presented with 90° rotations. The wild-type sequence is indicated. The BC (red), DE (green), and FG (blue) loops are highlighted.

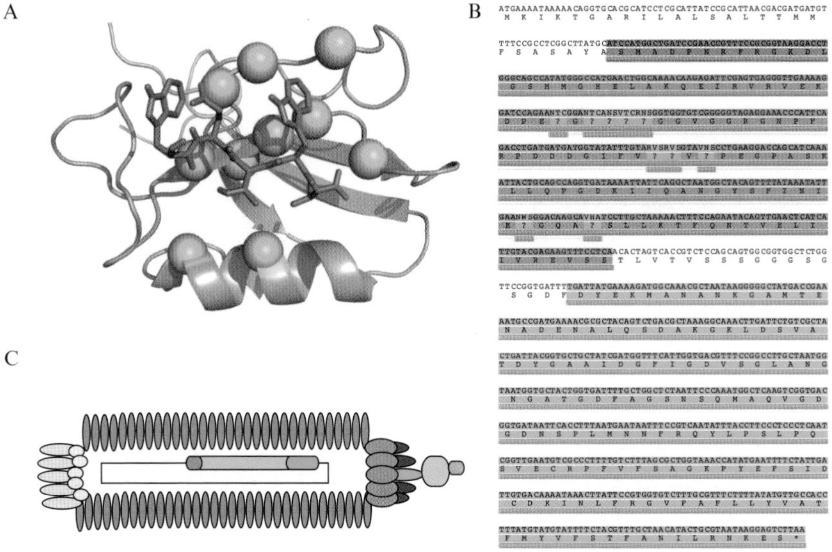

Megan E. McLaughlin and Sachdev S. Sidhu, Figure 15.1 Design and display of a combinatorial library of PDZ variants on phage. (A) Structure of the Erbin PDZ domain (PDB:1N7T) (*gray cartoon*) bound to C-terminal peptide $\mathrm{WETWV_{COOH}}$ (*magenta stick*). Ten domain residues predicted to determine binding specificity were diversified to generate a combinatorial library (*green spheres*). (B) Sequence of the recombinant phage coat protein p3 (*cyan*) displaying a PDZ variant (*gray with diversified residues in green*) with an N-terminal epitope tag (*orange*). Diversified codons are indicated (N = A/C/G/T, V = A/C/G, R = A/G, S = C/G, W = A/T, H = A/C/T). (C) Diagram of a phage particle, with enclosed phagemid genome encoding the displayed PDZ variant (*green PDZ variant, orange epitope tag, cyan coat protein p3*). Other major and minor coat proteins are shown as grey ellipses.

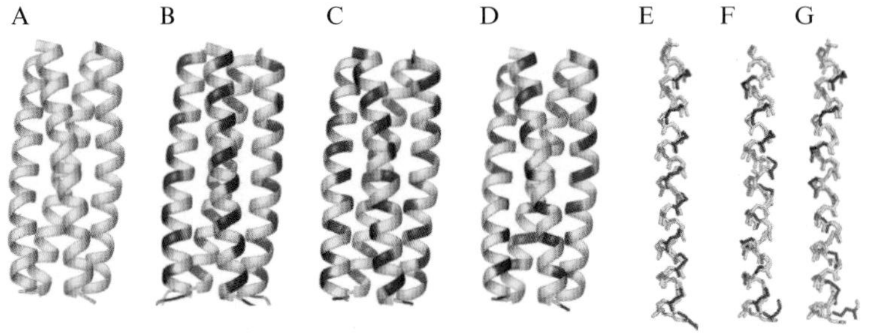

Lisa M. Johnson and Samuel H. Gellman, Figure 19.3 Helix bundles formed by α-peptide GCN4-pLI (A) (PDB ID: 1GCL; Harbury et al., 1993) and three α/β-peptide homologues with varying backbone patterns: (B) ααβαααβ (PDB ID: 2OXK), (C) ααβ (PDB ID: 3C3G), and (D) αααβ (PDB ID: 3C3F). Each image is based on a crystal structure. α Residues are shown in yellow, and β^3 residues are shown in blue. Backbone overlays between the α peptide GCN4-pLI and (E) ααβαααβ, (F) ααβ, and (G) αααβ homologues (Horne, Price, et al., 2008).

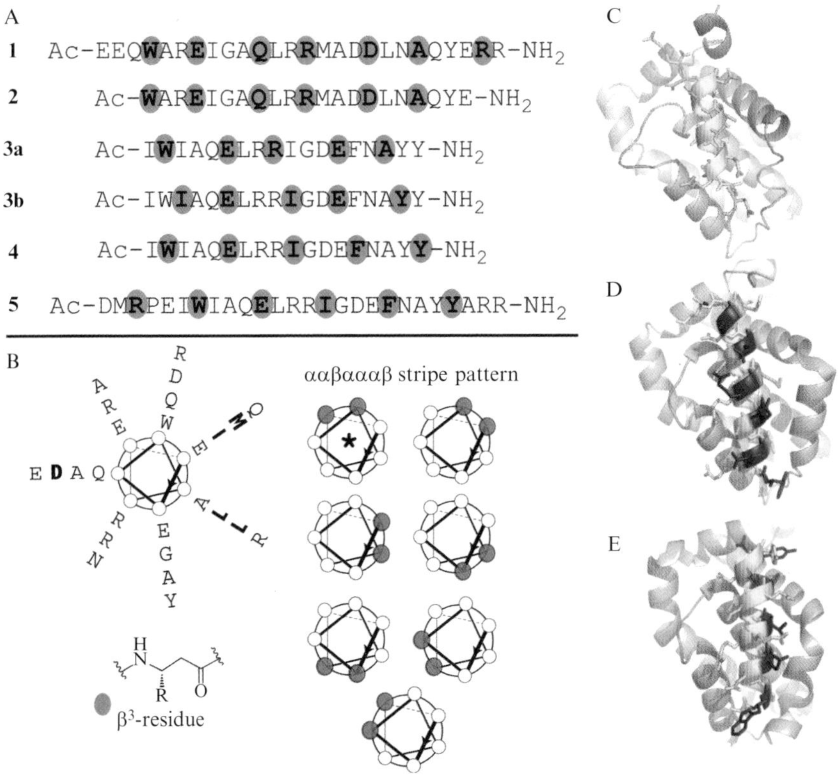

Lisa M. Johnson and Samuel H. Gellman, Figure 19.4 (A) α/β-Peptides with side chain sequences derived from the Puma BH3 domain (**1** and **2**) or the Bim BH3 domain (**3a-b**, **4**, and **5**). The standard single-letter code is used; letters covered with a blue dot indicate β^3-homologues of the α residue designated by the letter. Compounds **1**, **2**, and **3a-b** feature the ααβαααβ backbone repeat, while **4** and **5** have the αααβ backbone repeat. (B) Helical wheel diagram of the Puma BH3 domain with the four key hydrophobic residues and the key Asp in bold. The smaller helix-wheel diagrams show how the stripe of β^3 residues shifts incrementally around the helix periphery among the seven different versions of the ααβαααβ backbone repeat (Horne, Boersma, Windsor, & Gellman, 2008); the asterisk (*) indicates the α/β registry found in Puma-derived α/β-peptides **1** and **2**. (C–E) Images illustrating the complexes between the BH3 domain-derived α- or α/β-peptides (α-residues are yellow, β residues are blue) and Bcl-x_L (light purple): (C) Bak BH3 domain (NMR structure; PDB ID: 1BLX; Sattler et al., 1997); (D) α/β-peptide **2** (crystal structure; PDB ID: 2YJ1; Lee et al., 2011); (E) α/β-peptide **4** (crystal structure; PDB ID: 4A1W; Boersma et al., 2012).